Linear Integrated Circuits

Muhammad H Rashid, Ph.D.
Fellow IET (UK), Life Fellow IEEE (USA)
Electrical and Computer Engineering
University of West Florida

Cengage

Australia • Brazil • Canada • India • Mexico • Singapore • United Kingdom • United States

Cengage

Linear Integrated Circuits

Muhammad H Rashid

For product information and technology assistance, contact us at
asia.infoindia@cengage.com

For permission to use material from this text or product, submit all requests online at
www.cengage.com/permissions

ISBN-13: 978-81-315-2283-7
ISBN-10: 81-315-2283-0

Cengage Learning India Pvt. Ltd.
T-303, 3rd Floor
DAV Complex, Plot 3
Mayur Vihar Phase 1
Delhi 110091

Cengage is a leading provider of customized learning solutions with employees residing in nearly 40 different countries and sales in more than 125 countries around the world. Find your local representative at: **www.cengage.com**.

Cengage is committed to creating inclusive and diverse products, including perspectives that represent the diversity of people across all backgrounds, identities, opinions, and experiences. If you have any feedback related to this, please contact us at **support.cengage.com**.

To learn more about Cengage products, platforms and services, visit **www.cengage.co.in**.

Disclaimer:
This book contains link to resources which are a part of the imported edition. The same may not be available with the Indian edition. The buyer needs to purchase a copy of the imported edition to have full access to those resources. In no event the publisher shall be liable for any limitation or performance due to unavailability of these resources.

Printed by:
Saurabh Printers Pvt. Ltd.
67A-68, Ecotech Ext-I,
Greater Noida,-201308

Printed in India
Print Number: 03 Print Year: 2025

*To my parents,
my wife, Fatema,
my family, Faeza, Farzana, Hasan,
Hannah, Laith, Laila, and Nora*

CONTENTS

Chapter 2 Operational Amplifiers

Chapter 3 Introduction to Digital Electronics

Chapter 4 Integrated Analog Circuits and Applications

PREFACE

Semiconductor devices and integrated circuits (lCs) are the backbone of modern technology, and thus the study of electronicswhich deals with their characteristics and applicationsis an integral part of the undergraduate curriculum for students majoring in electrical, electronics, communications, or computer engineering. Op-amps can be used as the building blocks to design circuits for practical applications. It requires only the knowledge of external characteristics of the op-amp without knowing what is inside the op-amp. Thus, an electronics engineer must also know not only the applications of building blocks at the system level design, but also know the fundamental principles and the internal characteristics of the internal devices including their limitations.

Modern semiconductor technology has evolved to such an extent that many analog and digital circuits are available in the form of integrated circuit (IC) packages. Manufacturers of these packages provide application notes that can be used to implement circuit functions. Knowledge of the characteristics and operation of devices within the IC packages is essential, however, to understand the limitations of these ICs when they are interfaced as building blocks in circuit designs. Such knowledge also serves as the basis for developing future generations of IC packages. Although the trend in the technology suggests that discrete circuit design may disappear entirely in the future, transistor amplifiers (in large-scale or very large-scale integrated forms) will continue to be the building blocks of ICs.

An op-amp is made of semiconductor devices. Semiconductor diodes and transistors are the fundamental devices for electronic circuits and applications. The electronic devices are covered in the first course. A differential amplifier is the input stage of op-amps and is often covered to introduce internal parameters and characteristics of op-amps. This book focuses on the internal structures and characteristics of ICs and their non-linear applications. The objectives of this book are as follows:

- To develop an understanding of the internal structures and limitations of ICs.
- To develop analytical skills and design techniques of integrated circuits.
- To design integrated circuits to meet desired specifications.
- To introduce students to the various elements of the engineering design process, including formulation of specifications, analysis of alternative solutions, synthesis, decision making, iterations, consideration of cost factors, simulation, and tolerance issues.

Approach

The National Board of Accreditation (NBA) criteria and other engineering accreditation criteria that satisfy the international engineering educational standards under the Washington Accord graduate attributes (/) require the integration of design and computer usage throughout the curriculum. The lack of opportunities for design credits in engineering curricula is a common concern. Electronics is generally the first electrical engineering course well suited to the integration of design components and computer usage. This book is structured to permit design content to constitute at least 50% of the course, and it integrates computer usage through PSpice. Many design examples use PSpice to verify the design requirements, and the numerous computer-aided design examples illustrate the usefulness of personal computers as design tools, especially in cases in which design variables are subjected to component tolerances.

Student Course Learning Outcomes (CLOs or COs): Course Outcomes are narrower statements that describe what students are expected to know, and be able to do at the end of each course. These relate to the skills, knowledge, and behavior that students acquire in their matriculation through the course.

After successfully completing the course with a grade of C (2.0/4.0) or better, the student should be able to do the following:

- Identify the parameters influencing the differential and the common-mode gains of differential amplifiers.
- Analyze and evaluate cascode-connected transistors in order to obtain higher differential voltage gains.
- Describe the internal structure and types of op-amps (CMOS, bipolar, and BiCMOS) and the effects of amplifier configurations (difference and single-ended output) on the op-amp performance.
- Identify the circuit parameters of the internal design that affect the op-amp performance (offset voltage, offset current, and unity-gain frequency).
- Describe the definition of logic states and list the performance parameters of logic gates.
- List the relative advantages and disadvantages of logic families.
- Design and analyze simple logic gates.
- Apply some ICs and design circuits for waveform generation, conversion, detection, or signal processing.

Content and Organization

Differential amplifier, that is, the first stage of operational amplifiers, is covered first. The operating principles, analysis, and design of operational amplifiers are covered next. The digital electronics including CMOS inverters are fundamentals to the modern digital circuit. Op-amps find numerous applications in signal processing, and the common types of practical applications are covered. The chapters are organized in order of differential amplifiers, operational amplifiers, digital electronics, and integrated circuits and applications.

As a guide to the instructors for course planning, the table of contents of the book on *Electronic Devices and Circuits* by M.H. Rashid, Cengage Learning India (1/e, 2014), includes the following chapters:

Chapter I: Introduction to Electronics and Design
Chapter 2: Introduction to Amplifiers and Frequency Response
Chapter 3: Introduction to Operational Amplifiers and Applications
Chapter 4: Semiconductor Diodes
Chapter 5: Applications of Diodes
Chapter 6: Semiconductors and *pn* Junction Characteristics
Chapter 7: Metal Oxide Semiconductor Field-Effect Transistors
Chapter 8: Bipolar Junction Transistors and Amplifiers
Chapter 9: Differential Amplifiers

The table of contents of the book on *Electronic Circuits and Applications* by M.H. Rashid, Cengage Learning India (1/e, 2014) includes the following chapters:

Chapter 1: Differential Amplifiers
Chapter 2: Feedback Amplifiers
Chapter 3: Power Amplifiers
Chapter 4: Active Filters
Chapter 5: Oscillators

Pedagogy and Supplements

Mathematical derivations are kept to a minimum by using approximate circuit models of operational amplifiers, transistors, and diodes. The significance of these approximations is established by computer-aided analysis using PSpice. Important circuits are analyzed in worked-out examples in order to introduce the basic techniques and emphasize the effects of parameter variations. At the end of each chapter, review questions and problems test students' learning of the concepts developed in the chapter. The student learning outcomes (SLOs) are listed at the beginning of each chapter. Symbols and their meanings have been uniquely identified at the beginning of each chapter to serve as a quick reference to the students. Every chapter opens with an introduction that puts the content of the chapter in perspective of the field of microelectronics. Solved examples carry captions that identify the objective of the example. Notes interspersed through the text provide a link to other chapters and serve to guide students against common misconceptions and mistakes. Key points of most of the sections are summarized in a box in addition to an end-of-chapter summary. A list of references is included at the end of each chapter for those interested in further reading. End-of-chapter exercises are divided into Review Questions and Problems. Design problems and PSpice problems are identified by relevant symbols.

Support for Instructors

The solutions manual and slides of the figures in this book are available on request from Cengage Learning.

Teaching plans and suggested course outlines using this book are included just after this preface.

Acknowledgments

Thanks are due to the editorial team at Cengage Learning for their guidance and support.

I would also like to thank the following reviewers for their comments and suggestions:

Dr. Ezzat G. Bakhoum
University of West Florida

Dr. William T. Baumann
Virginia Polytechnic Institute and State University

Dr. Paul J. Benkeser
Georgia Institute of Technology

Dr. Alok K. Berry
George Mason University

Dr. Michael A. Bridgwood
Clemson University

Dr. Nadeem N. Bunni
Clarkson University

Dr. Wai-Kai Chen
University of Illinois at Chicago

Dr. Shirshak K. Dhali
Southern Illinois University

Dr. Constantine Hatziadoniu
Southern Illinois University

Dr. Bruce P. Johnson
University of Nevada-Reno

Dr. Frank Kornbaum
South Dakota State University

Dr. Oguz Kucur
Gebze Institute of Technology, Turkey

Dr. John A. McNeill
Worcester Polytechnic Institute

Dr. Bahram Nabet
Drexel University

Dr. Hemanshu R. Pota
Australian Defense Force Academy

Dr. Jack R. Smith
University of Florida

Dr. Robert D. Strattan
University of Tulsa

Finally, thanks to my family for their patience while I was occupied with this and other projects.

Any comments and suggestions regarding this book are welcome. They should be sent to the author at mrashidfl@gmail.com.

Muhammad H. Rashid
Web: http://uwf.edu/mrashid

TEACHING PLANS AND SUGGESTED COURSE OUTLINES

This text has more material than can be covered in a single-semester course. Instructors are advised to select topics that are appropriate to the course depending on the students' background preparations. A course on electronic devices and circuits is a pre-requisite for this course on integrated circuits and applications.

Third Electronics Course

This course usually covers (a) bipolar and MOS current sources; (b) differential amplifiers; (c) analog multipliers; (d) internal structures and characteristics of bipolar, MOS, and BiCMOS op-amps; (e) performance parameters of BJT and CMOS inverters; (f) BJT and CMOS logic circuits; (g) op-amp wave-shaping circuits; (h) voltage-controlled oscillators, 555 Timer, and phase-lock loops; (i) voltage-to-frequency and frequency-to-voltage converters; (j) sample-and-hold circuits; (k) digital-to-analog converters and analog-to-digital converters; and (l) IC voltage regulators. The suggested course topics are shown in Table 1.

TABLE 1 Suggested topics for third electronics course

Number of Weeks	Topics	Chapters	Sections
3	Differential Amplifiers	1	1.1–1.11
3	Operational Amplifiers	2	2.1–2.8
4	Introduction to Digital Electronics	3	3.1–3.10, 3.13–3.17
4	Integrated Analog Circuits and Applications	4	4.1–4.17

Types of Design Projects

Engineering design involves open-ended problems whose objectives are only partially defined. Problem definition and identification of constraints are needed to achieve a satisfactory solution. Often considerable ingenuity is required to find an acceptable solution from alternative pathways, and generally several iterations are needed to reach the solution. Also, it is necessary to verify the solution by simulation and/or testing to ensure that the design objectives have been satisfied. The design projects that appear in this text vary in complexity; the time required to complete them varies from 1 h to 1 month. The design projects can be classified into four categories, depending on the time required: short design projects, mini design projects, medium design projects, and large design projects.

Short Design Projects

Short design projects can be completed in 1 or 2 h. Some of the problems at the end of each chapter fall into this category, including the following:

1. Design of simple current source biasing circuits (Chapter 1)
2. MOS Differential Amplifiers (1)
3. BJT Differential Amplifiers (1)

Mini Design Projects

Mini design projects can be completed in approximately 1 week. Following are some of the end-of-chapter problems in the text that fall into this category:

1. Design of Schmitt trigger circuits (Chapter 4)
2. Design of sample and hold circuits (Chapter 4)
3. Design of timing circuits (Chapter 4)

Medium Design Projects

Medium design projects can be completed in approximately 2 to 3 weeks. Following are some of the end-of-chapter problems in the text that fall into this category:

1. Design of active current sources (Chapter 1)
2. Design of differential amplifiers with current source biasing (Chapter 1)
3. Design of electronic circuits using A/D and D/A converters (Chapter 3)
4. Design of electronic circuits using ICs of phase-lock loop (PLL) and voltage-controlled oscillators (VCO) (Chapter 4)

Large Design Projects

Large design projects can be completed in approximately 4 to 5 weeks. Following are some of the end-of-chapter problems in the text that fall into this category:

1. Design of logic gates (Chapter 3)

Suggested Format for Design Report

It is recommended that in your design reports you do the following:

- Give the complete design, including the ratings and values of each component.
- Justify the use of a particular circuit topology.
- Verify your design objectives by simulating your circuit using PSpice/SPICE/OrCAD or Electronics Workbench. Include a worst-case analysis (with 10% tolerances for all passive components, unless specified).
- Give a cost estimate. The project should be as inexpensive as possible.

A suggested format for design reports is as follows:

1. Title page (including your name, the course number, and the date)
   Include the following statement on the cover sheet:

I/we certify that this assignment is the result of my/our own efforts.
Signature(s): Date:

2. Design objectives and specifications
3. Design steps (including the circuit topology)
4. Design modifications
5. Computer simulation and design verification
6. Components and costs
7. Flowchart of the design process
8. Costs versus reliability and safety considerations
9. Conclusions

ABOUT THE AUTHOR

Muhammad H. Rashid is currently employed by the University of West Florida as a *Professor of Electrical and Computer Engineering*. Previously, he was employed by the University of Florida as *Professor and Director* of UF/UWF Joint Program. Rashid received B.Sc. degree in Electrical Engineering from the Bangladesh University of Engineering and Technology, and M.Sc. and Ph.D. degrees from the University of Birmingham in UK. Previously, he worked as Professor of Electrical Engineering and the Chair of the Engineering Department at Indiana University-Purdue University at Fort Wayne. He also worked as Visiting Assistant Professor of Electrical Engineering at the University of Connecticut, Associate Professor of Electrical Engineering at Concordia University (Montreal, Canada), Professor of Electrical Engineering at Purdue University Calumet, and Visiting Professor of Electrical Engineering at King Fahd University of Petroleum and Minerals (Saudi Arabia), as a design and development engineer with Brush Electrical Machines Ltd. (England, UK), a Research Engineer with Lucas Group Research Centre (England, UK), a Lecturer and Head of Control Engineering Department at the Higher Institute of Electronics (in Libya and Malta).

Dr. Rashid is actively involved in teaching, researching, and lecturing in electronics, power electronics, and professional ethics. He has published 18 books listed in the US Library of Congress and more than 160 technical papers. His books are adopted as textbooks all over the world. His book, *Power electronics* has translations in Spanish, Portuguese, Indonesian, Korean, Italian, Chinese, Persian, and Indian edition. His book *Microelectronics* has translations in Spanish in Mexico and in Spain, Italian, and Chinese.

He has received many invitations from foreign governments and agencies to give keynote lectures and consult, by foreign universities to serve as an external examiner for undergraduate, master's, and Ph.D. examinations, by funding agencies to review research proposals, and by US and foreign universities to evaluate promotion cases for professorship. Dr. Rashid has worked as a regular employee or consultant in Canada, Korea, the United Kingdom, Singapore, Malta, Libya, Malaysia, Saudi Arabia, Pakistan, and Bangladesh. Dr. Rashid has traveled to almost all States in the USA and many countries to lecture and present papers (Japan, China, Hong Kong, Indonesia, Taiwan, Malaysia, Thailand, Singapore, India, Pakistan, Turkey, Saudi Arabia, United Arab Emirates, Qatar, Libya, Jordan, Egypt, Morocco, Malta, Italy, Greece, United Kingdom, Brazil, and Mexico).

He is a *Fellow* of the Institution of Engineering and Technology (IET, UK) and a *Life Fellow* of the Institute of Electrical and Electronics Engineers (IEEE, USA). He was elected as an IEEE Fellow with the citation "*Leadership in power electronics education and contributions to the analysis and design methodologies of solid-state power converters*." Dr. Rashid is the recipient of the *1991 Outstanding Engineer Award* from The Institute of Electrical and Electronics Engineers (IEEE). He received the 2002 IEEE Educational Activity Award (EAB) Meritorious Achievement Award in Continuing Education with the following citation *"for contributions to the design and delivery of continuing education in power electronics and computer-aided simulation*." He is the recipient of the 2008 IEEE Undergraduate Teaching Award with citation "*For his distinguished leadership and dedication to quality undergraduate electrical engineering education, motivating students and publication of outstanding textbooks*." He is also the recipient of the IEEE 2013 Industry Applications Society *Outstanding Achievement Award*.

Dr. Rashid is an ABET program evaluator for electrical and computer engineering (also from 1995 to 2000) and was an engineering evaluator for the Southern Association of Colleges and Schools (SACS, USA). He is also an ABET program evaluator for (general) engineering program. He is the Series Editors of *Power Electronics and Applications*, and *Nanotechnology and Applications* with the CRC Press. He serves as the Editorial Advisor of *Electric Power and Energy* with Elsevier Publishing. He lectures and conducts workshops on Outcome-Based Education (OBE) and its implementations including assessments.

Dr. Rashid is a Distinguished Lecturer for the IEEE Education Society and a Regional Speaker (previously Distinguished Lecture) for the IEEE Industrial Applications Society. He also authored a book on *The Process of Outcome-Based EducationImplementation, Assessment and Evaluations,* 2012 UiTM Press, Malaysia.

CHAPTER 1

DIFFERENTIAL AMPLIFIERS

Learning Outcomes

After completing this chapter, students should be able to do the following:

- Design, analyze, and evaluate different types of MOSFET constant-current sources for biasing MOSFET amplifiers.
- Design, analyze, and evaluate different types of BJT constant-current sources for biasing BJT amplifiers.
- Describe and analyze the characteristics of differential amplifiers and their DC and small-signal characteristics.
- Identify the parameters influencing the differential and the common-mode gains of differential amplifiers.
- Analyze and evaluate cascode-connected transistors in order to obtain higher differential voltage gains.

Symbols and Their Meanings

Symbol	Meaning
A_d, A_c	Small-signal differential and common-mode voltage gains
CMRR	Common-mode rejection ratio of an amplifier
v_{id}, v_{ic}	Small-signal differential and common-mode signal
v_{o1}, v_{o2}	Small-signal output voltages due to input voltages at inverting and noninverting terminals

Symbol	Meaning
r_{o1}, r_{o2}	Small-signal output resistances of transistors
V_M, V_A	MOSFET channel modulation and BJT Early voltages
V_t, V_p	Threshold voltage of enhancement MOSFET and pinch-off voltage of depletion MOSFET

1.1 Introduction

Differential amplifiers are commonly used as an input stage in various types of analog ICs, such as operational amplifiers, voltage comparators, voltage regulators, video amplifiers, power amplifiers, and balanced modulators and demodulators [1–3]. A differential amplifier is a very important transistor stage and determines many of the performance characteristics of an IC. In ICs, including differential amplifiers, it is unnecessary to bias transistors by setting the values of biasing resistors. Because of variations in resistor values, power supply, and temperature, the quiescent point of transistors changes. Transistors can be used to generate the characteristics of DC constant-current sources. Transistors can also be used to produce an output voltage source that is independent of its load or, equivalently, of the output current. This chapter covers the operation, analysis, and characteristics of differential amplifiers using BJTs and MOSFETs. It also covers active current sources and voltage sources.

1.2 Internal Structure of Differential Amplifiers

A differential amplifier acts as an input stage; its output voltage is proportional to the difference between its two input voltages v_1 and v_2. It has a high voltage gain and is directly DC coupled to the input voltages and the load. As we will see later in this chapter, the voltage gain of a differential amplifier depends directly on the output resistance of the current source acting as an active load.

In amplifiers with discrete components, passive components such as resistors and capacitors are less expensive than active devices such as transistors (i.e., MOSFETs and BJTs); thus, in multistage amplifiers, interstage coupling is accomplished with capacitors. However, in monolithic circuits, the die area is the principal determining cost factor. Capacitors of the values and sizes used in amplifiers made with discrete components cannot be included in ICs and must be external to the chip. But using external capacitors increases the pin count of the package and the cost of the IC. A DC-coupled circuit is used to eliminate capacitors. The cheapest component in an IC is the one that can be fabricated within the least area, usually the transistor. The optimal IC has as little resistance as possible and more transistors.

1.2.1 Characteristics of Differential Amplifiers

The differential stage can be represented by an equivalent amplifier, as shown in Fig. 1.1(a). If the two input voltages are equal, a differential amplifier gives an output voltage of almost zero. Its voltage gain is very large, so the input voltage is low, typically less than 50 mV. Thus, we can consider the input voltages as small signals with zero DC components. That is, $v_{G1} = v_{g1}$ and $v_{G2} = v_{g2}$.

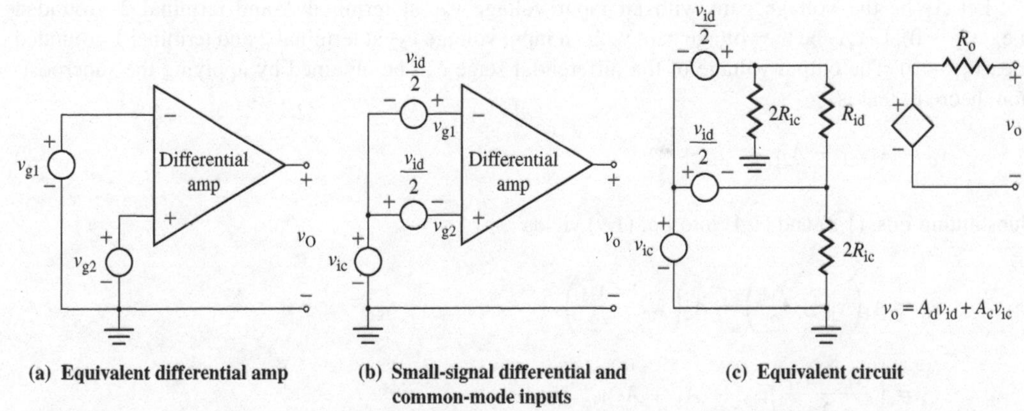

FIGURE 1.1 Small-signal equivalent circuit with differential and common-mode inputs

Let us define a differential voltage v_{id} as

$$v_{id} = v_{g1} - v_{g2} \tag{1.1}$$

and a common-mode voltage v_{ic} as

$$v_{ic} = \frac{v_{g1} + v_{g2}}{2} \tag{1.2}$$

From Eqs. (1.1) and (1.2), the two input voltages can be expressed as

$$v_{g1} = v_{ic} + \frac{v_{id}}{2} \tag{1.3}$$

and
$$v_{g2} = v_{ic} - \frac{v_{id}}{2} \tag{1.4}$$

By replacing the input signals with the equivalent differential and common-mode signals, we can represent the differential stage by an equivalent amplifier, as shown in Fig. 1.1(b). Let v_{o1} be the output voltage due to v_{g1} only, and let v_{o2} be the output voltage due to v_{g2} only. Then we can define a differential output voltage v_{od} as

$$v_{od} = v_{o1} - v_{o2} \tag{1.5}$$

and a common-mode output voltage as

$$v_{oc} = \frac{v_{o1} + v_{o2}}{2} \tag{1.6}$$

From Eqs. (1.5) and (1.6), the two output voltages can be expressed as

$$v_{o1} = v_{oc} + \frac{v_{od}}{2} \tag{1.7}$$

and
$$v_{o2} = v_{oc} - \frac{v_{od}}{2} \tag{1.8}$$

Let A_1 be the voltage gain with an input voltage v_{g1} at terminal 1 and terminal 2 grounded (i.e., $v_{g2} = 0$). Let A_2 be the voltage gain with an input voltage v_{g2} at terminal 2 and terminal 1 grounded (i.e., $v_{g1} = 0$). The output voltage of the differential stage can be obtained by applying the superposition theorem; that is,

$$v_o = A_1 v_{g1} + A_2 v_{g2} \tag{1.9}$$

Substituting Eqs. (1.3) and (1.4) into Eq. (1.9) yields

$$v_o = A_1\left(v_{ic} + \frac{v_{id}}{2}\right) + A_2\left(v_{ic} - \frac{v_{id}}{2}\right)$$

$$= \left(\frac{A_1 - A_2}{2}\right) v_{id} + (A_1 + A_2) v_{ic}$$

$$= A_d v_{id} + A_c v_{ic} \tag{1.10}$$

$$= A_d\left(v_{id} + \frac{A_c}{A_d} v_{ic}\right) \tag{1.11}$$

where $A_d = (A_1 - A_2)/2$ = differential voltage gain and $A_c = A_1 + A_2$ = common-mode voltage gain.

The output voltage v_o in Eq. (1.11) is due to a common-mode input voltage v_{ic} and a differential input voltage v_{id}. If A_d is much greater than A_c, the output voltage will be almost independent of the common-mode signal v_{ic}. A differential amplifier is expected to amplify the differential voltage as much as possible while rejecting (not amplifying) common-mode signals such as noise or other unwanted signals, which will be present in both terminals.

The ability of an amplifier to reject common-mode signals is defined by a performance criterion called the *common-mode rejection ratio* (CMRR), which is defined by

$$\text{CMRR} = \left|\frac{A_d}{A_c}\right| \tag{1.12}$$

$$= 20 \log\left|\frac{A_d}{A_c}\right| \quad \text{(in dB)} \tag{1.13}$$

Substituting Eq. (1.12) into Eq. (1.11) gives the output voltage

$$v_o = A_d\left(v_{id} + \frac{1}{\text{CMRR}} v_{ic}\right) \tag{1.14}$$

which shows that, to reduce the effect of v_{ic} on the output voltage v_o—that is, to get v_{oc} to approach zero—the value of CMRR must be very large, tending to infinity for an ideal amplifier. Thus, a differential amplifier should behave differently for common-mode and differential signals. The small-signal equivalent circuit is shown in Fig. 1.1(c). R_{id} and R_{ic} are the input resistances due to the differential and common-mode signals, respectively. The parameters of a differential amplifier are A_d (ideally ∞), A_c (0), CMRR (∞), R_{id} (∞), and R_{ic} (∞). In the following sections, we will determine the circuit elements affecting these parameters.

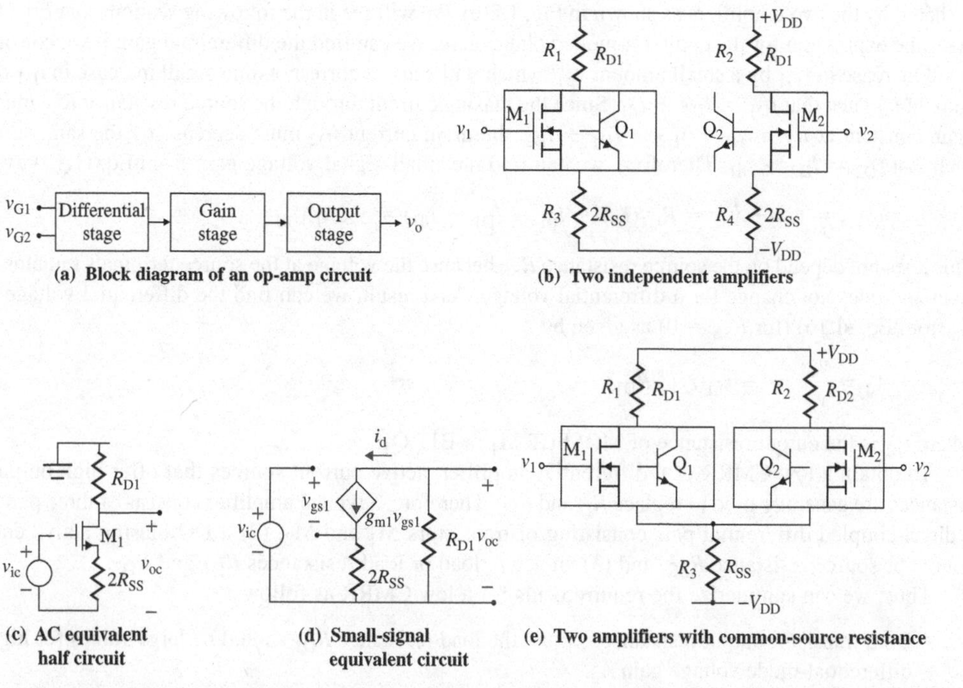

FIGURE 1.2 Basic structure of a differential amplifier

1.2.2 Internal Structure of Differential Amplifiers

A differential amplifier [4] acts as an input stage of an op-amp circuit as shown in Fig. 1.2(a). It serves as a direct DC-coupled differential stage. It may be viewed as consisting of two common-source (CS) amplifiers as shown in Fig. 1.2(b). If the two amplifiers are identical—that is, $R_{D1} = R_{D2}$—their DC biasing drain currents will be the same, and $i_{D1} = i_{D2}$. The DC output voltages are given by

$$v_{o1} = V_{DD} - R_{D1}i_{D1} \quad \text{and} \quad v_{o2} = V_{DD} - R_{D2}i_{D2}$$

The output voltage between the two output terminals of M_2 and M_1 is

$$v_{o(21)} = v_{o2} - v_{o1} = V_{DD} - R_{D2}i_{D2} - V_{DD} + R_{D1}i_{D1} = R_{D1}(i_{D2} - i_{D1}) \tag{1.15}$$

Therefore, we get $v_{o(21)} = v_{o(12)} = 0$ for identical amplifiers of the condition $i_{D1} = i_{D2}$. The small-signal AC equivalent half-circuit is shown in Fig. 1.2(c), which, after replacing the transistor by it transconductance model, becomes as shown in Fig. 1.2(d). Assume that the output of the transistor M_1 is $r_{o1} >> R_{D1}$. From Fig. 1.2(d), we get $v_{gs1} = v_{ic} - g_{m1}v_{gs1}\,(2R_{SS})$, $v_{gs1}(1 + g_{m1}\,2R_{SS}) = v_{ic}$, $v_{oc} = g_{m1}v_{gs1}R_{D1}$. Now we can find the small-signal common-mode voltage gain as given by

$$A_c = \frac{v_{oc}}{v_{ic}} = -\frac{g_{m1}R_{D1}}{1 + 2g_{m1}R_{SS}} \tag{1.16}$$

The derivation of Eq. (1.16) assumes that no interaction exists between the two amplifiers, but there is an interaction between them. The source (or emitter) resistance, whose value is $R_3 \parallel R_4 (= 2R_{SS} \parallel 2R_{SS} = R_{SS})$, is shared by the two amplifiers as shown in Fig. 1.2(e). We will see in the following sections that Eq. (1.16) gives the expression for the common-mode voltage gain. We can find the differential gain if we consider a small increase in v_{G1} by a small amount v_{g1}, which will cause a corresponding small increase in i_{D1} by an amount i_{d1} such that $i_{D1} = I_{D1} + i_{d1}$. Since the biasing current through the source resistance R_{SS} must remain constant at $I_O = i_{D1} + i_{D2} = I_{D1} + I_{D2}$, the drain current i_{D2} must decrease by the same amount such that $i_{D2} = I_{D2} - i_{d1}$. Therefore, we can find the small-signal voltage gain from Eq. (1.15) as

$$v_{o(21)} = v_{o2} - v_{o1} = R_{D1}(I_{D2} - i_{d1} - I_{D1} - i_{d1}) = -2R_{D1}i_{d1} \tag{1.17}$$

This does not depend on the source resistance R_{SS} because the voltage at the source terminals remains constant and does not change for a differential voltage. As a result, we can find the differential voltage gain A_d from Eq. (1.16) (for $R_{SS} = 0$) as given by

$$A_d = \frac{v_{od}}{v_{id}} = -g_{m1}(r_{o1} \| R_{D1}) \tag{1.18}$$

where r_{o1} is the output resistance of MOSFET M_1 or BJT Q_1.

To obtain a low CMRR of a differential amplifier, active current sources that offer high output resistances are generally used to replace R_D and R_{SS}. Therefore, a typical amplifier consists of three parts: (1) a direct-coupled differential pair consisting of transistors M_1 and M_2, (2) a DC biasing active current source or source resistance R_{SS}, and (3) an active load or load resistances R_{D1} and R_{D2}.

Thus, we can summarize the requirements for a low CMRR as follows:

- Both transistor output resistance r_{o1} and the load resistance R_{D1} should be large values for high differential-mode voltage gain A_d.
- The source resistance R should be large values for a low common-mode voltage gain A_c.

KEY POINTS OF SECTION 1.2

- A differential amplifier consists of an active biasing circuit, an active load, and a differential transistor pair.
- The performance of a differential amplifier is measured by a differential gain A_d that occurs in response to a differential voltage between two input terminals, a common-mode gain A_c that occurs in response to a voltage common to both input terminals, and a common-mode rejection ratio CMRR.
- The CMRR is the ratio of the differential gain to the common-mode gain, and it is a measure of the ability of an amplifier to amplify the differential signal and reject common-mode signals.

1.3 MOSFET Current Sources

An active current source has an inherent high output resistance. Thus, replacing the load resistance of a transistor with a current source increases the voltage gain significantly. Transistor current sources are widely used in analog ICs both as biasing elements and as loads for amplifying stages. Current sources are less sensitive to variations in DC power supply and temperature. Especially, for a small value of bias current, the current sources are more economical than resistors in terms of die area required for resistors. A current source can be designed by using either MOSFETs or BJTs [5].

MOSFET current sources are analogous to BJT current sources. We can convert a BJT current source to an equivalent MOSFET current source by assuming that the β_F of the BJTs is infinite. Since MOSFETs

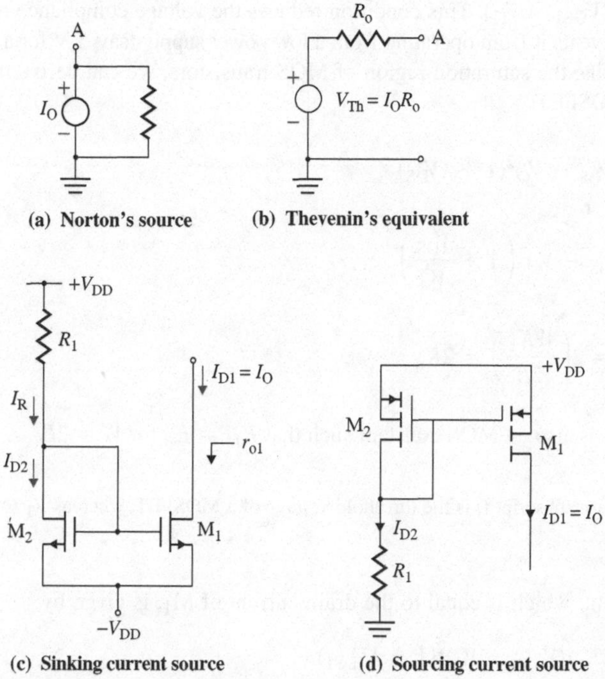

(a) Norton's source **(b) Thevenin's equivalent**

(c) Sinking current source **(d) Sourcing current source**

FIGURE 1.3 Basic MOSFET current source

do not draw any gate current, there is no need for gate current compensation as there is with BJTs. The choice of a BJT source or a MOSFET source generally depends on the type of integrated circuit involved (e.g., bipolar or MOS). BJT sources have some advantages over MOSFET sources, such as a wider compliance range and a higher output resistance. However, a higher output resistance can be obtained by cascode-like connections of MOSFETs.

Norton's equivalent of a current source I_O having an output resistance R_o is shown in Fig. 1.3(a), and its Thevenin's equivalent is in Fig. 1.3(b). An ideal current source should maintain a constant current at an infinite output resistance under all operating conditions. To achieve this goal, a number of current sources are developed. The commonly used current sources are the basic current source, the modified basic source, the cascoded current source, and the Wilson current source.

1.3.1 Basic Current Source

A basic MOSFET current source is shown in Fig. 1.3(c). The direction of the current flow in Fig. 1.3(c) is *into* the current source circuit; this type of constant-current source is often referred to as a *current sink.* Let us assume that the two transistors M_1 and M_2 are identical. Since their gate–source voltages are equal, their drain currents will be the same; that is, $I_{D1} = I_{D2}$. Thus, the output current I_O $(=I_{D1})$ will be the mirror of I_{D2}. Since $V_{DS2} = V_{GS2}$, M_2 will be in saturation. Let V_{t1} and V_{t2} be the threshold voltages of M_1 and M_2, respectively. For M_1 also to be in saturation, V_{DS1}, which is greater than or equal to $(V_{GS2} - V_{t2})$, must be greater than $(V_{GS1} - V_{t1})$. This condition reduces the voltage compliance range of the MOSFET current source and prevents it from operating from a low power supply (say, 1 V for a battery source). Using the equations that define the saturation region of MOS transistors, we can derive the drain current of an enhancement-type MOSFET:

$$I_D = \frac{WK_x}{2L}(V_{GS} - V_t)^2(1 + \lambda V_{DS}) \tag{1.19}$$

$$= \frac{K_m}{2}(V_{GS} - V_t)^2\left(1 + \frac{V_{DS}}{V_M}\right) \tag{1.20}$$

$$K_m = \frac{WK_x}{L} = 2\left(\frac{WK_x}{2L}\right) = 2K_n \tag{1.21}$$

where $K_n = WK_x/2L$ is also an MOS constant such that $K_m = K_n$ for $W = 2L$.

▶ **NOTE** V_t (lowercase subscript t) is the threshold voltage of a MOSFET, whereas V_T (capital subscript T) is the thermal voltage.

The output current, which is equal to the drain current of M_1, is given by

$$I_{D1} = I_O = K_{n1}(V_{GS1} - V_{t1})^2(1 + \lambda V_{DS1}) \tag{1.22}$$

Drain current I_{D2}, which is equal to the reference current I_R, is given by

$$I_{D2} = I_R = K_{n2}(V_{GS2} - V_{t2})^2(1 + \lambda V_{DS2}) \tag{1.23}$$

In practice, all the components of the current source are processed on the same IC, and hence all physical parameters such as K_x and V_t are identical for both devices. Thus, the ratio of I_O to I_R is given by

$$\frac{I_O}{I_R} = \frac{K_{n1}(1 + \lambda V_{DS1})}{K_{n2}(1 + \lambda V_{DS2})} = \frac{(W/L)_1}{(W/L)_2} \times \frac{(1 + \lambda V_{DS1})}{(1 + \lambda V_{DS2})} \tag{1.24}$$

In practice, $\lambda V_{DS} << 1$. Thus, Eq. (1.24) can be approximated by

$$\frac{I_O}{I_R} = \frac{(W/L)_1}{(W/L)_2} \tag{1.25}$$

By controlling the ratio W/L, therefore, we can change the output current. The gate length L is usually held fixed, and the gate width W is varied from device to device to give the desired current ratio I_O/I_R. By choosing identical transistors with $W_1 = W_2$ and $L_1 = L_2$, a designer can ensure that the output current I_O is almost equal to the reference current I_R.

Since $V_{GS2} = V_{DD} - R_1 I_R$ and $V_{DS2} = V_{GS2}$, the reference current I_R can be found approximately from Eq. (1.23); that is,

$$I_R = I_{D2} = K_{n2}(V_{DD} - R_1 I_R - V_{t2})^2 \tag{1.26}$$

can be solved for known values of V_{t2}, K_{n2}, V_{DD}, and R_1. The current source in Fig. 1.3(c) behaves as a current sink rather than a source. A current source equivalent to this current sink can be obtained by using *p*-channel MOS (PMOS) transistors as shown in Fig. 1.3(d).

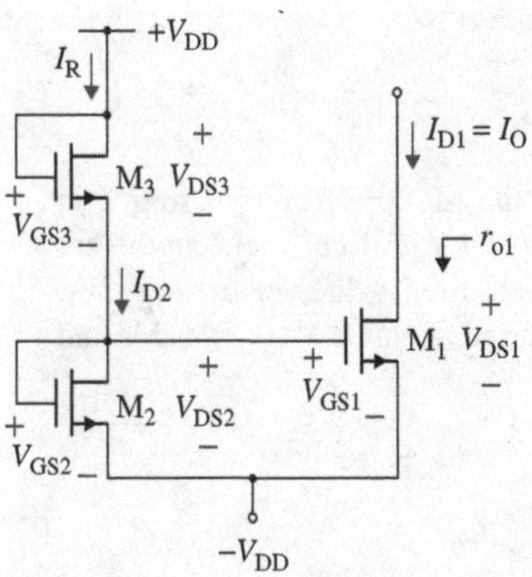

FIGURE 1.4 Basic MOSFET current source without resistance

1.3.2 Modified Basic Current Source

The reference resistance R_1 can be replaced by another MOSFET M_3, as shown in Fig. 1.4. Transistors M_2 and M_3 are used as voltage dividers to control the gate–source voltage of transistor M_1. If M_1 and M_2 are identical, the output current I_O exactly mirrors the drain current through M_2 and M_3. The value of V_{GS1} should be made as low as possible without taking M_1 out of the saturation region.

Since $V_{GS1} = V_{GS2}$, the drain current I_{D1} (= I_O) is equal to the drain current I_{D2} and is given by

$$I_{D1} = I_{D2} = I_R = K_{n1}(V_{GS1} - V_{t1})^2(1 + \lambda V_{GS1}) \tag{1.27}$$

Since $V_{DS2} = V_{GS2}$, the drain current I_{D2} is equal to the reference current I_R and is given by

$$I_{D2} = I_R = K_{n2}(V_{GS2} - V_{t2})^2(1 + \lambda V_{GS2}) \tag{1.28}$$

Since $V_{GS3} = V_{DD} - V_{GS2}$, the drain current of M_3 is given by

$$\begin{aligned} I_{D3} = I_R &= K_{n3}(V_{GS3} - V_{t3})^2(1 + \lambda V_{DS3}) \\ &= K_{n3}(V_{DD} - V_{GS2} - V_{t3})^2[1 + \lambda(V_{DD} - V_{GS2})] \end{aligned} \tag{1.29}$$

Since $I_{D2} = I_{D3} = I_R$, from Eqs. (1.28) and (1.29) we get

$$\frac{K_{n2}(V_{GS2} - V_{t2})^2(1 + \lambda V_{GS2})}{K_{n3}(V_{DD} - V_{GS2} - V_{t3})^2[1 + \lambda(V_{DD} - V_{GS2})]} = 1 \tag{1.30}$$

Thus, by controlling the constants K_{n2} and K_{n3}, we can obtain the desired value of $V_{GS2} = V_{GS1}$, which will give the desired output current.

Output Resistance R_o

The small-signal drain-source resistance r_{ds1} can be derived from Eq. (1.20):

$$\frac{1}{r_{ds1}} = \frac{\delta i_{D1}}{\delta v_{DS1}} = \frac{K_{n1}}{V_M}(V_{GS} - V_t)^2 \approx \frac{I_{D1}}{V_M} \tag{1.31}$$

Thus, the small-signal output resistance of the current source becomes

$$R_o = r_{ds1} = \frac{V_M}{I_{D1}} = \frac{1}{\lambda I_{D1}} = r_{o1} \tag{1.32}$$

which is relatively small. This small output resistance is a disadvantage of having only one MOSFET M_1 at the output side of a current source.

EXAMPLE 1.1

D

Designing a simple MOSFET current source The parameters of the MOSFET current source in Fig. 1.4 are $V_t = 1$ V, $I_O = 50$ μA, $I_R = 40$ μA, $V_{DD} = 10$ V, and $V_M = 10$ V. All channel lengths are equal, $L_1 = L_2 = L_3 = L = 10$ μm, and $K_x = 20$ μA/V^2. Calculate the required values of **(a)** K_{n1}, W_1, **(b)** K_{n2}, W_2, **(c)** K_{n3}, W_3, and **(d)** the output resistance R_o of the current source. Assume $V_{GS1} = 1.5$ V and $V_{DS1} = 5$ V.

SOLUTION

(a) From Eq. (1.22),

$$50 \times 10^{-6} = K_{n1}(1.5 - 1)^2\left(1 + \frac{5}{10}\right)$$

which gives $K_{n1} = 133.3$ μA/V^2. From Eq. (1.21),

$$133.3 \times 10^{-6} = \frac{W_1 \times 20 \times 10^{-6}}{10 \times 10^{-6}}$$

which gives $W_1 = 66.65$ μm.

(b) $V_{DS2} = V_{GS2} = V_{GS1} = 1.5$ V. From Eq. (1.28),

$$40 \times 10^{-6} = K_{n2}(1.5 - 1)^2\left(1 + \frac{1.5}{10}\right)$$

which gives $K_{n2} = 139.1$ μA/V^2. From Eq. (1.21),

$$139.1 \times 10^{-6} = \frac{W_2 \times 20 \times 10^{-6}}{10 \times 10^{-6}}$$

which gives $W_2 = 69.55$ μm.

(c) $V_{GS3} = V_{DS3} = V_{DD} - V_{DS2} = V_{DD} - V_{GS1} = 10 - 1.5 = 8.5$ V. From Eq. (1.29),

$$40 \times 10^{-6} = K_{n3}(8.5 - 1)^2\left(1 + \frac{8.5}{10}\right)$$

which gives $K_{n3} = 0.384$ μA/V^2. From Eq. (1.21),

$$0.384 \times 10^{-6} = \frac{W_3 \times 20 \times 10^{-6}}{10 \times 10^{-6}}$$

which gives $W_3 = 0.192$ μm. In practice, because of manufacturing limitations, the minimum value of L or W is 10 μm. Since the value of W_3 is smaller than 10 μm, V_M must be increased to make W_3 at least 10 μm.

(d) From Eq. (1.32), the output resistance R_o is

$$R_o = r_{ds1} = \frac{V_M}{K_{n1}(V_{GS1} - V_t)^2} = \frac{10}{133.3 \times 10^{-6} \times (1.5 - 1)^2} = 300 \text{ k}\Omega$$

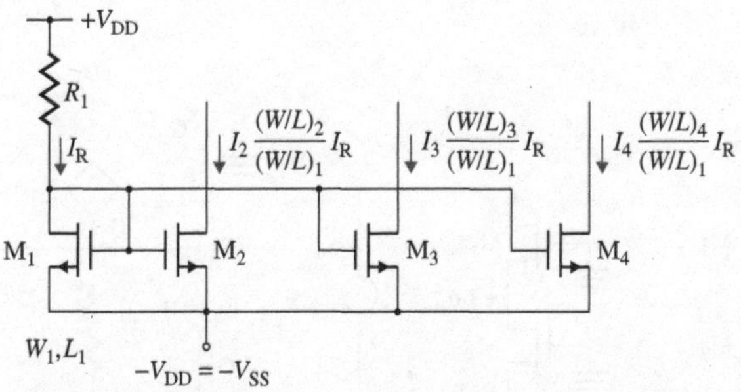

FIGURE 1.5 Multiple MOSFET current sources

1.3.3 Multiple Current Sources

Since there is no gate current in a MOSFET, a number of MOSFETs can be connected to a single reference MOSFET M_1, as shown in Fig. 1.5. Different output currents can be obtained by suitably adjusting the width-to-length ratios of MOSFETs (i.e., M_2, M_3, and M_4). In practice, the gate length L is normally kept constant, and the gate widths (W) of M_2, M_3, and M_4 are varied, to give the desired output currents. Thus, for equal L, Eq. (1.25) gives the relationship of the output currents I_2, I_3, and I_4 to I_R as

$$I_2 = \left(\frac{W_2}{W_1}\right) I_R$$

$$I_3 = \left(\frac{W_3}{W_1}\right) I_R$$

$$I_4 = \left(\frac{W_4}{W_1}\right) I_R$$

Output Resistance R_o

The small-signal output resistance of the current source is

$$R_o = r_{ds1} = \frac{V_M}{I_{D1}} = \frac{1}{\lambda I_{D1}} = r_{o1}$$

1.3.4 Cascode Current Source

The output resistance of the basic current source in Fig. 1.3(c) can be increased by adding two more MOSFETs in a cascodelike connection, as shown in Fig. 1.6(a). The analysis of the circuit is straightforward. The small-signal circuit for finding the output resistance is shown in Fig. 1.6(b), and its small-signal equivalent is shown in Fig. 1.6(c); r_{o2} is the output resistance of transistor M_2. Using KVL and the relation $v_{gs1} = -r_{o2}i_x$, we get

$$\begin{aligned} v_x &= r_{o1}i_1 + r_{o2}i_x \\ &= r_{o1}(i_x - g_{m1}v_{gs1}) + r_{o2}i_x \\ &= r_{o1}(i_x + g_{m1}r_{o2}i_x) + r_{o2}i_x \end{aligned}$$

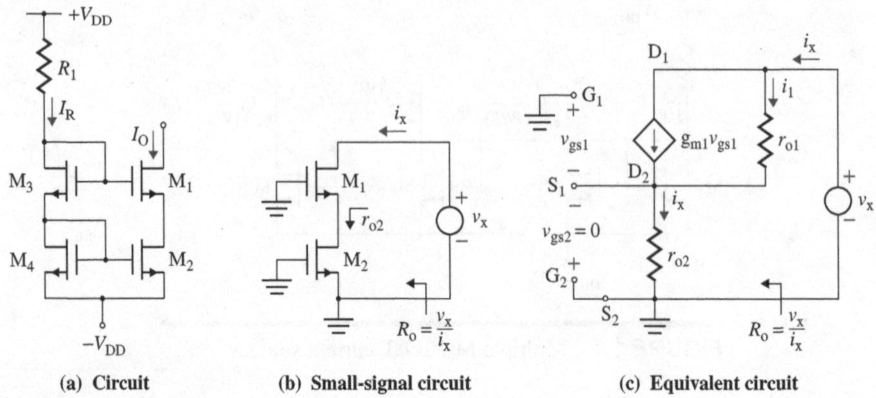

FIGURE 1.6 Cascode current source

which gives the output resistance R_o of the current source as

$$R_o = r_{o1}(1 + g_{m1}r_{o2}) + r_{o2} \tag{1.33}$$

For identical transistors, $r_{o1} = r_{o2} = r_o$, and R_o becomes

$$R_o = r_o(2 + g_{m1}r_o) \tag{1.34}$$

$$\approx g_{m1}r_o^2 \tag{1.35}$$

Thus, the output resistance can be significantly increased, to a level comparable to that of a BJT source. However, the voltage compliance range will be reduced because of the two drain-source voltages in series (i.e., $V_{DS1} + V_{DS2}$).

1.3.5 Wilson Current Source

The MOSFET version of the Wilson current source is shown in Fig. 1.7(a). The equivalent circuit for finding the output resistance R_o is shown in Fig. 1.7(b). We have

$$v_{gs3} = \frac{i_x}{g_{m2}}$$

and $\quad v_{gs1} + v_{gs3} = -g_{m3}v_{gs3}r_{o3}$

which can be simplified to relate v_{gs1} to v_{gs3} and v_{gs2} by

$$v_{gs1} = -(1 + g_{m3}r_{o3})v_{gs3} = -\frac{(1 + g_{m3}r_{o3})i_x}{g_{m2}}$$

Applying KVL to Fig. 1.7(b), we get

$$v_x = (i_x - g_{m1}v_{gs1})r_{o1} + \frac{i_x}{g_{m2}}$$

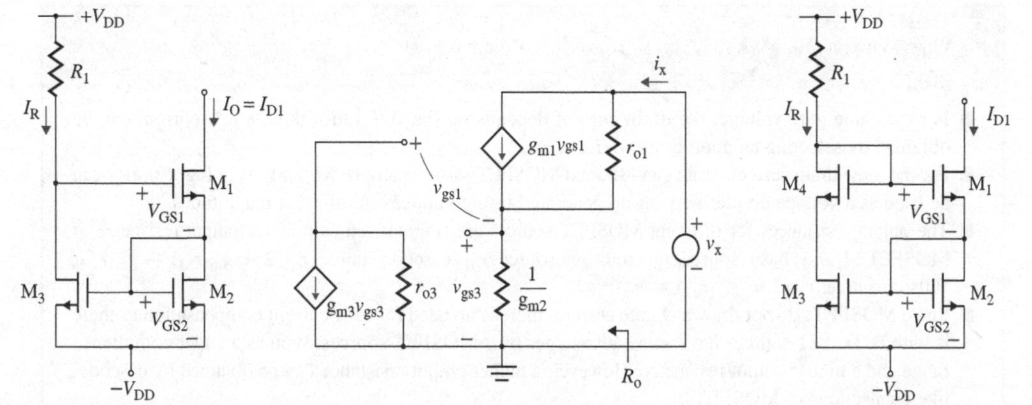

(a) **Wilson current source** (b) **Equivalent circuit for finding R_o** (c) **Modified Wilson current source**

FIGURE 1.7 Wilson current source

which, after substituting for v_{gs1} and simplifying, gives the output resistance R_o as

$$R_o = \frac{v_x}{i_x} = r_{o1} + \frac{1}{g_{m2}} + \frac{g_{m1}}{g_{m2}} r_{o1}(1 + g_{m3} r_{o3})$$

$$\approx r_{o1} + r_{o1}(1 + g_{m3} r_{o3}) \quad (\text{for } g_{m1} = g_{m2} = g_{m3}) \tag{1.36}$$

where r_{o2} and r_{o1} are the output resistances of transistors M_2 and M_1, respectively.

The problem with this circuit is that the drain voltages V_{D1} and V_{D3} of M_1 and M_3 are unequal. As a result, their drain currents I_{D1} and I_{D3} are also unequal. This problem can be solved by adding one diode-connected MOSFET, as shown in Fig. 1.7(c). This modification ensures that M_1 and M_3 have equal drain voltages and thus equal drain currents.

1.3.6 Design of Active Current Sources

The specifications for designing a current source will include the output current I_Q, the output resistance R_o, and the DC supply voltage V_{DD}. The design sequence is as follows:

Step 1. Determine the design specifications: output current and output resistance.

Step 2. Decide on the type of device to use—either BJTs or MOSFETs.

Step 3. Choose the circuit topology best suited to the specifications. Use simple transistor models for hand analysis to find the circuit-level solution, including component values and specifications of BJTs or MOSFETs.

Step 4. Use the standard values of components—for example, $R_1 = 5.6\ \text{M}\Omega \pm 5\%$ instead of $5.72\ \text{M}\Omega$, $R_2 = 30\ \text{k}\Omega \pm 5\%$ instead of $29.3\ \text{k}\Omega$, and $R_3 = 27\ \text{k}\Omega \pm 5\%$ instead of $27.5\ \text{k}\Omega$. Evaluate your design and modify the values, if necessary.

Step 5. Use PSpice/SPICE verification, employing complex circuit models to calculate the worst-case results due to component and parameter variations. Modify your design, if necessary.

KEY POINTS OF SECTION 1.3

- For the same gate voltage, the drain current depends on the W/L ratio; thus, a low current can be obtained by selecting an appropriate W/L ratio.
- For the same drain current, drain gate–shorted MOSFETs—for example, M_3 and M_4 in Fig. 1.6(a)—can be used as a voltage divider network to generate biasing voltages of different magnitudes.
- The output resistances for different MOSFET sources are summarized as follows: output resistance of MOSFET M_1, r_{o1}; basic source, r_{o1}; multiple source, r_{o1}; cascode source, $r_{o1}(2 + g_{m1}r_{o1}) \simeq g_{m1}r_{o1}^2$; Wilson source, $r_{o1}(2 + g_{m1}r_{o1}) \simeq g_{m1}r_{o1}^2$.
- Since MOSFETs do not draw any gate current, there is no need for base current compensation as there is with BJTs. BJT sources have some advantages over MOSFET sources, such as a wider compliance range and a higher output resistance. However, a higher output resistance can be obtained by cascode-like connections of MOSFETs.

1.4 MOS Differential Amplifiers

During the past few years, MOS technology has developed considerably. MOS transistors are being used increasingly in analog integrated circuits. It is relatively easy to connect MOS transistors in cascode form to control the drain current and give high output resistance. MOS differential pairs are the building blocks in MOS ICs.

1.4.1 NMOS Differential Pair

An n-channel NMOS pair is shown in Fig. 1.8. The DC biasing is normally done by a MOS current source. Although a resistor R_D is shown as the load in Fig. 1.8, a MOS active current mirror is normally used as the load.

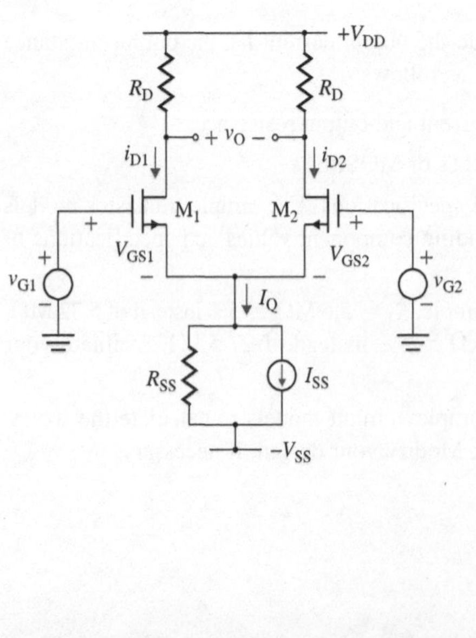

FIGURE 1.8 MOS differential pair

DC Transfer Characteristics

The following analysis is performed for an *n*-channel MOSFET pair, but it is equally applicable to a *p*-channel pair with appropriate sign changes. The analysis can be simplified by making the following assumptions:

1. The output resistances of the MOSFETs are infinite: $r_o = \infty$.
2. The MOSFETs are identical and operate in the saturation region. The threshold voltages are the same, $V_{t1} = V_{t2} = V_t$, and the constants are equal, $K_{n1} = K_{n2} = K_n$.
3. The output resistance of the transistor current source is infinite: $R_{SS} = \infty$.

Assuming that the drain current is related to v_{GS} by the approximate square law relationship in Eq. (G.14) in Appendix G, we can write

$$i_D = K_n(v_{GS} - V_t)^2 \tag{1.37}$$

Taking the square root of both sides of Eq. (1.37), we can write the square root of the drain currents as

$$\sqrt{i_{D1}} = \sqrt{K_n}(v_{GS1} - V_t) \tag{1.38}$$

$$\sqrt{i_{D2}} = \sqrt{K_n}(v_{GS2} - V_t) \tag{1.39}$$

We can subtract $\sqrt{i_{D2}}$ from $\sqrt{i_{D1}}$ to find a relation for the differential voltage $v_{id} = v_{GS1} - v_{GS2}$:

$$\begin{aligned}\sqrt{i_{D1}} - \sqrt{i_{D2}} &= \sqrt{K_n}(v_{GS1} - V_t) - \sqrt{K_n}(v_{GS2} - V_t) = \sqrt{K_n}(v_{GS1} - v_{GS2}) \\ &= \sqrt{K_n}v_{id}\end{aligned} \tag{1.40}$$

The sum of i_{D1} and i_{D2} must equal I_Q; that is,

$$I_Q = i_{D1} + i_{D2} \tag{1.41}$$

Substituting Eq. (1.41) into Eq. (1.40) and solving the resultant quadratic, we find the drain currents:

$$i_{D1} = \frac{I_Q}{2} + \sqrt{2K_nI_Q}\left(\frac{v_{id}}{2}\right)\left[1 - \frac{(v_{id}/2)^2}{(I_Q/2K_n)}\right]^{1/2} \tag{1.42}$$

$$i_{D2} = \frac{I_Q}{2} - \sqrt{2K_nI_Q}\left(\frac{v_{id}}{2}\right)\left[1 - \frac{(v_{id}/2)^2}{(I_Q/2K_n)}\right]^{1/2} \tag{1.43}$$

At the quiescent point $v_{id} = 0$, we get

$$i_{D1} = i_{D2} = \frac{I_Q}{2}$$

$$v_{GS1} = v_{GS2} = V_{GS}$$

$$I_Q = 2I_D = 2K_n(V_{GS} - V_t)^2$$

$$\sqrt{2K_nI_Q} = 2K_n\frac{(V_{GS} - V_t)^2}{(V_{GS} - V_t)} = \frac{I_Q}{(V_{GS} - V_t)}$$

$$\left(\frac{I_Q}{2K_n}\right) = (V_{GS} - V_t)^2$$

Substituting these relations into Eqs. (1.42) and (1.43), we can rewrite $i_{D1} = i_{D2}$ as

$$i_{D1} = \frac{I_Q}{2} + \left(\frac{I_Q}{V_{GS} - V_t}\right)\left(\frac{v_{id}}{2}\right)\left[1 - \left(\frac{v_{id}/2}{V_{GS} - V_t}\right)^2\right]^{1/2} \tag{1.44}$$

$$i_{D2} = \frac{I_Q}{2} - \left(\frac{I_Q}{V_{GS} - V_t}\right)\left(\frac{v_{id}}{2}\right)\left[1 - \left(\frac{v_{id}/2}{V_{GS} - V_t}\right)^2\right]^{1/2} \tag{1.45}$$

For $v_{id}/2 << (V_{GS} - V_t)$, i_{D1} and i_{D2} can be approximated by

$$i_{D1} = \frac{I_Q}{2} + \left(\frac{I_Q}{V_{GS} - V_t}\right)\left(\frac{v_{id}}{2}\right) \tag{1.46}$$

$$i_{D2} = \frac{I_Q}{2} - \left(\frac{I_Q}{V_{GS} - V_t}\right)\left(\frac{v_{id}}{2}\right) \tag{1.47}$$

Thus, the change in drain current from the quiescent value of $I_Q/2$ is given by

$$\Delta I_D = \left(\frac{I_Q}{V_{GS} - V_t}\right)\left(\frac{v_{id}}{2}\right) \tag{1.48}$$

which can be normalized with respect to the maximum value $I_Q/2$ as

$$\frac{\Delta I_D}{I_Q/2} = \frac{v_{id}}{V_{GS} - V_t} = \frac{v_{id}}{\sqrt{I_Q/2K_n}} \tag{1.49}$$

If v_{id} is sufficiently large, all of the biasing current I_D must flow through only one of the MOSFETs. The range of v_{id} for which both transistors conduct can be found from Eq. (1.49) under the condition $\Delta I_D = I_Q/2$. That is,

$$v_{id} \leq \left(\frac{I_Q}{2K_n}\right)^{1/2} \tag{1.50}$$

This equation gives the value of v_{id} for which the current I_Q is carried by one of the two transistors. Thus, outside the range defined by Eq. (1.50), currents i_{D1} and i_{D2} will be either zero or I_Q. The plots of the normalized currents i_{D1} and i_{D2} against the differential voltage v_{id}/v_n are shown in Fig. 1.9, where $v_n = \sqrt{I_Q/2K_n}$.

The output voltages of a MOSFET pair are as follows:

$$v_{O1} = V_{DD} - i_{D1}R_D \tag{1.51}$$

$$v_{O2} = V_{DD} - i_{D2}R_D \tag{1.52}$$

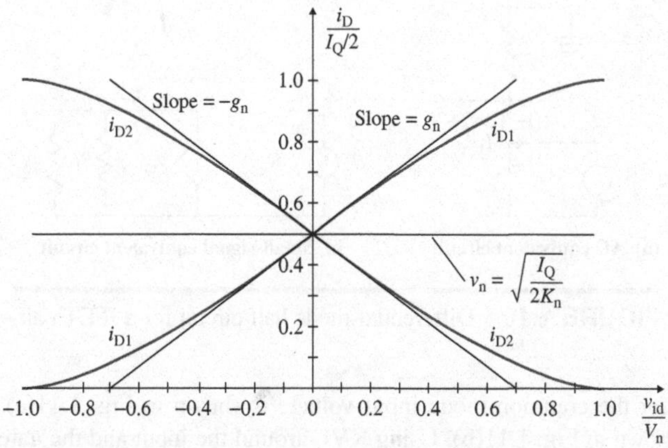

FIGURE 1.9 Normalized DC transfer characteristic of MOSFET pair

The differential DC output voltage is

$$v_{od} = v_{O1} - v_{O2} = R_D(i_{D2} - i_{D1}) = -R_D \Delta I_D$$

Substituting for ΔI_D from Eq. (1.47), we get

$$v_{od} = -R_D\left(\frac{I_Q}{V_{GS} - V_t}\right)\left(\frac{v_{id}}{2}\right) \tag{1.53}$$

$$= -R_D\left(\frac{K_n I_Q}{2}\right)^{1/2} v_{id}$$

which shows the relation between the output voltage v_{od} and the differential voltage v_{id}. If v_{id} is zero, v_{od} is also zero.

Small-Signal Analysis

The transconductance of the small-signal model is given by

$$g_m = 2K_n(V_{GS} - V_t) \tag{1.54}$$

$$= \sqrt{2K_n I_Q}$$

which shows that for a MOSFET a higher value of g_m requires a higher value of the biasing current I_Q.

The half circuit for differential input voltage is shown in Fig. 1.10(a), and its small-signal equivalent circuit is shown in Fig. 1.10(b). The $(r_{o1} \| R_D)$ differential voltage gain A_d for a single-ended output can easily be derived as

$$A_d = \frac{v_{od}/2}{v_{id}/2} = -g_m(r_{o1} \| R_D) \tag{1.55}$$

which, for $r_{o1} >> R_D$, can be approximated to $A_d = -g_m R_D$.

▶ **NOTE** If we define $v_{od} = v_{o2} - v_{o1}$ and $v_{gd} = v_{G1} - v_{G2}$, then $A_d = g_m(R_D \| r_{o1})$.

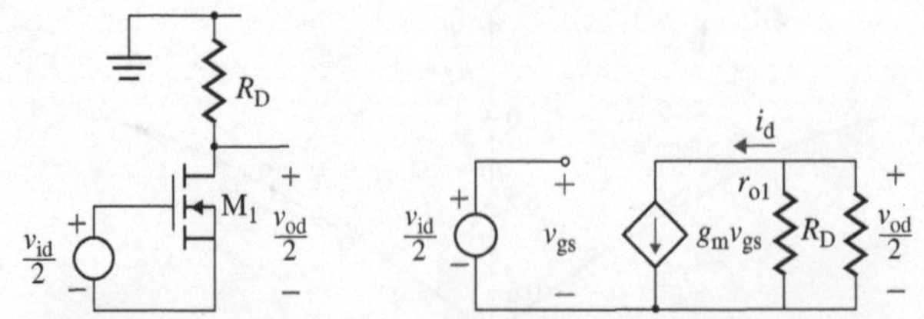

(a) AC equivalent circuit (b) Small-signal equivalent circuit

FIGURE 1.10 Differential-mode half circuit for a JFET pair

The half circuit for the common-mode input voltage is shown in Fig. 1.11(a), and its small-signal equivalent circuit is shown in Fig. 1.11(b). Using KVL around the input and the gate–source loop, we get

$$v_{gs} = v_{ic} - v_{gs} g_m 2R_{SS}$$

which leads to the following relationship between v_{gs} and v_{ic}:

$$v_{gs}(1 + g_m 2R_{SS}) = v_{ic} \tag{1.56}$$

The common-mode output voltage is given by

$$v_{oc} = -R_D i_d = -R_D g_m v_{gs}$$

Substituting v_{gs} from Eq. (1.56) into the above equation, we get

$$v_{oc} = v_{ic}\left[\frac{-g_m R_D}{1 + g_m 2R_{SS}}\right]$$

which gives the common-mode voltage gain A_c (for a single-ended output) as

$$A_c = \frac{v_{oc}}{v_{ic}} = \frac{-g_m R_D}{1 + g_m 2R_{SS}} \tag{1.57}$$

▶ **NOTE** To simplify the analysis, the effect of transistor output resistance r_{o1} is not included in Eq. (1.57).

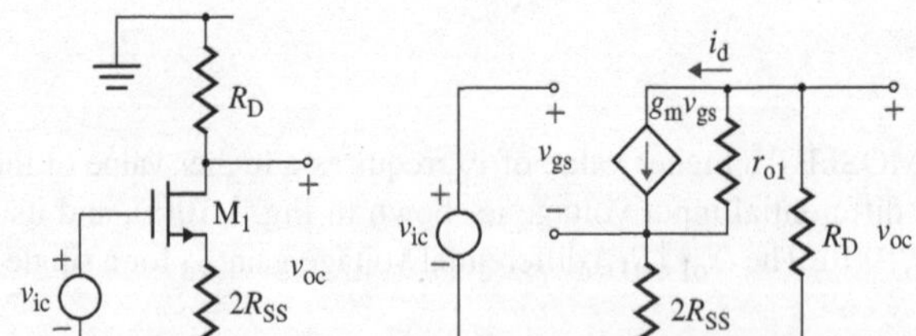

(a) AC equivalent circuit (b) Small-signal equivalent circuit

FIGURE 1.11 Common-mode half circuit for depletion MOSFET pair

From Eqs. (1.55) and (1.57) for $r_{o1} >> R_D$, we can find CMRR as

$$\text{CMRR} = \frac{A_d}{A_c} = 1 + 2g_m R_{SS} \tag{1.58}$$

which is valid only for single-ended output; that is, the output is taken out from one of the output terminals.

The common-mode and differential input resistances are given by

$$R_{ic} = R_{id} = \infty \tag{1.59}$$

▶ **NOTE** To simplify the analysis, the effect of transistor output resistance r_o is not included in Eqs. (1.57) and (1.58).

EXAMPLE 1.2

Analyzing a MOS differential pair with an active current source The parameters of the MOS differential pair in Fig. 1.8 are $R_{SS} = 50\ \text{k}\Omega$, $I_Q = 10$ mA, $V_{DD} = 30$ V, and $R_D = 5\ \text{k}\Omega$. The NMOSs are identical and have $K_n = 1.25\ \text{mA/V}^2$ and $V_t = 1.0$ V. Assume $V_M = 100$ V.

(a) Calculate the DC drain currents through the MOSFETs if $v_{id} = 10$ mV.

(b) Assuming $I_{D1} = I_{D2}$, calculate A_d, A_c, and CMRR; R_{id} and R_{ic}; and the small-signal output voltage if $v_{g1} = 10$ mV and $v_{g2} = 20$ mV.

(c) Find the drain voltage V_D.

SOLUTION

(a) For $v_{id} = 10$ mV, Eq. (1.42) gives the DC drain current i_{D1} for transistor M_1 as

$$i_{D1} = \frac{10\ \text{m}}{2} + \sqrt{2 \times 1.25\ \text{m} \times 10\ \text{m}} \times \left(\frac{10\ \text{m}}{2}\right)\left[1 - \frac{(10\ \text{m}/2)^2}{10\ \text{m}/(2 \times 1.25\ \text{m})}\right]^{1/2} = 5.25\ \text{mA}$$

$$i_{D2} = I_D - i_{D1} = 10\ \text{mA} - 5.25\ \text{mA} = 4.75\ \text{mA}$$

(b) We know that $I_{D1} = I_{D2} = I_Q/2 = 10\ \text{mA}/2 = 5$ mA. From Eq. (1.54),

$$g_m = \sqrt{2 \times K_n I_Q} = \sqrt{2 \times 1.25\ \text{m} \times 10\ \text{m}} = 5\ \text{mA/V}$$

$$r_{o1} = \frac{V_M}{I_D} = \frac{100}{5\ \text{m}} = 20\ \text{k}\Omega$$

From Eq. (1.55), the single-ended differential voltage gain A_d is

$$A_d = -g_m(R_D \| r_{o1}) = -5\ \text{m} \times (5\ \text{k} \| 20\ \text{k}) = -20\ \text{V/V}$$

From Eq. (1.57), the single-ended common-mode voltage gain A_c is

$$A_c = \frac{-g_m R_D}{1 + g_m 2R_{SS}} = \frac{-5\ \text{m} \times 5\ \text{k}}{1 + 5\ \text{m} \times 2 \times 50\ \text{k}} = -0.0399\ \text{V/V}$$

Thus, CMRR $= |A_d/A_c| = 20/0.0399 = 501$ (or 54 dB).

From Eq. (1.59),

$$R_{id} = R_{ic} = \infty$$

We know that

$$v_{id} = v_{g2} - v_{g1} = 20 \text{ mV} - 10 \text{ mV} = 10 \text{ mV}$$

and
$$v_{ic} = \frac{v_{g1} + v_{g2}}{2} = \frac{10 \text{ mV} + 20 \text{ mV}}{2} = 15 \text{ mV}$$

Using Eq. (1.10), we have

$$v_o = A_d v_{id} + A_c v_{ic} = -20 \times 10 \text{ mV} - 0.0399 \times 15 \text{ mV} = -201 \text{ mV}$$

(c) The DC drain voltage at the drain terminal of a transistor is

$$V_D = V_{DD} - I_D R_D = 30 \text{ V} - 5 \text{ mA} \times 5 \text{ k}\Omega = 5 \text{ V}$$

Thus, for $A_d = -20$, the maximum differential voltage will be $v_{id} = 5/20 = 250$ mV. Therefore, V_{DD} must be greater than $I_D R_D$ in order to allow output voltage swing due to the input voltages.

EXAMPLE 1.3

Analyzing an NMOS differential pair with an active current source Repeat Example 1.2 if the transistor current source is replaced by the resistance $R_{SS} = 50$ kΩ; that is, $I_{SS} = 0$. Assume $V_M = 100$ V.

SOLUTION

(a) The DC drain current and the gate–source voltage of the MOSFETs can be determined from the DC common-mode half circuit for $v_{g1} = v_{g2} = 0$, shown in Fig. 1.12:

$$V_{GS} + 2I_D R_{SS} = V_{SS}$$

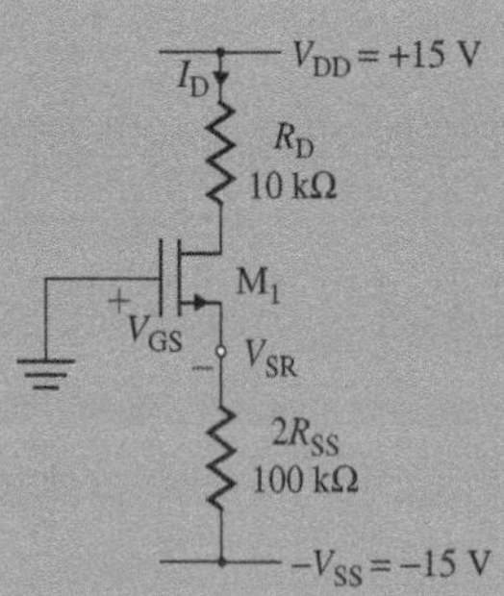

FIGURE 1.12 Common-mode half circuit for DC biasing of MOS pair

Substituting $i_D = I_D$ and $v_{GS} = V_{GS}$ from Eq. (1.37) into the preceding equation, we get

$$V_{GS} + 2R_{SS}K_n(V_{GS} - V_t)^2 = V_{SS}$$

Substituting the values, we get

$$V_{GS} + 2 \times 50\text{ k} \times (V_{GS} - 1)^2 = 15$$

This quadratic equation yields the solution for the gate–source voltage $V_{GS} = 1.478$ V, or 0.514 V. Since $V_{GS} > V_t$, the acceptable value is $V_{GS} = 1.478$ V, which gives the drain current as

$$I_{D1} = K_n(V_{GS} - V_t)^2 = 1.25\text{ m} \times (1.478 - 1)^2 = 285.6\ \mu\text{A}$$

Therefore, the voltage at the source with respect to the ground is

$$V_{SR} = -V_{GS} = -1.478\text{ V} \quad \text{and} \quad I_Q = 2I_D = 2 \times 285.6\ \mu = 571.2\ \mu\text{A}$$

(b) From Eq. (1.54),

$$g_m = \sqrt{2K_nI_Q} = \sqrt{2 \times 1.25\text{ m} \times 571.2\ \mu} = 1.194\text{ mA/V}$$

$$r_{o1} = \frac{V_M}{I_D} = \frac{100}{285.6\ \mu} = 350\text{ k}\Omega$$

From Eq. (1.55),

$$A_d = -g_m(R_D \| r_{o1}) = -1.194\text{ m} \times (5\text{ k} \| 350\text{ k}) = -5.89\text{ V/V}$$

From Eq. (1.57),

$$A_c = \frac{-g_mR_D}{1 + g_m \times 2R_{SS}} = \frac{-1.194\text{ m} \times 5\text{ k}}{1 + 1.194\text{ m} \times 2 \times 50\text{ k}} = -0.0489$$

Thus, CMRR $= |A_d/A_c| = 5.89/0.0489 = 120.4$, or 41.61 dB. From Eq. (1.59),

$$R_{id} = R_{ic} = \infty$$

We know that

$$v_{id} = v_{g2} - v_{g1} = 20\text{ mV} - 10\text{ mV} = 10\text{ mV}$$

$$v_{ic} = \frac{v_{g1} + v_{g2}}{2} = \frac{10\text{ mV} + 20\text{ mV}}{2} = 15\text{ mV}$$

Using Eq. (1.10), we have

$$v_o = A_dv_{id} + A_cv_{ic} = -5.887 \times 10\text{ mV} - 0.489 \times 15\text{ mV} = -59.6\text{ mV}$$

Thus, the output voltage and the voltage gain are much lower than with current-source biasing because the biasing current is low.

EXAMPLE 1.4

D

Designing a MOS differential pair with an active current source

(a) Design a MOS differential pair as shown in Fig. 1.13, in which one input terminal is grounded. The output is taken from the drain of transistor M_1. The DC biasing current is $I_Q = 10$ mA, and $V_{DD} = V_{SS} = 15$ V. The MOSFETs are identical and have $K_n = 1.25\ \text{mA/V}^2$ and $V_t = 1$ V. A small-signal voltage gain of $A_1 - 10$ V/V is required. Assume $V_M = \infty$.

(b) Calculate the design values of A_d, A_c, and CMRR.

(c) Find the maximum permissible value of R_D and the corresponding voltage gain A_d.

SOLUTION

(a) $I_Q = 10$ mA, and $I_{D1} = I_{D2} = I_Q/2 = 10\ \text{mA}/2 = 5$ mA. From Eq. (1.54),

$$g_m = \sqrt{2 \times K_n I_Q} = \sqrt{2 \times 1.25\ \text{m} \times 10\ \text{m}} = 5\ \text{mA/V}$$

From Eq. (1.37), the DC gate–source voltage is

$$V_{GS1} = -\left(\sqrt{\frac{I_{D1}}{K_n}} + V_t\right) = -\left(\sqrt{\frac{5\ \text{mA}}{1.25\ \text{m}}} + 1\right) = -2.236\ \text{V}$$

Therefore, the voltage at the source terminal with respect to the ground is $V_{SR} = -V_{GS} = -2.236$ V, and

$$R_{SS} = \frac{V_{SR} - V_{SS}}{I_{SS}} = \frac{(-2.236 + 15)\ \text{V}}{10\ \text{mA}} = 1.3\ \text{k}\Omega$$

(b) $v_{G1} = V_{G1} + v_{g1}$, $V_{G1} = 0$, and $v_{G2} = v_{g2} = 0$. Then

$$v_{id} = v_{g1} - v_{g2} = v_{g1}$$

$$v_{ic} = \frac{v_{g1} + v_{g2}}{2} = \frac{v_{g1}}{2}$$

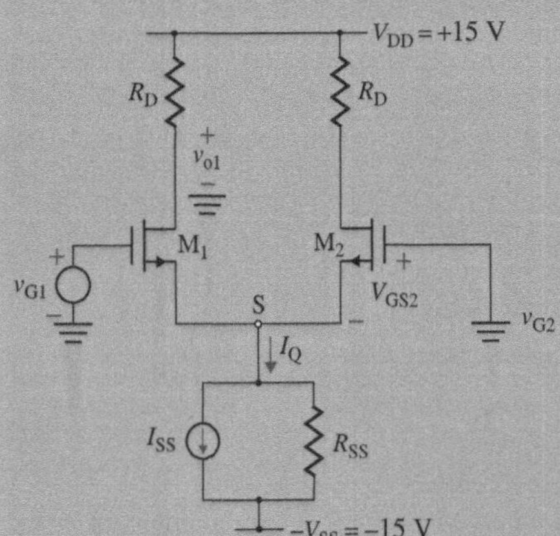

FIGURE 1.13 MOS differential pair with a single input

From Eq. (1.7),

$$v_{o1} = v_{oc} + \frac{v_{od}}{2} = A_c v_{ic} + A_d v_{id} = A_c \frac{v_{g1}}{2} + A_d v_{g1} = \frac{v_{g1}}{2}(A_c + 2A_d)$$

Substituting for A_d and A_c from Eqs. (1.55) and (1.57) gives the voltage gain A_1 as

$$A_1 = \frac{v_{o1}}{v_{g1}} = \frac{1}{2}(A_c + 2A_d) = -\frac{1}{2}\left[\frac{g_m R_D}{1 + 2g_m R_{SS}} + 2g_m R_D\right]$$

Substituting $A_1 = -10$, $g_m = 5$ mA/V, and $R_{SS} = 1.3$ kΩ into the above equation gives $R_D = 1.93$ kΩ.

$$r_{o1} = \frac{V_M}{I_{D1}} = \infty$$

From Eq. (1.55),

$$A_d = -g_m(R_D \| r_{o1}) = -5\text{ m} \times (1.93\text{ k} \| 20\text{ k}) = -8.8\text{ V/V}$$

From Eq. (1.57),

$$A_c = \frac{-g_m R_D}{1 + g_m \times 2R_{SS}} = \frac{-5\text{ m} \times 1.93\text{ k}}{1 + 5\text{ m} \times 2 \times 1.7\text{ k}} = -0.64\text{ V/V}$$

Thus, CMRR = $|A_d/A_c| = 8.8/0.64 = 13.76$, or 22.78 dB.

(c) For $I_D = I_Q/2 = 5$ mA and $V_{DD} = 15$ V, the maximum value of R_D is

$$R_{D(max)} = \frac{V_{DD}}{I_D} = \frac{15\text{ V}}{5\text{ mA}} = 3\text{ k}\Omega$$

which gives the maximum value of

$$A_{d(max)} = -g_m R_{D(max)} = -5\text{ mA} \times 3\text{ k}\Omega = -15\text{ V/V}$$

1.4.2 MOS Differential Pair with Active Load

MOS differential amplifiers are normally used with current mirror active loads. A commonly used configuration is shown in Fig. 1.14. The current mirror consists of transistors M_3 and M_4. Increasing the input voltage to M_1 by $v_{id}/2$ will cause the drain current of M_1 to increase by an amount $g_m v_{id}/2$. This increase will cause a similar increase in the drain current of M_4 due to the current mirror effect and also a decrease in the drain current of M_2. Since M_1 and M_2 are NMOS transistors and their complements (M_3 and M_4) are PMOS types, this configuration is known as a *CMOS amplifier*; the manufacturing process by which the amplifiers are produced is known as *CMOS technology*.

The small-signal equivalent of the output side of the CMOS amplifier is shown in Fig. 1.15. The output resistance R_o is the parallel combination of r_{o2} and r_{o4}. That is,

$$R_o = r_{o2} \| r_{o4} \tag{1.60}$$

where

$$r_{o2} = \frac{2V_M}{I_Q} = \text{output resistance of transistor } M_2$$

$$r_{o4} = \frac{2V_M}{I_Q} = \text{output resistance of transistor } M_4$$

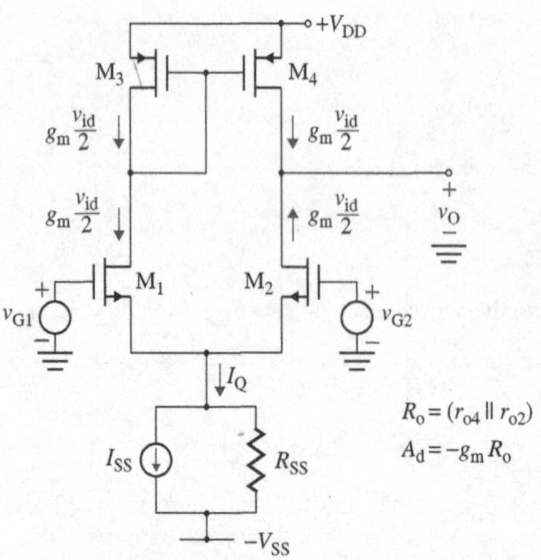

FIGURE 1.14 CMOS differential amplifier

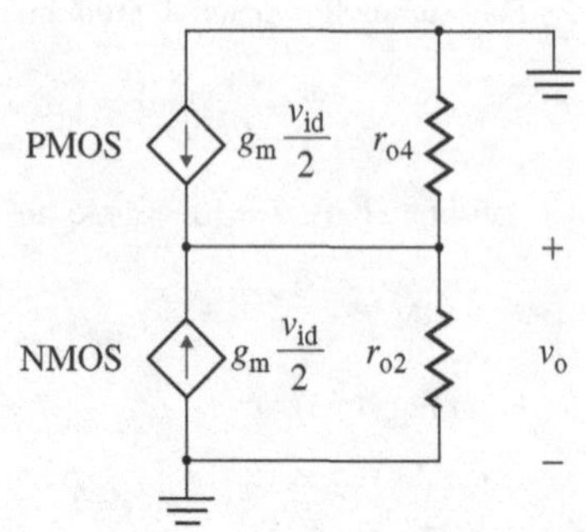

FIGURE 1.15 Small-signal equivalent of the CMOS amplifier

Using Eq. (1.55), we find the differential voltage gain A_d:

$$A_d = \frac{v_o}{v_{id}} = -g_m(r_{o2} \| r_{o4}) \tag{1.61}$$

where g_m, which is the transconductance of MOSFET M_2, is given by Eq. (1.54). Substituting $r_{o4} = r_{o2} = 2V_M/I_Q$ and $g_m = \sqrt{2K_n I_Q}$ and simplifying, we get

$$A_d = -\sqrt{\frac{2K_n}{I_Q}} V_M \tag{1.62}$$

which gives a higher voltage gain for a lower value of I_Q.

EXAMPLE 1.5

D

Designing a CMOS amplifier

(a) Design the CMOS amplifier shown in Fig. 1.16 by determining the *W/L* ratios of the MOSFETs and the threshold voltage V_t. The differential voltage gain should be $A_d = 60$ V/V at biasing current $I_Q = 10$ μA. Assume identical transistors whose channel modulation voltage is $V_M = 20$ V, channel constant is $K_x = 20$ μA/V², and channel length is $L = 10$ μm. The *W/L* ratio of the current source is 2, and $K_x = 10$ μA/V². Assume $V_{DD} = V_{SS} = 5$ V.

(b) Use PSpice/SPICE to find the small-signal differential voltage gain A_d for $v_{G1} = 1$ mV and $v_{G2} = 0$.

SOLUTION

(a) $A_d = 60$, $I_Q = 10$ μA, $V_M = 20$ V, $K_x = 20$ μA/V², and $L = 10$ μm. From Eq. (1.62), we find the MOS constant K_n as

$$K_{n2} = \left(\frac{I_Q}{2}\right)\left(\frac{A_d}{V_M}\right)^2 = \left(\frac{10\ \mu\text{A}}{2}\right)\left(\frac{60}{20}\right)^2 = 45\ \mu\text{A/V}^2$$

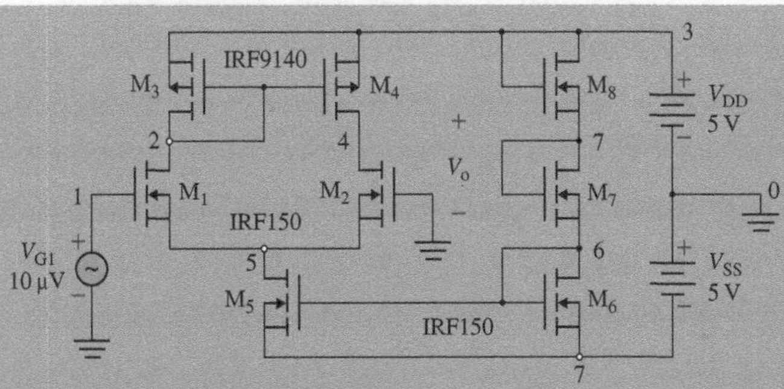

FIGURE 1.16 CMOS amplifier

From Eq. (1.21), we find the W/L ratio as

$$\frac{W}{L} = \frac{K_n}{K_x} = \frac{45\ \mu}{20\ \mu} = 2.25$$

Since $L = 10\ \mu\text{m}$, the desired value is $W = 2.25 \times 10\ \mu\text{m} = 22.5\ \mu\text{m}$. For identical transistors for the current source,

$$V_{GS6} = \frac{V_{DD} + V_{SS}}{3} = \frac{(5 + 5)\text{ V}}{3} = 3.33\text{ V}$$

Since $W = 2L$ for the MOSFETs of the current source, the MOS constant is

$$K_{n6} = \frac{W_6 K_x}{2L_6} = \frac{2 \times 10}{2}\ \mu\text{A/V}^2 = 10\ \mu\text{A/V}^2$$

The biasing current is given by

$$I_Q = K_{n6}(V_{GS6} - V_t)^2$$

which, for $V_{GS6} = 3.33$ V, $I_Q = 10\ \mu$A, and $K_{n6} = 10\ \mu\text{A/V}^2$, gives $V_t = 2.33$ V. Then

$$r_{o2} = r_{o4} = \frac{2V_M}{I_Q} = \frac{2 \times 20}{10\ \mu} = 4\text{ M}\Omega$$

$$R_o = r_{o2} \| r_{o4} = 2\text{ M}\Omega$$

$$r_{o5} = \frac{V_M}{I_Q} = \frac{20}{10\ \mu} = 2\text{ M}\Omega$$

(b) The CMOS amplifier for PSpice simulation is shown in Fig. 1.17. The plot of the output voltage is shown in Fig. 1.18, which gives a voltage gain 81.72 V/V (expected value 60). Note there is no phase shift of 180° because the output is taken from the drain terminal of the pair.

The results of simulation (.TF analysis) are as follows. (The expected values are listed on the right.)

`****  SMALL-SIGNAL CHARACTERISTICS`	
`V(4)/VID1=8.189E+01=81.89`	($A_{id} = 60$ V/V)
`INPUT RESISTANCE AT VG1=1.000E+20`	($R_{id} = \infty$)
`OUTPUT RESISTANCE AT V(4)=2.221E+06=2.221 MΩ`	($R_{od} = 2$ MΩ)
`ID1=5.44E-06=5.44 μA`	($I_{D1} = 5.44\ \mu$A)
`ID=ID5=1.09E-05=10.9 μA`	($I_Q = 10\ \mu$A)

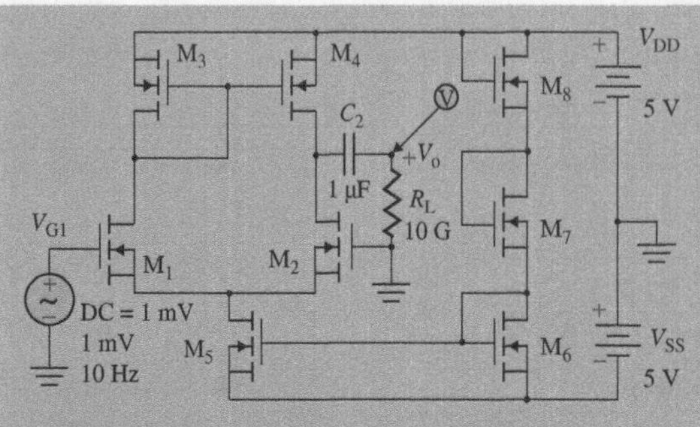

FIGURE 1.17 CMOS amplifier for a PSpice simulation for Example 1.5

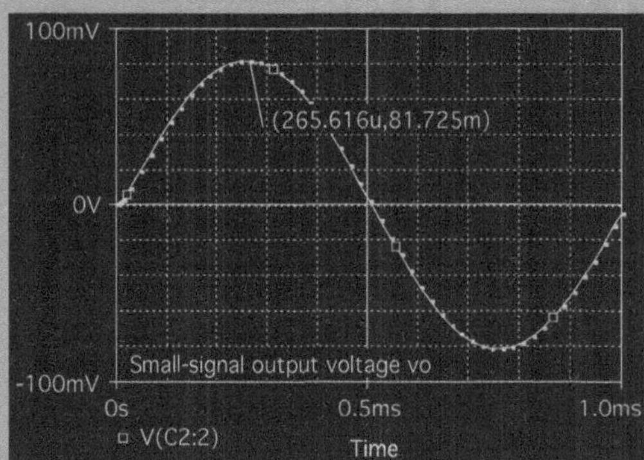

FIGURE 1.18 PSpice plot of small-signal output voltage for Example 1.5

NOTE: As expected, the PSpice results depend on the values of the *W/L* ratio for the MOSFETs. If you run PSpice from the schematic, you will need to change the model parameters of the MOSFETs; otherwise, the results will be different from those shown above.

1.4.3 Cascoded MOS Differential Amplifier

We can notice from Eq. (1.61) that the differential gain increases with the output resistance R_o of the active amplifier load. MOSFETs can be connected in cascode configuration as shown in Fig. 1.6(a) to increase the output resistance and to improve the frequency response. A common modification to Fig. 1.14(a) is shown in Fig. 1.19(a). Transistors M_1 and M_2 are connected in a common-source configuration and form a common-source differential pair, whereas M_3 and M_4 are connected in a common-gate configuration and form a common-gate differential stage. M_6 and M_8 act as the load. For identical MOSFETs, the DC biasing currents are equal, $I_{D1} = I_{D2} = I_{D4} = I_{D6} = I_{D8} = I_Q/2$. All MOSFETs have equal output resistances, $r_{o1} = r_{o2} = r_{o4} = r_{o6} = r_{o8} = V_M/I_{D1}$, and equal transconductances, $g_{m1} = g_{m2} = g_{m4} = g_{m6} = g_{m8} = 2K_n(V_{GS1} - V_t)$. The small-signal half-circuit is shown in Fig. 1.19(b). Using Eq. (1.34), we can find the effective load resistance offered by M_6 and M_8 as given by

$$r'_{o6} = r_{o6}(2 + g_{m1}r_{o6}) \tag{1.63}$$

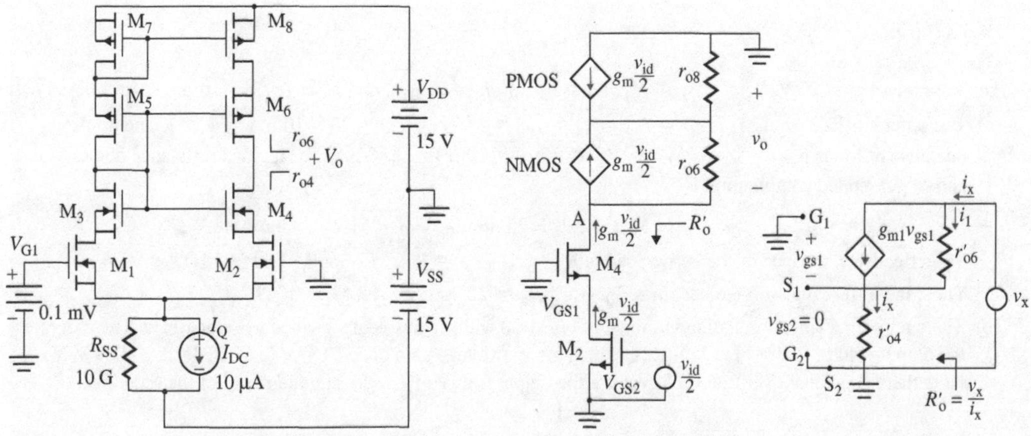

(a) **Cascoded MOS amplifier** (b) **Small-signal half-circuit** (c) **Equivalent circuit**

FIGURE 1.19 Cascoded MOS differential amplifier

Similarly, we can find the effective resistance offered by M_2 and M_4 as given by

$$r'_{o4} = r_{o4}(2 + g_{m1}r_{o4}) \tag{1.64}$$

Therefore, the effective output resistance at the output terminal as shown in Fig. 1.19(c) is given by

$$R_o = r'_{o4} \| r'_{o6} \tag{1.65}$$

Applying Eq. (1.61) we can find the small-signal differential output voltage as given by

$$A_d = \frac{v_{od}}{v_{id}} = -g_{m1}R_o = -g_{m1}(r'_{o4} \| r'_{o6}) \tag{1.66}$$

This should give a much larger differential gain compared to that of Fig. 1.14(a). For example, if $|V_M| = 20$ V, $I_{D1} = 5\ \mu$A, $g_{m1} = 31.5\ \mu$A/V, $r_{o4} = r_{o6} = 5$ MΩ, $r'_{o6} = r'_{o8} = 514$ MΩ, and $R_o = 257$ MΩ, we get $A_d = 8126$ V/V.

EXAMPLE 1.6

Analyzing cascoded MOS amplifiers The DC biasing current of the MOS amplifier shown in Fig. 1.19(a) is kept constant at $I_Q = 10\ \mu$A. All MOS transistors are identical: $V_M = 20$ V, $K_n = 25\ \mu$A/V^2, $W = 30\ \mu$m, and $L = 10\ \mu$m.

(a) Determine the differential voltage gain A_d for single-ended output at the drain terminal of M_4.

(b) Use PSpice to verify the result.

SOLUTION

You are given $V_M = 20$ V, $K_n = 25\ \mu\text{A/V}^2$, $W = 30\ \mu$m, $L = 10\ \mu$m, and $I_D = I_Q/2 = 10\ \mu\text{A}/2 = 5\ \mu$A.

(a) The output resistance of M_2 is $r_{o2} = r_{o4} = r_{o6} = r_{o8} = 2V_M/I_Q = 2 \times 20\ \text{V}/10\ \mu\text{A} = 4\ \text{M}\Omega$. The transconductance of M_2 is $g_{m2} = \sqrt{2K_n I_Q} = \sqrt{2 \times 25\ \mu \times 10\ \mu} = 22.36\ \mu$A/V. The load resistance due to M_2 and M_4 cascoded combination is

$$r'_{o4} = r'_{o6} = r_{o4}(2 + g_{m2}r_{o4}) = 4\ \text{M}\Omega \times (2 + 22.36\ \mu\text{A/V} \times 4\ \text{M}\Omega) = 366\ \text{M}\Omega.$$

The effective resistance of the active load is $R_o = r'_{o4} \parallel r'_{o6} = 366\ \text{M} \parallel 366\ \text{M} = 183\ \text{M}\Omega$.

Thus, the differential voltage becomes $A_d = g_{m2}R_o = 22.36\ \mu \times 183\ \text{M} = 4089$ V/V.

(b) The circuit for PSpice simulation is shown in Fig. 1.20 with a sinusoidal signal of 1 μV at 10 Hz. The plot of the output voltage is shown in Fig. 1.21, which gives a voltage gain of 5370 V/V (expected value 4089 V/V). Note there is no phase shift of 180° because the output is taken from the drain terminal of the other pair.

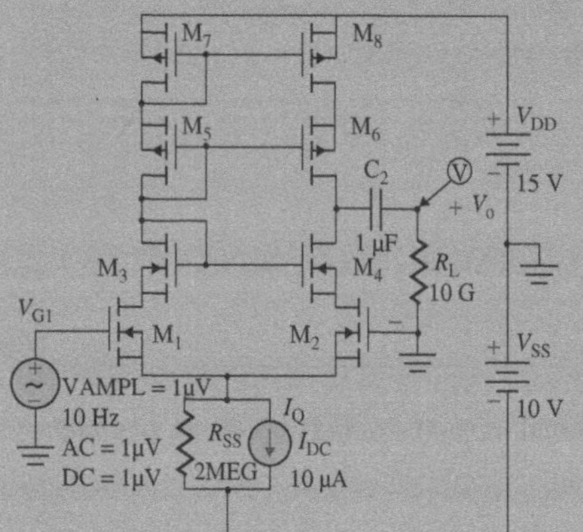

FIGURE 1.20 PSpice schematic for Example 1.6

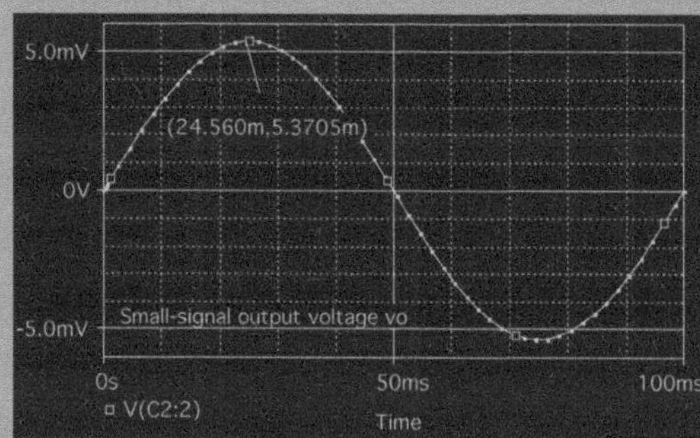

FIGURE 1.21 PSpice plot of small-signal output voltage for Example 1.6

KEY POINTS OF SECTION 1.4

- A MOS amplifier exhibits a linear DC characteristic and has a very high input resistance, tending to infinity.
- It is relatively easy to connect MOS transistors in cascode form in order to control the drain current and give high output resistance. A high-voltage gain can be obtained with a cascode connection.

1.5 Depletion MOS Differential Amplifiers

The depletion MOS amplifiers are similar to the enhancement MOSs, except that the gate–source voltage of a depletion MOS can range from a negative value to a positive value. The drain current becomes the maximum $i_{D(max)} = I_{DSS}$ when $v_{GS} = 0$.

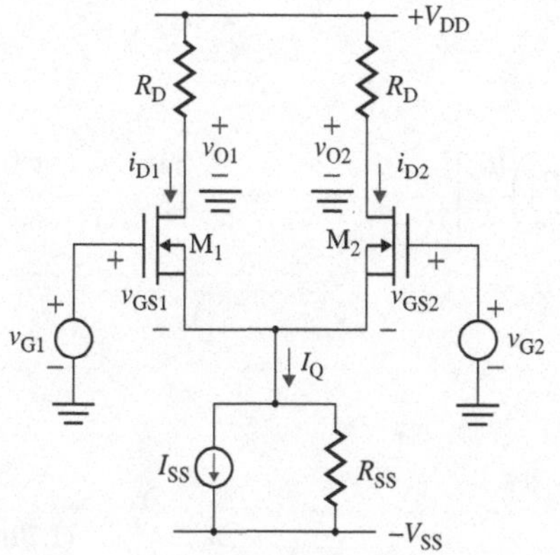

FIGURE 1.22 Depletion MOSFET differential pair

1.5.1 Depletion MOS Differential Pair with Resistive Load

A depletion NMOS source-coupled pair is shown in Fig. 1.22. Although the DC biasing circuit can be either a simple resistor (in which case the equivalent current generator will be zero) or a transistor current source, a current source is generally used.

DC Transfer Characteristics

The following analysis is performed for an *n*-channel MOS pair, but it is equally applicable to a *p*-channel pair with appropriate sign changes. The analysis can be simplified by making the following assumptions:

1. The output resistances of the MOSFETs are infinite: $r_d = \infty$.
2. The MOSFETs are identical and operate in the saturation region. The pinch-down voltages are the same, $V_{p1} = V_{p2} = V_p$, and the drain currents (with drain $V_{GS} = 0$) are equal, $I_{DSS1} = I_{DSS2} = I_{DSS}$.
3. The output resistance of the transistor current source is infinite: $R_{SS} = \infty$.

Using KVL around the loop formed by the two input voltages and two gate–source junctions, we get

$$v_{G1} - v_{GS1} + v_{GS2} - v_{G2} = 0 \tag{1.67}$$

Assuming that the drain current is related to v_{GS} by the approximate square law relationship in Eq. (G.16) in Appendix G, we get

$$i_D = I_{DSS}\left(1 - \frac{v_{GS}}{V_p}\right)^2$$

from which we can find

$$\frac{v_{GS}}{V_p} = 1 - \left(\frac{i_D}{I_{DSS}}\right)^{1/2} \tag{1.68}$$

Substituting Eq. (1.68) into Eq. (1.67) yields

$$\begin{aligned} -v_{G1} + v_{G2} &= -v_{GS1} + v_{GS2} \\ &= -V_p\left[1 - \left(\frac{i_{D1}}{I_{DSS}}\right)^{1/2}\right] + V_p\left[1 - \left(\frac{i_{D2}}{I_{DSS}}\right)^{1/2}\right] \\ &= V_p\left(\frac{i_{D1}}{I_{DSS}}\right)^{1/2} - V_p\left(\frac{i_{D2}}{I_{DSS}}\right)^{1/2} \end{aligned} \tag{1.69}$$

which gives

$$-\frac{v_{G1} + v_{G2}}{V_p} = \frac{v_{id}}{V_p} = \left(\frac{i_{D1}}{I_{DSS}}\right)^{1/2} - \left(\frac{i_{D2}}{I_{DSS}}\right)^{1/2} \tag{1.70}$$

where $v_{id} = v_{G2} - v_{G1}$ is the differential DC voltage.

Using KCL at the source nodes of the transistors gives

$$I_Q = i_{D1} + i_{D2} \tag{1.71}$$

Substituting Eq. (1.71) into Eq. (1.70) and solving the resultant quadratic, we get

$$i_{D1} = \frac{I_Q}{2} + \frac{I_Q}{2}\left(\frac{v_{id}}{V_p}\right)\left[2\left(\frac{I_{DSS}}{I_Q}\right) - \left(\frac{v_{id}}{V_p}\right)^2\left(\frac{I_{DSS}}{I_Q}\right)^2\right]^{1/2} \tag{1.72}$$

$$i_{D2} = \frac{I_Q}{2} - \frac{I_Q}{2}\left(\frac{v_{id}}{V_p}\right)\left[2\left(\frac{I_{DSS}}{I_Q}\right) - \left(\frac{v_{id}}{V_p}\right)^2\left(\frac{I_{DSS}}{I_Q}\right)^2\right]^{1/2} \tag{1.73}$$

If v_{id} is sufficiently large, all of the biasing current I_Q must flow through only one of the MOSFETs. The range of v_{id} for which both transistors conduct can be found from Eq. (1.72) with the condition $i_{D2} = 0$; that is,

$$\frac{I_Q}{2} - \frac{I_Q}{2}\left(\frac{v_{id}}{V_p}\right)\left[2\left(\frac{I_{DSS}}{I_Q}\right) - \left(\frac{v_{id}}{V_p}\right)^2\left(\frac{I_{DSS}}{I_Q}\right)^2\right]^{1/2} = 0$$

which gives

$$\left|\frac{v_{id}}{V_p}\right| \le \sqrt{\frac{I_Q}{I_{DSS}}} \tag{1.74}$$

This equation gives the value of v_{id} for which the current I_Q is carried by one of the two transistors. Typical drain currents for various values of I_Q are shown in Fig. 1.23. The range of v_{id} for which the circuit exhibits a linear characteristic is approximately equal to the pinch-down voltage $|V_p|$ (typically 2 V to 5 V), compared to V_T (26 mV) for BJTs.

The output voltages for a MOS differential pair are

$$v_{O1} = V_{DD} - i_{D1}R_D \tag{1.75}$$

$$v_{O2} = V_{DD} - i_{D2}R_D \tag{1.76}$$

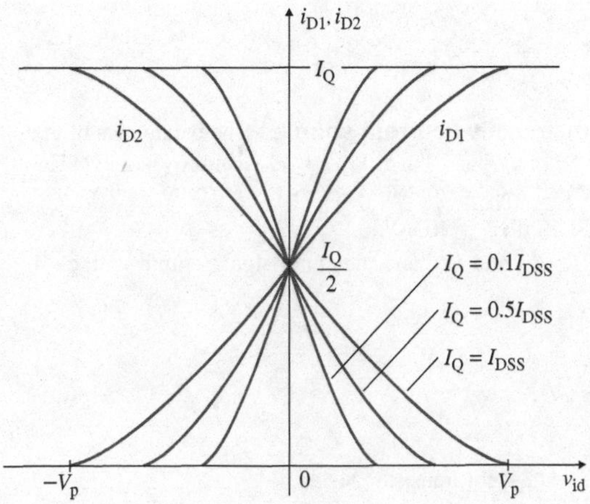

FIGURE 1.23 DC transfer characteristic of depletion NMOS pairs

The differential DC output voltage is

$$v_{od} = v_{O1} - v_{O2} = R_D(i_{D2} - i_{D1})$$

Substituting i_{D1} and i_{D2} from Eqs. (1.72) and (1.73) into the above equation and simplifying, we get

$$v_{od} = -\frac{I_Q R_D}{V_p} v_{id}\left[2\left(\frac{I_{DSS}}{I_Q}\right) - \left(\frac{v_{id}}{V_p}\right)^2\left(\frac{I_{DSS}}{I_Q}\right)^2\right]^{1/2} \tag{1.77}$$

As with the any differential circuit, if v_{id} is zero, v_{od} is also zero. A MOS-coupled pair allows direct coupling of cascoded stages without introducting DC offsets.

Small-Signal Analysis

The drain current is $i_D = \dfrac{I_{DSS}}{V_p{}^2}(V_{GS} - V_p)^2$ and the transconductance is given by

$$g_m = \frac{di_D}{dv_{GS}}$$

$$= \frac{2}{|V_p|}[|I_D I_{DSS}|]^{1/2} \tag{1.78}$$

The small-signal analyses for differential and common-mode gains are the same as the enhancement MOS pair. From Eqs. (1.55) and (1.57), we get

$$A_d = -g_m(R_D \,\|\, r_{o1}) \tag{1.79}$$

$$A_c = \frac{-g_m R_D}{1 + g_m 2R_{SS}} \tag{1.80}$$

▶ **NOTE** To simplify the analysis, the effect of transistor output resistance r_o is not included in Eq. (1.80).

EXAMPLE 1.7

Analyzing a depletion MOS differential pair with an active current source The parameters of the depletion MOS differential pair in Fig. 1.22 are $R_{SS} = 50\ k\Omega$, $I_Q = 10$ mA, $V_{DD} = 30$ V, and $R_D = 5\ k\Omega$. The depletion MOSFETs are identical and have $V_p = -4$ V and $I_{DSS} = 20$ mA. Assume $V_M = 100$ V.

(a) Calculate the DC drain currents through the MOSFETs if $v_{id} = 10$ mV.

(b) Assuming $I_{D1} = I_{D2}$, calculate A_d, A_c, and CMRR; R_{id} and R_{ic}; and the small-signal output voltage if $v_{g1} = 10$ mV and $v_{g2} = 20$ mV.

(c) Find the drain voltage V_D.

SOLUTION

(a) For $v_{id} = 10$ mV, Eq. (1.72) gives the DC drain current i_{D1} for transistor M_1 as

$$i_{D1} = \frac{10\text{ m}}{2}\left\{1 + \frac{100\text{ m}}{-4}\left[2\left(\frac{20\text{ m}}{10\text{ m}}\right) - \left(\frac{100\text{ m}}{-4}\right)^2\left(\frac{20\text{ m}}{10\text{ m}}\right)^2\right]^{1/2}\right\} = 5.25\text{ mA}$$

$$i_{D2} = I_Q - i_{D1} = 10\text{ mA} - 5.25\text{ mA} = 4.75\text{ mA}$$

(b) We know that $I_{D1} = I_{D2} = I_Q/2 = 10\text{ mA}/2 = 5$ mA. From Eq. (1.78),

$$g_m = \frac{2[|I_{D1}I_{DSS}|]^{1/2}}{|V_p|} = \left(\frac{2}{4}\right) \times \sqrt{5\text{ mA} \times 20\text{ mA}} = 5\text{ mA/V}$$

$$r_{o1} = \frac{V_M}{I_D} = \frac{100}{5\text{ m}} = 20\ k\Omega$$

From Eq. (1.79), the single-ended differential voltage gain A_d is

$$A_d = -g_m(R_D \| r_{o1}) = -5\text{ m} \times (5\text{ k} \| 20\text{ k}) = -20\text{ V/V}$$

From Eq. (1.80), the single-ended common-mode voltage gain A_c is

$$A_c = \frac{-g_m R_D}{1 + g_m 2R_{SS}} = \frac{-5\text{ m} \times 5\text{ k}}{1 + (5\text{ m} \times 2 \times 50\text{ k})} = -0.0399\text{ V/V}$$

Thus, CMRR $= |A_d/A_c| = 20/0.0399 = 501$ (or 54 dB). From Eq. (1.59),

$$R_{id} = R_{ic} = \infty$$

We know that

$$v_{id} = v_{g2} - v_{g1} = 20\text{ mV} - 10\text{ mV} = 10\text{ mV}$$

and $$v_{ic} = \frac{v_{g1} + v_{g2}}{2} = \frac{10\text{ mV} + 20\text{ mV}}{2} = 15\text{ mV}$$

Using Eq. (1.10), we have

$$v_o = A_d v_{id} + A_c v_{ic} = -20 \times 10\text{ mV} - 0.0399 \times 15\text{ mV} = -250.6\text{ mV}$$

(c) The DC drain voltage at the drain terminal of a transistor is

$$V_D = V_{DD} - I_D R_D = 30\text{ V} - 5\text{ mA} \times 5\ k\Omega = 5\text{ V}$$

Thus, for $A_d = -20$, the maximum differential voltage will be $v_{id} = 5/20 = 250$ mV. Therefore, V_{DD} must be greater than $I_D R_D$ in order to allow output voltage swing due to the input voltages.

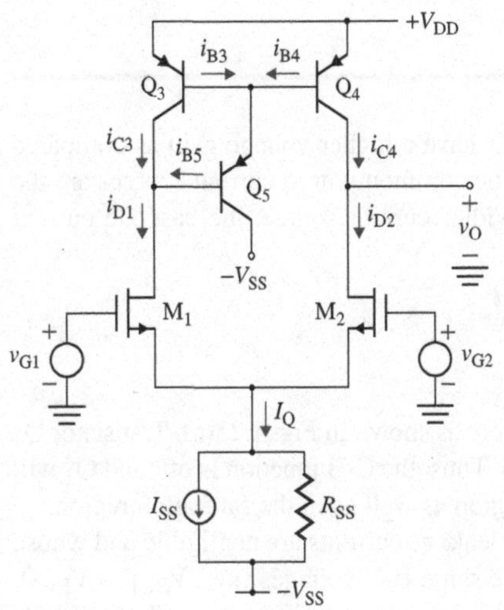

FIGURE 1.24 Depletion MOS differential amplifier with current mirror active load

1.5.2 Depletion MOS Differential Pair with Active Load

Like MOS differential amplifiers, depletion MOS differential amplifiers use a current mirror active load to achieve a large voltage gain. The active load can be made either with MOSFETs similar to that in Fig. 1.24 or with BJTs. A depletion MOS differential amplifier with a basic current source as the BJT active load is shown in Fig. 1.24. From Eq. (1.60), we know that the output resistance R_o is the parallel combination of r_{o2} and r_{o4}; that is,

$$R_o = r_{o2} \| r_{o4} \tag{1.81}$$

where $r_{o2} = 2V_M/I_Q$ = output resistance of transistor M_2 and $r_{o4} = 2V_A/I_Q$ = output resistance of transistor Q_4.

Using Eq. (1.61), we find the differential voltage gain A_d:

$$A_d = \frac{v_o}{v_{id}} = -g_m(r_{o2} \| r_{o4}) \tag{1.82}$$

where g_m, which is the transconductance of MOSFET M_2, is given by Eq. (1.78).

KEY POINTS OF SECTION 1.5

- A MOSFET amplifier has a very high input resistance, in the range of $10^9\ \Omega$ to $10^{12}\ \Omega$.
- An active load considerably increases the differential gain of a MOSFET amplifier.
- The range of v_{id} for which the circuit exhibits a linear DC characteristic is much higher for MOSFETs than for BJTs. For MOSFETs, this range is approximately equal to the pinch-down voltage $|V_p|$ (typically 2 V to 5 V), compared to V_T (26 mV) for BJTs.

1.6 BJT Current Sources

The BJT current sources are similar to MOS sources. Since BJTs have a higher voltage gain as compared to MOSFETs, BJT sources can offer higher output resistances. The commonly used current sources are the basic current sources, the modified basic current source, the Widlar current source, the cascode current source, and the Wilson current source.

1.6.1 Basic Current Source

The simplest current source consists of a resistor and two transistors as shown in Fig. 1.25(a). Transistor Q_1 is diode connected, and its C-B voltage is forced to zero: $v_{CB} = 0$. Thus, the C-B junction is off, and Q_1 will operate in the active region. Transistor Q_2 can be in the active region as well as in the saturation region.

Let us assume that Q_1 and Q_2 are identical transistors whose leakage currents are negligible and whose output resistances are infinite. Since the two transistors have the same B-E voltages (i.e., $V_{BE1} = V_{BE2}$), the collector and base currents are equal: $I_{C1} = I_{C2}$ and $I_{B1} = I_{B2}$. Applying KCL at the collector of Q_1 gives the reference current:

$$I_R = I_{C1} + I_{B1} + I_{B2} = I_{C1} + 2I_{B1}$$

Since $I_{C1} = \beta_F I_{B1}$,

$$I_R = I_{C1} + 2I_{B1} = I_{C1} + \frac{2I_{C1}}{\beta_F}$$

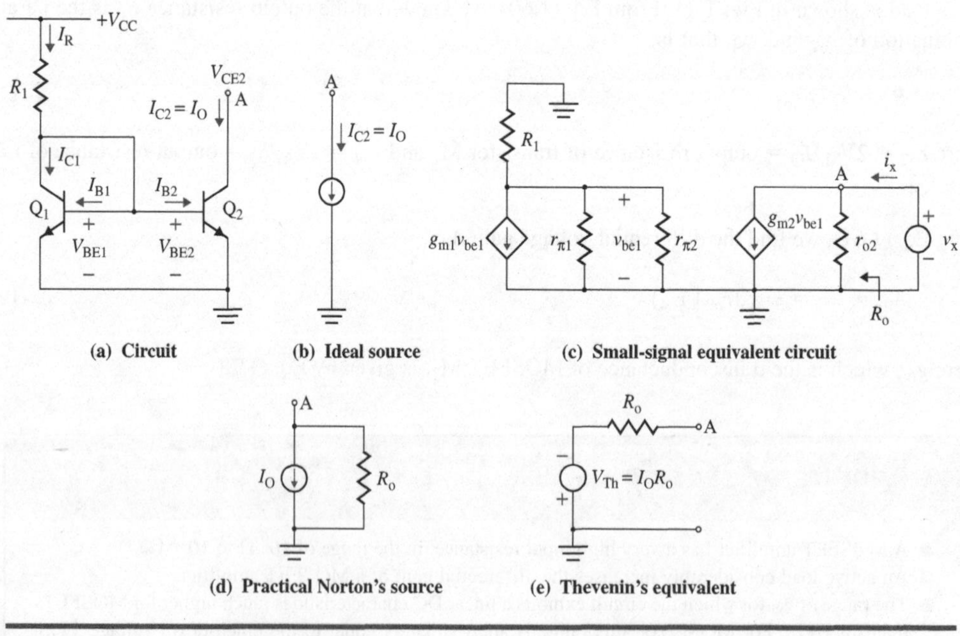

FIGURE 1.25 Basic current source

which gives the collector current I_{C1} as

$$I_{C1} = I_{C2} = \frac{I_R}{1 + 2/\beta_F} \tag{1.83}$$

$$= \frac{V_{CC} - V_{BE1}}{R_1} \times \frac{1}{1 + 2/\beta_F} \tag{1.84}$$

If the DC current gain $\beta_F >> 2$, which is usually the case, Eq. (1.84) is reduced to

$$I_{C1} = I_{C2} \approx I_R$$

Thus, for two identical transistors, the reference and output currents are almost equal. I_{C2}, which is the mirror image of I_{C1}, is known as the *mirror current* of I_{C1}. For transistors with small values of β_F, the current ratio will not be unity. In practice, however, the transistors may not be identical, and the two collector currents will have a constant ratio. The equivalent circuit for the ideal current source is shown in Fig. 1.25(b).

For a transistor with finite output resistance, the effect of the Early voltage V_A should be taken into account, and the collector current in Eq. (G.19) in Appendix G can be modified to

$$I_C = I_S\left[\exp\left(\frac{V_{BE}}{V_T}\right) - 1\right]\left(1 + \frac{V_{CE}}{V_A}\right) \tag{1.85}$$

If we take into account the variation in the collector current due to the C-E voltage, the ratio of the two collector currents can be found from

$$\frac{I_{C2}}{I_{C1}} = \frac{1 + V_{CE2}/V_A}{1 + V_{CE1}/V_A} \tag{1.86}$$

Output Resistance R_o

The small-signal AC equivalent circuit for determining the output resistance is shown in Fig. 1.25(c). Output resistance R_o is the same as r_{o2}; that is,

$$R_o = \frac{v_x}{i_x} = r_{o2} = \frac{V_A}{I_{C2}} \tag{1.87}$$

Thevenin's equivalent voltage is given by

$$V_{Th} = I_O R_o = I_{C2} R_o = I_{C2}\frac{V_A}{I_{C2}} = V_A \tag{1.88}$$

Norton's and Thevenin's equivalents of the current source are shown in Fig. 1.25[(d) and (e)]. If the output of the current source is open-circuited, a voltage of $-V_{Th}$ is expected to appear across transistor Q_1. However, this will not actually happen because transistor Q_1 will saturate when the voltage across the current source (i.e., the C-E voltage of Q_1) reaches zero.

The current source in Fig. 1.25(a) behaves as a current sink rather than a source. A current source equivalent to this current sink can be obtained by using *pnp* transistors and a negative power supply. This arrangement is shown in Fig. 1.26.

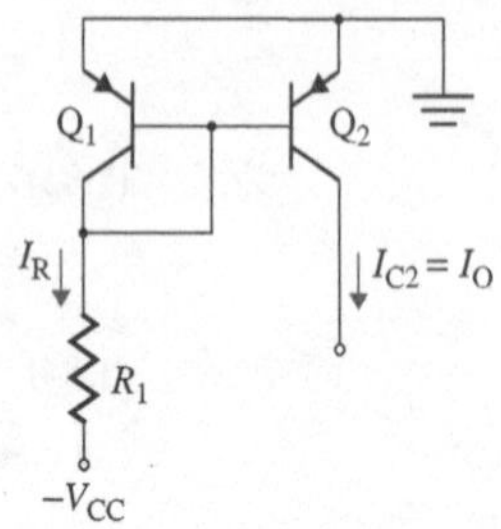

FIGURE 1.26 Current source using *pnp* transistors

EXAMPLE 1.8

D

Designing a simple basic current source

(a) Design the basic current source in Fig. 1.25(a) to give an output current of $I_O = 5\ \mu A$. The transistor parameters are $\beta_F = 100$, $V_{CC} = 30$ V, $V_{BE1} = V_{BE2} = V_{CE1} = 0.7$ V, and $V_A = 150$ V.

(b) Calculate the output resistance R_o, Thevenin's equivalent voltage V_{Th}, and the collector current ratio if $V_{CE2} = 20$ V.

SOLUTION

$V_{BE1} = V_{BE2} = V_{CE1} = 0.7$ V, and $V_A = 150$ V.

(a) From Eq. (1.83),

$$I_O = I_{C2} = I_{C1} = \frac{I_R}{1 + 2/\beta_F}$$

which, for $I_O = 5\ \mu A$, gives $I_R = (5\ \mu A)(1 + 2/\beta_F) = 5.1\ \mu A$. From Eq. (1.84),

$$R_1 = \frac{V_{CC} - V_{BE1}}{I_R} = \frac{30\text{ V} - 0.7\text{ V}}{5.1\ \mu\text{A}} = 5.75\text{ M}\Omega$$

(b) From Eq. (1.87),

$$R_o = \frac{V_A}{I_{C2}} = \frac{150\text{ V}}{5\ \mu\text{A}} = 30\text{ M}\Omega$$

From Eq. (1.88),

$$V_{Th} = V_A = 150\text{ V}$$

From Eq. (1.86),

$$\frac{I_{C2}}{I_{C1}} = \frac{1 + V_{CE2}/V_A}{1 + V_{CE1}/V_A} = \frac{1 + 20/150}{1 + 0.7/150} = 1.128$$

Thus, $I_{C2} = 1.128 \times I_{C1} = 1.128 \times 5\ \mu A = 5.64\ \mu A$, which agrees, with a degree of error, with the desired value of $I_O = 5\ \mu A$.

▶ **NOTES**

1. For a current source of 5 μA, a resistor of 5.75 MΩ would be required. Resistors of such high value are very costly in terms of the die area. Resistors over 50 kΩ are generally avoided for IC applications. Thus, this current source is not suitable for generating a current of less than about 0.6 mA at $V_{CC} = 30$ V and 0.3 mA at $V_{CC} = 15$ V.
2. If the output of the current source is open-circuited, a voltage of $-V_{Th} = -150$ V will not appear across transistor Q_1. Rather, the voltage will be $V_{CE1(sat)} \approx 0.2$ V.

1.6.2 Modified Basic Current Source

Notice from Eq. (1.83) that the collector current I_{C2} $(=I_{C1})$ differs from the reference current I_R by a factor of $(1 + 2/\beta_F)$. For low-gain transistors (especially *pnp* types), I_{C2} can differ significantly from I_R. The error can be reduced by adding another transistor so that I_{C2} becomes less dependent on the transistor parameter β_F. This type of circuit is shown in Fig. 1.27(a). Applying KCL at the emitter of transistor Q_3 gives

$$I_{E3} = I_{B1} + I_{B2} = \frac{I_{C1}}{\beta_F} + \frac{I_{C2}}{\beta_F} \tag{1.89}$$

Since $V_{BE1} = V_{BE2}$, it follows that $I_{C1} = I_{C2}$. Thus, Eq. (1.89) becomes

$$I_{E3} = \frac{2}{\beta_F} I_{C2}$$

The base current of Q_3 is related to I_{E3} by

$$I_{B3} = \frac{I_{E3}}{1 + \beta_F} = \frac{2}{\beta_F(1 + \beta_F)} I_{C2} \tag{1.90}$$

Using KCL at the collector of Q_1 gives

$$I_R = I_{C1} + I_{B3} = I_{C1} + \frac{2}{\beta_F(1 + \beta_F)} I_{C2} \tag{1.91}$$

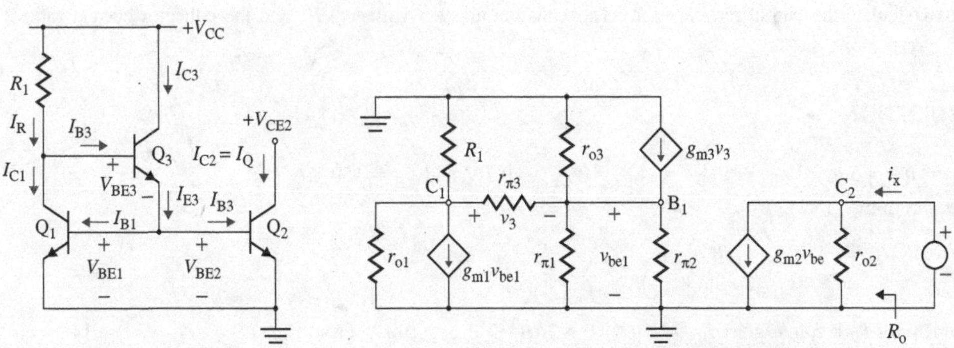

(a) Circuit (b) Small-signal equivalent circuit

FIGURE 1.27 Modified basic current source

Since $I_{C1} = I_{C2}$, Eq. (1.91) gives the output current I_O as

$$I_O = I_{C2} = \frac{I_R}{1 + 2/(\beta_F^2 + \beta_F)} \tag{1.92}$$

which indicates that the reference current I_R is related to the output current I_O by a factor of only $[1 + 2/(\beta_F^2 + \beta_F)]$. The reference current can be found from

$$I_R = \frac{V_{CC} - V_{BE1} - V_{BE3}}{R_1} \tag{1.93}$$

In the derivation of Eq. (1.92), the output resistances of the transistors are ignored. However, for a finite transistor output resistance, the collector current ratio in Eq. (1.86) is applicable.

Output Resistance R_o

The small-signal equivalent circuit for determining the output resistance is shown in Fig. 1.27(b). The output resistance R_o is the same as r_{o2}; that is,

$$R_o = \frac{v_x}{i_x} = r_{o2} = \frac{V_A}{I_{C2}} \tag{1.94}$$

Thevenin's equivalent voltage is given by

$$V_{Th} = I_O R_o = I_{C2} R_o = I_{C2} \frac{V_A}{I_{C2}} = V_A \tag{1.95}$$

EXAMPLE 1.9

D

Designing a simple modified basic current source

(a) Design the modified basic current source in Fig. 1.27(a) to give an output current of $I_O = 5\ \mu A$. The transistor parameters are $\beta_F = 100$, $V_{CC} = 30$ V, $V_{BE1} = V_{BE2} = V_{BE3} = 0.7$ V, and $V_A = 150$ V.

(b) Calculate the output resistance R_o, Thevenin's equivalent voltage V_{Th}, and the collector current ratio if $V_{CE2} = 20$ V.

SOLUTION

$I_O = I_{C2} = 5\ \mu A$, $V_{BE1} = V_{BE2} = V_{BE3} = V_{CE1} = 0.7$ V, and $V_A = 150$ V.

(a) From Eq. (1.92),

$$I_O = I_{C2} = \frac{I_R}{1 + 2/(\beta_F^2 + \beta_F)}$$

which, for $I_O = 5\ \mu A$, gives $I_R = (5\ \mu A)[1 + 2/(\beta_F^2 + \beta_F)] = 5\ \mu A$. From Eq. (1.93),

$$R_1 = \frac{V_{CC} - V_{BE1} - V_{BE3}}{I_R} = \frac{30 - 0.7 - 0.7}{5\ \mu A}$$

which gives the required value of the resistor as $R_1 = 5.72\ \text{M}\Omega$.

(b) From Eq. (1.94),

$$R_o = \frac{V_A}{I_{C2}} = \frac{150}{5\ \mu\text{A}} = 30\ \text{M}\Omega$$

From Eq. (1.95),

$$V_{Th} = V_A = 150\ \text{V}$$

From Eq. (1.86),

$$\frac{I_{C2}}{I_{C1}} = \frac{1 + V_{CE2}/V_A}{1 + (V_{BE1} + V_{BE3})/V_A} = \frac{1 + 20/150}{1 + (0.7 + 0.7)/150} = 1.123$$

Neglecting I_{B3}, we have $I_{C1} \approx I_O = 5\ \mu\text{A}$. Thus, $I_{C2} = 1.123 \times I_{C1} = 1.123 \times 5\ \mu\text{A} = 5.62\ \mu\text{A}$, which agrees, with a degree of error, with the desired value of $I_O = 5\ \mu\text{A}$.

1.6.3 Widlar Current Source

Biasing currents of low magnitudes, typically on the order of 5 μA, are required in a variety of applications. Currents of low magnitude can be obtained by inserting a resistance of moderate value in series with the emitter Q_2 in Fig. 1.25(a). A circuit with this modification, as shown in Fig. 1.28(a), is known as a *Widlar current source*. As a result of the addition of R_2 into the circuit, I_{C2} is no longer equal to I_R, and the value of I_{C2} can be made much smaller than that of I_{C1}. This circuit can give currents in the microampere range, with acceptable circuit resistance values of less than 50 kΩ.

Using KVL around the B-E loop in Fig. 1.28(a) gives

$$V_{BE1} - V_{BE2} - (I_{C2} + I_{B2})R_2 = 0$$

Since $I_{C2} >> I_{B2}$,

$$V_{BE1} - V_{BE2} - I_{C2}R_2\left(1 + \frac{1}{\beta_F}\right) = 0 \tag{1.96}$$

Assuming that the transistors have infinite output resistances, $V_A = \infty$, Eq. (1.85) becomes

$$I_C = I_S \exp\left(\frac{V_{BE}}{V_T}\right)$$

which gives

$$V_{BE} = V_T \ln\left(\frac{I_C}{I_S}\right) \tag{1.97}$$

Applying V_{BE} from Eq. (1.97) in Eq. (1.96), we get

$$V_T \ln\left(\frac{I_{C1}}{I_{S1}}\right) - V_T \ln\left(\frac{I_{C2}}{I_{S2}}\right) - I_{C2}R_2 = 0 \quad (\text{for } \beta_F >> 1)$$

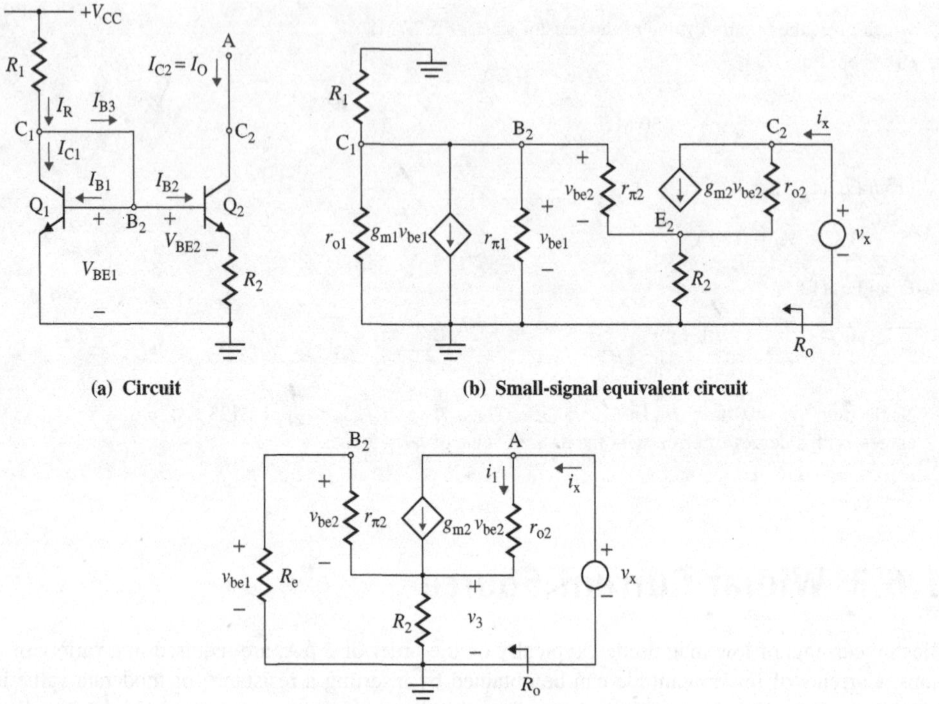

(a) Circuit

(b) Small-signal equivalent circuit

(c) Equivalent circuit

FIGURE 1.28 Widlar current source

which, for identical transistors with $I_{S1} = I_{S2}$, can be simplified to

$$V_T \ln\left(\frac{I_{C1}}{I_{S1}} \times \frac{I_{S2}}{I_{C2}}\right) - I_{C2}R_2 = 0$$

This equation gives the relation of I_{C2} to I_{C1} and R_2 as

$$V_T \ln\left(\frac{I_{C1}}{I_{C2}}\right) = I_{C2}R_2 \tag{1.98}$$

The reference current I_R can be found from

$$I_R = \frac{V_{CC} - V_{BE1}}{R_1} \tag{1.99}$$

which is also related to the base and collector currents; that is,

$$I_R = I_{C1} + I_{B1} + I_{B2}$$

$$= I_{C1}\left(1 + \frac{1}{\beta_F}\right) + \frac{I_{C2}}{\beta_F} \tag{1.100}$$

Substituting I_{C1} from Eq. (1.98) into the preceding equation, we get I_R in terms of I_{C2} and R_2 as

$$I_R = \frac{1 + \beta_F}{\beta_F} I_{C2} \exp\left(\frac{I_{C2}R_2}{V_T}\right) + \frac{I_{C2}}{\beta_F} \tag{1.101}$$

Thus, I_{C2} is a nonlinear function of I_R and R_2. To find I_{C2}, we must solve this transcendental equation by trial and error, using known values of I_R and R_2. However, for design purposes, I_{C2} and I_R are known, and it is necessary only to find the value of R_2.

Output Resistance R_o

The small-signal equivalent circuit for determining the output resistance is shown in Fig. 1.28(b). This circuit can be reduced to Fig. 1.28(c), where R_e is the parallel equivalent of $r_{\pi 1}$, $1/g_{m1}$, r_{o1}, and R_1; that is,

$$R_e = r_{\pi 1} \left\| \frac{1}{g_{m1}} \right\| R_1 \| r_{o1} \tag{1.102}$$

Since transconductance is $g_m = \dfrac{I_C}{V_T} = \dfrac{\beta_E}{r_\pi}$, we can relate the input resistance $r_{\pi 1}$ to the transconductance g_{m1} and the small-signal current gain $\beta_f(\approx\beta_F)$:

$$r_{\pi 1} = \frac{\beta_f(\approx\beta_F)}{g_{m1}}$$

Since $\beta_f >> 1$, $r_{\pi 1} >> 1/g_{m1}$, and in general $R_1 >> 1/g_{m1}$, Eq. (1.102) can be approximated by

$$R_e \approx \frac{1}{g_{m1}} \tag{1.103}$$

The collector current flows through the parallel combination of R_2 and $(r_{\pi 2} + R_e)$, so

$$v_3 = i_x[R_2 \| (r_{\pi 2} + R_e)] \tag{1.104}$$

The voltage v_3 is related to v_2 by

$$v_2 = -\frac{v_3 r_{\pi 2}}{r_{\pi 2} + R_e} = -\frac{i_x r_{\pi 2}[R_2 \| (r_{\pi 2} + R_e)]}{r_{\pi 2} + R_e} \tag{1.105}$$

The current i_1 thus becomes

$$i_1 = i_x - g_{m2}v_2$$

The test voltage v_x is

$$v_x = v_3 + r_{o2}i_1 = v_3 + r_{o2}i_x - r_{o2}g_{m2}v_2$$

$$= v_3 + i_x r_{o2} + \frac{r_{o2}g_{m2}v_3 r_{\pi 2}}{r_{\pi 2} + R_e} \tag{1.106}$$

Substituting v_3 from Eq. (1.104) into Eq. (1.106) and simplifying, we get the resistance at the collector of the transistor:

$$R_o = \frac{v_x}{i_x} = R_2 \| (r_{\pi 2} + R_e) + r_{o2}\left[1 + \frac{g_{m2}r_{\pi 2}}{r_{\pi 2} + R_e} R_2 \| (r_{\pi 2} + R_e)\right] \tag{1.107}$$

Since r_{o2} is large, the first term is much smaller than the second one. If the first term is neglected, Eq. (1.107) can be reduced to

$$R_o \approx r_{o2}\left[1 + \frac{g_{m2}r_{\pi 2}}{r_{\pi 2} + R_e} R_2 \,\|\, (r_{\pi 2} + R_e)\right] \tag{1.108}$$

Since transconductance is $g_m = \dfrac{I_C}{V_T} = \dfrac{\beta_E}{r_\pi}$, we get

$$\frac{1}{g_{m1}} = \frac{1}{I_{C1}/V_T.} = \frac{V_T}{I_{C1}} \tag{1.109}$$

$$r_{\pi 2} = \frac{\beta_F}{g_{m2}} = \frac{\beta_F}{I_{C2}/V_T} = V_T\frac{\beta_F}{I_{C2}} = \beta_F \frac{V_T}{I_{C1}} \times \frac{I_{C1}}{I_{C2}} = \frac{\beta_F}{g_{m1}} \times \frac{I_{C1}}{I_{C2}} \tag{1.110}$$

Since $I_{C1} >> I_{C2}$, $r_{\pi 2} >> 1/g_{m1}$ and $r_{\pi 2} >> R_e$; that is,

$$R_e + r_{\pi 2} \approx r_{\pi 2} \tag{1.111}$$

Substituting R_e from Eq. (1.103) into Eq. (1.107) gives

$$R_o \approx r_{o2}[1 + g_{m2}(R_2 \,\|\, r_{\pi 2})] \tag{1.112}$$

Since $\beta_F >> 1$ and $r_{\pi 2} = \beta_F/g_{m2}$, Eq. (1.112) can be rewritten as

$$R_o = r_{o2}\frac{g_{m2}R_2(1 + \beta_F) + \beta_F}{g_{m2}R_2 + \beta_F} = r_{o2}\left[\frac{1 + g_{m2}R_2}{1 + g_{m2}R_2/\beta_F}\right] \tag{1.113}$$

Generally, $\beta_F >> g_{m2}R_2$, so Eq. (1.113) can be approximated by

$$R_o \approx r_{o2}(1 + g_{m2}R_2) \tag{1.114}$$

$$\approx r_{o2}\left(1 + \frac{I_{C2}R_2}{V_T}\right) \tag{1.115}$$

Thus, R_o depends on $I_{C2}R_2$, which is the DC voltage drop across R_2. The larger this drop is made, the higher the output resistance becomes. For the Widlar source, $I_{C2}R_2$ is limited to several hundred millivolts for practical current ratios, and the corresponding value of V_{Th} is limited to about $10V_A$.

EXAMPLE 1.10

D

Designing a Widlar current source

(a) Design the Widlar current source in Fig. 1.28(a) to give $I_O = 5$ μA and $I_R = 1$ mA. The parameters are $V_{CC} = 30$ V, $V_{BE1} = 0.7$ V, $V_T = 26$ mV, $V_A = 150$ V, and $\beta_F = 100$.

(b) Calculate the output resistance R_o and Thevenin's voltage V_{Th}. Assume $V_T = 26$ mV.

SOLUTION

$I_{C2} = 5\ \mu A$, and $V_T = 26$ mV.

(a) From Eq. (1.99),

$$1\ \text{mA} = \frac{30 - 0.7}{R_1}$$

so $R_1 = 29.3\ \text{k}\Omega$. From Eq. (1.100),

$$1\ \text{mA} = I_{C1}\left(1 + \frac{1}{100}\right) + \frac{5\ \mu A}{100}$$

which gives $I_{C1} \approx 990\ \mu A$. From Eq. (1.98),

$$26\ \text{mV} \times \ln\left(\frac{990\ \mu A}{5\ \mu A}\right) = 5\ \mu A \times R_2$$

so $R_2 = 27.5\ \text{k}\Omega$.

(b) $r_{o2} = V_A/I_{C2} = 150/(5\ \mu A) = 30\ \text{M}\Omega$. From Eq. (1.110),

$$r_{\pi 2} = \frac{V_T\beta_F}{I_{C2}} = 26\ \text{m} \times \frac{100}{5\ \mu} = 520\ \text{k}\Omega$$

Also from Eq. (1.109),

$$g_{m2} = \frac{I_{C2}}{V_T} = \frac{5\ \mu A}{26\ \text{mV}} = 192.3\ \mu A/V$$

From Eq. (1.112),

$$R_o \approx 30\ \text{M}\Omega \times [1 + 192.3\ \mu A/V \times (27.5\ \text{k}\Omega \parallel 520\ \text{k}\Omega)] = 180.68\ \text{M}\Omega$$

Using the approximation in Eq. (1.115), we have

$$R_o \approx 30\ \text{M}\Omega \times \left(1 + \frac{5\ \mu A \times 27.5\ \text{k}\Omega}{26\ \text{mV}}\right) = 188.66\ \text{M}\Omega$$

and $V_{Th} = R_o I_{C2} = 188.65\ \text{M}\Omega \times 5\ \mu A = 943.3$ V

NOTE: The Widlar current source gives a low output current at high output resistance, and Thevenin's equivalent voltage is very high.

1.6.4 Cascode Current Source

The emitter resistance R_2 in the Widlar current source in Fig. 1.28(a) can be replaced by a basic current source consisting of two transistors Q_3 and Q_4. This arrangement, shown in Fig. 1.29, will give a larger output resistance. In a cascodelike connection, two or more transistors are connected in series so that their collector biasing currents are almost identical (e.g., transistors Q_1 and Q_3), whereas in a cascodelike connection, the transistors operate in parallel fashion so that one transistor drives the other (e.g., Q_3 and Q_4). According to Eq. (1.55), the larger the output resistance, the greater the voltage gain of an amplifier

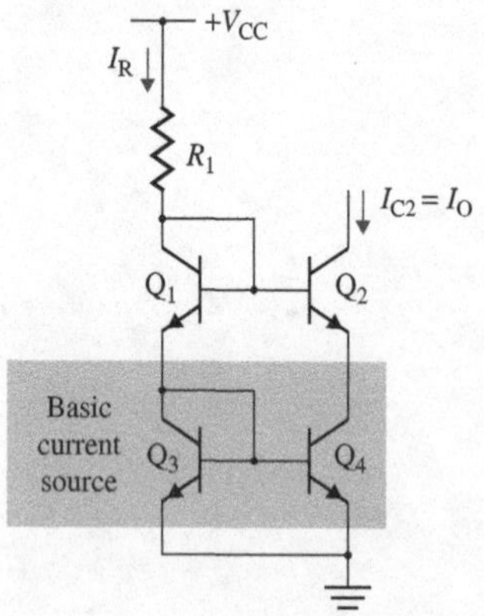

FIGURE 1.29 Cascode current source

becomes. Substituting the output resistance r_{o4} of transistor Q_4 for R_2 in Eq. (1.112) gives the output resistance of this cascode source:

$$R_o \approx r_{o2}[1 + g_{m2}(r_{o4} \| r_{\pi 2})] = r_{o2}(1 + g_{m2}r_{\pi 2}) = r_{o2}(1 + \beta_F) \tag{1.116}$$

1.6.5 Wilson Current Source

A Wilson current source, shown in Fig. 1.30(a), also gives a high output resistance. However, the output current is approximately equal to the reference current. Base current I_{B2}, which is the difference between the reference current I_R and the collector current I_{C1}, is multiplied by $(1 + \beta_F)$ to give I_{E2}, which flows in the diode-connected transistor Q_3 and causes a collector current of the same magnitude to flow in Q_1. There is a feedback path that regulates I_{C1}, which is approximately equal to I_{E2} and I_{C2}. Thus, I_{C2} remains very nearly equal to I_{C1} and nearly constant, giving a high output resistance:

$$I_{E2} = (1 + \beta_F)I_{B2} = (I_R - I_{C1})(1 + \beta_F)$$

Assuming that the transistors have infinite output resistances, $V_A = \infty$, and that the transistors are identical so that $I_{C1} = I_{C3}$, we can write

$$\begin{aligned} I_{E2} &= I_{C3} + I_{B3} + I_{B1} = I_{C3}\left(1 + \frac{1}{\beta_F}\right) + \frac{I_{C1}}{\beta_F} \\ &= I_{C3}\left(1 + \frac{2}{\beta_F}\right) \end{aligned} \tag{1.117}$$

Using Eq. (1.117), we can relate I_{C2} to I_{E2} and I_{C3}:

$$I_{C2} = I_{E2}\frac{\beta_F}{1 + \beta_F} = I_{C3}\left(1 + \frac{2}{\beta_F}\right)\frac{\beta_F}{1 + \beta_F} = I_{C3}\frac{2 + \beta_F}{1 + \beta_F}$$

which gives

$$I_{C3} = I_{C2}\frac{1 + \beta_F}{2 + \beta_F} \tag{1.118}$$

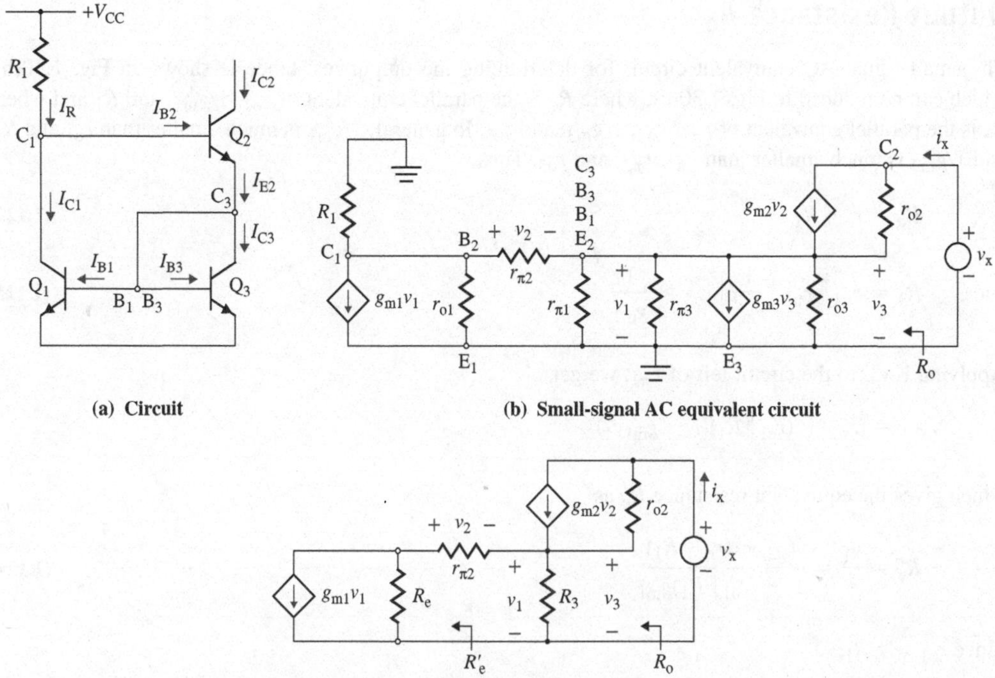

(a) Circuit

(b) Small-signal AC equivalent circuit

(c) Equivalent circuit

FIGURE 1.30 Wilson current source

I_{C1} is related to the reference current I_R by

$$I_{C1} = I_R - I_{B2} = I_R - \frac{I_{C2}}{\beta_F} \tag{1.119}$$

Since $I_{C3} = I_{C1}$, equating Eq. (1.118) to Eq. (1.119) yields

$$I_{C1} = I_{C3} = I_{C2}\frac{1 + \beta_F}{2 + \beta_F} = I_R - \frac{I_{C2}}{\beta_F}$$

which gives the output current $I_O = I_{C2}$ as

$$I_O = I_{C2} = I_R\left[\frac{(2 + \beta_F)\beta_F}{\beta_F^2 + 2\beta_F + 2}\right] = I_R\left[1 - \frac{2}{\beta_F^2 + 2\beta_F + 2}\right] \tag{1.120}$$

Since $\beta_F >> 1$, Eq. (1.120) can be approximated by

$$I_O \approx I_R \tag{1.121}$$

Thus, the output current almost equals the reference current and is less sensitive to the current gain β_F, which varies in response to temperature changes.

Output Resistance R_o

The small-signal AC equivalent circuit for determining the output resistance is shown in Fig. 1.30(b), which can be reduced to Fig. 1.30(c), where R_e is the parallel equivalent of r_{o1}, $1/g_{m1}$, and R_1 and where R_3 is the parallel equivalent of $r_{\pi 1}$, $r_{\pi 3}$, $1/g_{m3}$, and r_{o3}. In general, $1/g_{m1}$ is much smaller than r_{o1} and R_1, and $1/g_{m3}$ is much smaller than $r_{\pi 1}$, $r_{\pi 3}$, and r_{o3}. Thus,

$$R_e = r_{o1} \| R_1 \tag{1.122}$$

and
$$R_3 = r_{\pi 1} \| r_{\pi 3} \| \frac{1}{g_{m3}} \| r_{o3} \approx \frac{1}{g_{m3}} \tag{1.123}$$

Applying KVL to the circuit left of $r_{\pi 1}$, we get

$$v_1 = i r_{\pi 2} + (r_{o1} \| R_1)(i - g_{m1} v_1)$$

which gives the equivalent resistance R'_e as

$$R'_e = \frac{v_1}{i} = \frac{r_{\pi 2} + (r_{o1} \| R_1)}{1 + (r_{o1} \| R_1) g_{m1}} \tag{1.124}$$

Since $r_{o1} >> R_1$,

$$R'_e = \frac{r_{\pi 2} + R_1}{1 + g_{m1} R_1}$$

which, for $R_1 >> r_{\pi 2}$ and $g_{m1} R_1 >> 1$, can be simplified to

$$R'_e \approx \frac{1}{g_{m1}}$$

The output resistance can be found from Eq. (1.107) by substituting R_3 for R_2 and R'_e for $(r_{\pi 2} + R_e)$. That is,

$$R_o = \frac{v_x}{i_x} = R_3 \| R'_e + r_{o2}\left[1 + \frac{g_{m2} r_{\pi 2}}{R'_e}(R_2 \| R'_e)\right] \tag{1.125}$$

Since r_{o2} is large, the first term is much smaller than the second one and can be ignored. Equation (1.124) can be approximated by

$$R_o \approx r_{o2}\left[1 + \frac{g_{m2} r_{\pi 2}}{R'_e}(R_3 \| R'_e)\right] \tag{1.126}$$

Assuming $(R_3 \| R'_e)/R'_e = 1/2$, $R'_e = 1/g_{m1}$, and $g_{m1} = g_{m3}$, Eq. (1.126) becomes

$$R_o \approx r_{o2}\left(1 + \frac{g_{m2} r_{\pi 2}}{2}\right)$$

$$\approx r_{o2}\left(1 + \frac{\beta_F}{2}\right) \tag{1.127}$$

EXAMPLE 1.11

D

Designing a Wilson current source

(a) Design the Wilson current source in Fig. 1.30(a) to give $I_O = 5\ \mu A$. The parameters are $V_{CC} = 30$ V, $V_{BE1} = V_{BE2} = V_{BE3} = 0.7$ V, $V_T = 26$ mV, $V_A = 150$ V, and $\beta_F = 100$. Assume that all transistors are identical.

(b) Calculate the output resistance R_o and Thevenin's equivalent voltage V_{Th}.

(c) Use PSpice/SPICE to calculate the output current, the output resistance, and the reference current for $\beta_F = 100$ and 400. Assume that all transistors are identical, and $V_{CE} = 10$ V.

SOLUTION

$I_{C2} = 5\ \mu A$, and $V_T = 26$ mV.

(a) From Eq. (1.120),

$$5\ \mu A = I_R\left[1 - \frac{2}{(100^2 + 2 \times 100 + 2)}\right]$$

so $I_R \approx 5\ \mu A$. The reference current is

$$I_R = \frac{V_{CC} - V_{BE1} - V_{BE2}}{R_1}$$

That is,

$$5\ \mu A = \frac{30 - 0.7 - 0.7}{R_1}$$

which gives $R_1 = 5.72\ M\Omega$.

(b) From Eq. (1.118),

$$I_{C3} = I_{C1} = \frac{5\ \mu A \times (1 + 100)}{2 + 100} = 4.95\ \mu A$$

Then $r_{o1} = r_{o3} = \dfrac{V_A}{I_{C1}} = \dfrac{150}{4.95\ \mu} = 30.3\ M\Omega$

$$r_{o2} = \frac{V_A}{I_{C2}} = \frac{150}{5\ \mu} = 30\ M\Omega$$

Since $1/g_{m1} = 1/g_{m3} = V_T/I_{C1} = 26\ mV/4.95\ \mu A = 5.25\ k\Omega$,

$$g_{m1} = g_{m3} = 190.4\ \mu A/V$$

Since $1/g_{m2} = 26\ mV/5\ \mu A = 5.2\ k\Omega$,

$$g_{m2} = 192.3\ \mu A/V$$

Then $r_{\pi 1} = r_{\pi 3} = \dfrac{\beta_F}{g_{m1}} = 100 \times 5.25\text{ k} = 525\text{ k}\Omega$

$$r_{\pi 2} = 100 \times 5.2\text{ k} = 520\text{ k}\Omega$$

From Eq. (1.124),

$$R'_e = \frac{520\text{ k}\Omega + (30.3\text{ M}\Omega \parallel 5.72\text{ M}\Omega)}{[1 + (30.3\text{ M}\Omega \parallel 5.72\text{ M}\Omega) \times 190.4\ \mu\text{A/V}]}$$

$$= 5.81\text{ k}\Omega$$

From Eq. (1.123),

$$R_3 = 525\text{ k} \parallel 525\text{ k} \parallel 5.25\text{ k} \parallel 30\text{ M} \approx 5.15\text{ k}\Omega$$

From Eq. (1.126),

$$R_o = (30\text{ M})\left[1 + \frac{192.3\ \mu \times 520\text{ k}}{5.81\text{ k}} \times 5.15\text{ k} \parallel 5.81\text{ k}\right] = 1.44\text{ G}\Omega$$

and $V_{Th} = 1.44\text{ G} \times 5\ \mu = 7.2\text{ kV}$

NOTE: If we use Eq. (1.127),

$$R_o \approx (30\text{ M})\left(1 + \frac{100}{2}\right) = 1.53\text{ G}\Omega$$

(c) The Wilson current source for PSpice simulation is shown in Fig. 1.31. We will use parametric sweep for the model parameter β_F of the transistors. The voltage source V_y acts as an ammeter for the output current.

The results of simulation (.TF analysis) are as follows. (The hand calculations are shown in parentheses.) For $\beta_F = 100$, the simulation gives

```
VOLTAGE SOURCE CURRENTS
NAME     CURRENT
VCC      -5.022E-06
```

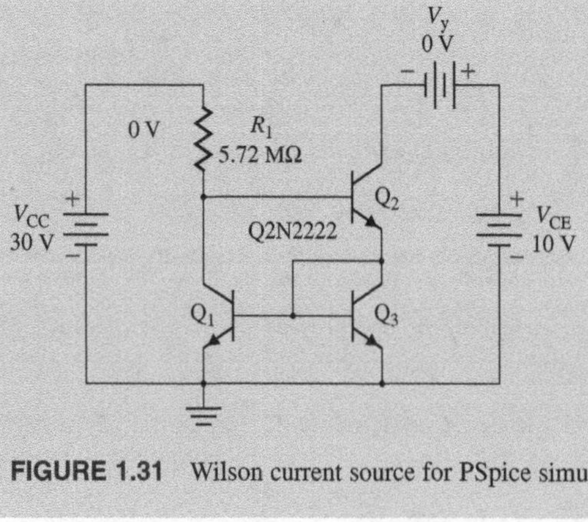

FIGURE 1.31 Wilson current source for PSpice simulation

```
VX        5.022E-06       (I_R = 5 μA)
VY        5.006E-06       (I_R = 5 μA)
VCE      -5.006E-06
**** BIPOLAR JUNCTION TRANSISTORS
NAME      Q1          Q2          Q3
IB        4.95E-08    4.73E-08    4.95E-08
IC        4.98E-06    5.01E-06    4.95E-06
VBE       6.37E-01    6.36E-01    6.37E-01
VBC      -6.36E-01   -8.73E+00    0.00E+00
VCE       1.27E+00    9.36E+00    6.37E-01
BETADC    1.00E+02    1.06E+02    1.00E+02
GM        1.92E-04    1.94E-04    1.92E-04
RPI       5.22E+05    5.47E+05    5.22E+05    (rπ1 = 525 kΩ, rπ2 = 520 kΩ, rπ3 = 525 kΩ)
RO        3.03E+07    3.17E+07    3.03E+07    (ro1 = 30.3 MΩ, ro2 = 30 MΩ, ro3 = 30.3 MΩ)
**** SMALL-SIGNAL CHARACTERISTICS
I(VY)/VCC=1.739E-07
INPUT RESISTANCE AT VCC=5.730E+06             (R1 = 5.72 MΩ)
OUTPUT RESISTANCE AT I(VY)=1.602E+09          (Ro = 1.44 GΩ)
```

For $\beta_F = 400$, the simulation gives

```
I(VY)/VCC=1.738E-07
INPUT RESISTANCE AT VCC=5.730E+06
OUTPUT RESISTANCE AT I(VY)=5.441E+09          (Ro = 6.03 GΩ)
```

NOTE: R_o changes from 1.602E+09 Ω (for $\beta_F = 100$) to 5.441E+09 Ω (for $\beta_F = 400$ Ω) and depends on β_F, as expected.

1.6.6 Multiple Current Sources

A DC reference current can be generated in one location and reproduced in another location for biasing amplifier circuits in ICs. A group of current sources with only one reference current is shown in Fig. 1.32. This is an extension of the modified basic current source in Fig. 1.25. Transistor Q_1 and resistor R_1 serve as the reference for current-sink transistors Q_3 through Q_6. Transistor Q_2 supplies the total base currents for the transistors and makes the collector current of Q_1 almost equal to the reference current I_R; that is,

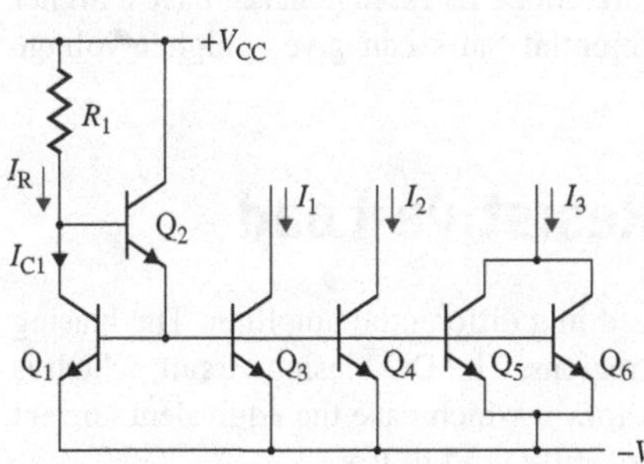

FIGURE 1.32 Multiple current sources

TABLE 1.1 Comparison of BJT current sources

Current Source Type	Output Resistance R_o	Beta (β_F) Dependency of I_o	Comments
Basic source	r_o	$-2/\beta_F$	Not suitable for low currents (i.e., typically less than 0.3 mA)
Modified source	r_o	$-2/(\beta_F^2 + \beta_F)$	Not suitable for low currents (i.e., typically less than 0.3 mA)
Widlar source	$r_{o2}(1 + g_{m2}R_2)$	Nonlinear	Current as low as 5 μA; higher output resistance
Cascode source	$r_{o2}(1 + \beta_F)$	$-2/\beta_F$	Not suitable for low currents; higher output resistance
Wilson source	$r_{o2}(1 + \beta_F/2)$	$-2/(\beta_F^2 + 2\beta_F + 2)$	Not suitable for low currents; higher output resistance

$I_R \approx I_{C1}$. The collector currents I_1 and I_2 will be mirrors of current I_R. Since two transistors Q_5 and Q_6 are connected in parallel, I_3 will be two times I_R (i.e., $I_3 = 2I_R$). The parallel combination of Q_5 and Q_6 should be equivalent to a single transistor whose E-B junction has double the area of Q_1. Therefore, the emitter areas of transistors can be scaled in ICs so as to provide multiples of the reference current simply by designing the transistors so that they have an area ratio equal to the desired multiple.

KEY POINTS OF SECTION 1.6

- Transistors can generate the characteristics of a constant-current source. An ideal current source has a very high output resistance, and its output current is not sensitive to the transistor parameter β_F.
- Table 1.1 compares various BJT current sources. The output resistances for different BJT sources are also summarized in the table.

1.7 BJT Differential Amplifiers

The BJT differential pairs are analogous to MOS differential pairs. Since BJTs, in general, have a higher transconductance compared to identical MOSFETs, BJT differential pairs can give a higher voltage gain, but they have limitations of a lower input resistance.

1.7.1 BJT Differential Pair with Resistive Load

An emitter-coupled pair, as shown in Fig. 1.33, is commonly used in a differential amplifier. The biasing current should be such that the transistors operate in the active regions. The DC biasing circuit, which is shown as a constant-current source, can be either a simple resistor, in which case the equivalent current generator will be zero, or a transistor current source, which is generally used in ICs.

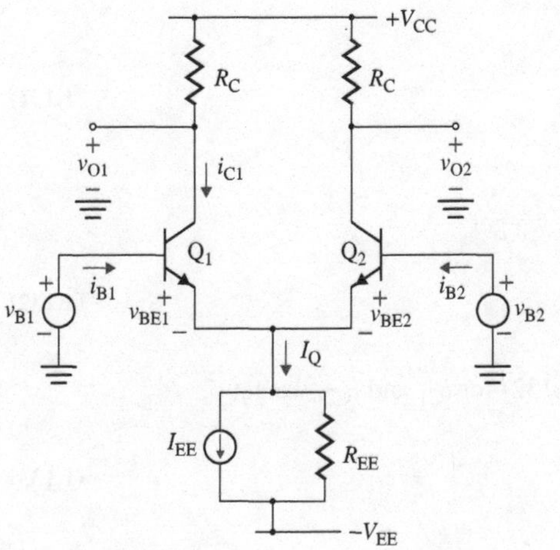

FIGURE 1.33 Emitter-coupled differential pair

DC Transfer Characteristics

The DC transfer characteristic, which gives the relation between the input and output voltages, can be determined from the large-signal analysis, and it should be linear over a wide range. The analysis can be simplified by making the following assumptions:

1. The output resistances of the transistors are infinite: $r_o = \infty$.
2. The output resistance of the transistor current source is infinite: $R_{EE} = \infty$.

Using KVL around the loop formed by the two input voltages and the two B-E junctions, we get

$$v_{B1} - v_{BE1} + v_{BE2} - v_{B2} = 0 \tag{1.128}$$

Assuming v_{BE1}, $v_{BE2} >> V_T$ and equal leakage currents $I_{S1} = I_{S2} = I_S$ and using the transistor current equations, we get

$$v_{BE1} = V_T \ln\left(\frac{i_{C1}}{I_{S1}}\right) = V_T \ln\left(\frac{i_{C1}}{I_S}\right) \tag{1.129}$$

$$v_{BE2} = V_T \ln\left(\frac{i_{C2}}{I_{S2}}\right) = V_T \ln\left(\frac{i_{C2}}{I_S}\right) \tag{1.130}$$

Substituting Eqs. (1.129) and (1.130) into Eq. (1.128) gives

$$v_{B1} - V_T \ln\left(\frac{i_{C1}}{I_S}\right) + V_T \ln\left(\frac{i_{C2}}{I_S}\right) - v_{B2} = 0$$

That is,

$$v_{B1} - v_{B2} = V_T\left[\ln\left(\frac{i_{C1}}{I_S}\right) - \ln\left(\frac{i_{C2}}{I_S}\right)\right]$$

which can be simplified to

$$\frac{i_{C1}}{i_{C2}} = \exp\left(\frac{v_{B1} - v_{B2}}{V_T}\right) = \exp\left(\frac{v_{id}}{V_T}\right) \tag{1.131}$$

where $v_{id} = v_{B1} - v_{B2}$ is the differential DC input voltage.

Using KCL at the emitter terminal of the transistors, we get

$$I_Q = \frac{1}{\alpha}(i_{C1} + i_{C2}) \tag{1.132}$$

where $\alpha = \beta_F/(1 + \beta_F) \approx 1$. Solving Eqs. (1.131) and (1.132) for i_{C1} and i_{C2}, we get

$$i_{C1} = \frac{\alpha I_Q}{1 + \exp(-v_{id}/V_T)} \tag{1.133}$$

$$\approx \alpha I_Q \quad \text{for } v_{id} >> V_T \tag{1.134}$$

$$i_{C2} = \frac{\alpha I_Q}{1 + \exp(v_{id}/V_T)} \tag{1.135}$$

$$\approx 0 \quad \text{for } v_{id} >> V_T \tag{1.136}$$

Thus, if i_{C1} increases, i_{C2} decreases such that $i_{C1} + i_{C2} = \alpha I_Q = \alpha I_{EE}$ remains constant.

The plots of the two collector currents are shown as a function of v_{id} in Fig. 1.34(a). Notice that, for $v_{id} >> V_T$, i_{C1} and i_{C2} become independent of v_{id} and all the currents flow through one of the transistors. For $v_{id} \leq V_T$, i_{C1} and i_{C2} have an approximately linear relation. The differential voltage change Δv_{id} required to shift the current distribution from $i_{C1} = 0.9I_Q$ and $i_{C2} = 0.1I_Q$ to the opposite case, $i_{C1} = 0.1I_Q$ and $i_{C2} = 0.9I_Q$, is called the *transition voltage*; it will have a value of approximately $2V_T = 52.6$ mV.

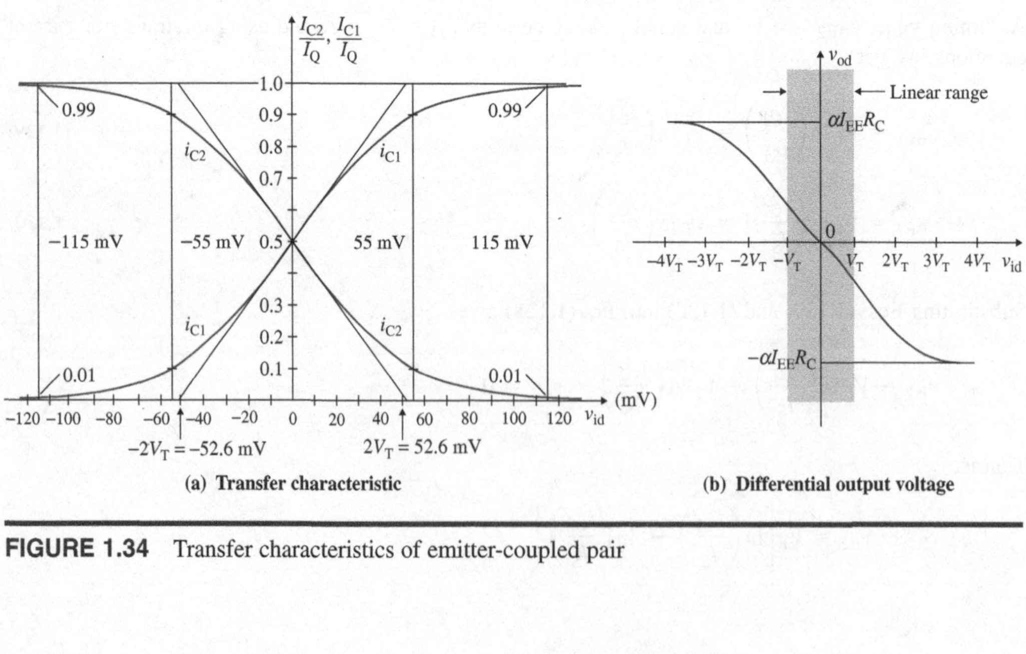

FIGURE 1.34 Transfer characteristics of emitter-coupled pair

The DC output voltages are

$$v_{O1} = V_{CC} - i_{C1}R_C \tag{1.137}$$

$$v_{O2} = V_{CC} - i_{C2}R_C \tag{1.138}$$

The differential DC output voltage is

$$v_{od} = v_{O1} - v_{O2} = R_C(i_{C2} - i_{C1})$$

Substituting Eqs. (1.133) and (1.135) into the above equation and simplifying yields

$$v_{od} = \alpha I_{EE}R_C \tanh\left(-\frac{v_{id}}{2V_T}\right) \tag{1.139}$$

Since tanh $x \approx x$ for a small value of x, Eq. (1.139) can be approximated by

$$v_{od} \approx -\alpha I_{EE}R_C\left(\frac{v_{id}}{2V_T}\right) \tag{1.140}$$

The plot of v_{od} as a function of v_{id} is shown in Fig. 1.34(b). If v_{id} is zero, v_{od} is also zero, and this feature allows direct coupling of cascoded stages without introducing DC offsets. Thus, the amplifier is a true differential, or difference, amplifier, responding only to the difference in the voltages applied to the two input terminals. If $v_{B1} = -v_{B2}$, then $v_{ic} = (v_{B1} + v_{B2})/2$ is zero and there will be only a differential voltage. If, on the other hand, $v_{B1} = v_{B2}$, then v_{id} is zero and there will be a pure common-mode voltage (i.e., no output voltage). However, the range of the differential input voltage v_{id} over which the emitter-coupled pair exhibits a linear characteristic is very small, typically two or three times V_T. This range can be extended by inserting emitter degeneration resistors, as shown in Fig. 1.35(a); in this case, the range over which the characteristic is linear is approximately equal to I_QR_E. The factor by which the voltage gain is reduced is approximately the same as the factor by which the input range is increased. The variations of v_{od} with several values of R_E are shown in Fig. 1.35(b).

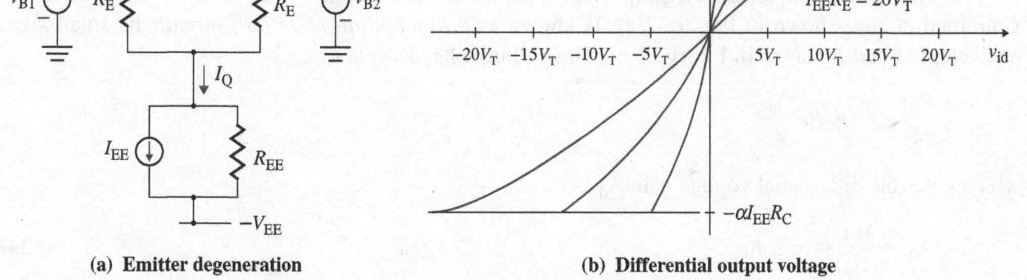

FIGURE 1.35 Emitter-coupled pair with degeneration resistors

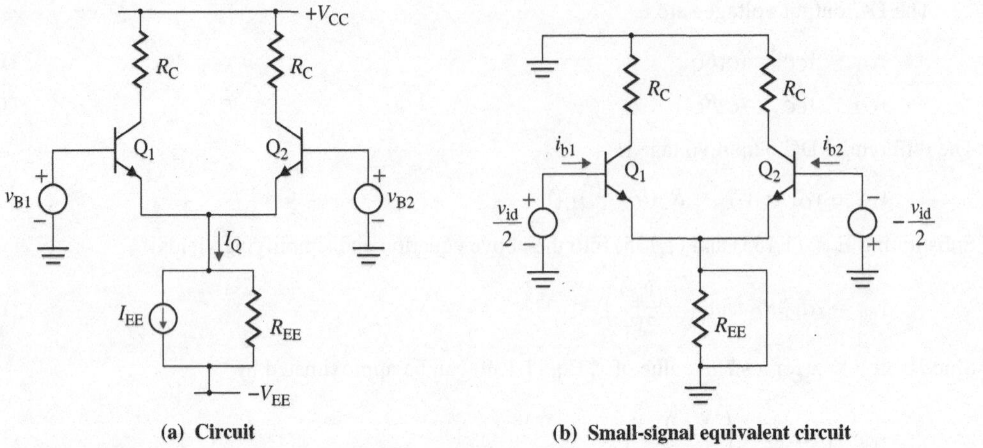

FIGURE 1.36 Emitter-coupled pair with differential input

Small-Signal Analysis

In studying a BJT differential amplifier, it is often of interest to examine the small-signal behavior for small DC differential voltages near zero, when the amplifier operates in the linear portion of the transfer characteristic. Circuit properties such as differential voltage gain A_d, common-mode gain A_c, common-mode rejection ratio (CMMR), common-mode input resistance R_{ic}, and differential mode input resistance R_{id} can be determined from the small-signal analysis.

Small Differential Signal Let us assume that the common-mode signal is zero, $v_{ic} = 0$, and only the differential input voltage v_{id} is applied. This situation is shown in Fig. 1.36(a). If the DC differential biasing is removed, we get the small-signal equivalent circuit shown in Fig. 1.36(b). The input voltage to one transistor is $+v_{id}/2$, and that to the other transistor is $-v_{id}/2$. Assuming that the two transistors are identical and the circuit is balanced, the increase in voltage at the emitter junction due to $+v_{id}/2$ will be compensated for by an equal decrease in voltage due to $-v_{id}/2$. As a result, the voltage at the emitters of the transistors will not vary at all. The emitter junction, which experiences no voltage variation, can be regarded as the ground potential. Thus, the resistor R_{EE} may be replaced by a short circuit, as shown in Fig. 1.36(b), which shows two identical sides.

The characteristic of a balanced amplifier can be determined from only one side of the amplifier. This simplified circuit, shown in Fig. 1.37(a), is known as a *differential-mode half circuit*; its small-signal equivalent circuit is shown in Fig. 1.37(b). The output voltage v_{od} is given by

$$\frac{v_{od}}{2} = -g_m R_C \frac{v_{id}}{2}$$

which gives the differential voltage gain A_d as

$$A_d = \frac{v_{od}}{v_{id}} = -g_m R_C \tag{1.141}$$

▶ **NOTE** If we include the output resistance r_{o1} of Q_1, Eq. (1.141) becomes $A_d = -g_m(R_C \parallel r_{o1})$.

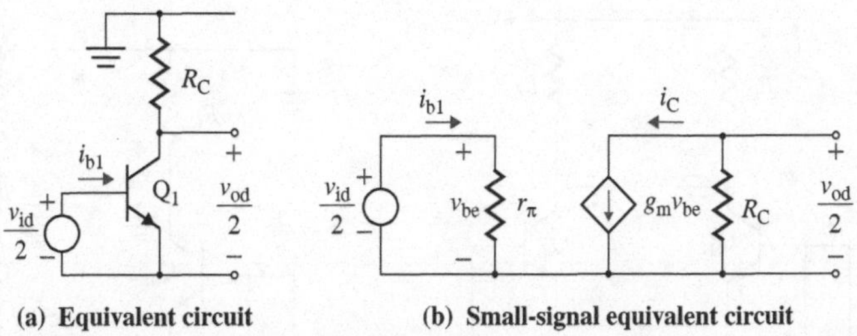

FIGURE 1.37 Differential-mode half circuit

High values of R_C and g_m are required to give a high value of A_d. An active load (discussed in Sec. 1.6) rather than a discrete resistance ensures a high value of R_C and hence of A_d. From Fig. 1.37(b), we can write

$$\frac{v_{id}}{2} = i_{bl} r_\pi$$

which gives the differential input resistance R_{id} as

$$R_{id} = \frac{v_{id}}{i_{bl}} = 2r_\pi \tag{1.142}$$

A high value of r_π $(=V_T/I_C)$—that is, a low collector biasing current ($I_{C1} = I_{C2} = I_Q/2$)—is required to achieve a higher value of R_{id}.

Small Common-Mode Signal An emitter-coupled pair with only common-mode input v_{ic} is shown in Fig. 1.38(a). If the DC common-mode biasing is removed, the result is the small-signal equivalent circuit shown in Fig. 1.38(b). Assuming that the two transistors are identical, the collector currents must be identical, and the voltage at the emitter junction will increase by the same amount in response to inputs at both transistors. Since the voltage across R_{EE} will be the same for both inputs, the resistor R_{EE} can be split into two parallel resistors, each of value $2R_{EE}$, as shown in Fig. 1.38(c). As a result of symmetry, no current will flow through the lead that connects the two sides, and $i_x = 0$. Thus, this lead can be disconnected without affecting the circuit behavior; as shown in Fig. 1.38(d), the two half circuits may be considered to be completely independent. The common-mode behavior can be determined from only one side, as shown in Fig. 1.39(a). The small-signal equivalent circuit is shown in Fig. 1.39(b), from which we get

$$v_{ic} = i_{b1} r_\pi + i_b(1 + \beta_F)2R_{EE} \tag{1.143}$$

which gives the common-mode input resistance R_{ic} as

$$R_{ic} = \frac{v_{ic}}{i_{b1}} = r_\pi + (1 + \beta_F)2R_{EE} \tag{1.144}$$

Equation (1.144) will give a high value for R_{ic}, so R_{ic} should be calculated using r_μ in the BJT model. The current through r_μ can be found from

$$i_\mu = \frac{v_{ic} - v_o}{r_\mu} = \frac{v_{ic} + \beta_F R_C i_{b1}}{r_\mu}$$

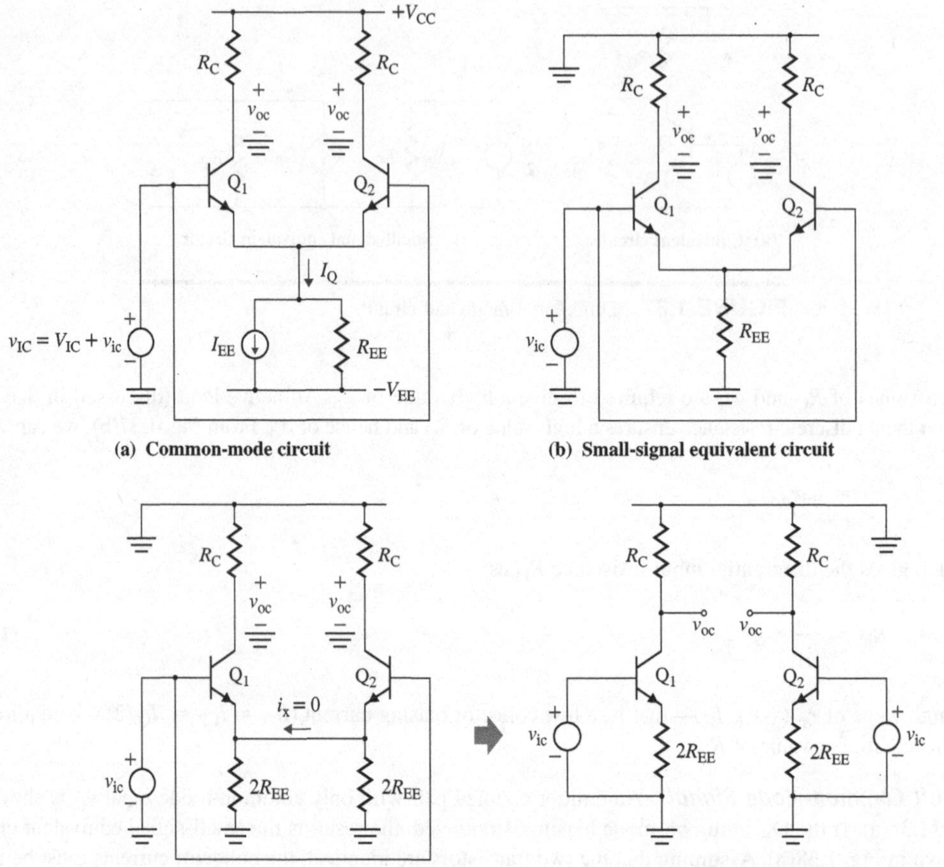

(a) **Common-mode circuit** (b) **Small-signal equivalent circuit**

(c) **Small-signal equivalent circuit with split R_{EE}** (d) **Independent half circuits**

FIGURE 1.38 Emitter-coupled pair with common-mode input

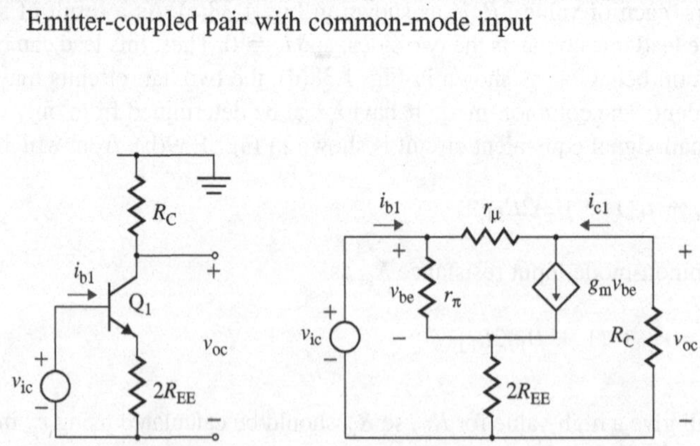

(a) **Common-mode half circuit** (b) **Small-signal equivalent circuit**

FIGURE 1.39 Common-mode half circuit

which, after substitution of $i_{b1} = v_{ic}/[r_\pi + (1 + \beta_F)2R_{EE}]$, gives the effective resistance as

$$R_{i\mu} = \frac{r_\mu}{1 + \beta_F R_C/[r_\pi + (1 + \beta_F)2R_{EE}]}$$

Thus, the common-mode resistance R_{ic} becomes

$$R_{ic} = [r_\pi + 2R_{EE}(1 + \beta_F)] \parallel \frac{r_\mu}{r_\pi + (1 + \beta_F)2R_{EE}}$$

The common-mode output voltage v_{oc} is

$$v_{oc} = -R_C i_{c1} = -R_C \beta_F i_{b1} = v_{ic}\left[\frac{-\beta_F R_C}{r_\pi + (1 + \beta_F)2R_{EE}}\right]$$

which gives the common-mode voltage gain A_c as

$$A_c = \frac{v_{oc}}{v_{ic}} = \frac{-\beta_F R_C}{r_\pi + (1 + \beta_F)2R_{EE}} \tag{1.145}$$

$$= \frac{-g_m R_C}{1 + 2g_m R_{EE}(1 + 1/\beta_F)} \tag{1.146}$$

Small-Signal CMRR From Eqs. (1.141) and (1.146), we can find the CMRR as

$$\text{CMRR} = \left|\frac{A_d}{A_c}\right| = 1 + 2g_m R_{EE}\left(1 + \frac{1}{\beta_F}\right) \tag{1.147}$$

$$\approx 1 + 2g_m R_{EE} \quad \text{for } \beta_F >> 1 \tag{1.148}$$

which indicates that a high output resistance R_{EE} on the biasing current source will improve the CMRR; that is, the value of R_{EE} should be as large as possible. To obtain a high value of g_m $(=I_C/V_T)$, the collector biasing current ($I_{C1} = I_{C2} = I_Q/2 = I_{EE}/2$) should be made large.

The small-signal input currents, which will flow when both v_{id} and v_{ic} are applied, can be found by superposition. Since R_{ic} is common to both i_{b1} and i_{b2}, we get

$$i_{b1} = \frac{v_{ic}}{2R_{ic}} + \frac{v_{id}}{R_{id}} \tag{1.149}$$

$$i_{b2} = \frac{v_{ic}}{2R_{ic}} + \frac{v_{id}}{R_{id}} \tag{1.150}$$

The input resistance can thus be represented by the π-equivalent circuit of Fig. 1.40(a), where R_{ic} is assumed to be much larger than R_{id}. The T-equivalent circuit is shown in Fig. 1.40(b), and its values are shown in Fig. 1.40(c).

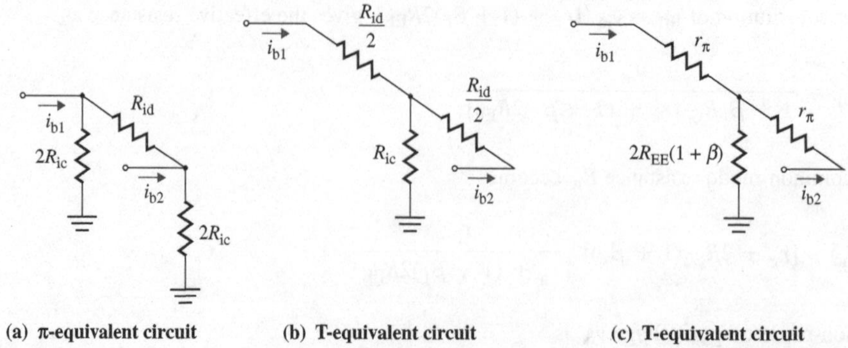

FIGURE 1.40 π- and T-equivalent circuits

EXAMPLE 1.12

Finding the performance parameters of an emitter-coupled pair The parameters of the emitter-coupled pair in Fig. 1.36(a) are $\beta_F = 100$, $R_{EE} = 50\ \text{k}\Omega$, $I_Q = 1$ mA, $V_{CC} = 15$ V, and $R_C = 10\ \text{k}\Omega$.

(a) Calculate the DC collector currents through the transistors if $v_{id} = 5$ mV.

(b) Assuming $i_{C1} = i_{C2}$, calculate A_d, A_c, and CMRR; R_{id} and R_{ic}; and the small-signal output voltage if $v_{B1} = 20$ mV and $v_{B2} = 10$ mV. Assume $V_T \approx 26$ mV.

SOLUTION

$\alpha = \beta_F/(1 + \beta_F) = 100/(1 + 100) = 0.99$.

(a) From Eq. (1.133),

$$i_{C1} = \frac{\alpha I_Q}{1 + \exp(-v_{id}/V_T)} = \frac{0.99 \times 1\,\text{mA}}{1 + \exp(-5\text{ m}/26\text{ m})} = 0.543\text{ mA}$$

and $i_{C2} = 1\text{ m} - 0.543\text{ m} = 0.457$ mA

(b) We know that $i_{C1} = i_{C2} = 1\text{ mA}/2 = 0.5$ mA. Thus,

$$g_m = \frac{i_{C1}}{V_T} = \frac{0.5\text{ mA}}{26\text{ mV}} = 19.23\text{ mA/V}$$

From Eq. (1.141),

$$A_d = -g_m R_C = -19.23\text{ m} \times 10\text{ k} = -192.3\text{ V/V}$$

From Eq. (1.146),

$$A_c = \frac{-g_m R_C}{1 + 2g_m R_{EE}(1 + 1/\beta_F)}$$

$$= \frac{-19.23\text{ mA/V} \times 10\text{ k}\Omega}{1 + 2 \times 19.23\text{ mA/V} \times 50\text{ k}\Omega \times (1 + 1/100)} = -0.099\text{ V/V}$$

Thus, CMRR = $|A_d/A_c|$ = 192.3/0.099 = 1942.4 (or 65.77 dB). From Eq. (1.148),

$$\text{CMRR} \approx 1 + 2g_m R_{EE} = 1 + 2 \times 19.23\text{ m} \times 50\text{ k} = 1924$$

We know that $r_\pi = \beta_F/g_m = 100/19.23\text{ m} = 5.2\text{ k}\Omega$. From Eq. (1.142),

$$R_{id} = 2r_\pi = 2 \times 5.2\text{ k} = 10.4\text{ k}\Omega$$

From Eq. (1.144),

$$R_{ic} = r_\pi + (1 + \beta_F)2R_{EE} = 5.2\text{ k} + (1 + 100) \times 2 \times 50\text{ k} = 10.1\text{ M}\Omega$$

We know that $v_{id} = 20 - 10 = 10$ mV, and $v_{ic} = (20 + 10)/2 = 15$ mV. From Eq. (1.10),

$$v_o = A_d v_{id} + A_c v_{ic} = -192.3 \times 10\text{ mV} - 0.099 \times 15\text{ mV} = -1924.5\text{ mV}$$

▶ **NOTE** To apply Eq. (1.10) and other equations in Sec. 1.2, we must have $|v_{id}| \leq V_T$. The DC collector voltage of a transistor is

$$V_C = V_{CC} - I_C R_C = 15\text{ V} - 0.5\text{ mA} \times 10\text{ k}\Omega = 10\text{ V}$$

Thus, for $|A_d| = 192.3$, the maximum differential voltage will be $v_{id} = 10/192.3 = 52$ mV. Therefore, to allow output voltage swing due to the input voltages, V_{CC} must be greater than $I_C R_C$.

EXAMPLE 1.13

D

Designing an emitter-coupled pair

(a) Design an emitter-coupled pair as shown in Fig. 1.41 in which one input terminal is grounded. The output is taken from the collector of transistor Q_1. The biasing current is $I_{EE} = 1$ mA, and $V_{CC} = V_{EE} = 15$ V. The transistors are identical. Assume $V_{BE} = 0.7$ V, $V_T = 26$ mV, $V_A = \infty$, and $\beta_F = 100$. A small-signal voltage gain of $A_1 = -250$ V/V is required.

(b) Calculate the design values of A_d, A_c, and CMRR.

SOLUTION

(a) $I_{EE} = 1$ mA. Since $v_{B2} = 0$, we can write

$$R_{EE} I_{EE} + V_{BE} = V_{EE} = 15\text{ V}$$

Thus,

$$R_{EE} = \frac{15 - 0.7}{1\text{ m}} = 14.3\text{ k}\Omega$$

We have

$$i_{C1} = i_{C2} = \frac{I_{EE}}{2} = \frac{1\text{ mA}}{2} = 0.5\text{ mA}$$

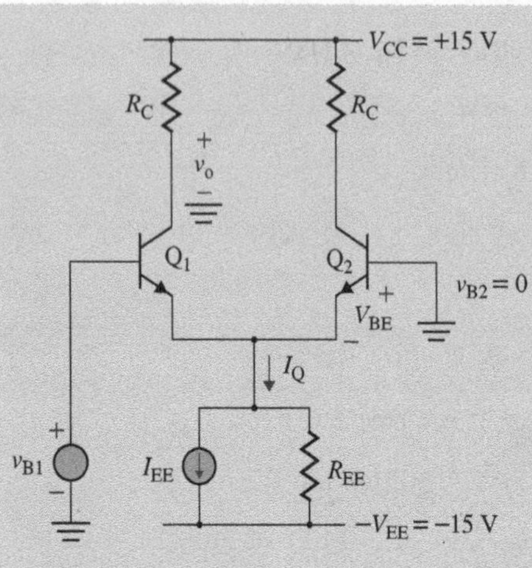

FIGURE 1.41 Emitter-coupled pair with single input

and $$g_m = \frac{i_{C1}}{V_T} = \frac{0.5 \text{ mA}}{26 \text{ mV}} = 19.23 \text{ mA/V}$$

Since $v_{B2} = V_{B2} + v_{b2} = 0$, $v_{b2} = 0$. Then

$$v_{id} = v_{b1} - v_{b2} = v_{b1}$$

and $$v_{ic} = \frac{v_{b1} + v_{b2}}{2} = \frac{v_{b1}}{2}$$

From Eq. (1.10), we get

$$v_{o1} = A_c v_{ic} + \frac{A_d v_{id}}{2} = A_c \frac{v_{b1}}{2} + \frac{A_d v_{b1}}{2} = \frac{v_{b1}}{2}(A_c + A_d) \quad \textbf{(1.151)}$$

Substituting for A_d and A_c from Eqs. (1.141) and (1.146) gives the voltage gain A_1 as

$$A_1 = \frac{v_{o1}}{v_{b1}} = \frac{1}{2}(A_c + A_d) = -\frac{1}{2}\left[\frac{g_m R_C}{1 + 2g_m R_{EE}(1 + 1/\beta_F)} + g_m R_C\right] \quad \textbf{(1.152)}$$

Substituting $A_1 = -250$, $g_m = 19.23$ mA/V, $R_{EE} = 14.3$ kΩ, and $\beta_F = 100$ into Eq. (1.152) gives $R_C = 25.95$ kΩ.

(b) From Eq. (1.141),

$$A_d = -g_m R_C = -19.23 \text{ m} \times 25.95 \text{ k} = -499.01$$

From Eq. (1.146),

$$A_c = \frac{-19.23 \text{ mA/V} \times 25.95 \text{ k}\Omega}{1 + 2 \times 19.23 \text{ mA/V} \times 14.3 \text{ k}\Omega \times (1 + 1/100)} = -0.897$$

Thus, CMRR $= |A_d/A_c| = 499.01/0.897 = 556.3$ (or 54.91 dB).

1.7.2 BJT Differential Amplifiers with Basic Current Mirror Active Load

We saw in Sec. 1.7.1 that the differential gain of a differential pair with a resistive load R_C is $-g_m R_C = -R_C I_C/V_T = -R_C I_Q/2V_T$. In differential amplifiers, a very small value of biasing current I_Q, in the microampere range, is often used. As a result, a very large value of R_C, on the order of megaohms, will be required to give a substantial voltage gain. However, a large value of R_C will cause a large DC voltage drop, reducing the collector voltage to $V_{CC} - R_C I_Q/2$, which will be substantially less than V_{CC}. This low collector voltage will reduce the allowable input voltage range of the amplifier.

Active devices such as transistors occupy much less silicon area than medium-size or large resistors. In practical amplifiers, the load resistor R_C is normally replaced by a constant-current source, which offers a very high load resistance to the amplifier and hence can give a high voltage gain. This type of load, known as an *active load*, has a small voltage drop, typically 0.7 V, and hence allows a wider input voltage range.

A differential amplifier with a basic current source as the active load is shown in Fig. 1.42. The active load consists of transistors Q_3 and Q_4. Since their B-E voltages are the same, their collector currents will be equal. That is, $i_{C3} \approx i_{C4}$. Thus, the current through Q_4 will be the mirror of the current through Q_3. Under quiescent conditions, the differential amplifiers will be balanced such that $I_{C1} = I_{C2}$. Since $I_{C1} = I_{C2}$ and $I_{C3} = I_{C4}$, we can find the quiescent load current:

$$I_O = I_{B3} + I_{B4} = \frac{I_{C3} + I_{C4}}{\beta_F} \approx \frac{I_{C1} + I_{C2}}{\beta_F} = \frac{\alpha I_Q}{\beta_F} = \frac{I_Q}{1 + \beta_F}$$

Since $\beta_F >> 1$, I_O will be very small.

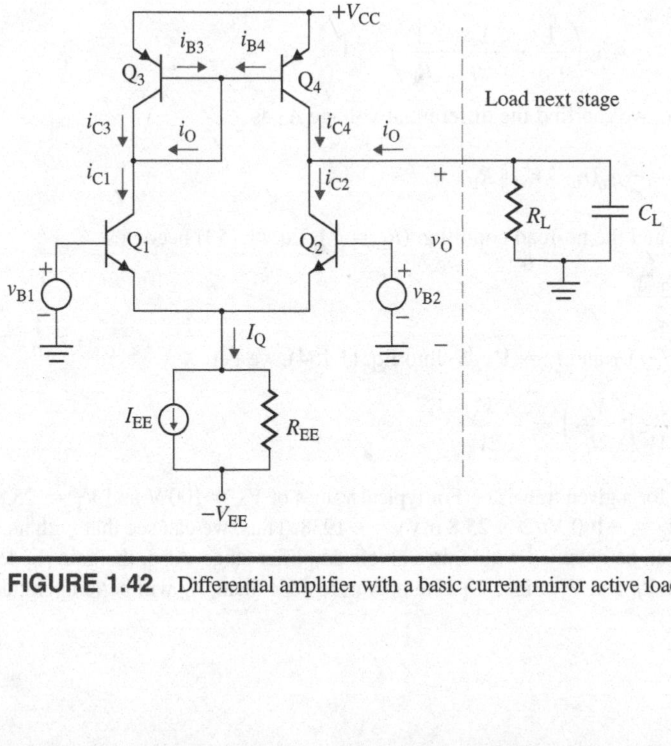

FIGURE 1.42 Differential amplifier with a basic current mirror active load

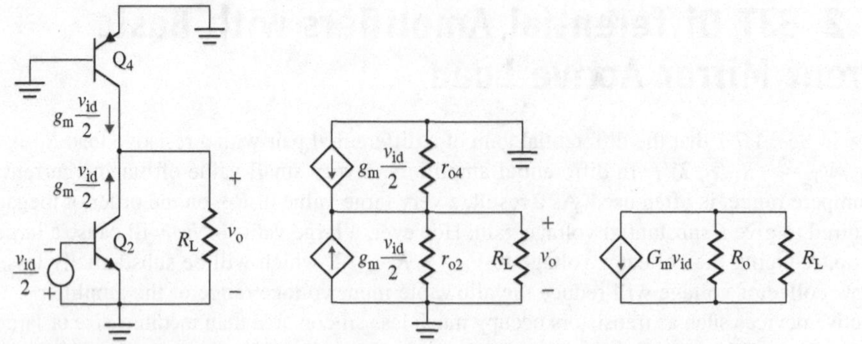

(a) Equivalent half circuit (b) Small-signal equivalent circuit (c) Transconductance representation

FIGURE 1.43 Equivalent basic current mirror circuit

Small-Signal Analysis

Small Differential Signal If the input voltages change by a small differential amount v_{id}, the collector current of Q_1 will change by a small amount $g_m v_{id}/2$, and the collector current of Q_3 will change by the same amount. Since i_{C4} is a mirror of i_{C3}, the collector current of Q_4 will change by $g_m v_{id}/2$. The B-E voltage of Q_2 will decrease by $v_{id}/2$, causing its collector current to change by an amount $-g_m v_{id}/2$. The equivalent half circuit is shown in Fig. 1.43(a), and its small-signal equivalent appears in Fig. 1.43(b).

Using KCL at the collector of transistors Q_2 and Q_4, we can find the output voltage v_o in terms of v_{id}:

$$\frac{-2 \times g_m v_{id}}{2} = \frac{v_o}{r_{o2}} + \frac{v_o}{r_{o4}} + \frac{v_o}{R_L}$$

$$= v_o\left(\frac{1}{r_{o2}} + \frac{1}{r_{o4}} + \frac{1}{R_L}\right)$$

After simplification, we can find the differential voltage A_d as

$$A_d = \frac{v_o}{v_{id}} = -g_m(r_{o2} \parallel r_{o4} \parallel R_L) \tag{1.153}$$

For $r_{o2} = r_{o4} = r_o$ and the no-load condition ($R_L = \infty$), Eq. (1.153) becomes

$$A_d = -\frac{g_m r_o}{2} \tag{1.154}$$

Substituting $g_m = I_C/V_T$ and $r_o = V_A/I_C$ into Eq. (1.154), we get

$$A_d = -\left(\frac{I_C}{V_T}\right)\left(\frac{V_A}{2I_C}\right) = -\frac{V_A}{2V_T} \tag{1.155}$$

which is a constant for a given transistor. For typical values of $V_A = 100$ V and $V_T = 25.8$ mV, the differential gain becomes $A_d = -100$ V/(2 × 25.8 mV) = −1938. Thus, we can see that with an active load, a very large voltage gain can be obtained with only a single amplifier stage. Also, the gain A_d depends only on the physical parameters V_A and V_T. Since V_T is temperature dependent, A_d will be temperature dependent too.

The differential input resistance R_{id} is given by

$$R_{id} = \frac{v_{id}}{i_d} = 2r_\pi \quad (\text{for } I_C > 0) \tag{1.156}$$

Equation (1.156) is not valid for the no-load condition—that is, when $I_C = 0$. Substituting $r_\pi = \beta_F V_T / I_C$ into Eq. (1.156) gives

$$R_{id} = 2\frac{\beta_F V_T}{I_C} = 2 \times \frac{\beta_F V_T}{I_Q/2} = \frac{4\beta_F V_T}{I_Q} \tag{1.157}$$

which indicates that a lower biasing current I_Q will give a higher value of R_{id} while still maintaining a high voltage gain A_d. A very low value of I_Q, however, will affect the frequency and transient responses of the amplifier, which is undesirable. If a low biasing current is desired to achieve a high input resistance, then a MOSFET differential amplifier is preferable because the amplifier can be operated at a relatively higher value of biasing current without affecting the frequency and transient responses.

The output resistance R_o is the parallel combination of r_{o2} and r_{o4}; that is,

$$R_o = r_{o2} \parallel r_{o4} \tag{1.158}$$

$$= \frac{r_o}{2}$$

(since $r_{o2} = r_{o4}$).

A differential amplifier is normally followed by other stages. The input resistance R_L of the next stage acts as a load of the amplifier and hence influences the overall voltage gain. The amplifier is often represented as a transconductance amplifier so that the short-circuit current and the effect of load resistance on the output voltage can be determined easily. This arrangement is shown in Fig. 1.43(c). The total current is $2g_m v_{id}/2 = g_m v_{id}$, which gives the effective G_m as

$$G_m = g_m = \frac{I_C}{V_T} = \frac{I_Q}{2V_T} \tag{1.159}$$

Small Common-Mode Signal The common-mode input resistance R_{ic} can be found from Eq. (1.144). The approximate value of the common-mode gain A_c can be found from Eq. (1.145) by replacing R_C by R_{o4}. For practical amplifiers, A_c is generally very small and can be ignored in finding the output voltage.

1.7.3 Differential Amplifier with Modified Current Mirror

Let us use the modified basic current shown in Fig. 1.27(a) as the active load. This arrangement is shown in Fig. 1.44. The active load consists of transistors Q_3, Q_4, and Q_5. The addition of Q_5 makes the ratio i_{C4}/i_{C3} independent of current gain β_F, and i_{C4} approximates a true mirror of i_{C3}. Transistors Q_6 and Q_7 belong to the second stage and act as the load of the differential amplifier. Q_6 and Q_7 form a compound transistor (Darlington pair), in which the emitter current of Q_6 becomes the base current of Q_7. As a result, the effective current gain becomes β_F^2 such that $i_{C7} = \beta_F^2 i_{B6}$.

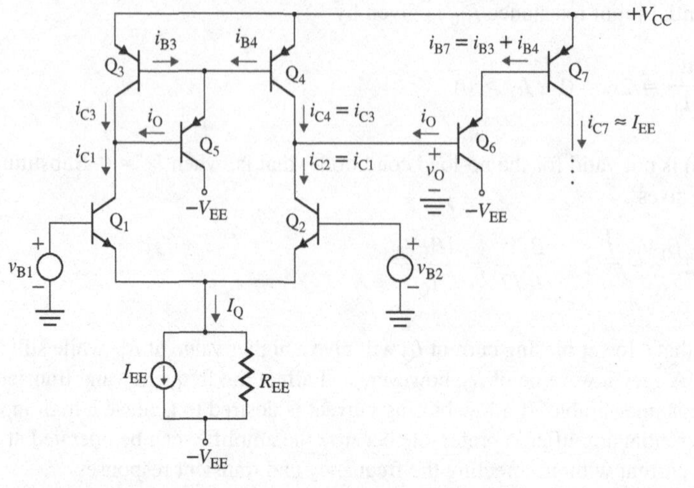

FIGURE 1.44 Differential amplifier with a modified current mirror active load

Under quiescent conditions with identical transistors, $I_{C1} + I_{C2} = I_Q$, $I_{C1} = I_{C2}$, and $I_{C3} = I_{C4}$. Thus, $I_{B5} = I_{B6}$, so $I_{E5} = I_{E6}$; that is,

$$I_{B7} = I_{E6} = I_{E5} = I_{B3} + I_{B4} = \frac{I_{C3} + I_{C4}}{\beta_F} = \frac{I_Q}{\beta_F}$$

Thus, the collector current of Q_7 becomes $I_{C7} = \beta_F I_{B7} = I_Q$. This current mirror causes the quiescent current of the next stage to have the same value as the differential amplifier while having the same performance as the amplifier of Fig. 1.42.

EXAMPLE 1.14

Analyzing a BJT differential amplifier with a current mirror active load The parameters of the differential amplifier in Fig. 1.44 are $\beta_F = 100$, $I_Q = 20\ \mu\text{A}$, and $V_{CC} = 15$ V. Calculate A_d, R_{id}, R_o, and the overall voltage gain with load $A_{d(load)}$. Assume $V_T = 26$ mV and $V_A = 100$ V.

SOLUTION

We have

$$I_{C1} = I_{C2} = I_{C4} = \frac{20\ \mu\text{A}}{2} = 10\ \mu\text{A}$$

$$I_{B6} = \frac{I_Q}{\beta_{F7}\beta_{F8}} = \frac{20\ \mu\text{A}}{100 \times 100} = 2\ \text{nA}$$

$$g_m = \frac{I_{C2}}{V_T} = \frac{10\ \mu A}{26\ mV} = 387.6\ \mu A/V$$

$$r_{o2} = r_{o3} = r_o = \frac{V_A}{I_C} = \frac{100\ V}{10\ \mu A} = 10\ M\Omega$$

$$r_\pi = \frac{\beta_F V_T}{I_C} = \frac{100 \times 26\ mV}{10\ \mu A} = 260\ k\Omega$$

From Eq. (1.155),

$$A_d = -\frac{V_A}{2V_T} = -\frac{100\ V}{2 \times 26\ mV} = -1923\ V/V$$

From Eq. (1.156),

$$R_{id} = 2r_\pi = 2 \times 260\ k\Omega = 520\ k\Omega$$

and $$R_o = r_{o2} \| r_{o4} = \frac{r_o}{2} = \frac{10\ M\Omega}{2} = 5\ M\Omega$$

Since Q_6 and Q_7 form a compound transistor, its effective B-E voltage is that of two B-E junctions in series. Thus,

$$R_L = 2r_{\pi 6} = \frac{2V_T}{I_{B6}} = \frac{2 \times 26\ mV}{2\ nA} = 26\ M\Omega$$

The effective transconductance is $G_m = g_m = 387.6\ \mu A/V$. Thus, the overall voltage gain with load $A_{d(load)}$ is

$$A_{d(load)} = -G_m(R_o \| R_L) = -387.6\ \mu A/V \times (5\ M\Omega \| 26\ M\Omega) = -1625\ V/V$$

1.7.4 Cascode Differential Amplifier

Notice from Eq. (1.154) that the differential gain increases with the output resistance R_o of the differential amplifier. Transistors are often connected in cascode configurations to increase the output resistance and also to improve the frequency response. A common modification to Fig. 1.44 is shown in Fig. 1.45. Transistors Q_5 and Q_6 are connected in a common-base configuration and form a common-base differential stage. Note that Q_5 and Q_6 must be biased by connecting the base terminals of Q_5 and Q_6 to the DC supply V_{CC} through a potential divider.

The equivalent half circuit is shown in Fig. 1.46(a), which can be simplified to Fig. 1.46(b). Q_4 is replaced by its equivalent circuit. R'_o, which is the equivalent output resistance of the Q_2 and Q_6 combination, can be determined from the test circuit shown in Fig. 1.46(c). The emitter resistance of Q_6 is the parallel combination of $r_{\pi 6}$ and r_{o2}; that is,

$$R'_E = r_{\pi 6} \| r_{o2} \approx r_{\pi 6} \tag{1.160}$$

(since $r_{o2} >> r_{\pi 6}$). Using Eq. (1.114), we get R'_o:

$$\begin{aligned} R'_o &= r_{o6}(1 + g_{m6}R'_E) = r_{o6}(1 + g_{m6}r_{\pi 6}) \\ &= r_{o6}(1 + \beta_{F6}) \approx \beta_{F6} r_{o6} \end{aligned} \tag{1.161}$$

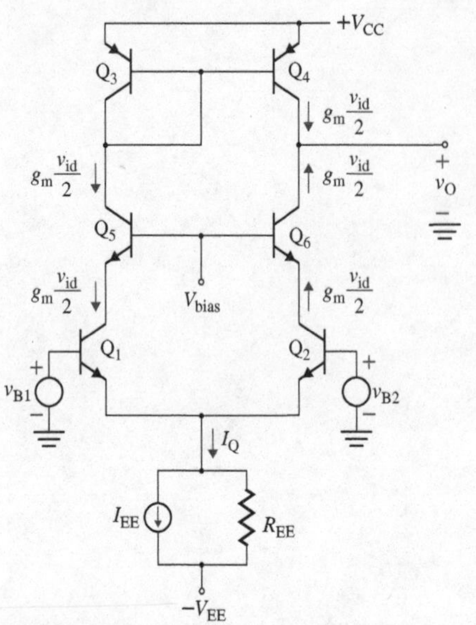

FIGURE 1.45 Cascode differential amplifier

Since all devices are biased at the same current $r_{o6} = r_{o3} = r_{o2} = r_o$, all β_F are the same. Thus,

$$R'_o = \beta_F r_o \tag{1.162}$$

The output resistance of the amplifier becomes

$$R_o = r_o \parallel R'_o = r_o \parallel \beta_F r_o \approx r_o \tag{1.163}$$

The differential gain becomes

$$A_d = -g_m R_o = -g_m r_o \tag{1.164}$$

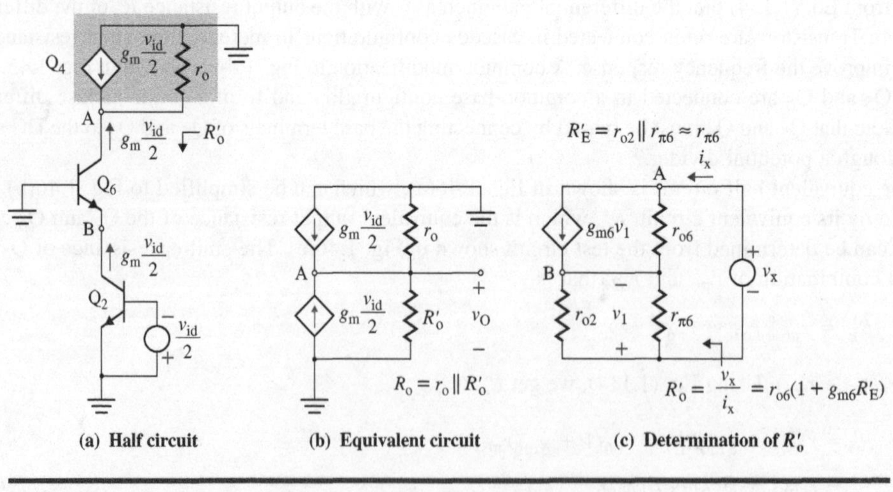

FIGURE 1.46 Circuit for determination of output resistance R'_o

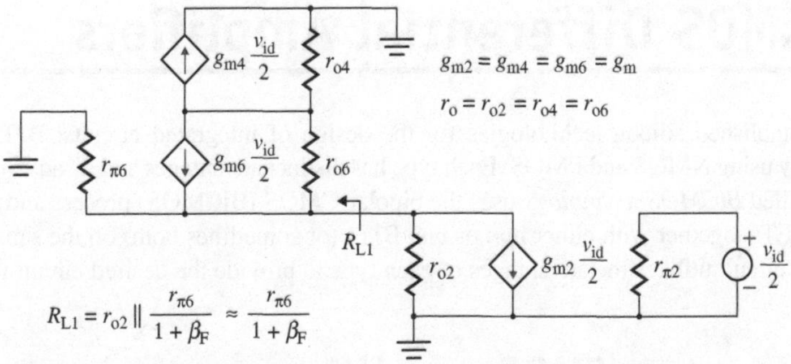

FIGURE 1.47 Circuit for determination of load resistance R_{L1} at the collector of Q_2

Substituting $g_m = I_C/V_T$ and $r_o = V_A/I_C$ into Eq. (1.164), we get

$$A_d = -\frac{V_A}{V_T} \tag{1.165}$$

Thus, the differential gain is twice that of the modified current mirror amplifier in Fig. 1.44. Transistor Q_6 offers a very high resistance seen from the output side.

Transistor Q_6 offers a load resistance R_{L1} seen from the collector of Q_2. R_{L1} can be determined from the test circuit shown in Fig. 1.47; it is the parallel combination of r_{o2}, r_{o6}, and $r_{\pi 6}$ (referred to as the emitter of transistor Q_6). That is,

$$R_{L1} = r_{o2} \| \frac{r_{\pi 6}}{1+\beta_F} \| r_{o6} \approx \frac{r_{\pi 6}}{1+\beta_F} \tag{1.166}$$

which will have a low value. Thus, transistor Q_5 (or Q_6) acts as the *current buffer*, accepting the signal current $(g_m v_{id}/2)$ from the collector of Q_1 at a low resistance R_{L1} and delivering an almost equal current $(g_m v_{id}/2)$ to the load at a very high resistance R_o'. The low resistance at the collector of Q_2 improves the frequency response. Additionally, the internal capacitances of Q_6 (from collector to base and emitter to base) are connected to the ground, and there is no Miller multiplication effect. As a result, the high cutoff frequency is increased.

KEY POINTS OF SECTION 1.7

- The DC transfer characteristic of a BJT differential pair is nonlinear. However, a BJT differential pair is normally operated in the linear region where $v_{id} < V_T$.
- The values of the load resistance R_C and the current source resistance R_{EE} should be large for large values of differential gain and CMRR, respectively. A discrete resistor limits the maximum differential input voltage range.
- The voltage gain is $g_m R_C$ for the difference output and $g_m R_C/2$ for the single-sided output. A small biasing current I_Q increases the transconductance and the voltage gain; however, the differential input resistance is reduced.
- An active load increases the differential gain considerably. The gain is directly proportional to the ratio V_A/V_T. Also, a cascodelike connection increases the gain.

1.8 BiCMOS Differential Amplifiers

There are two established silicon technologies for the design of integrated circuits: BJT technology and CMOS technology using NMOS and PMOS. Each type has distinct advantages and disadvantages. An emerging technology called *BiCMOS technology* uses the bipolar-CMOS (BiCMOS) process and combines *n*- and *p*-channel MOSFETs together with either *npn* or *pnp* BJTs (or sometimes both) on the same semiconductor chip. A BiCMOS circuit utilizes the advantages of each type to provide the desired circuit functions.

1.8.1 BJT versus CMOS Amplifiers

We will begin by considering the basic BJT and CMOS amplifiers. Figure 1.48(a) shows a BJT amplifier with active load that is similar to the half circuit of a BJT differential pair. The output resistance is given by

$$R_o = r_o = \frac{V_A}{I_C} \tag{1.167}$$

where V_A is the Early voltage. Typically, $V_A = 50$ V and $I_C = 5$ μA, so $r_o = 50$ V/(5 μA) = 10 MΩ. Assuming that the current source load has infinite resistance, the voltage gain is given by

$$\begin{aligned} A_d &= -g_m R_o = -g_m r_o \\ &= -\frac{I_C}{V_T} \times \frac{V_A}{I_C} = -\frac{V_A}{V_T} \end{aligned} \tag{1.168}$$

which is independent of the biasing current I_C. Since typically $V_A = 50$ V and $V_T = 25.8$ mV at room temperature, the intrinsic gain of a BJT amplifier is $A_d = -1938$ V/V. The input resistance is

$$R_i = r_\pi = \beta_F \frac{V_T}{I_C} \tag{1.169}$$

which is generally low. Typically, $\beta_F = 60$ and $I_C = 5$ μA, so $R_i = 60 \times 25$ mV/5 μA = 300 kΩ. Although lowering the value of I_C will increase R_i, it will lower the g_m of the transistor and hence the upper frequency limit of the amplifier f_T, which is directly proportional to g_m.

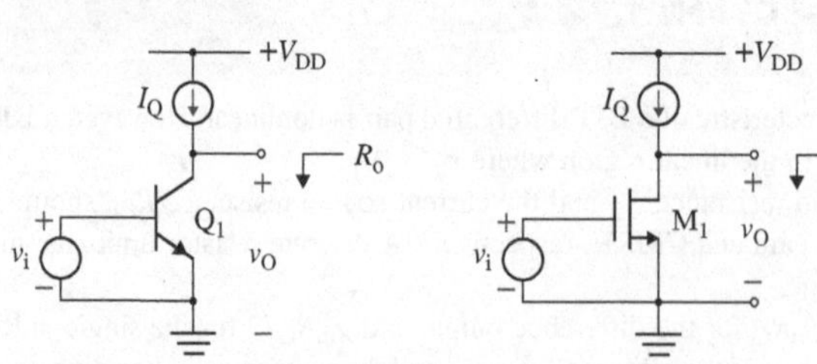

FIGURE 1.48 Basic BJT and CMOS amplifiers

Now consider the MOSFET amplifier shown in Fig. 1.48(b). The output resistance is given by

$$R_o = r_o = \frac{V_M}{I_D} \tag{1.170}$$

where V_M is the channel modulation voltage. Typically, $V_A = 20$ V and $I_D = 5$ μA, so $r_o = 20$ V/(5 μA) = 4 MΩ. Assuming that the current source load has infinite resistance, the voltage gain is given by

$$\begin{aligned} A_d &= -g_m R_o = -g_m r_o \\ &= -2\sqrt{I_D K_n}\left(\frac{V_M}{I_D}\right) = -2\left(\frac{K_n}{I_D}\right)^{1/2} V_M \end{aligned} \tag{1.171}$$

Thus, the gain is inversely proportional to $\sqrt{I_D}$ and will increase as the biasing current is lowered. Decreasing the DC biasing current, however, reduces the amplifier bandwidth. For example, if $I_D = 5$ μA, $V_M = 20$ V, and $K_n = 25$ μA/V^2, $A_d = -89$ V/V.

In summary, for the same value of biasing current I_D, the values of g_m and r_o are much larger for a BJT amplifier than for a MOSFET amplifier—typically 2.5 times as large. The V_A of a BJT amplifier (typically 50 V) is greater than the V_M of a MOSFET (typically 20 V). The voltage gain of a BJT amplifier is greater than that of a MOSFET amplifier by a factor of about 10. However, a MOSFET amplifier has a practically infinite input resistance.

1.8.2 BiCMOS Amplifiers

A BiCMOS amplifier combines the best features of BJT and MOSFET amplifiers. It consists of cascode-like connections of BJTs and MOSFETs. The basic half circuit for a BiCMOS configuration is shown in Fig. 1.49(a). The MOSFET M_1 acts as the driving device, and the BJT Q_1 acts as the load. The cascode configuration is shown in Fig. 1.49(b). The BJT Q_1 acts as the driving device, and the MOSFET M_1 and

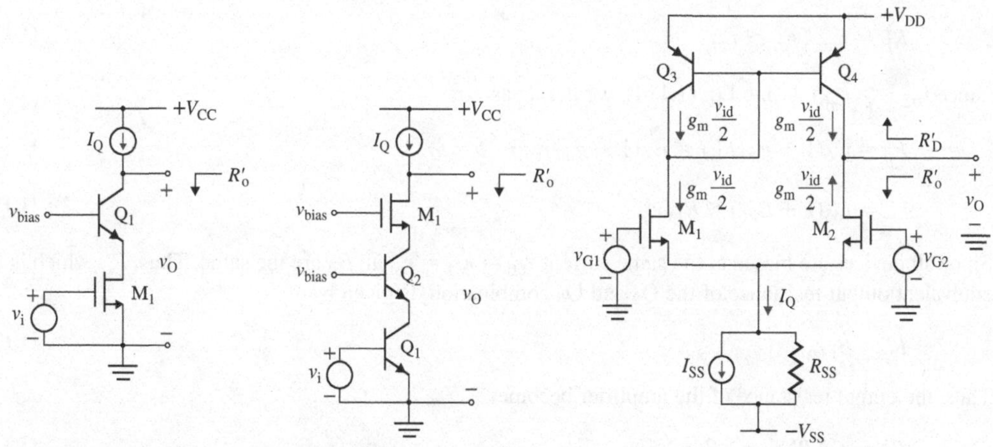

(a) Basic BiCMOS configuration **(b) Cascode BiCMOS configuration** **(c) BiCMOS amplifier**

FIGURE 1.49 BiCMOS amplifier

the BJT Q_2 act as the load. A BiCMOS amplifier is shown in Fig. 1.49(c); this amplifier is identical to the CMOS amplifier in Fig. 1.14. As its abbreviation indicates, the current mirror in a BiCMOS is bipolar rather than unipolar, as in MOS devices. Transistors M_1 and M_2 are the amplifying devices. Transistor Q_4 acts as the load of transistor M_2. The output resistance R_o is the parallel combination of the output resistance $r_{o2} = 2V_M/I_Q$ for transistor M_2 and the output resistance $r_{o4} = 2V_A/I_Q$ for transistor Q_4; that is,

$$R_o = R'_o \| R'_D = r_{o2} \| r_{o4} \tag{1.172}$$

$$= \frac{2V_A V_M}{I_Q(V_A + V_M)} \tag{1.173}$$

The differential voltage gain A_d is given by

$$A_d = \frac{v_o}{v_{id}} = -g_m(r_{o2} \| r_{o4}) \tag{1.174}$$

where g_m, which is the transconductance of the driving MOSFET M_2, is given by Eq. (1.54). Substituting $(r_{o4} \| r_{o2}) = 2V_A V_M/I_Q(V_A + V_M)$ and $g_m = \sqrt{2K_n I_Q}$ into Eq. (1.174), we get

$$A_d = -\sqrt{2K_n I_Q}\frac{2V_M V_A}{I_Q(V_A + V_M)} \tag{1.175}$$

Since $r_{o4} > r_{o2}$, R_o and A_d will be greater for a BiCMOS amplifier than for a CMOS amplifier.

1.8.3 Cascode BiCMOS Amplifiers

Like the cascode BJT amplifier in Fig. 1.45, BiCMOS amplifiers can use cascodelike transistors to increase the voltage gain. This arrangement is shown in Fig. 1.50. Transistors Q_5 and Q_6 are connected in a CB configuration and form a CB differential stage. As shown in Fig. 1.46(c), the emitter resistance R'_E of Q_6 is the parallel combination of $r_{\pi 6}$ and r_{o2}; that is,

$$R'_E = r_{\pi 6} \| r_{o2} \approx r_{\pi 6} \tag{1.176}$$

(since $r_{o2} >> r_{\pi 6}$). Using Eq. (1.114), we get R'_o as

$$R'_o = r_{o6}(1 + g_{m6}R'_E) = r_{o6}(1 + g_{m6}r_{\pi 6})$$

$$= r_{o6}(1 + \beta_{F6}) \approx \beta_{F6}r_{o6} \tag{1.177}$$

Since all devices are biased at the same current $r_{o6} = r_{o4} = r_o$, all β_F are the same. Thus, R'_o, which is the equivalent output resistance of the Q_2 and Q_6 combination, is given by

$$R'_o = \beta_F r_o \tag{1.178}$$

Thus, the output resistance of the amplifier becomes

$$R_o = r_o \| R'_o = r_o \| \beta_F r_o \approx r_o \tag{1.179}$$

The differential voltage gain becomes

$$A_d = -g_m R_o = -g_{m1} r_o \tag{1.180}$$

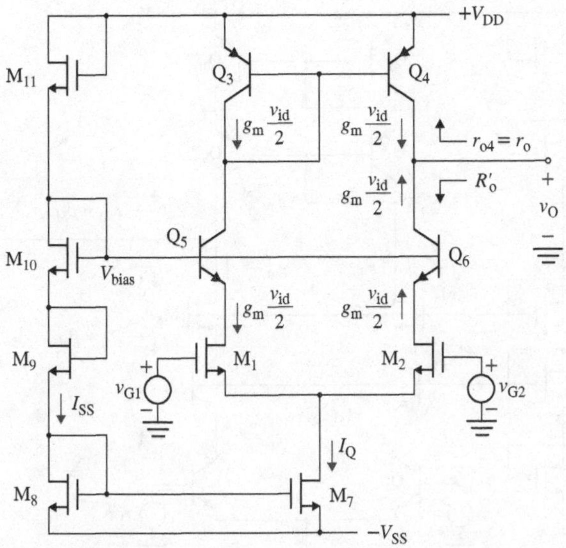

FIGURE 1.50 Cascode BiCMOS amplifier

Substituting $g_m = \sqrt{2K_n I_Q}$ and $r_o = 2V_A/I_Q$ into the preceding equation, we get

$$A_d = -\sqrt{2K_n I_Q}\,\frac{2V_A}{I_Q} = -2V_A\sqrt{\frac{2K_n}{I_Q}} \tag{1.181}$$

which gives a greater voltage gain than Eq. (1.175).

1.8.4 Double-Cascode BiCMOS Amplifiers

A very high gain can be obtained by double cascoding, as shown in Fig. 1.51. There are two cascode connections: the first connects BJTs Q_1, Q_2, Q_3, and Q_4, and the second connects MOSFETs M_5 and M_6. The resistance R'_S looking from the collector of Q_4 will be

$$R'_S = r_{o4}(1 + \beta_{F4}) \simeq \beta_{F4} r_{o4} \tag{1.182}$$

which will be the source resistance of M_6. Using Eq. (1.33), we can find the output resistance R'_o of M_6 as

$$R'_o \simeq r_{o6}(1 + g_{m6} R'_S) = r_{o6}(1 + g_{m6}\beta_{F4} r_{o4}) \tag{1.183}$$

$$\simeq r_{o6} g_{m6} \beta_{F4} r_{o4} \tag{1.184}$$

Like Q_2 and Q_4, transistors Q_8 and Q_{10} are also connected in cascode, and their equations will be similar to those for Q_2 and Q_4. From Eq. (1.177), the resistance looking from the collector of Q_8 is

$$R'_D = r_{o8}(1 + \beta_{F8}) \simeq \beta_{F8} r_{o8} \tag{1.185}$$

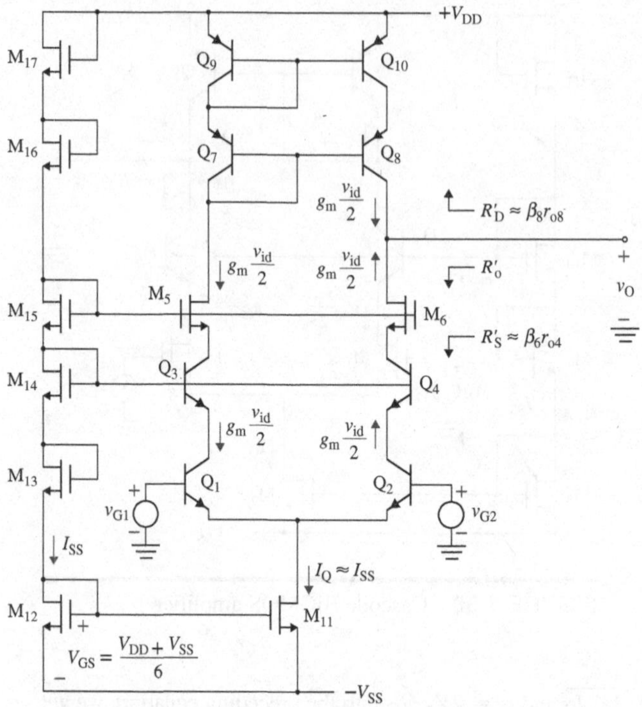

FIGURE 1.51 Double-cascode BiCMOS amplifier

Thus, the output resistance of the amplifier becomes

$$R_o = R'_D \parallel R'_S = (r_{o6} g_{m6} \beta_{F4} r_{o4}) \parallel (\beta_{F8} r_{o8}) \tag{1.186}$$

$$= \beta_F r_o \quad (\text{for } r_{o4} = r_{o8} = r_o,\ \beta_{F8} = \beta_F) \tag{1.187}$$

The differential voltage gain becomes

$$A_d = -g_m R_o = -g_m (r_{o6} g_{m6} \beta_{F4} r_{o4}) \parallel (\beta_{F8} r_{o8}) \tag{1.188}$$

$$= -g_m \beta_F r_o \tag{1.189}$$

where $g_m = I_Q/2V_T$. Equation (1.189) shows a considerable increase in the voltage gain.

EXAMPLE 1.15

Analyzing BiCMOS amplifiers The DC biasing current of a BiCMOS amplifier is kept constant at $I_Q = 10\ \mu A$. All bipolar transistors are identical, with $V_A = 50$ V and $\beta_F = 40$. Also, the MOS transistors are identical, with $V_M = 20$ V, $K_n = 25\ \mu A/V^2$, $W = 30\ \mu m$, and $L = 10\ \mu m$. Assume $V_T = 25.8$ mV.

Determine the differential voltage gain A_d for single-ended output **(a)** for the BiCMOS amplifier in Fig. 1.49(c), **(b)** for the cascode BiCMOS amplifier in Fig. 1.50, and **(c)** for the double-cascode BiCMOS amplifier in Fig. 1.51.

SOLUTION

$V_A = 50$ V, $\beta_F = 40$, $V_M = 20$ V, $K_n = 25\ \mu A/V^2$, $W = 30\ \mu m$, $L = 10\ \mu m$, and $I_D = I_C = I_Q/2 = 10\ \mu A/2 = 5\ \mu A$.

(a) We have

$$r_{o2} = \frac{2V_M}{I_Q} = \frac{2 \times 20\ \text{V}}{10\ \mu\text{A}} = 4\ \text{M}\Omega$$

$$r_{o4} = \frac{2V_A}{I_Q} = \frac{2 \times 50\ \text{V}}{10\ \mu\text{A}} = 10\ \text{M}\Omega$$

$$R_o = r_{o2} \| r_{o4} \equiv 4\ \text{M}\Omega \| 10\ \text{M}\Omega = 2.86\ \text{M}\Omega$$

$$g_{m2} = \sqrt{2K_n I_Q} = \sqrt{2 \times 25\ \mu \times 10\ \mu} = 22.36\ \mu\text{A/V}$$

Thus, the differential voltage becomes

$$A_d = -g_{m2}R_o = -22.36\ \mu \times 2.86\ \text{M} = -63.9$$

(b) We have

$$r_{o4} = r_{o6} = \frac{2V_A}{I_Q} = \frac{2 \times 50\ \text{V}}{10\ \mu\text{A}} = 10\ \text{M}\Omega$$

$$R_o' = \beta_F r_{o6} = 40 \times 10\ \text{M}\Omega = 400\ \text{M}\Omega$$

$$R_o = r_{o4} \| R_o' = 10\ \text{M} \| 400\ \text{M} = 10\ \text{M}\Omega$$

$$g_{m2} = \sqrt{2K_n I_Q} = \sqrt{2 \times 25\ \mu \times 10\ \mu} = 22.36\ \mu\text{A/V}$$

Thus, the differential voltage becomes

$$A_d = -g_{m2}R_o = -22.36\ \mu \times 10\ \text{M} = -223.6\ \text{V/V}$$

(c) From Eq. (1.187),

$$R_o = \beta_F r_{o4} = 40 \times 10\ \text{M} = 400\ \text{M}\Omega$$

$$g_{m2} = \frac{I_Q}{2V_T} = \frac{10\ \mu\text{A}}{2 \times 25.8\ \text{mV}} = 193.8\ \mu\text{A/V}$$

Thus, the differential voltage gain becomes

$$A_d = -g_{m2}R_o = -193.8\ \mu\text{A/V} \times 400\ \text{M}\Omega = -77{,}520\ \text{V/V}$$

which is considerably larger than for the other two configurations.

KEY POINTS OF SECTION 1.8

- A BiCMOS amplifier combines the advantages of BJT and MOS technologies to achieve the desirable circuit functions of infinite input resistance and a very large voltage gain and CMRR.
- BJT and MOS transistors can be connected in cascode to give an extremely large output resistance and voltage gain.

1.9 Frequency Response of Differential Amplifiers

The transistors (either MOSFETs or BJTs) in differential amplifiers have capacitance—hence, the gain of such amplifiers will be frequency dependent [5–7]. In this section, we consider only the differential gain with passive and active loads. R_C represents the output resistance of the active source.

1.9.1 Frequency Response with Resistive Load

Let us first consider the differential half circuit shown in Fig. 1.52(a). Replacing the transistor by its frequency model gives the circuit in Fig. 1.52(b). The transistor can be either a BJT or a MOSFET. We can apply the same model and analysis for MOSFETs by substituting $r_\pi = \infty$, $C_\pi \approx C_{gs}$ and $C_\mu \approx C_{gd}$ in the derivations. Since the input signal is $v_{id}/2$, the r_π, C_π, and C_μ parameters of the transistor model are sealed by a factor of 2. Since $g_m(R_C \| r_{o1}) >> 1$, C_μ will dominate the high cutoff frequency ω_H. If we replace $C_\mu/2$ by its effective Miller capacitance C_M, we get the equivalent circuit shown in Fig. 1.52(c) and Eq. (G.3) in Appendix G; that is,

$$C_M = \left(\frac{C_\mu}{2}\right)[1 + g_m(R_C \| r_{o1})] \tag{1.190}$$

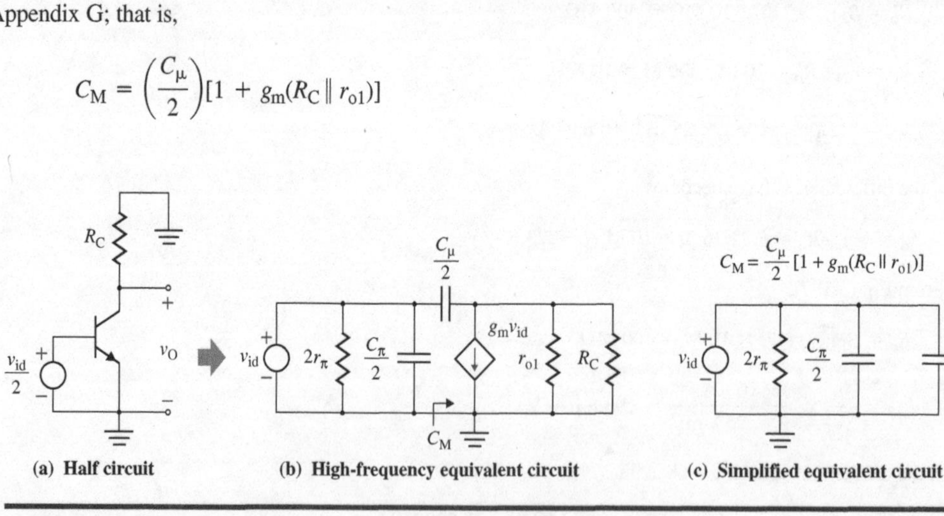

(a) Half circuit **(b) High-frequency equivalent circuit** **(c) Simplified equivalent circuit**

FIGURE 1.52 Differential half circuit and high-frequency equivalent

Thus, the high cutoff frequency ω_H is given by

$$\omega_H = \frac{1}{2r_\pi(C_\pi/2 + C_M)} = \frac{1}{2r_\pi\{C_\pi/2 + (C_\mu/2)[1 + g_m(R_C \| r_{o1})]\}} \tag{1.191}$$

which, for $R_C = r_{o1} = r_o$ and $g_m(R_C \| r_{o1}) >> 1$, can be approximated by

$$\omega_H = \frac{1}{2r_\pi(C_\mu/2)g_m r_o/2} = \frac{2}{r_\pi C_\mu g_m r_o} = \frac{2}{r_\pi C_\mu g_m R_C} \tag{1.192}$$

The signal source resistance $R_s = 0$. The effective resistance for $C_\pi/2$ will be $R_{C\pi} = (2r_\pi \| R_s) = 0$. Thus, the break frequency due to $C_\pi/2$ will be at infinity and will not influence ω_H in Eq. (1.192). Therefore, the frequency-dependent differential gain is given by

$$A_d(j\omega) = \frac{A_{do}}{1 + j\omega/\omega_H} \tag{1.193}$$

where $A_{do} = g_m(r_{o2} \| r_{o4})$ is the low-frequency differential gain with differential output voltage.

1.9.2 Frequency Response with Active Load

Now let us consider the output circuit for the amplifier with an active load, as shown in Fig. 1.42. The AC equivalent of that circuit is shown in Fig. 1.53(a). Replacing the transistors by their frequency model, we get Fig. 1.53(b). Since the effective transconductance is $2g_m$, the Miller capacitance C_M [from Eq. (G.3) in Appendix G] becomes

$$C_M = \left(\frac{C_\mu}{2}\right)[1 + 2g_m(r_{o2} \| r_{o4})] \tag{1.194}$$

Thus, the high cutoff frequency ω_H is given by

$$\omega_H = \frac{1}{2r_\pi\{C_\pi/2 + (C_\mu/2)[1 + 2g_m(r_{o2} \| r_{o4})]\}} \tag{1.195}$$

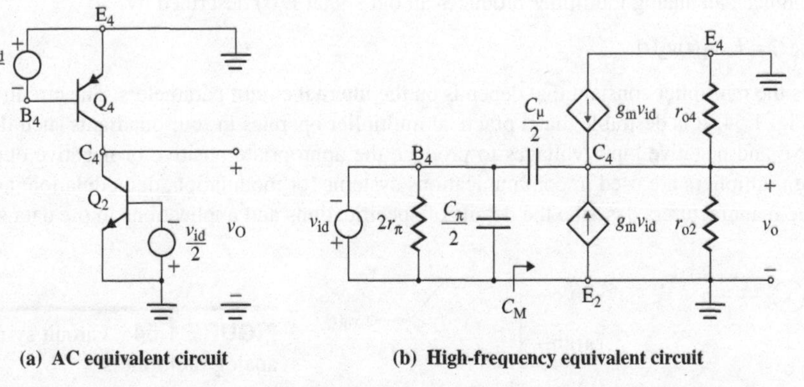

FIGURE 1.53 Differential amplifier with active load and high-frequency equivalent

which, for $r_{o2} = r_{o4} = r_o$ and $2g_m(r_{o2} \| r_{o4}) >> 1$, can be approximated by

$$\omega_H = \frac{1}{2r_\pi/(C_\mu/2)g_m r_o} = \frac{1}{r_\pi C_\mu g_m r_o} \tag{1.196}$$

Therefore, the frequency-dependent differential gain is given by

$$A_d(j\omega) = \frac{A_{do}}{1 + j\omega/\omega_H} \tag{1.197}$$

where $A_{do} = -2g_m(r_{o2} \| r_{o4})$ is the low-frequency differential gain with an active load and a single-sided output.

Notice from Eq. (1.192) that the high cutoff frequency is double that for an amplifier with an active load and single-sided output. However, this high cutoff frequency for an amplifier with a passive load is obtained at the expense of the output voltage gain. For this reason, the first stage of a wide-band amplifier often uses a balanced circuit with differential voltage when the CMRR is not the prime consideration.

KEY POINTS OF SECTION 1.9

- The transistors in differential amplifiers have capacitance; hence, the gain of such amplifiers will be frequency dependent and exhibit a high-pass characteristic.
- The high cutoff frequency for an amplifier with a passive load is obtained at the expense of the output voltage gain. For this reason, the first stage of a wide-band amplifier often uses a balanced circuit with differential voltage.

1.10 Analog Multipliers

In analog-signal processing, there is often a need for a circuit that takes two analog inputs and produces an output proportional to their product. Such circuits are termed *analog multipliers*. If $v_1(t)$ and $v_2(t)$ are the two input signals, an analog multiplier produces an out signal $v_o(t)$ described by

$$v_o(t) = K_c v_1(t) v_2(t) \tag{1.198}$$

where K_c is the multiplier constant that depends on the internal circuit parameters. The circuit symbols are shown in Fig. 1.54. It is desirable that a practical multiplier operates in four quadrants such that it can accept positive and negative input voltages to produce the appropriate positive or negative output voltage. The analog multipliers are used in communications systems for modulation, demodulation, and phase detection. The manufacturers describe the details of specifications and applications in the data sheets [9].

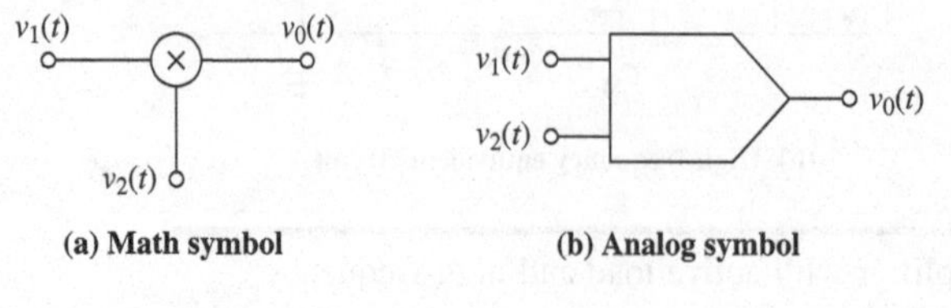

(a) Math symbol (b) Analog symbol

FIGURE 1.54 Circuit symbol for analog multipliers

1.10.1 Gilbert Multiplier Cell

The differential output voltage v_{od} of a MOS differential pair in Fig. 1.8 is proportional to the differential input voltage v_{id} as shown in Eq. (1.53) and the source biasing current I_Q. The relationship is practically linear for a small value of v_{id} as shown in Fig. 1.9. Similarly, the differential output voltage v_{od} of a BJT differential pair in Fig. 1.33 is proportional to the differential input voltage v_{id} as shown in Eq. (1.140) and the emitter biasing current I_Q. The relationship is practically linear for a small value of v_{id} as shown in Fig. 1.34. Since the collector (or drain) current of a BJT (or a MOSFET) of a differential pair depends on the biasing current I_Q, differential pairs are connected in cascade to perform an analog multiplication.

The Gilbert cell [5, 8] is the fundamental block for virtually all analog multipliers as shown in Fig. 1.55. It consists of three emitter-coupled pairs, Q_1–Q_2, Q_3–Q_4, and Q_5–Q_6. The emitter biasing current I_Q is provided by a Widlar current source consisting of Q_7 and Q_8. All resistors and transistors are matched or identical in the fabrication process. The Gilbert cell can also be implemented by replacing the BJTs by MOSFETs. Using Eq. (1.101), the biasing current I_Q of the Widlar current source as shown in Eq. (1.101) can be found approximated from

$$I_Q = I_{EE} = \frac{V_T}{R_2}\ln\left(\frac{I_{C7}}{I_{C8}}\right) \tag{1.199}$$

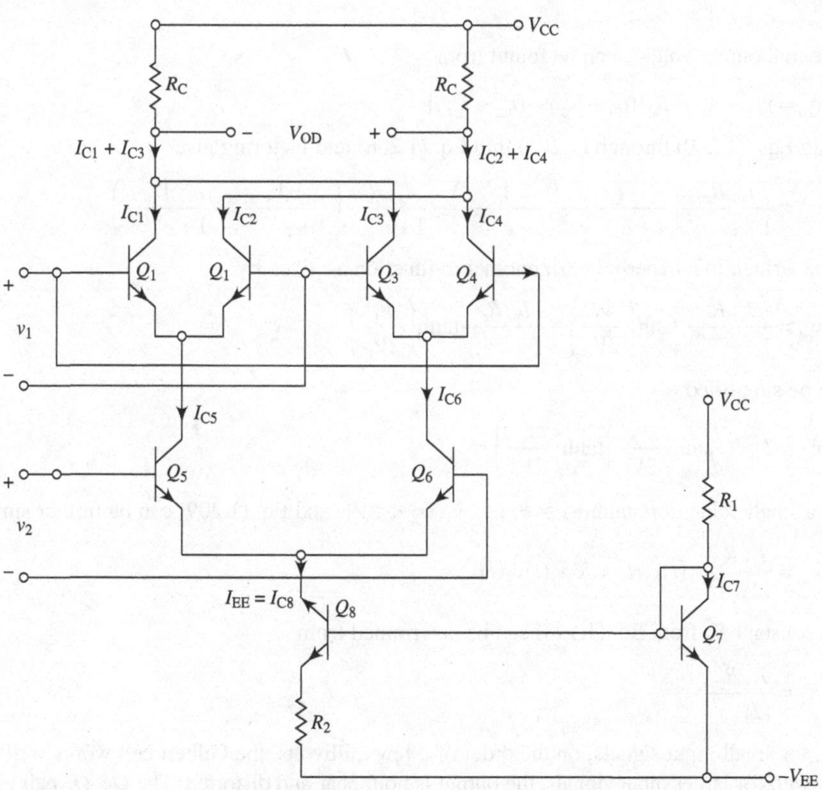

FIGURE 1.55 Gilbert cell multiplier

Assuming that the beta (β_F) of the BJTs is large such that the forward current gain $\alpha_F = \beta_F/(1 + \beta_F) \cong 1$, Eqs. (1.133) and (1.135) can be applied to relate the collector currents to the emitter currents of the differential pairs as given by

$$i_{C1} = \frac{i_{C5}}{1 + e^{-v_1/V_T}} \tag{1.200}$$

$$i_{C2} = \frac{i_{C5}}{1 + e^{v_1/V_T}} \tag{1.201}$$

$$i_{C3} = \frac{i_{C6}}{1 + e^{v_1/V_T}} \tag{1.202}$$

$$i_{C4} = \frac{i_{C6}}{1 + e^{-v_1/V_T}} \tag{1.203}$$

$$i_{C5} = \frac{I_Q(= I_{EE})}{1 + e^{-v_2/V_T}} \tag{1.204}$$

$$i_{C6} = \frac{I_Q(= I_{EE})}{1 + e^{v_2/V_T}} \tag{1.205}$$

The differential output voltage can be found from

$$v_{od} = v_{C2} - v_{C1} = R_C[(i_{C1} + i_{C3}) - (i_{C2} + i_{C4})] \tag{1.206}$$

Substituting Eqs. (1.200) through (1.205) into Eq. (1.206) and factoring give

$$v_{od} = \frac{I_{EE}R_C}{1 + e^{-v_1/V_T}}\left(\frac{1}{1 + e^{-v_2/V_T}} - \frac{1}{1 + e^{v_2/V_T}}\right) - \frac{I_{EE}R_C}{1 + e^{v_1/V_T}}\left(\frac{1}{1 + e^{-v_2/V_T}} - \frac{1}{1 + e^{v_2/V_T}}\right) \tag{1.207}$$

This can be written in a hyperbolic trigonometric function as given by

$$v_{od} = \frac{I_{EE}R_C}{1 + e^{-v_1/V_T}}\tanh\left(\frac{v_2}{2V_T}\right) - \frac{I_{EE}R_C}{1 + e^{v_1/V_T}}\tanh\left(\frac{v_2}{2V_T}\right) \tag{1.208}$$

which can be simplified to

$$v_{od} = I_{EE}R_C \tanh\left(\frac{v_2}{2V_T}\right)\tanh\left(\frac{v_1}{2V_T}\right) \tag{1.209}$$

Since, for a small value of θ, $\tanh(\theta) \approx \theta$, i.e., $v_1, v_2 << V_T$, and Eq. (1.209) can be further simplified to

$$v_{od} = \frac{I_{EE}R_C}{(2V_T)^2} v_1(t)v_2(t) = K_c v_1(t)v_2(t) \tag{1.210}$$

where the constant K_c from Eq. (1.198) can be determined from

$$K_c = \frac{I_{EE}R_C}{(2V_T)^2} \tag{1.211}$$

Therefore, for small input signals, on the order of a few millivolts, the Gilbert cell works well as a multiplier. However, for larger input signals, the output is nonlinear and distorted. The Q_5–Q_6 pair can be operated as an emitter-coupled pair with degeneration resistors as shown in Fig. 1.35(a). This increases the linear range as shown in Fig. 1.35(b) for input signal v_2. But this cannot be done with the cross-coupled pairs Q_1–Q_4 because the degeneration resistors removes the required nonlinear relation between I_C and V_{BE} in these devices.

1.10.2 Hyperbolic Tangent Transfer Characteristic

The output of the multiplier can be linearized by introducing a nonlinearity that pre-distorts the input signals and produces an inverse hyperbolic tangent characteristic. By pre-distorting the input signal by performing the $\tanh^{-1}(v/2V_T)$ operation, Eq. (1.209) becomes

$$v_{od} = I_{EE}R_C \tanh\left[\tanh^{-1}\left(\frac{v_1}{2V_T}\right)\right]\tanh\left[\tanh^{-1}\left(\frac{v_2}{2V_T}\right)\right]$$

$$= I_{EE}R_C\left(\frac{v_1}{2V_T}\right)\left(\frac{v_2}{2V_T}\right) \tag{1.212}$$

An inverse hyperbolic tangent circuit [5, 8] is shown in Fig. 1.56(a). The voltage–current converter within the box develops a differential output current that is linearly related to the input voltage v_1. It can be implemented as differential pair consisting of Q_{11}–Q_{12} as shown in Fig. 1.56(b). The differential output voltages Δv_{o1} of the converter across the emitter terminals of the diode-connected transistors (Q_9 and Q_{10}) is given by

$$\Delta v_{o1} = v_{C12} - v_{C11} = v_{BE9} - v_{BE10} \tag{1.213}$$

Using the diode relation $V_{BE} = \eta V_T \ln\left(\frac{i_D}{I_S}\right)$, Eq. (1.213) becomes

$$\Delta v_{o1} = V_T\ln\left(\frac{i_{C9}}{I_{S9}}\right) - V_T\ln\left(\frac{i_{C10}}{I_{S10}}\right) = V_T\ln\left(\frac{i_{C9}}{i_{C10}}\right) = V_T\ln\left(\frac{i_{C11}}{i_{C12}}\right) \tag{1.214}$$

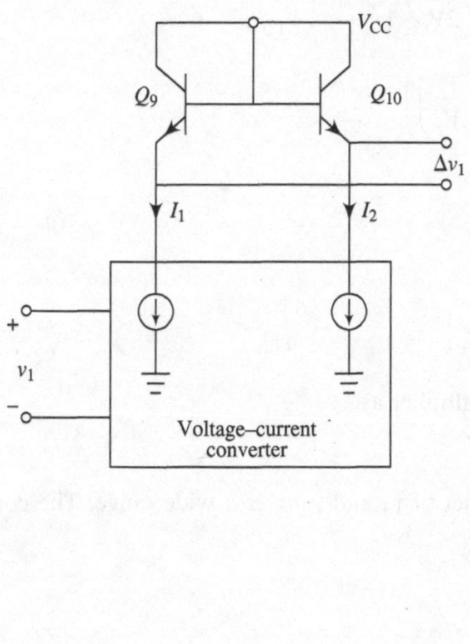

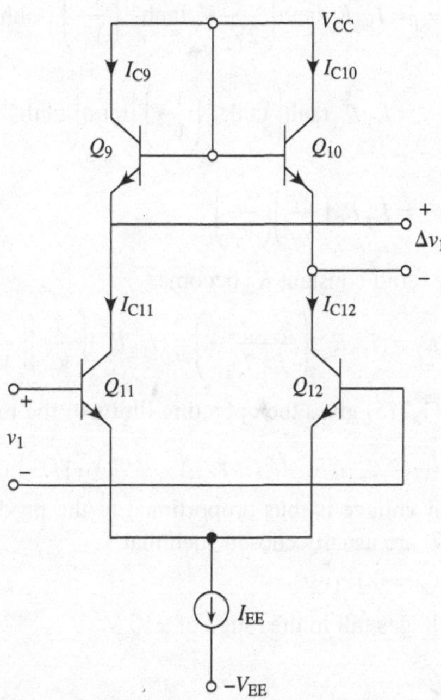

FIGURE 1.56 Inverse hyperbolic tangent circuit

If the input signal v_1 changes by a small amount, the collector currents of Q_{11} and Q_{12} vary from the dc biasing conditions of $I_{C11} = I_{C12} = I_{EE}/2$. Applying Eq. (1.214), the output voltage of the circuit is given by

$$v_{od1} = \Delta v_{o1} = V_T \ln\left(\frac{I_{C11} + g_{m11}v_1}{I_{C12} - g_{m12}v_1}\right) = V_T \ln\left(\frac{1+(g_{m11}v_1/I_{C11})}{1-(g_{m12}v_1/I_{C12})}\right) \quad \textbf{(1.215)}$$

where $g_{m11} = g_{m12} = g_m$ is the trans-conductance of transistor Q_{11} or Q_{12}. Applying the hyperbolic trigonometric identity,

$$\tanh^{-1}(x) = \frac{1}{2}\ln\left(\frac{1+x}{1-x}\right) \quad \textbf{(1.216)}$$

into Eq. (1.215), we get

$$v_{od1} = 2V_T \tanh^{-1}\left(\frac{g_{m11}v_1}{I_{C11}}\right) = 2V_T \tanh^{-1}\left(\frac{v_1}{V_T}\right) \quad \textbf{(1.217)}$$

This is an inverse hyperbolic tangent function. If the input signals v_1 and v_2 are passed through two such circuits, the output of the Gilbert cell multiplier operates as a multiplier of both large and small signals.

1.10.3 Four-Quadrant Gilbert Cell Multiplier

The complete Gilbert cell multiplier with two inverse hyperbolic tangent circuits [8] is shown in Fig. 1.57. The circuit consisting of transistors Q_{13}–Q_{14} and Q_{15}–Q_{16} generates the inverse hyperbolic tangent function for input signal v_2. Substituting Δv_{o1} from Eq. (1.217) for v_1 and Δv_{o2} for v_2 into Eq. (1.209), the overall transfer characteristic becomes

$$\begin{aligned} v_{od} &= I_{EE}R_C \tanh\left(\frac{2}{2V_T}V_T \tanh^{-1}\left(\frac{v_2}{V_T}\right)\right)\tanh\left(\frac{2}{2V_T}V_T \tanh^{-1}\left(\frac{v_1}{V_T}\right)\right) \\ &= I_{EE}R_C \tanh\left(\tanh^{-1}\left(\frac{v_2}{V_T}\right)\right)\tanh\left(\tanh^{-1}\left(\frac{v_1}{V_T}\right)\right) \\ &= I_{EE}R_C\left(\frac{v_1}{V_T}\right)\left(\frac{v_2}{V_T}\right) \end{aligned} \quad \textbf{(1.218)}$$

This, the overall constant K_c becomes

$$K_c = (I_{EE}R_C)\left(\frac{g_{m11}g_{m15}}{I_{C11}I_{C15}}\right) = (I_{EE}R_C)\left(\frac{1}{V_T}\right)\left(\frac{1}{V_T}\right) \quad \textbf{(1.219)}$$

Equation (1.215) gives the operating limits of the multiplier as

$$I_{C11} > g_{m11}v_1; \quad I_{C15} > g_{m15}v_2 \quad (\text{or } v_1, v_2 < V_T) \quad \textbf{(1.220)}$$

The output voltage is thus proportional to the product of v_1 and v_2 over a wide range. The constants in Eq. (1.219) are usually chosen such that

$$v_{od} = 0.1v_1v_2 \quad \textbf{(1.221)}$$

and all voltages fall in the range of ±10 V.

EXAMPLE 1.16

Finding the parameters of a Gilbert cell

The circuit parameters of the Gilbert cell shown in Fig. 1.55 are $V_{CC} = V_{EE} = 15$ V, $I_{C1} = I_{C2} = I_{C3} = I_{C4} = 50\ \mu$A, and $R_{C1} = R_{C2} = 47$ kΩ. Determine (a) the values of R_1 and R_2, (b) the multiplier constant K_c, and (c) the maximum value of the input voltages if $v_1 = v_2$ and the output voltage v_o is limited to 5 V. Assume the transistor current gain $\beta_F >> 1$ and the reference current $I_{C7} = 1$ mA, $V_{BE} = 0.7$ V, and $V_T = 25.8$ mV.

SOLUTION

$$I_{EE} = I_{C8} = I_{C1} + I_{C2} + I_{C3} + I_{C4} = 4 \times 50\ \mu\text{A} = 200\ \mu\text{A}$$

(a) $$R_1 = \frac{V_{CC} - (-V_{EE}) - V_{BE}}{I_{C7}} = \frac{15 + 15 - 0.7}{1 \times 10^{-3}} = 29.3\ \text{k}\Omega$$

$$R_2 = \frac{V_T}{I_{C8}} \ln\left(\frac{I_{C7}}{I_{C8}}\right) = \frac{25.8 \times 10^{-3}}{200 \times 10^{-6}} \ln\left(\frac{1 \times 10^{-3}}{200 \times 10^{-6}}\right) = 207.62\ \Omega$$

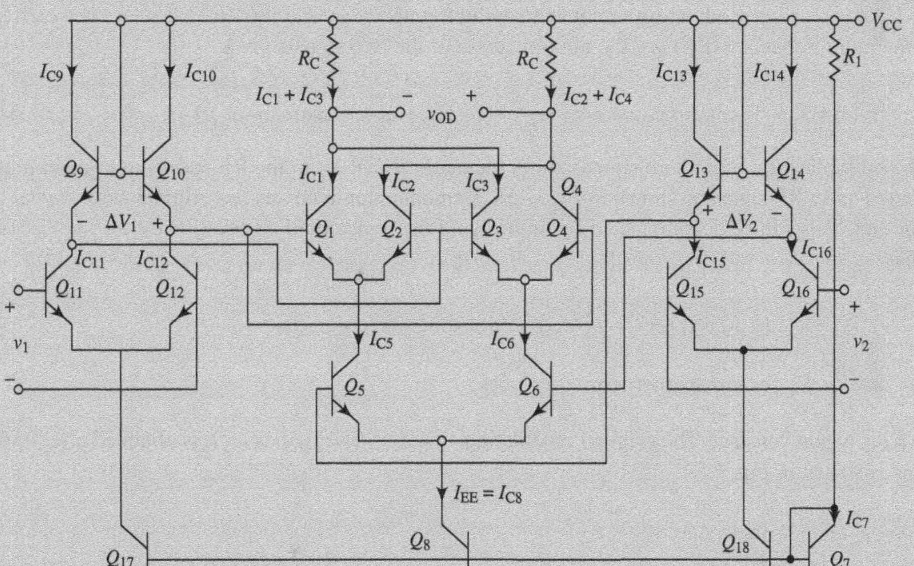

FIGURE 1.57 Four-quadrant Gilbert cell multiplier

(b) Using Eq. (1.219),

$$K_c = I_{C8}R_{C1}\left(\frac{1}{V_T}\right)^2 = 200\times10^{-6}\times47\times10^3\left(\frac{1}{25.8\times10^{-3}}\right)^2 = 14.1\times10^3$$

(c) For $v_1 = v_2$,

$$v_1 = v_2 = \sqrt{\frac{v_o}{K_c}} = \sqrt{\frac{5}{14.1\times10^3}} = 18.82\ \text{mV}$$

NOTE: If it requires a higher output voltage, the circuit values must be selected carefully to ensure $v_1 = v_2 < V_T$.

1.10.4 Multipliers as Frequency Modulation and Demodulation Signal

In this type of applications, the input signals are small and the transistors operate in the active region. If $v_1(t) = V_m \cos \omega_1 t$ is a modulating signal at an audio frequency ω_1 and $v_2(t) = V_c \cos \omega_2 t$ is a carrier signal in the radio frequency (RF) ω_2, the multiplication of the two signals gives

$$v_o(t) = K_c V_m V_c \cos \omega_1 t \cos \omega_2 t = \frac{K_c V_m V_c}{2}[\cos(\omega_2 + \omega_1)t + \cos(\omega_2 - \omega_1)t] \tag{1.222}$$

The multiplication of the two signals shifts the audio signal ω_1 to the RF spectrum ω_2, and it can be transmitted as an RF signal as shown in Fig. 1.58. Demodulation recovers the original audio signal at the receiver end. This can be accomplished by multiplying the transmitted signal $v_o(t)$ by $v_2(t) = V_c \cos \omega_2 t$ to obtain

$$v_o'(t) = V_m \cos \omega_2 t v_o(t) = \frac{K_c V_m V_c^2}{2}\cos \omega_2 t[\cos(\omega_2 + \omega_1)t + \cos(\omega_2 - \omega_1)t]$$

$$= K_1 \cos \omega_1 t + \text{other terms} \tag{1.223}$$

where K_1 is a gain constant. The original modulating signal, $v_1(t) = v_m \cos \omega_1 t$, is obtained after low-pass filtering as shown in Fig. 1.59.

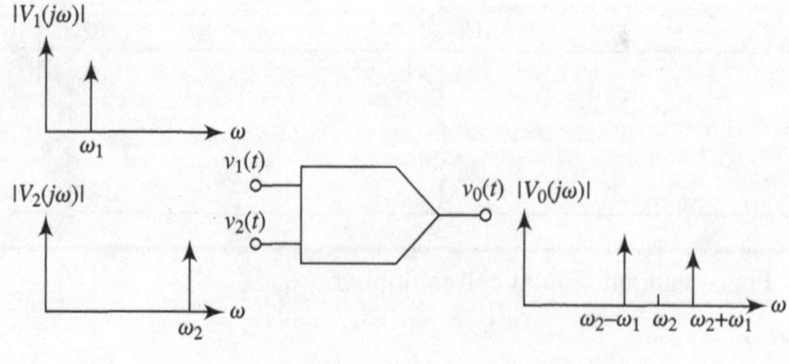

FIGURE 1.58 Multiplier for frequency modulation

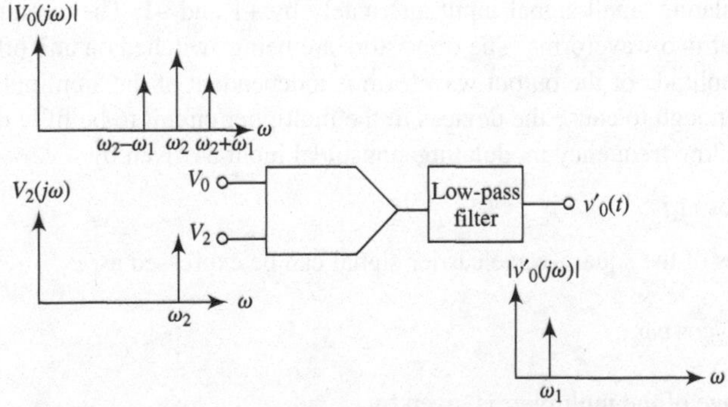

FIGURE 1.59 Multiplier for frequency demodulation

1.10.5 Multipliers as Balanced and Amplitude Modulation

We have seen in Sec. 1.10.4 that the modulation or the demodulation can be accomplished by multiplying two signals at two different frequencies. However, the multiplication is relatively slow. The multiplication of a continuously varying signal by a square wave is often needed in communications systems. The multiplier circuit can perform this function by applying a sufficiently large signal (as compared to $2V_T$) directly to the cross-coupled pair such that two of the four transistors are alternately turned-off completely. The other two transistors conduct all the currents. Since the transistors in the circuit do not enter saturation, this process can be accomplished at high speed.

A typical waveform of a modulating input signal is shown in Fig. 1.60(a), and the square-waver carrier signal is shown in Fig. 1.60(b). The output voltage as shown in Fig. 1.60(c) is produced by

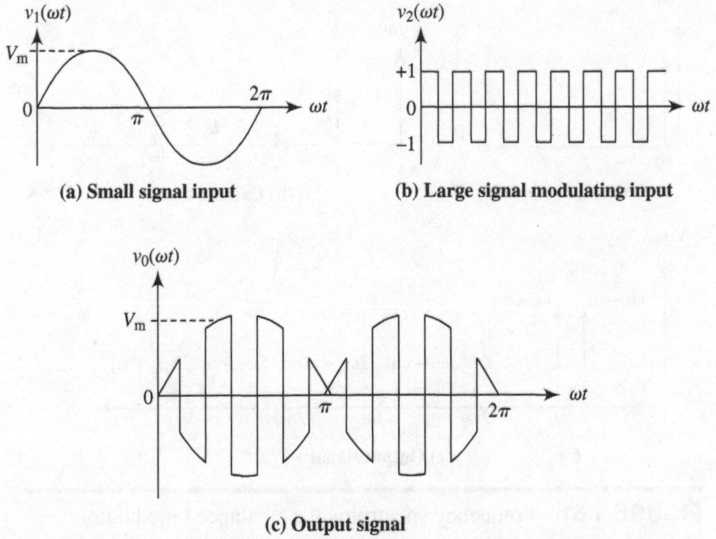

FIGURE 1.60 Input and output waveforms of a multiplier

multiplying the modulating small-signal input alternately by +1 and –1. The circuit does not perform a linear multiplication of two waveforms. The transistors are being switched on and off by the square-wave carrier signal. The amplitude of the output waveform is independent of the amplitude of the square wave as long as it is large enough to cause the devices in the multiplier circuit to be fully on or fully off.

Let us consider a low-frequency modulating sinusoidal input as given by

$$v_1(t) = V_m \cos \omega_1 t \tag{1.224}$$

The Fourier series of the square-wave carrier signal can be expressed as

$$v_2(t) = \sum_{n=1}^{\infty} A_n \cos n\omega_2 t \tag{1.225}$$

Thus, the output voltage of the multiplier is given by

$$v_o(t) = K_c v_1(t) v_2(t) = K_c \sum_{n=1}^{\infty} V_m A_n \cos \omega_1 t \cos n\omega_2 t$$

$$= K_c \sum_{n=1}^{\infty} \frac{V_m A_n}{2} [\cos(n\omega_2 + \omega_1)t + \cos(n\omega_2 - \omega_1)t] \tag{1.226}$$

where K_c is the multiplier gain. The spectrum has components located at frequencies ω_1 above and below each of the harmonics of ω_2, but no component at the carrier frequency ω_2 or its harmonics. The frequency spectrums are shown in Fig. 1.61(c). The output does not contain any component at the carrier frequency and this can be filtered so that only the components near ω_2 are retained.

A dc component is often introduced intentionally to provide amplitude modulation or it can be the result of offset voltages in the devices within the modulator. When the dc component is added to the modulating input, it introduces a signal component in the output at the carrier frequency and its harmonics. The modulating signal in Eq. (1.224) becomes as given by

$$V_1(t) = V_m (1 + M \cos \omega_1 t) \tag{1.227}$$

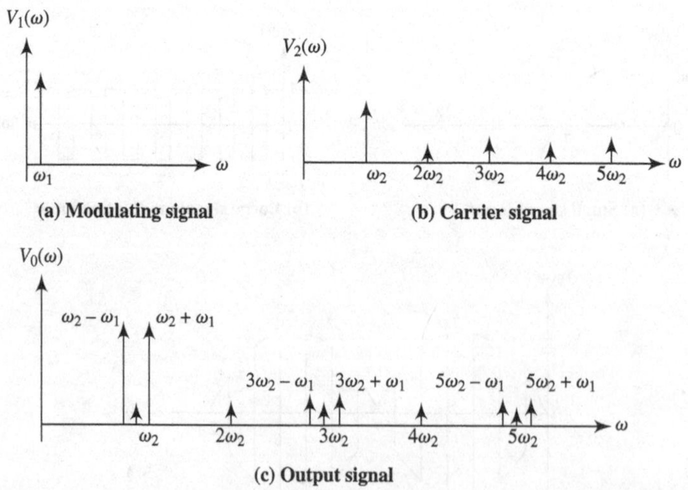

FIGURE 1.61 Frequency spectrums for a balanced modulator

where the parameter M is called the modulation index, $0 < M \le 1$. The output signal is given by

$$v_o(t) = K_c \sum_{n=1}^{\infty} V_m A_n \left[\cos(n\omega_2 t) + \frac{M}{2}\cos(n\omega_2 + \omega_1)t + \frac{M}{2}\cos(n\omega_2 - \omega_1)t \right] \quad \textbf{(1.228)}$$

Information contained within the modulating signal $v_1(t)$ was originally concentrated at the modulating frequency ω_1. The modulator has translated this information such that it is contained in spectral components located near the harmonics of the high-frequency carrier signal $v_2(t)$. The demodulation does the extraction of the information from the frequency band near the carrier and the retranslation of the information back to low frequencies.

1.10.6 Multipliers as a Phase Detector

If two signals of identical frequency ω_o are applied to the two inputs of the multiplier, it behaves as phase detector and produces an output whose dc component is proportional to the phase difference between the two inputs. Let us consider the two input waveforms in Fig. 1.62(a) and (b) which are applied to the Gilbert

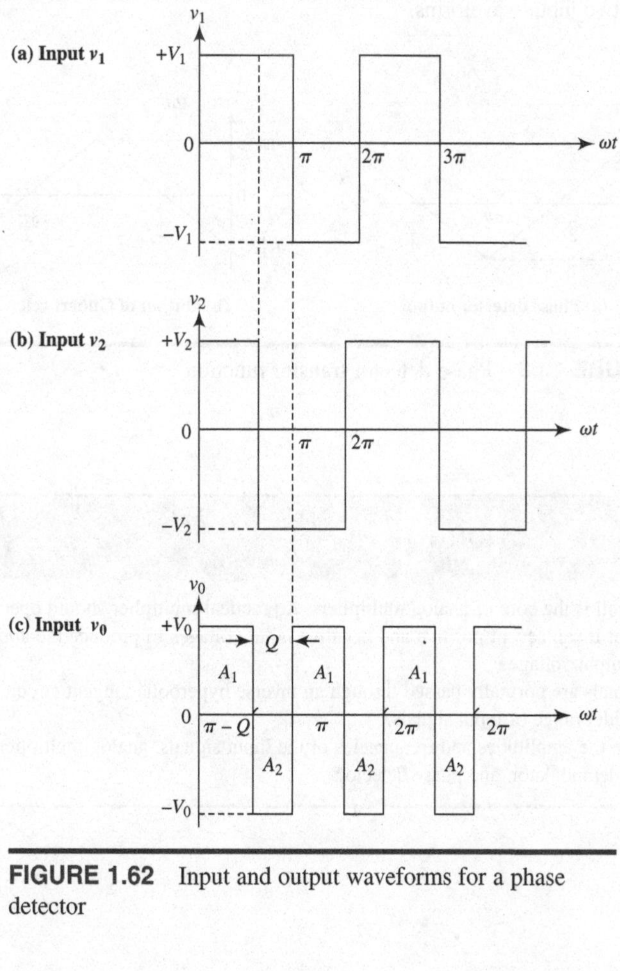

FIGURE 1.62 Input and output waveforms for a phase detector

cell in Fig. 1.55. Let us also assume that both inputs are large in magnitudes so that all the transistors are behaving as switches. The output waveform as shown in Fig. 1.62(c) has a frequency of twice the input frequency ω_O.

The average output voltage is given by

$$V_{av}(\theta) = \frac{1}{2\pi}\int_0^{2\pi} v_o(t)d\theta = \frac{1}{2\pi}(A_1 - A_2 + A_1 - A_2) = \frac{1}{\pi}(A_1 - A_2)$$

$$= \frac{1}{\pi}[V_O(\pi - \theta) - V_O\theta] = \frac{V_O}{\pi}[\pi - 2\theta] = \frac{V_O}{\pi}\left[1 - \frac{2\theta}{\pi}\right] \quad \textbf{(1.229)}$$

where $V_O = -I_{EE}RI_C$ for the Gilbert cell in Fig. 1.55. For $R_C = 47$ kΩ and $I_{EE} = 200$ μA in Example 1.16, $V_O = 9.4$ V. The plot of Eq. (1.229) is shown in Fig. 1.63(a). The output of the Gilbert cell across the terminals of Q_1 and Q_4 is shown in Fig. 1.63(b). This phase demodulation technique is widely used in phase-locked loops. If the input signal amplitude is large, the actual waveform shape is unimportant since the multiplier simply switches from one state to the other at the zero crossings of the waveform. For the case in which the amplitude of one or both the input signals has amplitude comparable to or small than V_T, the circuit still acts as a phase detector. However, the output voltage then depends on the phase difference and the amplitude of the two input waveforms.

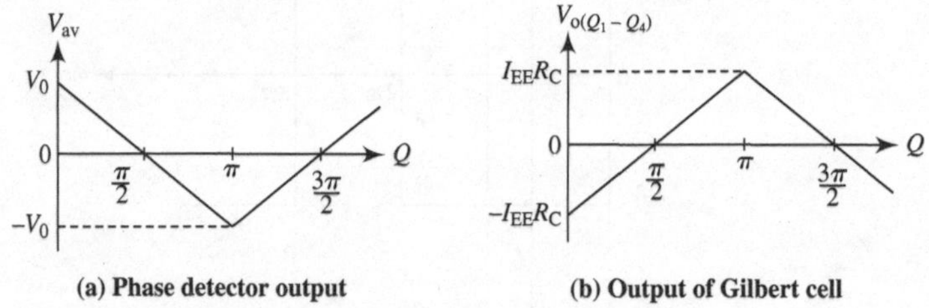

FIGURE 1.63 Phase detector transfer function

KEY POINTS OF SECTION 1.10

- The Gilbert cell is the core of analog multipliers. A practical multiplier should operate in four quadrants such that it can accept positive and negative input voltages to produce the appropriate positive or negative output voltage.
- The input signals are normally passed through an inverse hyperbolic tangent circuit for linear operation over a wider range of input signals.
- Depending on the amplitudes and frequencies of the input signals, analog multipliers can be used as a modulator, demodulator, and phase detector.

1.11 Design of Differential Amplifiers

Differential amplifiers are used as the input stage and are designed for a high differential gain, a high CMRR, and high differential and common-mode input resistances. The design of a differential stage involves the following steps:

Step 1. Identify the specifications: the differential gain A_d, the CMRR, the input resistance R_{id}, and the DC supply voltages V_{CC} and V_{EE} (or V_{DD} and V_{SS}).

Step 2. Select the type of differential amplifier (BJT, CMOS, or BiCMOS).

Step 3. Determine the biasing current I_Q required for the desired differential gain and input resistance.

Step 4. Choose the type of current source (BJT or MOSFET) and determine its component ratings. Use the standard values of components. (See Appendix E.)

Step 5. For an active load with a current mirror, choose the type of current source (BJT or MOSFET) needed to obtain the desired voltage gain and determine its component ratings.

Step 6. Determine the voltage, current, and power ratings of active and passive components.

Step 7. Analyze and evaluate the complete differential amplifier to check for the desired specifications.

Step 8. Use PSpice/SPICE to simulate and verify your design, using the standard values of components with their tolerances.

1.12 Electronic Noise

Noise is generated in electronic circuits, that is, an electronic circuit will have an output even without any input signal. Noise is generated by resistances and semiconductor devices. The capacitors and inductors do not generate noise.

1.12.1 Classification of Noise

Noise can be classified as one of five types:

Thermal Noise

Thermal noise is generated in resistors due to random motion produced by the thermal agitation of electrons. This noise is dependent on temperature. The equivalent circuit for thermal noise in resistors is shown in Fig. 1.64 [10]. The mean-square value of the noise generator is expressed as

$$V_t^2 = 4kTR\Delta f \quad \text{V}^2 \tag{1.230}$$

$$I_t^2 = 4kT\frac{\Delta f}{R} \quad \text{A}^2 \tag{1.231}$$

where k is the Boltzmann's constant (1.3806×10^{-23} J/K); T is the temperature, K (273 + T°C); R is the resistance, ohms; Δf is the noise bandwidth, Hz.

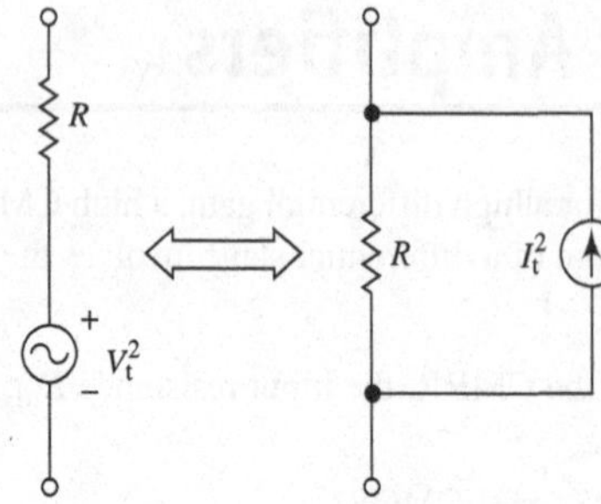

FIGURE 1.64 Equivalent circuit for noise in resistors

Shot Noise

Shot noise is generated by random fluctuations in the number of charged carriers that are emitted from a surface or diffused from a junction. This noise is always associated with a direct current flow and is present in bipolar transistors. The mean-square value of the noise current is expressed by

$$I_s^2 = 2qI_D\Delta f \quad A^2 \tag{1.232}$$

where Δf is the noise bandwidth, Hz; q is the electron charge (1.6022×10^{-16} C); I_D is the dc current, A.

Flicker Noise

Flicker noise is generated due to surface imperfections resulting from the emission. This noise is associated with all active devices and some discrete passive elements such as carbon resistors. The mean square value of the noise current is expressed by

$$I_f^2 = K_f \frac{I_D^a}{f}\Delta f \quad A^2 \tag{1.233}$$

where Δf is the noise bandwidth, Hz; f is the operating frequency, Hz; I_D is the dc current, A; K_f is the flicker constant for a particular device in the range 0 to 1; a is the flicker-exponent constant in the range of 0.5 to 2.

Burst Noise

Burst noise is generated due to the presence of heavy metal ion contamination and is found in some integrated circuits and discrete transistors. The repetition rate of noise pulses is in the audio frequency range (a few kilohertz or less) and produces a "popping" sound when played through a speaker. This noise is also known as *popcorn noise*. The mean-square value of the noise current is expressed as

$$I_b^2 = K_b \frac{I_D^c}{1+(f/f_c)}\Delta f \quad A^2 \tag{1.234}$$

where Δf is the noise bandwidth, Hz; I_D is the direct current; K_b is the burst constant for a particular device; c is the burst-exponent constant in the range of 0.5 to 2; f_c is a particular frequency for a given noise in the range of 20 Hz to 15 kHz.

Avalanche Noise

Avalanche noise is produced by Zener or avalanche breakdown in *pn* junctions. The holes and electrons in the depletion region of a reverse-biased *pn* junction acquire sufficient energy to create hole–electron pairs by collision. This process is cumulative, resulting in the production of a random series of large noise spikes. This noise is associated with direct current and is much greater than shot noise for the same current. Zener diodes are normally avoided in circuits requiring low noise.

1.12.2 Noise in Diodes

The equivalent circuit for noise in diodes is shown in Fig. 1.65 [10]. There are two noise generators. The voltage generator is due to thermal noise in the resistance of the silicon. The current source is due to the shot noise and flicker noise. The mean-square values of the noise voltage and current are given by

$$V_s^2 = 4kTr_s\Delta f \quad \text{V}^2 \tag{1.235}$$

$$I_d^2 = 2qI_D\Delta f + K_f\frac{I_D^a}{f}\Delta f \quad \text{A}^2 \tag{1.236}$$

where I_D is the forward direct current, A; r_s is the resistance of the silicon, ohms; K_f is the flicker constant for a particular device in the range 0 to 1; a is the flicker-exponent constant in the range of 0.5 to 2.

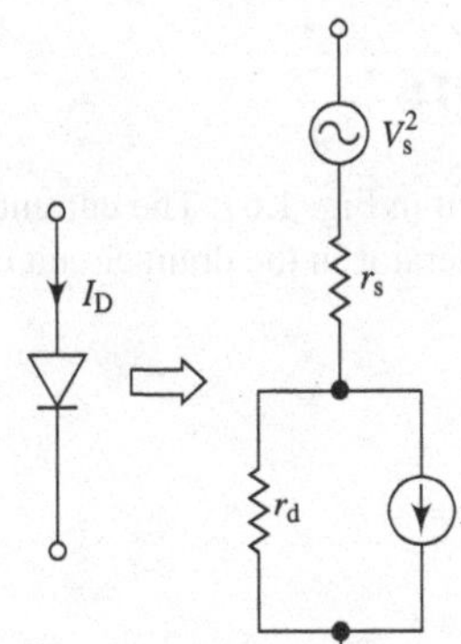

FIGURE 1.65 Equivalent circuit for noise in diodes

1.12.3 Noise in Bipolar Transistors

The equivalent circuit for noise in bipolar transistors is shown in Fig. 1.66 [10]. The current generator in the collector is due to shot noise. The noise voltage generator in the base circuit is due to thermal noise in the base resistance. The current generator in the base circuit consists of shot noise, flicker noise, and burst noise. The mean-square values of the noise voltage and currents are expressed by

$$V_b^2 = 4kTr_b\Delta f \quad \text{V}^2 \tag{1.237}$$

$$I_c^2 = 2qI_C\Delta f \quad \text{A}^2 \tag{1.238}$$

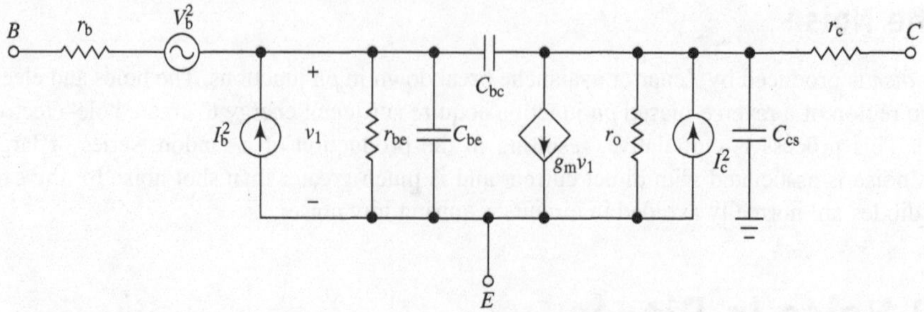

FIGURE 1.66 Equivalent circuit for noise in bipolar transistors

$$I_{\mathrm{b}}^{2}=2qI_{\mathrm{B}}\Delta f+K_{\mathrm{f}}\frac{I_{\mathrm{B}}^{\mathrm{a}}}{f}\Delta f+K_{\mathrm{b}}\frac{I_{\mathrm{B}}^{\mathrm{c}}}{1+(f/f_{\mathrm{c}})}\Delta f\ \ \mathrm{A}^{2} \tag{1.239}$$

where I_B is the base bias current, A; I_C is the collector bias current, A; r_b is the resistance at the transistor base, ohms; K_b is the burst constant for a particular device in the range 0 to 1; c is the burst-exponent constant in the range of 0.5 to 2; f_c is the a particular frequency for a given noise in the range of 20 Hz to 15 kHz; k_f is the flicker constant for a particular device in the range 0 to 1; a is the flicker-exponent constant in the range of 0.5 to 2.

1.12.4 Noise in Field-Effect Transistors

The equivalent circuit for noise in field-effect transistors (FETs) is shown in Fig. 1.67. The current generator in the gate is due to shot noise, which is very small. The current generator in the drain circuit consists of thermal and flicker noise. The noise current is expressed by

$$I_{\mathrm{g}}^{2}=2qI_{\mathrm{G}}\Delta f\ \ \mathrm{A}^{2} \tag{1.240}$$

$$I_{\mathrm{d}}^{2}=4kT\left(\frac{2}{3}g_{\mathrm{m}}\right)\Delta f+K_{\mathrm{f}}\frac{I_{\mathrm{D}}^{\mathrm{a}}}{f}\Delta f\ \ \mathrm{A}^{2} \tag{1.241}$$

where I_D is the drain bias current, A; I_G is the gate leakage current, A; g_m is the trans-conductance at the bias point, A/V; r_o is the resistance at the transistor output, ohms; K_f is the flicker constant for a particular device in the range 0 to 1; a is the flicker-exponent constant in the range of 0.5 to 2.

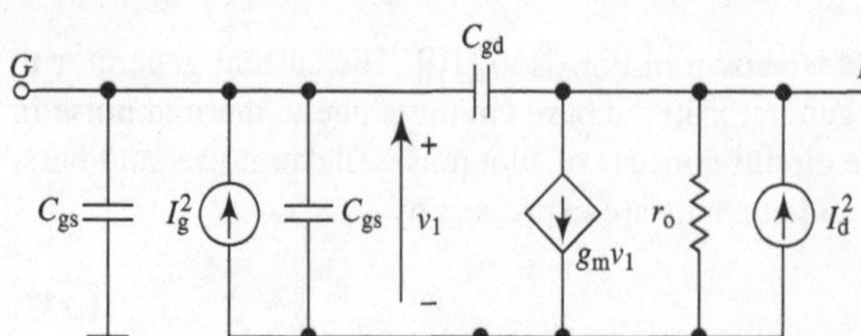

FIGURE 1.67 Equivalent circuit for noise in field-effect transistors

1.12.5 Equivalent Input Noise

Each noise generator contributes to the output of a circuit (Fig. 1.68). The effect of all the noise generators can be found by summing the mean-square value (but not the root-mean square value) of individual noise contributions. Once the total mean square noise output voltage is found, all the noises can be represented by an equivalent effective output voltage by the gain. The gain is the output noise voltage with respect to a defined input. It is time-consuming for hand calculations to determine the total noise-generated voltages and currents in electronic circuits. SPICE software can calculate the output noise and the equivalent input noise by the NOISE command [10–12]. It may be noted that PSpice calculates the noise in $V/\sqrt{Hz}$ or $A/\sqrt{Hz}$. Dividing the mean-square value of the noise output voltage $V_{out(noise)}$ by the noise bandwidth gives the output noise spectrum V_{out}, i.e.,

$$V_{out} = \frac{V_{out(noise)}}{\Delta f} \ \ V/\sqrt{Hz} \tag{1.242}$$

Dividing the output noise spectrum by the gain yields the equivalent input noise spectrum:

$$V_{in} = \frac{V_{out}}{G_v} \ \ V/\sqrt{Hz} \tag{1.243}$$

where G_V is the voltage gain with respect to the equivalent input source.

If the equivalent input is a current source, the equivalent input current noise spectrum becomes

$$I_{in} = \frac{V_{out}}{R_i} \ \ V/\sqrt{Hz} \tag{1.244}$$

where R_i is the trans-resistance with respect to the equivalent input source.

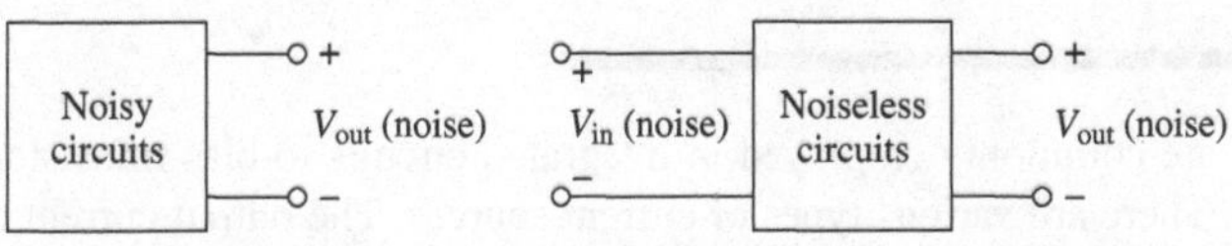

FIGURE 1.68 Equivalent noise input

EXAMPLE 9.16

Determining the noise-generated voltages and currents The noise parameters are $q = 1.6022 \times 10^{-19}$, $k = 1.3806 \times 10^{-23}$, $T = 25°C$, $K_f = 0.5$, $K_b = 0.5$, $f = 1$ Hz, $f_c = 2.5$ kHz, $a = 1$, and $c = 1$. Determine the noise-generated voltages and currents for (a) the resistance in Fig. 1.64 if $R = 10$ kΩ, (b) the diode in Fig. 1.65 if $I_D = 1$ mA, and $r_s = 40$ Ω, and (c) the transistor in Fig. 1.66 if $I_C = 1$ mA, $I_B = 10$ μA, and $r_b = 40$ Ω.

SOLUTION

(a) $R = 10\ \text{k}\Omega$. From Eqs. (1.230) and (1.231), we get

$$V_t = \sqrt{4kTR\Delta f} = \sqrt{4\times 1.3608\times 10^{-23}\times(273+25)\times 10\times 10^3\times 1} = 1.283\times 10^{-8}\ \text{V}$$

$$I_t = \sqrt{4kT\frac{\Delta f}{R}} = \sqrt{4\times 1.3608\times 10^{-23}\times(273+25)\times\frac{1}{10\times 10^3}} = 1.283\times 10^{-12}\ \text{A}$$

(b) $I_D = 1$ mA and $r_s = 40\ \Omega$. From Eqs. (9.235) and (9.236), we get

$$V_s = \sqrt{4kTr_s\Delta f} = \sqrt{4\times 1.3608\times 10^{-23}\times(273+25)\times 40\times 1} = 8.113\times 10^{-10}\ \text{V}$$

$$I_d = \sqrt{2qI_D\Delta f + K_f\frac{I_D^a}{f}\Delta f} = \sqrt{2\times 1.6022\times 10^{-19}\times 1\times 10^{-3}\times 1 + 0.5\times\frac{1\times 10^{-3}}{10}\times 1} = 7.071\times 10^{-3}\ \text{A}$$

(c) $I_C = 1$ mA, $I_B = 10\ \mu$A, and $r_b = 40\ \Omega$. From Eqs. (9.237), (9.238), and (9.239), we get

$$V_b = \sqrt{4kTr_b\Delta f} = \sqrt{4\times 1.3608\times 10^{-23}\times(273+25)\times 40\times 1} = 8.113\times 10^{-10}\ \text{V}$$

$$I_c = \sqrt{2qI_C\Delta f} = \sqrt{2\times 1.6022\times 10^{-19}\times 1\times 10^{-3}\times 1} = 1.79\times 10^{-11}\ \text{A}$$

$$I_b = \sqrt{2qI_C\Delta f + K_f\frac{I_B^a}{f}\Delta f + K_b\frac{I_B^c}{1+(f/f_c)}\Delta f}$$

$$= \sqrt{2\times 1.6022\times 10^{-19}\times 1\times 10^{-3}\times 1 + 0.5\times\frac{1\times 10^{-3}}{10}\times 1 + 0.5\times\frac{10\times 10^{-6}}{1+(10/2.5\times 10^3)^2}\times 1} = 2.345\times 10^{-3}\ \text{A}$$

Summary

Active current sources are commonly employed in integrated circuits to bias transistor circuits at appropriate operating points. There are various types of current sources. The output current of a good source is independent of the transistor parameters and offers a high output resistance. The design of BJT current sources can be simplified by ignoring the base currents of transistors and by assuming a constant voltage drop between the base and the emitter. Diode-connected BJTs and MOSFETs rather than diodes are normally used in current sources to give matching mirror characteristics.

A current source can be either a source or a sink, depending on how it is connected. Current mirrors are often used as an active load to increase the output resistance and the voltage gain of a differential amplifier. Also, cascodelike connections increase the voltage gain.

Emitter-coupled (or source-coupled) amplifiers offer the advantage of direct coupling of cascoded stages. They are usually used as the input stage of differential amplifiers to give a high input resistance and a high CMRR. However, the range of the differential input voltage is very small (typically two to three times V_T for the BJT amplifier). The CMRR depends on the output resistance of the biasing current source. Thus, a current source with a high value of output resistance is highly desirable. Source-coupled differential pairs offer a higher input resistance than emitter-coupled pairs, but they have a low CMRR.

The Gilbert cell is the core of analog multipliers. A practical multiplier operates in four quadrants such that it can accept positive and negative input voltages to produce the appropriate positive or negative output voltage. Depending on the amplitudes and frequencies of the input signals, analog multipliers can be used as a modulator, demodulator, and phase detector.

Noise is generated in electronic circuits, that is, an electronic circuit will have an output even without any input signal. Noise is generated by resistances and semiconductor devices. The capacitors and inductors do not generate noise.

References

1. P. E. Allen and D. R. Holberg, *CMOS Analog Circuit Design*. New York: Oxford University Press, 1996.
2. P. M. Chirlian, *Analysis and Design of Integrated Electronic Circuits*. New York: Wiley, 1986.
3. L. J. Giacoletto, *Differential Amplifiers*. New York: Wiley, 1970.
4. V. H. Grinich and H. G. Jackson, *Introduction to Integrated Circuits*. New York: McGraw-Hill, 1975.
5. P. R. Gray, P. J. Hurst, S. H. Lewis, and R. G. Meyer, *Analysis and Design of Analog Integrated Circuits*. New York: Wiley, 2001.
6. R. C. Jaeger and T. Blalock, *Microelectronic Circuit Design*. New York: McGraw-Hill, 2008.
7. B. Razavi, *Design of Analog CMOS Integrated Circuits.* New York: McGraw-Hill, 2001.
8. S. G. Burns and P. R. Bond, *Principles of Electronic Circuits*. Boston, MA: International Thomson Publishing Company, 2nd ed., 1997.
9. Datasheet Catalog, http://www.DatasheetCatalog.com. A datasheet source for electronic components and semiconductors. Accessed November 2012.
10. M. H. Rashid, *Introduction to PSpice. Using OrCAD for Circuits and Electronics*. Upper Saddle River, NJ: Prentice Hall Inc., 3rd ed., 2003.
11. P. Antognetti and G. Massobrio, *Semiconductor Device Modeling with SPICE*. New York, NY: McGraw-Hill, 1988.
12. I. Getreu, *Modeling the Bipolar Transistor*, Tektronix, Inc. part# 062-2841-00.

Review Questions

1. What are the advantages and disadvantages of a basic current source?
2. What are the advantages and disadvantages of a modified current source?
3. What are the advantages and disadvantages of a Widlar current source?
4. What are the advantages and disadvantages of a cascode current source?
5. What are the advantages and disadvantages of a Wilson current source?
6. What are the differences between Widlar and Wilson current sources?
7. What is a current mirror load?
8. What is the common-mode rejection ratio (CMRR)?
9. What are the advantages of an emitter-coupled pair?
10. What is the DC characteristic of an emitter-coupled pair?
11. What design criteria will yield a large value of CMRR in an emitter-coupled pair?
12. What are the advantages of a source-coupled pair?
13. What is the DC characteristic of a source-coupled pair?
14. What design criteria will yield a large value of CMRR in a source-coupled pair?
15. What is the purpose of cascodelike connections of transistors?
16. What is a Gilbert cell multiplier?
17. What is the function of an inverse hyperbolic tangent circuit in analog multipliers?
18. What is frequency modulation?
19. What is demodulation?
20. What is amplitude modulation?

21. What is modulation index?
22. What is a phase detector?
23. What is a thermal noise?
24. What is a shot noise?
25. What is a flicker noise?
26. What is a burst noise?
27. What is an avalanche noise?

Problems

The symbol **D** indicates that a problem is a design problem. The symbol **P** indicates that you can check the solution to a problem using PSpice/SPICE.

Assume that the device parameters are as follows: diodes, $I_S = 10^{-13}$ A and $I_{D(min)} = 1$ mA to ensure conduction; transistors, $\beta_F = h_{fe} = 50$, $V_{BE} = 0.7$ V, $I_S = 10^{-14}$ A, and $V_{CE(sat)} = 0.2$ V.

1.3 MOSFET Current Sources

1.1 The parameters of the MOSFET current source in Fig. 1.4 are $V_t = 1$ V, $I_O = 20$ μA, $I_R = 20$ μA, $V_{DD} = 15$ V, and $V_M = 40$ V. The channel lengths are $L_1 = L_2 = 10$ μm and $L_3 = 100$ μm, and $K_x = 20$ μA/V^2. Calculate the required values of **(a)** K_{n1}, W_1, **(b)** K_{n2}, W_2, **(c)** K_{n3}, W_3, and **(d)** the output resistance R_o of the current source. Assume $V_{GS1} = 1.5$ V and $V_{DS1} = 5$ V.

1.2 **a.** Design the cascode current source in Fig. 1.6(a) to give $I_O = 10$ μA. Assume $V_{DD} = 10$ V. All MOSFETs are identical and have $L = 20$ μm, $W = 60$ μm, $V_t = 1$ V, $K_x = 20$ μA/V^2, and $V_M = 40$ V.

b. Calculate the output resistance R_o and Thevenin's equivalent voltage V_{Th}.

1.3 **a.** Design the modified Wilson current source in Fig. 1.7(c) to give $I_O = 10$ μA. Assume $V_{DD} = 10$ V and $V_M = 40$ V. All MOSFETs are identical and have $L = 10$ μm, $W = 40$ μm, $K_x = 20$ μA/V^2, and $V_M = 40$ V.

D **b.** Calculate the output resistance R_o and Thevenin's equivalent voltage V_{Th}.

1.4 Design a MOSFET current source so that $R_o \geq 50$ kΩ at an output current of $I_O = 1$ mA.

D

1.5 Design a MOSFET current source so that $R_o \geq 500$ kΩ at an output current of $I_O = 0.1$ mA.

1.6 **D** The Widlar current source shown in Fig. P1.6 has $I_{ref} = 50$ μA, $R = 2$ kΩ, and $V_{DD} = 12$ V. The MOS parameters are $K_n = 100$ μA/V^2, $V_t = 1$ V, $|V_M| = 100$ V, and $(W/L)_1 = (W/L)_2 = 20$. Determine **(a)** the output current I_o, **(b)** the output resistance r_{o2}, and **(c)** the value of R_{ref}.

FIGURE P1.6

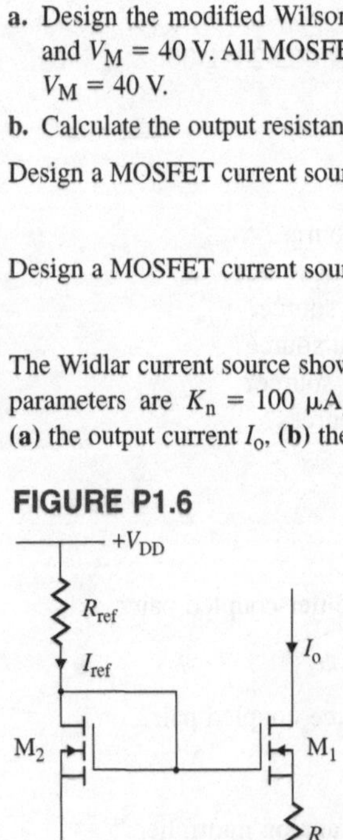

1.7 Design the modified Wilson current source in Fig. P1.7 to give $I_O = 10\ \mu A$ by determining the resistance R_{ref}. Assume $V_{DD} = 12$ V. All MOSFETs are identical and have $L = 10\ \mu m$, $W = 40\ \mu m$, $K_x = 50\ \mu A/V^2$, and $V_M = 100$ V. Calculate the output resistance R_o and Thevenin's equivalent voltage V_{Th}.

D

FIGURE P1.7

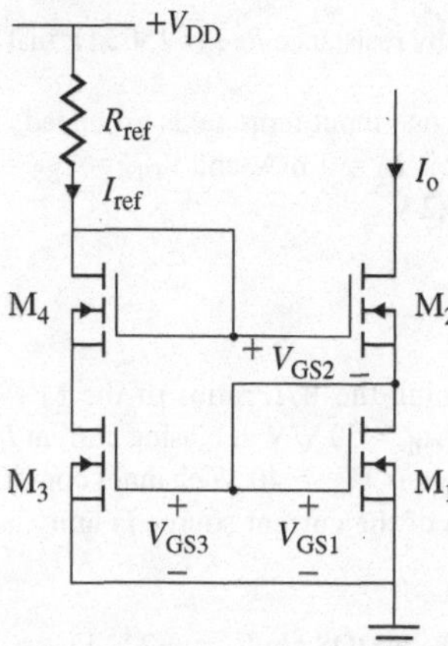

1.8 The current source shown in Fig. P1.8 has $I_{ref} = 150\ \mu A$ and $V_{DD} = 12$ V. The NMOS parameters are $K_n = 200\ \mu A/V^2$, $V_t = 1$ V, $V_M = -100$ V, and $(W/L)_1 = (W/L)_2 = (W/L)_3 = 20$. Determine **(a)** the gate–source voltages (V_{GS1}, V_{GS2}, V_{GS3}), **(b)** the drain–source voltages (V_{DS1}, V_{DS2}, V_{DS3}) if the devices are to operate in the active region, and **(c)** the value of R_{ref}.

FIGURE P1.8

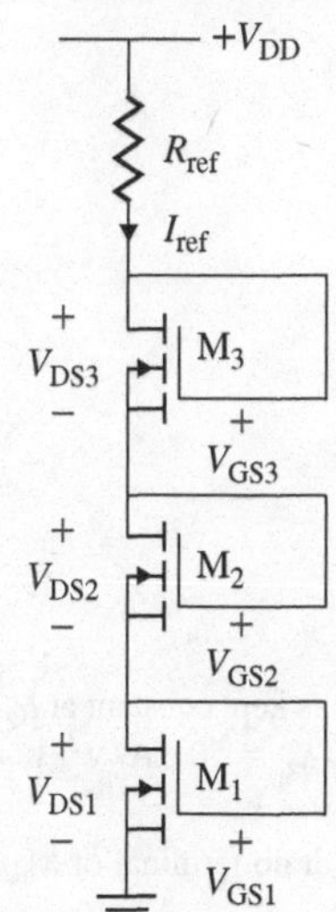

1.9 Repeat Prob. 1.8 if the current source in Fig. P1.8 has only two MOSFETs.

1.4 MOS Differential Amplifiers

1.10 The parameters of the MOS differential pair in Fig. 1.8 are $R_{SS} = 150\ \text{k}\Omega$, $I_Q = 1$ mA, $V_{DD} = 15$ V, and $R_D = 5\ \text{k}\Omega$. The MOSFETs are identical and have $K_n = 1.5\ \text{mA/V}^2$, $W/L = 1$, and $V_t = 1.2$ V.

a. Calculate the DC drain currents through the MOSFETs if $v_{id} = 10$ mV.

b. Assuming $I_{D1} = I_{D2}$, calculate A_d, A_c, and CMRR; R_{id} and R_{ic}; the small-signal output voltage if $v_{g1} = 5$ mV and $v_{g2} = 10$ mV; and the drain voltage V_D.

1.11 Repeat Prob. 1.10 if the active biasing current source is replaced by resistance $R_{SS} = 150\ \text{k}\Omega$; that is, $I_Q = 0$.

1.12 D Design a MOS differential pair as shown in Fig. 1.13 in which one input terminal is grounded. The output is taken from the drain of transistor M_1. The DC biasing current is $I_Q = 1$ mA, and $V_{DD} = V_{SS} = 15$ V. The MOSFETs are identical and have $K_n = 1.5\ \text{mA/V}^2$ and $V_t = 1.2$ V.

a. For a small-signal voltage gain of $A_1 = -20$ V/V.

b. Calculate the design values of A_d, A_c, and CMRR.

1.13 D Design the CMOS amplifier shown in Fig. 1.16 by determining the *W/L* ratios of the MOSFETs and the threshold voltage V_t. The differential voltage gain should be $A_d = 50$ V/V at biasing current $I_Q = 1$ mA. Assume identical transistors whose channel modulation voltage is $V_M = 40$ V, channel constant is $K_x = 10\ \mu\text{A/V}^2$, and channel length is $L = 10\ \mu\text{m}$. The *W/L* ratio of the current source is unity, and $V_{DD} = V_{SS} = 10$ V.

1.14 A CMOS amplifier is shown in Fig. P1.14. The parameters for the NMOS are $V_t = +2$ V, $V_M = -40$ V, and $V_{GS} = +4$ V at $I_D = 1$ mA; the parameters for the PMOS are $V_t = -3$ V, $V_M = 40$ V, and $V_{GS} = -6$ V at $I_D = 1$ mA. Calculate **(a)** A_d, A_c, and CMRR and **(b)** R_{id} and R_{ic}.

FIGURE P1.14

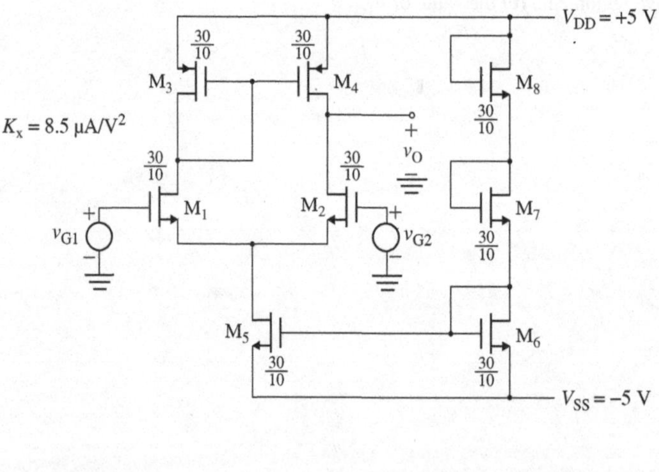

1.15 D The DC biasing current of the cascoded MOS amplifier shown in Fig. 1.19(a) is kept constant at $I_Q = 50\ \mu\text{A}$ and $V_{DD} = V_{SS} = 5$ V. All MOS transistors are identical and have $V_M = 50$ V, $K_n = 50\ \mu\text{A/V}^2$, $W = 50\ \mu\text{m}$, and $L = 10\ \mu\text{m}$.

a. Determine the differential voltage gain A_d for single-ended output at the drain terminal of M_4.

b. Use PSpice to verify the result.

1.16 The MOS parameters of the CMOS amplifier shown in Figure P1.16 are $K_n = 100\ \mu A/V^2$, $V_{tN} = 1$ V, $V_{tP} = 1.5$ V, $|V_M| = 100$ V, $L = 10\ \mu m$, $(W/L)_1 = 60$, $(W/L)_7 = (W/L)_8 = 120$, and $(W/L)_2 = (W/L)_3 = (W/L)_4 = (W/L)_5 = (W/L)_6 = 30$. The DC supply voltages are $V_{DD} = V_{SS} = 5$ V, the reference current is to be set at $I_{ref} = 100\ \mu A$, and $C_L \approx \infty$.

a. Determine the value of R_{ref} and V_{GS6}.

b. Calculate A_d, A_c, and CMRR.

c. Determine the value of capacitor C_x to limit the upper frequency at $f_H = 500$ kHz.

FIGURE P1.16

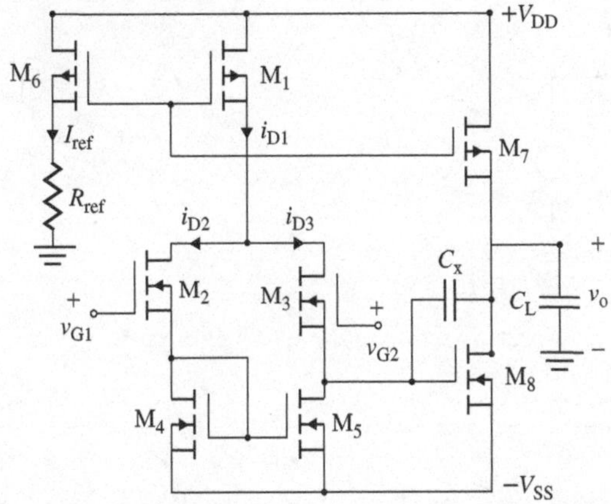

1.17 **D** Design a CMOS amplifier as shown in Fig. P1.16 by determining the W/L ratios of the MOSFETs and the threshold voltage V_t. The small-signal voltage gain should be $A_d = 250$ V/V at biasing current $I_{ref} = 100\ \mu A$. Assume identical transistors whose channel modulation voltage is $V_M = 70$ V, channel constant is $K_x = 10\ \mu A/V^2$, and channel length $L = 10\ \mu m$. Also, the width $(W/L)_6 = 1$ and $V_{DD} = V_{SS} = 5$ V.

1.5 *Depletion MOS Differential Amplifiers*

1.18 The parameters of the depletion MOS differential pair in Fig. 1.22 are $R_{SS} = 50\ k\Omega$, $I_Q = 1$ mA, $V_{DD} = V_{SS} = 30$ V, and $R_D = 2\ k\Omega$. The MOSFETs are identical and have $V_p = -4$ V and $I_{DSS} = 20$ mA.

a. Calculate the DC drain currents through the MOSFETs if $v_{id} = 30$ mV.

b. Assuming $I_{D1} = I_{D2}$, calculate A_d, A_c, and CMRR; R_{id} and R_{ic}; and the small-signal output voltage if $v_{g1} = 50$ mV and $v_{g2} = 20$ mV.

1.19 Repeat Prob. 1.18 if the transistor current source is replaced by resistance $R_{SS} = 100\ k\Omega$; that is, $I_{SS} = 0$.

1.20 **D** **P** **a.** Design a depletion MOS differential pair as shown in Fig. 1.13 in which one input terminal is grounded. The output is taken from the drain of transistor M_2. The DC biasing current is $I_Q = 5$ mA, and $V_{DD} = V_{SS} = 15$ V. The MOSFETs are identical and have $V_p = -4$ V and $I_{DSS} = 20$ mA. A small-signal voltage gain of $|A_2| = 20$ V/V is required.

b. Calculate the design values of A_d, A_c, and CMRR.

1.21 A depletion MOS amplifier is shown in Fig. P1.21. The MOS parameters are $V_P = 4$ V, $I_{DSS} = -400$ μA, $V_M = 40$ V, $I_Q = 200$ μA, and $V_{DD} = V_{SS} = 15$ V.

a. Calculate the DC drain currents through the MOSFETs if $v_{id} = 10$ mV.

b. Assuming $I_{D1} = I_{D2}$, calculate A_d, A_c, and CMRR; R_{id} and R_{ic}; and the small-signal output voltage if $v_{g1} = 20$ mV and $v_{g2} = 10$ mV.

FIGURE P1.21

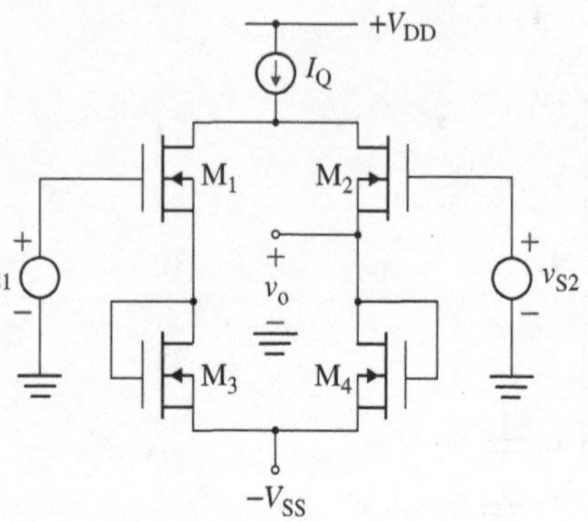

1.6 *BJT Current Sources*

1.22 The parameters of the basic current source in Fig. 1.25(a) are $\beta_F = 150$, $R_1 = 20$ kΩ, $V_{CC} = 15$ V, $V_{BE1} = V_{BE2} = 0.7$ V, and $V_A = 100$ V. Calculate **(a)** the output current $I_O = I_{C2}$, **(b)** the output resistance R_o, **(c)** Thevenin's equivalent voltage V_{Th}, and **(d)** the collector current ratio I_{C2}/I_{C1} if $V_{CE2} = 30$ V.

1.23 D P **a.** Design the basic current source in Fig. 1.25(a) to give an output current of $I_O = 200$ μA. The transistor parameters are $\beta_F = 100$, $V_{CC} = 30$ V, $V_{BE1} = V_{BE2} = V_{CE1} = 0.7$ V, and $V_A = 150$ V.

b. Calculate the output resistance R_o, Thevenin's equivalent voltage V_{Th}, and the collector current ratio if $V_{CE2} = 30$ V.

1.24 The parameters of the modified current source in Fig. 1.27(a) are $\beta_F = 150$, $R_1 = 10$ kΩ, $V_{CC} = 15$ V, $V_{BE1} = V_{BE2} = V_{BE3} = 0.7$ V, and $V_A = 100$ V. Calculate **(a)** the output current $I_O = I_{C2}$, **(b)** the output resistance R_o, **(c)** Thevenin's equivalent voltage V_{Th}, and **(d)** the collector current ratio I_{C2}/I_{C1} if $V_{CE2} = 30$ V.

1.25 D P **a.** Design the modified basic current source in Fig. 1.27(a) to give an output current of $I_O = 50$ μA. The transistor parameters are $\beta_F = 150$, $V_{CC} = 30$ V, $V_{BE1} = V_{BE2} = V_{BE3} = 0.7$ V, and $V_A = 100$ V.

b. Calculate the output resistance R_o, Thevenin's equivalent voltage V_{Th}, and the collector current ratio if $V_{CE2} = 20$ V.

1.26 The multiple transistors of the current source in Fig. P1.26 have $\beta_F = 150$, $R_1 = 10$ kΩ, $V_{CC} = 15$ V, and $V_A = 100$ V. The B-E voltages are equal, $V_{BE} = 0.7$ V. Calculate **(a)** the output current I_O, **(b)** the output resistance R_o, **(c)** Thevenin's equivalent voltage V_{Th}, and **(d)** the collector current ratio if $V_{CE2} = 15$ V.

FIGURE P1.26

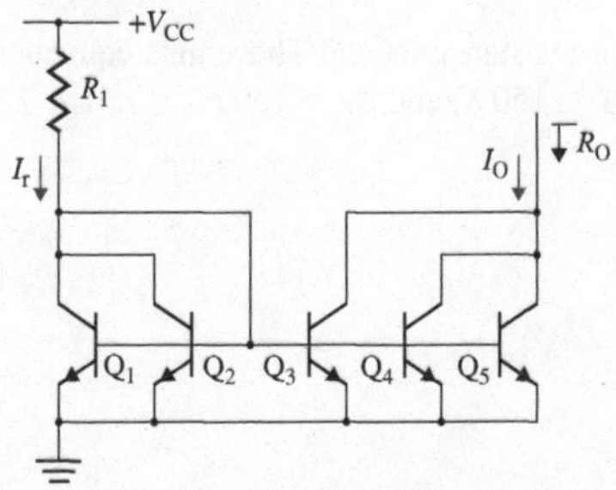

1.27 **a.** Design the Widlar current source in Fig. 1.28(a) to give $I_O = 10\ \mu A$ and $I_R = 2$ mA. The parameters are $V_{CC} = 30$ V, $V_{BE1} = 0.7$ V, $V_T = 26$ mV, $V_A = 150$ V, and $\beta_F = 100$.

D P **b.** Calculate the output resistance R_o and Thevenin's equivalent voltage V_{Th}.

1.28 Design a Widlar current source as shown in Fig. 1.28(a) to produce a 10-μA output current. Assume $\beta_F = 100$, $V_{CC} = 30$ V, and $R_1 = 30$ kΩ. Calculate the output resistance R_o.

D P

1.29 Determine the output current I_O and the output resistance R_o of the current source circuit in Fig. P1.29. Assume $V_{CC} = 30$ V, $R_1 = 20$ kΩ, $R_2 = 10$ kΩ, $V_{BE} = 0.7$ V, $V_A = 150$ V, and $\beta_F = 100$.

FIGURE P1.29

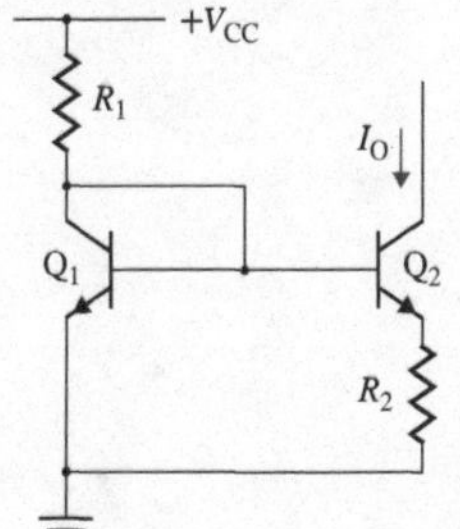

1.30 **a.** Design the Wilson current source in Fig. 1.30(a) to give $I_O = 10\ \mu A$. The parameters are $V_{CC} = 30$ V, $V_{BE} = 0.7$ V, $V_T = 26$ mV, $V_A = 100$ V, and $\beta_F = 150$.

D P **b.** Calculate the output resistance R_o and Thevenin's equivalent voltage V_{Th}.

1.31 For the Wilson current source in Fig. P1.31, determine the output current I_O and the output resistance R_o. Assume $V_{CC} = 20$ V, $V_{BE} = 0.7$ V, $V_T = 26$ mV, $V_A = 150$ V, and $\beta_F = 150$.

FIGURE P1.31

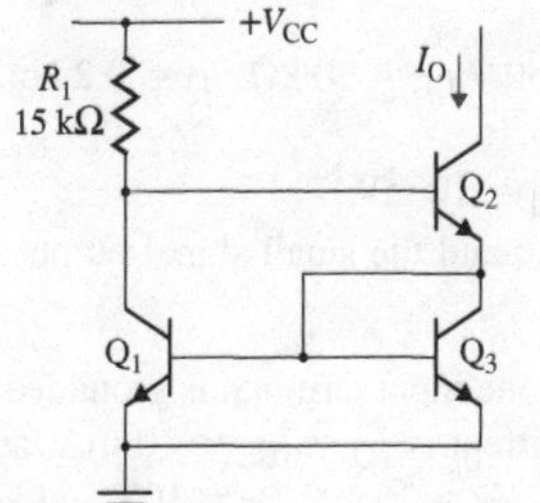

1.32 Repeat Prob. 1.31 for $V_{CC} = 30$ V.

1.33 For the current source in Fig. P1.33, determine the output resistance R_o and Thevenin's equivalent voltage V_{Th}. Assume $V_{CC} = 30$ V, $V_{BE} = 0.7$ V, $V_T = 26$ mV, $V_A = 150$ V, and $\beta_F = 150$.

FIGURE P1.33

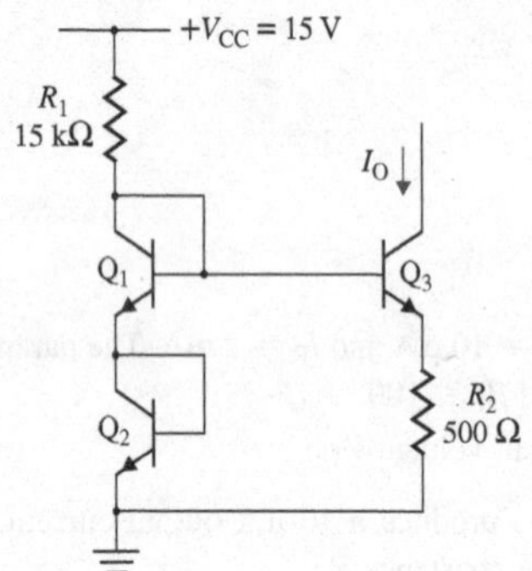

1.34 Determine the sensitivity S of output current I_O to supply voltage V_{CC} for the circuit in Fig. P1.34. S is defined as

$$S = \frac{V_{CC}/I_O}{\delta I_O/\delta V_{CC}}$$

FIGURE P1.34

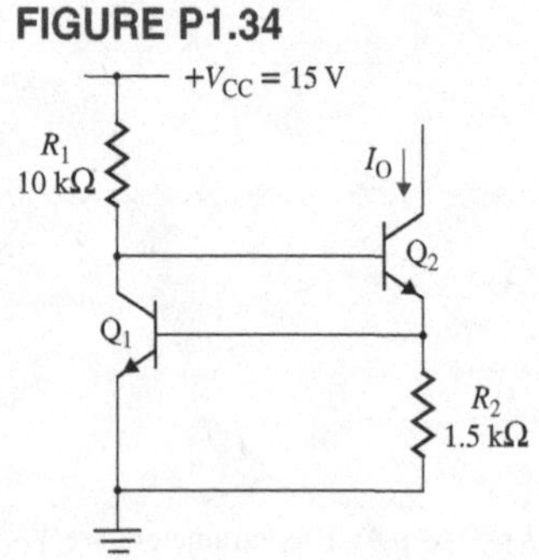

1.35 Design a BJT current source so that $R_o \geq 50$ kΩ at an output current of $I_O = 1$ mA.
D

1.36 Design a BJT current source so that $R_o \geq 500$ kΩ at an output current of $I_O = 1$ mA.
D

1.7 *BJT Differential Amplifiers*

1.37 The parameters of the emitter-coupled pair in Fig. 1.33 are $\beta_F = 150$, $R_{EE} = 20$ kΩ, $I_Q = 0.25$ mA, $V_{CC} =$
P 12 V, and $R_C = 10$ kΩ.

a. Calculate the DC collector currents through the transistors if $v_{id} = 10$ mV.

b. Assuming $I_{C1} = I_{C2}$, calculate A_d, A_c, and CMRR; R_{id} and R_{ic}; and the small-signal output voltage if $v_{B1} = 30$ mV and $v_{B2} = 20$ mV. Assume $V_T = 26$ mV.

1.38 **a.** Design an emitter-coupled pair as shown in Fig. 1.41, in which one input terminal is grounded. The out-
D put is taken from the collector of transistor Q_2. The biasing current is $I_Q = I_{EE} = 10$ mA, and $V_{CC} = -V_{EE} = 12$ V. The transistors are identical. Assume $V_{BE} = 0.7$ V, $V_T = 26$ mV, $\beta_F = 100$, and $V_A = 40$ V. A small-signal voltage gain of $A_1 = -150$ V/V is required.

b. Calculate the design values of A_d, A_c, and CMRR.

1.39 A differential amplifier is shown in Fig. P1.39. The transistors are identical. Assume $V_{BE} = 0.7$ V, $V_T = 26$ mV, $\beta_F = 50$, and $V_A = 40$ V. Calculate the values of A_d, R_{id}, A_c, R_{ic}, and CMRR.

FIGURE P1.39

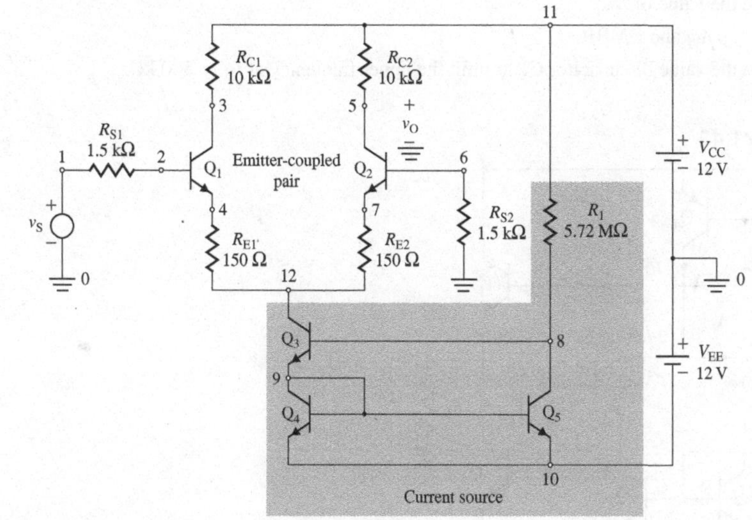

1.40 The parameters of the differential amplifier in Fig. 1.44 are $\beta_{F(npn)} = 100$, $\beta_{F(pnp)} = 50$, $V_A = 40$ V, $I_Q = 10$ μA, and $V_{CC} = 10$ V. Calculate A_d, R_{id}, R_o, and the overall voltage gain with load $A_{d(load)}$. Assume $V_T = 26$ mV.

1.41 The parameters of the differential amplifier in Fig. 1.45 are $\beta_{F(npn)} = 100$, $\beta_{F(pnp)} = 50$, $V_A = 40$ V, $I_Q = 5$ μA, and $V_{CC} = 10$ V. Calculate A_d, R_{id}, R_o, and the overall voltage gain with load $A_{d(load)}$. Assume $V_T = 26$ mV.

1.42 A differential amplifier is shown in Fig. P1.42. The transistors are identical. Assume $V_{BE} = 0.7$ V, $V_T = 26$ mV, $\beta_{F(npn)} = 100$, $\beta_{F(pnp)} = 50$, $V_A = 40$ V, and $V_{CC} = 10$ V. Calculate the values of R_1, R_2, A_d, R_{id}, A_c, R_{ic}, and CMRR.

FIGURE P1.42

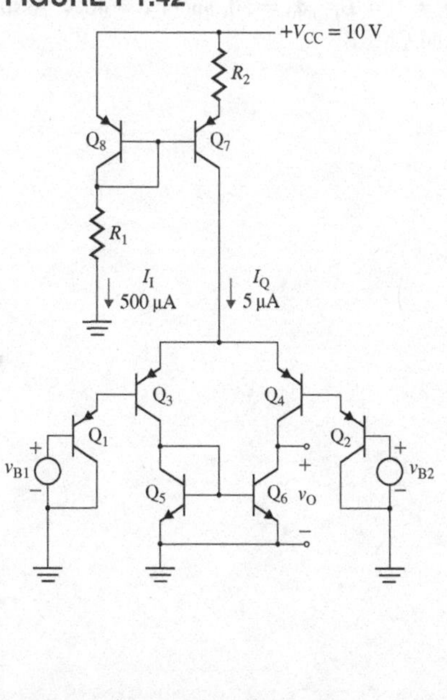

1.43 The parameters of the BJT amplifier shown in Fig. P1.43 are $\beta_{FN} = 100$, $\beta_{FP} = 120$, $|V_A| = 100$ V, and $V_{BEN} = -V_{BEP} = 0.65$ V. The DC supply voltages are $V_{CC} = V_{EE} = 5$ V, the reference current is to be set at $I_{ref} = 100$ μA, and $R_L = 50$ kΩ.

a. Determine the value of R_{ref}.

b. Calculate A_d, A_c, and CMRR.

c. Determine the value of capacitor C_x to limit the upper frequency at $f_H = 500$ kHz.

FIGURE P1.43

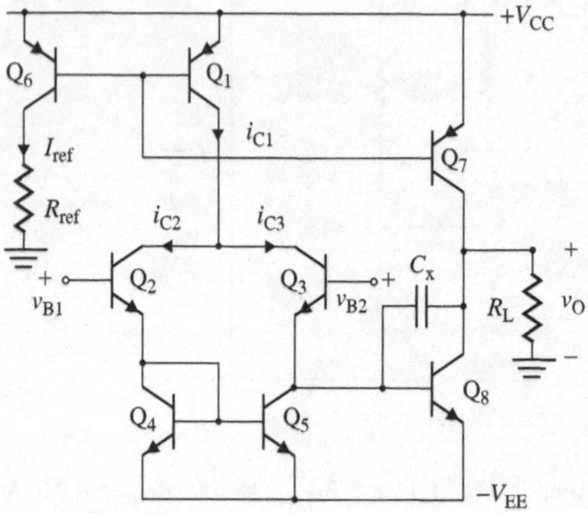

1.44 **D** Design a differential amplifier as shown in Fig. 1.44 to satisfy the following conditions: a differential voltage gain of $A_o \geq 1500$ V/V, CMRR of ≥ 3500, and load of $R_L = 50$ kΩ, with a coupling capacitor of $C_L = \infty$. Use all BJTs and the cascoded current source, similar to the one in Fig. 1.29. Use PSpice to verify your design with a $v_{id} = 5$ μV and $v_{ic} = 10$ mV. Supply voltages are $V_{CC} = -V_{EE} = 15$ V.

1.8 *BiCMOS Differential Amplifiers*

1.45 A BiCMOS amplifier is shown in Fig. P1.45. The depletion PMOS parameters are $V_p = 4$ V, $I_{DSS} = 400$ μA, and $V_M = 100$ V. The BJT parameters are $\beta_{F(npn)} = 100$, $\beta_{F(pnp)} = 50$, and $V_A = 40$ V. Assume $V_{DD} = V_{EE} = 15$ V and $I_Q = 200$ μA. Calculate A_d, A_c, and CMRR.

FIGURE P1.45

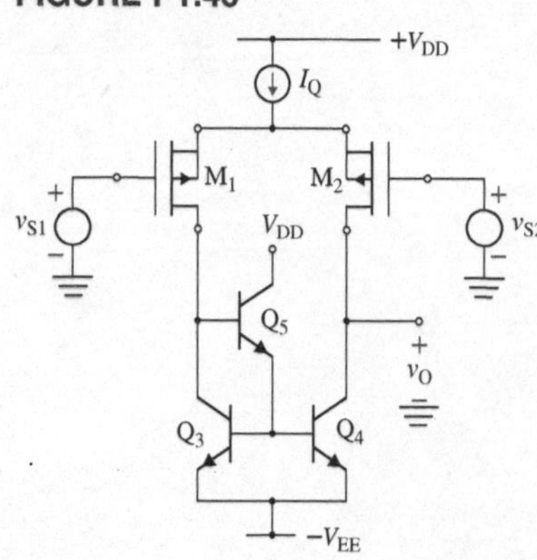

1.46 The DC biasing current of a BiCMOS amplifier is kept constant at $I_Q = 5\ \mu A$. All bipolar transistors are identical, with $\beta_{F(npn)} = 100$, $\beta_{F(pnp)} = 50$, and $V_A = 40$ V. Also, the MOS transistors are identical, with $|V_M| = 20$ V. For the NMOS, $V_t = +2$ V and $V_{GS} = +4$ V at $I_D = 1$ mA; for the PMOS, $V_t = -3$ V and $V_{GS} = -6$ V at $I_D = 1$ mA. Determine the differential voltage gain A_d for single-ended output **(a)** for the basic BiCMOS amplifier in Fig. 1.49(c), **(b)** for the cascode BiCMOS amplifier in Fig. 1.50, and **(c)** for the double-cascode BiCMOS amplifier in Fig. 1.51.

1.47 A BiCMOS amplifier is shown in Fig. P1.47. The PMOS parameters are $V_t = -3$ V and $V_{GS} = -6$ V at $I_D = 1$ mA. The BJT parameters are $\beta_{F(npn)} = 100$, $\beta_{F(pnp)} = 50$, and $V_A = 40$ V. Assume $V_{DD} = -V_{EE} = 15$ V and $I_Q = 200\ \mu A$. Calculate A_d, A_c, and CMRR.

FIGURE P1.47

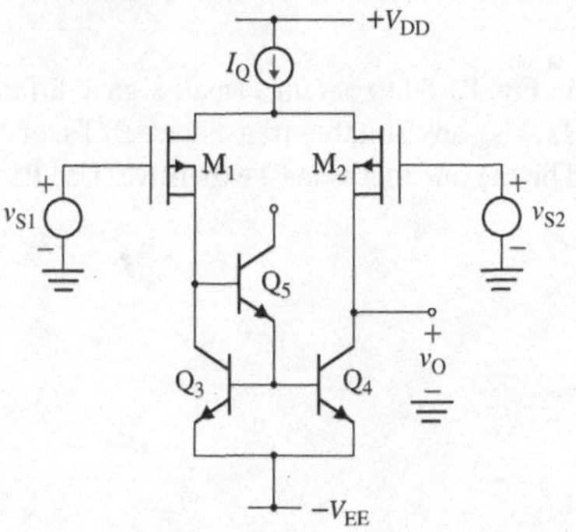

1.48 (D) Design a differential amplifier as shown in Fig. 1.51 to satisfy the following conditions: a differential voltage gain of $A_o \geq 1500$ V/V, CMRR of ≥ 3500, and load of $R_L = 50\ k\Omega$, with a coupling capacitor of $C_L = \infty$. Use all BJTs and NMOSs. Use PSpice to verify your design with $v_d = 5\ \mu V$ and $v_c = 10$ mV. Supply voltages are $V_{CC} = V_{EE} = 12$ V.

1.9 *Frequency Response of Differential Amplifiers*

1.49 The emitter-coupled pair in Fig. 1.33 has $R_{EE} = 20\ k\Omega$, $I_Q = 5$ mA, $V_{CC} = 12$ V, and $R_C = 10\ k\Omega$. The small-signal transistor parameters are $C_\pi = 5$ pF, $C_\mu = 2$ pF, $\beta_{F(npn)} = 100$, and $\beta_{F(pnp)} = 50$. Find the frequency-dependent gains $A_d(j\omega)$ and $A_c(j\omega)$.

1.50 The emitter-coupled pair in Fig. P1.50 has $I_Q = 5$ mA, $V_{CC} = -V_{EE} = 15$ V, and $R_C = 10\ k\Omega$. The small-signal transistor parameters are $C_\pi = 5$ pF, $C_\mu = 2$ pF, $\beta_{F(npn)} = 100$, and $\beta_{F(pnp)} = 50$. Find the frequency-dependent gains $A_d(j\omega)$ and $A_c(j\omega)$.

FIGURE P1.50

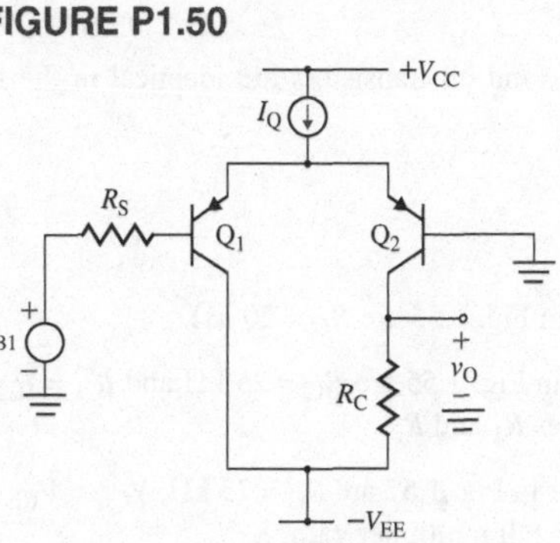

1.51 The differential amplifier in Fig. 1.42 has $V_A = 40$ V, $I_Q = 10$ μA at $R_{EE} = 50$ kΩ, and $V_{CC} = -V_{EE} = 15$ V. The small-signal transistor parameters are $C_\pi = 5$ pF, $C_\mu = 2$ pF, $\beta_{F(npn)} = 100$, and $\beta_{F(pnp)} = 50$. Find the frequency-dependent gains $A_d(j\omega)$ and $A_c(j\omega)$.

1.52 The source-coupled pair in Fig. 1.8 has $R_{SS} = 50$ kΩ, $I_Q = 10$ mA, $V_{DD} = 30$ V, and $R_D = 5$ kΩ. The MOSFETs are identical and have $K_n = 1.5$ mA/V^2, $W/L = 1$, $|V_M| = 100$ V, and $V_t = 1.2$ V. The small-signal MOS parameters are $C_{gs} = 5$ pF and $C_{gd} = 2$ pF. Find the frequency-dependent gains $A_d(j\omega)$ and $A_c(j\omega)$.

1.53 The MOS amplifier shown in Fig. 1.14 has $R_{SS} = 50$ kΩ, $I_Q = 1$ mA, and $V_{DD} = 15$ V. The MOSFETs are identical and have $K_n = 1.5$ mA/V^2, $W/L = 1$, $|V_M| = 100$ V, and $V_t = 1.2$ V. The small-signal MOS parameters are $C_{gs} = 5$ pF and $C_{gd} = 2$ pF. Find the frequency-dependent gains $A_d(j\omega)$ and $A_c(j\omega)$.

1.54 D P Design a differential amplifier configuration as shown in Fig. P1.54 to obtain a small-signal differential gain, with $A_d = 250$ V/V, CMRR $\geq 10^3$, and $f_b \geq 250$ kHz. Use any suitable transistors, BJTs, or MOSFETs. Assume $V_{CC} = V_{EE} = 12$ V. The configuration should be the one that is least expensive. Use PSpice to verify your design and give the component estimates.

FIGURE P1.54

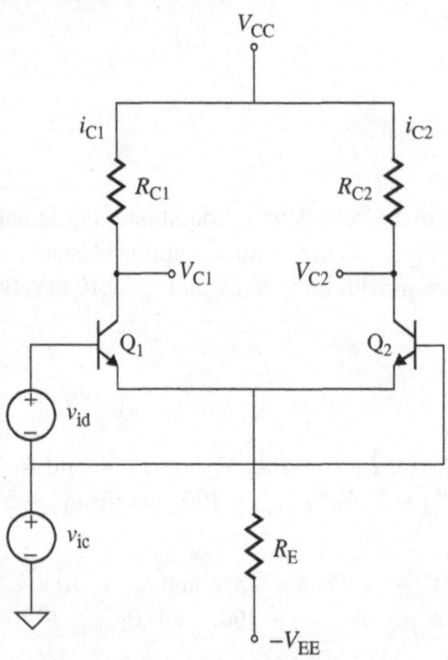

NOTE: Assume thermal constant $V_T = 25.6$ mV, $\beta_F >> 1$, and all transistors are identical in the following problems.

1.10 Analog Multipliers

1.55 The circuit parameters of the Gilbert cell multiplier in Fig. 1.55 are $R_C = 20$ kΩ.

1.56 The circuit parameters of the Gilbert cell multiplier in Fig. 1.55 are $R_C = 25$ kΩ and $I_{C1} = I_{C2} = I_{C3} = I_{C4} = 100$ μA. Determine the values of biasing resistances R_1 and R_2.

1.57 The circuit parameters of the four-quadrant multiplier in Fig. 1.57 are $R_C = 25$ kΩ, $V_{CC} = V_{EE} = 15$ V, $R_1 = 15$ kΩ, and $R_2 = R_3 = R_4 = 110$ Ω. Determine the overall multiplier gain, K_c.

1.58 If the amplitude of the modulating signal in Fig. 1.62(a) is $V_1 = 1.5$ V, what is peak amplitude of the fundamental component for $n = 1$.

1.59 The circuit parameters of the Gilbert cell phase detector in Fig. 1.55 are $R_C = 25$ kΩ, $V_{CC} = V_{EE} = 15$ V, and $I_{EE} = 400$ μA. If the phase delay is $\theta = \pi/4$, determine the average output voltage $V_{(O)av}$.

1.60 The circuit parameters of the Gilbert cell phase detector in Fig. 1.55 are $R_C = 20$ kΩ, $V_{CC} = V_{EE} = 15$ V, and $I_{EE} = 500$ μA. If the average output voltage $V_{(O)av} = 2$ V, determine the phase delay θ.

1.12 *Electronic Noise*

1.61 The noise parameters are $q = 1.6022 \times 10^{-19}$, $k = 1.3806 \times 10^{-23}$, $T = 25$°C, $K_f = 0.5$, $K_b = 0.5$, $f = 1$ Hz, $f_c = 2.5$ kHz, $a = 1$, and $c = 1$. Determine the noise-generated voltage for a resistance in Fig. 1.64 if $R = 10$ MΩ.

1.62 The noise parameters are $q = 1.6022 \times 10^{-19}$, $k = 1.3806 \times 10^{-23}$, $T = 25$°C, $K_f = 0.5$, $K_b = 0.5$, $f = 1$ Hz, $f_c = 2.5$ kHz, $a = 1$, and $c = 1$. Determine the noise-generated voltages and currents for a diode in Fig. 1.65 if $I_D = 10$ mA and $r_s = 50$ Ω.

1.63 The noise parameters are $q = 1.6022 \times 10^{-19}$, $k = 1.3806 \times 10^{-23}$, $T = 25$°C, $K_f = 0.5$, $K_b = 0.5$, $f = 1$ Hz, $f_c = 2.5$ kHz, $a = 1$, and $c = 1$. Determine the noise-generated voltages and currents for a BJT transistor in Fig. 9.66 if $I_C = 10$ mA, $I_B = 100$ μA, and $r_b = 100$ Ω.

1.64 The noise parameters are $q = 1.6022 \times 10^{-19}$, $k = 1.3806 \times 10^{-23}$, $T = 25$°C, $K_f = 0.5$, $K_b = 0.5$, $f = 1$ Hz, $f_c = 2.5$ kHz, $a = 1$, and $c = 1$. Determine the noise-generated voltages and currents for a FET in Fig. 1.67 if $I_D = 1$ mA, $I_G = 100$ nA, and $g_m = 1$ mA/V.

1.65 The noise parameters are $q = 1.6022 \times 10^{-19}$, $k = 1.3806 \times 10^{-23}$, $T = 25$°C, $K_f = 0.5$, $K_b = 0.5$, $f = 1$ Hz, $f_c = 2.5$ kHz, $a = 1$, and $c = 1$. Determine the noise-generated voltages and currents for a FET in Fig. 1.67 if $I_D = 10$ mA, $I_G = 100$ nA, and $g_m = 1.5$ mA/V.

CHAPTER 2

OPERATIONAL AMPLIFIERS

Learning Outcomes

After completing this chapter, students should be able to do the following:

- Describe the internal structure and types of op-amps (CMOS, bipolar, and BiCMOS) and the effects of amplifier configurations (difference and single-ended output) on the op-amp performance.
- Identify the circuit parameters of the internal design that affect the op-amp performance (offset voltage, offset current, and unity-gain frequency).
- List the typical values of performance parameters for different types of op-amps.
- Analyze op-amp circuits to determine the DC-biasing conditions and the performance parameters.

Symbols and Their Meanings

Symbol	Meaning
i_{B1}, i_{B2}	Instantaneous base currents of transistors 1 and 2
i_{Bn}, i_{Cn}, i_{En}	Base, collector, and emitter currents of an nth BJT Q_n
G_m	Transconductance of an amplifier
I_{Sn}	Saturation current of an nth BJT Q_n
K_{pn}, V_{tn}	MOS constant and threshold voltage of an nth MOSFET M_n
r_{on}	Output resistance of an nth BJT Q_n or MOSFET M_n

Symbol	Meaning
R_i, R_o	Input and output resistances of an amplifier stage
v_{B1}, v_{B2}	Instantaneous voltages at the base of transistors 1 and 2
V_{io}, V_{oo}	Input and output offset voltages
$V_{CC}(V_{DD})$, $V_{EE}(V_{SS})$	Positive and negative DC supply voltages
v_{GSn}, i_{Dn}	Gate–source voltage and drain current of an nth MOSFET M_n
$(W/L)_n$, $(W/L)_p$	Width-to-length ratios of an NMOS and a PMOS
β_{Fn}, g_{mn}	Current gain of an nth BJT Q_n and transconductance of an nth MOSFET M_n

2.1 Introduction

So far we have discussed separately the analysis and design of transistor amplifiers, differential amplifiers, and output stages. An operational amplifier normally consists of these stages. In this chapter, we examine the internal circuitry of 10 commercially available op-amps. Much of the circuitry will be closely related to that of other ICs. We analyze in detail one of the oldest but most popular amplifiers, the LM741.

2.2 Internal Structure of Op-Amps

The general configuration of an op-amp is shown in Fig. 2.1 [1]. All stages are direct coupled—that is, there are no coupling or bypass capacitors. Since capacitors and resistors of over 50 kΩ occupy large areas on IC chips and exhibit parasitic effects, they are usually avoided in ICs. Therefore, op-amp circuits are designed using transistors with matching characteristics. Mismatches do exist, however, and cause offset voltages.

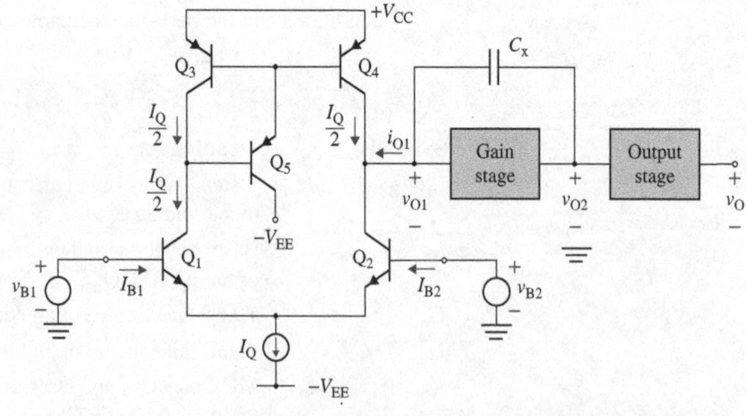

FIGURE 2.1 General configuration of an op-amp

The voltage gain of each of the differential and gain stages is normally in the range from 300 to 1000. The output stage is an emitter follower, and its gain is unity. Thus, the overall open-loop voltage gain of an op-amp is on the order of 10^5 to 10^6. The output stage provides a low-output impedance so that it can drive a load with relatively low values of load resistance. It is normally operated in class AB mode to reduce crossover distortion. In general, BJT op-amps have a larger voltage gain, whereas MOSFET op-amps have higher input resistances.

For stability of the op-amp, the phase shift of each stage is kept to a minimum. Since each stage contributes to the phase shift, the total number of stages is generally limited to three. A compensation capacitor C_X is normally connected across the second stage.

▶ **NOTE** Unless specified, we will assume that the base–emitter voltage is $V_{BE} = 0.6$ V.

KEY POINT OF SECTION 2.2

- In general, an op-amp circuit consists of a differential input stage, a gain stage, an output stage, and protection circuitry. Each stage uses active biasing and an active load.

2.3 Parameters and Characteristics of Practical Op-Amps

Most op-amps contain a differential-coupled pair as an input stage. Practical op-amps exhibit characteristics that deviate significantly from the ideal characteristics [2, 3]. Here we consider some parameters that depend on the internal design of the op-amp. The effects of deviations from the ideal can be incorporated in the equivalent circuit of the op-amp, as shown in Fig. 2.2 where $R_{cm} = R_{ic}$ is the common-mode input resistance.

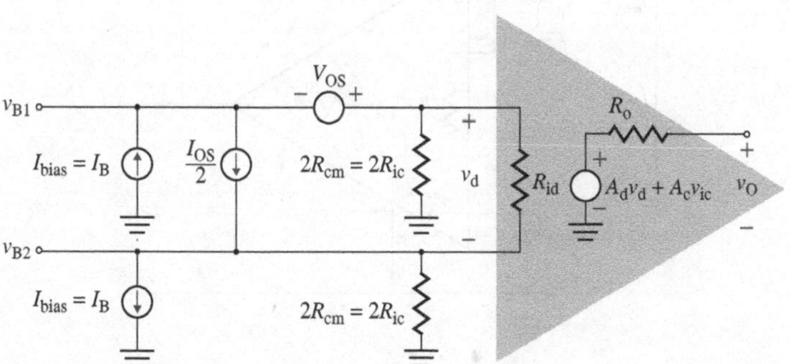

FIGURE 2.2 Equivalent circuit of an op-amp

2.3.1 Input Biasing Current

The input biasing current I_B is defined as the average of the two DC input currents to the bases of Q_1 and Q_2; that is,

$$I_B = \frac{I_{B1} + I_{B2}}{2} \tag{2.1}$$

$$\approx \frac{I_C}{\beta_F} = \frac{I_Q}{2\beta_F} \qquad \text{(for BJT input stage only)} \tag{2.2}$$

The polarity of I_B is shown for an *npn* transistor input stage in Fig. 2.2. For a *pnp* input stage, I_B would flow out of the amplifier terminals. Transistors in an ideal op-amp are assumed to have a large value of β_F, tending to infinity, and to draw zero DC input current. In a practical op-amp, however, β_F has a finite value. Thus, I_B has a small but finite value. The typical magnitudes of the biasing currents are 10 pA to 100 nA for BJT input devices and 1 nA to 10 pA for MOSFET input devices. This input biasing current will cause a small DC-output voltage when the external input voltage is zero.

The effect of biasing current I_B can be determined for the inverting or noninverting amplifier, as shown Fig. 2.3. Assuming $I_{B1} = I_{B2} = I_B$, then $v_d = R_i\,(I_{B1} - I_{B2}) = 0$. Since $v_d = 0$ due to the feedback of V_{ob}, there will be no current flowing through R_1; that is, $i_S = 0$, and the biasing current I_B will flow through R_F. Thus,

$$v_d = 0 = -I_B R_F + V_{ob}$$

which gives the output voltage due to input biasing current I_B as

$$V_{ob} = R_F I_B \tag{2.3}$$

Thus, the output offset voltage due to the input biasing current I_B depends directly on the feedback resistance R_F. To minimize the effect of I_B, the value of R_F should be small. However, the ratio R_F/R_1 determines the voltage gain v_O/v_S. The output voltage due to I_B does not depend on whether the input voltage source v_S is connected. Thus, Eq. (2.3) is applicable for both inverting and noninverting amplifiers.

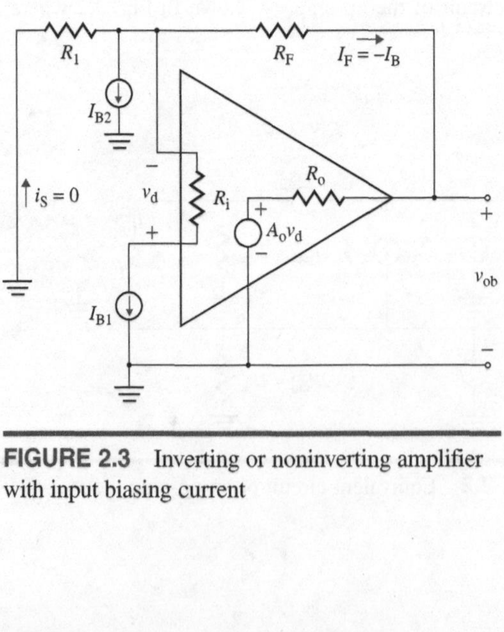

FIGURE 2.3 Inverting or noninverting amplifier with input biasing current

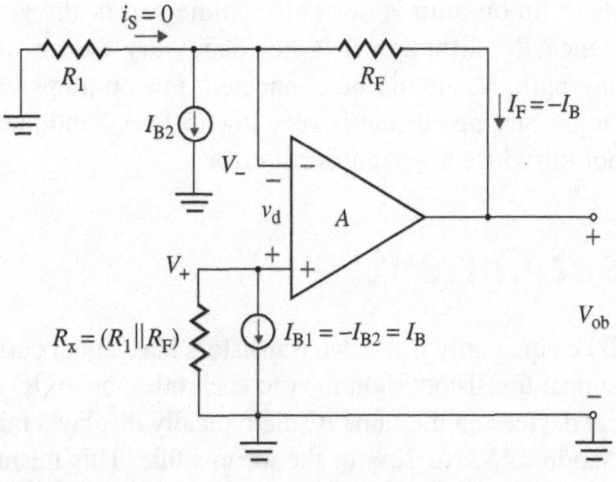

FIGURE 2.4 Effect of offset-minimizing resistance R_x

The effect of I_B on the output voltage can be eliminated or minimized by making Thevenin's equivalent resistance at the (−) terminal equal to that at the (+) terminal. This arrangement can be implemented by connecting a resistance R_x to the noninverting terminal, as shown in Fig. 2.4, such that

$$V_+ = V_- \quad \text{or} \quad R_{Th+} = R_{Th-}$$

which gives the *offset-minimizing resistance* R_x as

$$R_x = \frac{R_1 R_F}{R_1 + R_F} = R_1 \parallel R_F \tag{2.4}$$

For example, if $R_F = 100\text{ k}\Omega$ and $I_B = 500\text{ nA}$, then $V_{ob} = 100\text{ k}\Omega \times 500 \times 10^{-9} = 50\text{ mV}$ without R_x, and $V_{ob} = 0$ if $R_x = (10\text{ k}\Omega \parallel 100\text{ k}\Omega) = 9.091\text{ k}\Omega$. Therefore, by connecting resistance R_x, which is equal to the parallel combination of R_1 and R_F, we can minimize the output voltage due to the input biasing current. Since I_{B1} and I_{B2} are not exactly equal, V_{ob} will be minimized but not completely eliminated. The connections of the input voltage v_S and the input/output relations are shown in Fig. 2.5 for inverting and noninverting amplifiers.

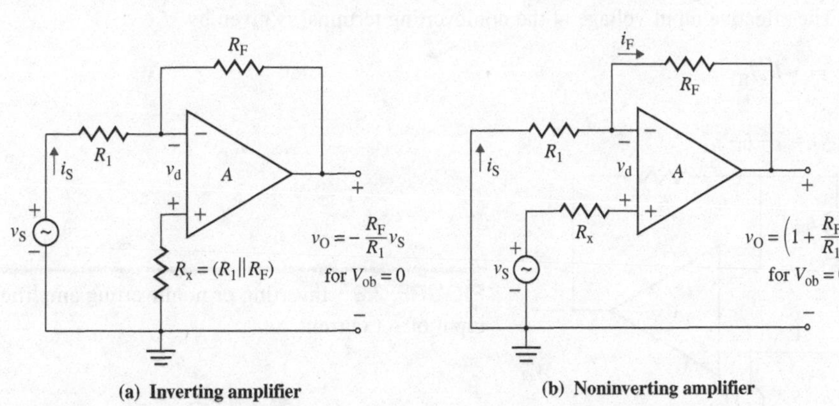

FIGURE 2.5 Amplifiers with offset-minimizing resistance R_x

One of the input terminals of an op-amp is normally connected to the ground terminal through an offset compensating resistance R_X, although it is not necessary for an ideal op-amp. However, while an op-amp circuit is being built, R_X should be connected. For op-amps with an FET input stage (e.g., the LF411 op-amp), the input biasing current is very low (50 pA), and the absence of offset-minimizing resistance R_X should not introduce any significant error.

2.3.2 Input Offset Current

The DC input base currents will be equal only if the two transistors have equal current gains (betas). However, even two theoretically identical transistors right next to each other on an IC chip will not be exactly identical. Geometrically identical devices on the same IC die typically display a mismatch that is normally distributed with a standard deviation of 5% to 10% of the mean value. This mismatch in the two biasing currents is random from circuit to circuit and is described by the input offset current I_{OS}, which is defined as the difference between the two base currents of the transistors:

$$I_{OS} = |I_{B1} - I_{B2}| \tag{2.5}$$

I_{OS} arises from mismatches in the area and β_F of the transistors. Beta and collector current values typically deviate by 10% and 1%, respectively. Thus, I_{OS} can be found approximately from

$$I_{OS} = \frac{(0.1 + 0.01)I_C}{\beta_F} = \frac{0.11 I_Q}{2\beta_F} = \frac{0.055 I_Q}{\beta_F} \tag{2.6}$$

The value of the bias-minimizing resistance R_X in Eq. (2.4) was derived by assuming equal input-biasing currents: $I_{B1} = I_{B2} = I_B$. In practice, these currents are not equal because of internal imbalances within the op-amp circuit. The input offset current I_{io} is a measure of the degree of mismatching, and it is defined by

$$I_{io} = I_{B1} - I_{B2} \tag{2.7}$$

For the μA741C op-amp, I_{io} is quoted as having a maximum absolute value of 200 nA and a typical value of 30 nA; it can be either positive or negative. An amplifier with offset-minimizing resistance R_X is shown in Fig. 2.6. The effective input voltage at the noninverting terminal is given by

$$V_+ = -R_X I_{B1} \tag{2.8}$$

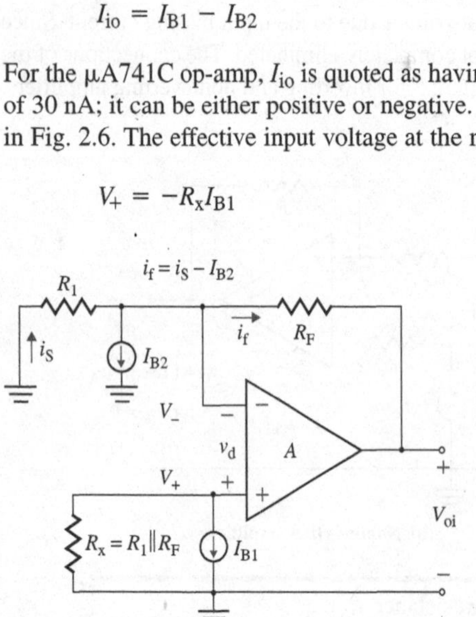

FIGURE 2.6 Inverting or noninverting amplifier with input offset current

Since the differential voltage between the input terminals is $v_d \approx 0$, $V_- = V_+$. Assuming the op-amp draws negligible current and applying KCL at the inverting terminal, we get

$$\frac{V_+}{R_1} + I_{B2} + \frac{V_+ - V_{oi}}{R_F} = 0$$

which gives

$$\frac{V_{oi}}{R_F} = \frac{V_+}{R_1 \parallel R_F} + I_{B2} \tag{2.9}$$

Substituting $V_+ = -R_x I_{B1}$, we get the output offset voltage

$$V_{oi} = R_F\left(\frac{-R_x}{R_1 \parallel R_F} I_{B1} + I_{B2}\right) \tag{2.10}$$

Substituting R_x $(=R_1 \parallel R_F)$ from Eq. (2.4) into Eq. (2.10) gives the output voltage as

$$V_{oi} = R_F(I_{B2} - I_{B1}) = -R_F I_{io} \tag{2.11}$$

For example, if $R_F = 100\ \text{k}\Omega$ and $I_{io} = \pm 200$ nA, then

$$V_{oi} = \pm 100 \times 10^3 \times 200 \times 10^{-9} = \pm 20\ \text{mV} \quad \text{(DC)}$$

To minimize the effect of I_{io}, the value of R_F should be small.

▶ **NOTE** If R_x is zero or is absent, Eq. (2.8) is to be used. If $R_x = R_1 \parallel R_F$, Eq. (2.11) is to be used.

EXAMPLE 2.1

Finding the effects of offsets on the output of an op-amp integrator The inverting integrator in Fig. 2.7 has $R_1 = 1\ \text{k}\Omega$, $R_x = 1\ \text{k}\Omega$, $C_F \doteq 0.1\ \mu\text{F}$, $V_{CC} = 15$ V, $-V_{EE} = -15$ V, and maximum saturation voltage $= \pm 14$ V. The op-amp parameters are $V_{io} = 6$ mV, $I_B = 500$ nA, and $I_{io} = 200$ nA at 25°C.

(a) Determine the total output offset voltage v_{of}.

(b) Repeat part (a) if $R_x = 0$

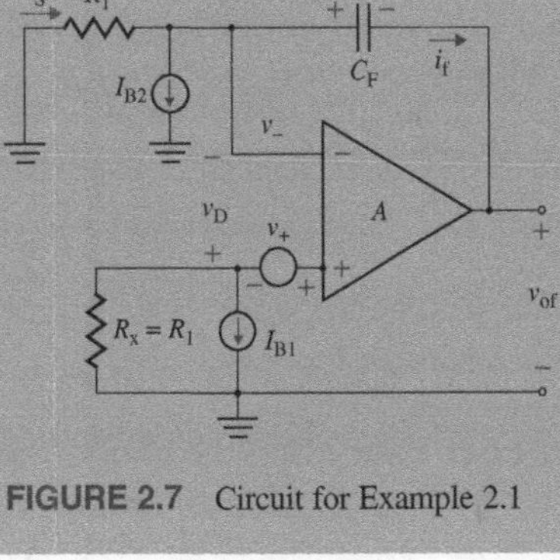

FIGURE 2.7 Circuit for Example 2.1

SOLUTION

$R_1 = 1\ \text{k}\Omega$, $R_x = 1\ \text{k}\Omega$, $V_{io} = 6$ mV, $I_B = 500$ nA, and $I_{io} = 200$ nA. The equivalent circuit of the integrator with input offset voltage and input biasing currents is also shown in Fig. 2.7.

(a) $R_x = R_1$. The output voltage will be due to V_{io} and I_{io}. With $V_{io} = 0$, we get

$$v_+ = -R_x I_{B1} = -R_1 I_{B1}$$

$$v_- = v_+ = -R_1 I_{B1}$$

$$i_S = -\frac{v_-}{R_1}$$

$$i_f = i_S - I_{B2} = I_{B1} - I_{B2}$$

With $I_{B1} = I_{B2} = 0$, we get

$$i_f = \frac{-V_{io}}{R_1}$$

Applying the superposition theorem, we get for the current flowing through the capacitor C_F

$$i_f = i_S + I_{B1} - I_{B2} = \frac{-V_{io}}{R_1} + I_{io} \tag{2.12}$$

where $I_{io} = (I_{B1} - I_{B2})$ is the input offset current. The total output offset voltage due to the capacitor current i_f can be found from

$$-v_{of} = \frac{1}{C_F}\int i_f dt + (-V_{io} + R_x I_{B1}) + v_C(t=0) \tag{2.13}$$

where $v_C(t = 0)$ is the initial capacitor voltage. Substituting i_f from Eq. (2.12) into Eq. (2.13) gives the total output offset voltage as

$$-v_{of} = \frac{1}{C_F}\int\left(\frac{-V_{io}}{R_1} + I_{io}\right)dt + (-V_{io} + R_x I_{B1}) + v_C(t=0)$$

$$v_{of} = \frac{V_{io}}{C_F R_1}t - \frac{I_{io}}{C_F}t + V_{io} - R_x I_{B1} - v_C(t=0) \tag{2.14}$$

This equation indicates that the output offset voltage will rise linearly until the output reaches the saturation voltage of the amplifier, which is ±14 V in this example. If the power supplies are turned on and enough time is allowed, the output will build up to the saturation voltage, even without any external input signal to the integrator. For this reason, this is not a practical circuit; it needs a DC feedback resistor R_F. After the power supply is switched on, the time required for the total output offset voltage to reach the saturation level $v_{of} = 14$ V can be found from Eq. (2.14) with $v_C(t = 0) = 0$:

$$14 = \left(\frac{6\times10^{-3}}{0.1\times10^{-6}\times1\times10^{3}} - \frac{200\times10^{-9}}{0.1\times10^{-6}}\right)t + 6\times10^{-3} - 1\ \text{k}\Omega\times600\times10^{-9} - 0$$

which gives $t = 241.3$ ms.

(b) If $R_x = 0$, the total output offset voltage will be due to the input offset voltage, the input biasing current, $-I_{B2} = -I_B + (I_{io}/2)$, and the input offset current, and it can be found from Eq. (2.14) by replacing I_{io} by I_{B2}. That is,

$$v_{of} = \left(\frac{V_{io}}{C_F R_1} + \frac{I_B - I_{io}/2}{C_F}\right)t + V_{io} - v_C(t = 0) \tag{2.15}$$

The time required for the output voltage to reach the saturation level $v_{of} = 14$ V can be found from Eq. (2.15) with $v_C(t = 0) = 0$:

$$14 = \left(\frac{6 \times 10^{-3}}{0.1 \times 10^{-6} \times 1 \times 10^{3}} + \frac{400 \times 10^{-9}}{0.1 \times 10^{-6}}\right)t + 6 \times 10^{-3} + 0$$

which gives $t = 218.7$ ms.

2.3.3 Input Offset Voltage

If the input terminals of an op-amp are tied together and connected to the ground, as shown in Fig. 2.8(a), a certain DC voltage exits at the output. This voltage is called the *output offset voltage* V_{oo}. The input offset voltage is the differential input voltage that exists between two terminals without any external inputs applied. In other words, it may be regarded as the input voltage that should be applied between the input terminals to force the output voltage to zero, as shown in Fig. 2.8(b). If V_{oo} is divided by the voltage gain A_o of the op-amp, the result is the *input offset voltage* V_{io}. Assuming that the output voltage is not saturated, V_{io} can be determined from

$$V_{io} = \frac{V_{oo}}{A_o} \quad \text{for } |V_{oo}| \le |V_{sat}| \tag{2.16}$$

Input offset voltage is quoted as an absolute value, and it may be positive or negative. The maximum value of V_{io} for the μA741C is ±6 mV. If the output voltage is saturated to V_{sat}, as is usually the case, Eq. (2.16) is valid only if $|V_{oo}| \le |V_{sat}|$. The polarity of V_{io} is unpredictable, and so is the output offset voltage V_{oo}.

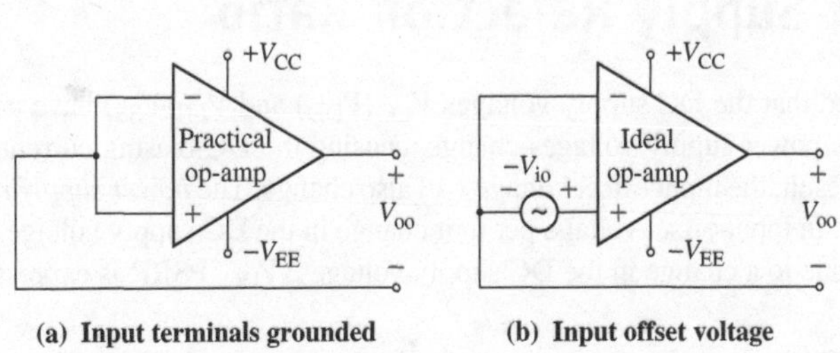

FIGURE 2.8 Input offset voltage

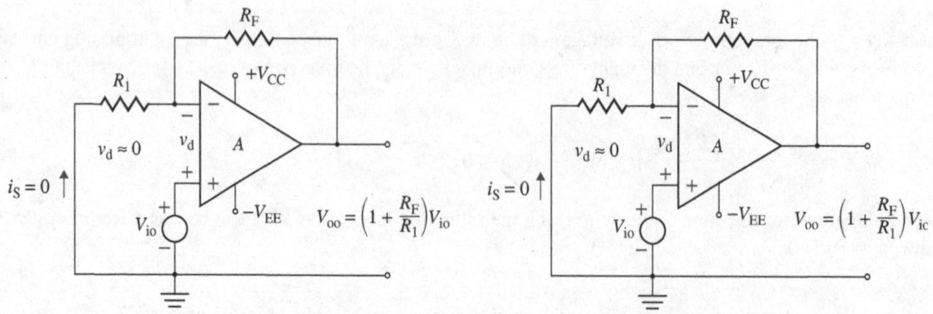

FIGURE 2.9 Inverting and noninverting amplifiers with input offset voltage

The output offset voltage is caused by internal mismatching in the input stage. A simple differential pair, as shown in Fig. 1.2(b), consists of two transistors Q_1 and Q_2. Any differential signal between the input terminals is amplified and gives the output voltage V_{oo}. In practice, the characteristics of the two transistors will not be exactly the same; therefore, the collector-biasing currents I_{C1} and I_{C2} will differ. As a result, even without any input voltages, there could be a differential output voltage, which is amplified in subsequent stages and possibly aggravated by more mismatching.

The effect of the input offset voltage can be determined for the inverting and noninverting amplifiers in Fig. 2.9. For both configurations, V_{io} may be considered as the input to the noninverting terminal, since there is no other input signal. Since the closed-loop gain of a non-inverting amplifier is $(1 + R_F/R_1)$, the offset output voltage is given by

$$V_{oo} = \left(1 + \frac{R_F}{R_1}\right)V_{io} \tag{2.17}$$

For example, if $R_1 = 10\ \text{k}\Omega$, $R_F = 100\ \text{k}\Omega$, and $V_{io} = \pm 6$ mV, then

$$V_{oo} = \pm\left(1 + \frac{100\ \text{k}\Omega}{10\ \text{k}\Omega}\right) \times 6\ \text{mV} = \pm 66\ \text{mV}$$

That is, the output voltage V_{oo} can be ± 66 mV (DC) without any external input signal v_S applied.

2.3.4 Power Supply Rejection Ratio

So far we have assumed that the DC supply voltages $V_{CC}(V_{DD})$ and $V_{EE}(V_{SS})$ have no effect on the output voltage. In practice, the power supply voltages change, causing the DC-biasing currents of the internal transistors to change. As a result, the input offset voltage will also change. The *power supply rejection ratio* (PSRR) is defined as the change in input offset voltage per unit change in the DC supply voltage. If ΔV_{io} is the change in input offset voltage due to a change in the DC supply voltage ΔV_{DC}, PSRR is expressed as

$$\text{PSRR} = \frac{\Delta V_{io}}{\Delta V_{DC}} \tag{2.18}$$

$$= 20 \log\left|\frac{\Delta V_{io}}{\Delta V_{DC}}\right| \tag{2.19}$$

This ratio is also known as the *supply voltage rejection ratio* (SVRR) or the *power supply sensitivity* (PSS). The maximum value of PSRR for the μA741C op-amp is 150 μV/V. For example, if the DC supply voltages change from $V_{DC} = \pm 15$ V to ± 12 V and PSRR = 150 μV/V, then

$$\Delta V_{DC} = 2 \times 15 - 2 \times 12 = 6 \text{ V}$$

and $$\Delta V_{io} = \text{PSRR}\,\Delta V_{DC} = 150\ \mu\text{V} \times 6 = 900\ \mu\text{V}$$

2.3.5 Thermal Voltage Drift

In previous sections, we assumed that the input offset voltage V_{io}, input biasing current I_B, and input offset current I_{io} remain constant. A practical op-amp consists of devices such as diodes and transistors whose parameters change with temperature. *Thermal drift* is a measure of the change in an offset parameter due to a unit change in temperature. *Thermal voltage drift* is defined as the rate of change of input offset voltage V_{io} per unit change in temperature, and it is expressed as

$$D_v = \frac{\Delta V_{io}}{\Delta T} \quad (\text{V}/^\circ\text{C}) \tag{2.20}$$

Thermal-biasing current drift is defined as the rate of change of input biasing current I_B per unit change in temperature, and it is expressed as

$$D_b = \frac{\Delta I_B}{\Delta T} \quad (\text{A}/^\circ\text{C}) \tag{2.21}$$

Thermal input offset current drift is defined as the rate of change of input offset current I_{io} per unit change in temperature, and it is expressed as

$$D_i = \frac{\Delta I_{io}}{\Delta T} \quad (\text{A}/^\circ\text{C}) \tag{2.22}$$

Thus, the output voltage due to drifts can be found from

$$V_{od} = \left(1 + \frac{R_F}{R_1}\right)\Delta V_{io} + R_F \Delta I_{io} = \left(1 + \frac{R_F}{R_1}\right) D_v \Delta T + R_F D_i \Delta T \tag{2.23}$$

$$= \left(1 + \frac{R_F}{R_1}\right) D_v \Delta T + R_F D_b \Delta T \quad \text{for } R_x = 0 \tag{2.24}$$

EXAMPLE 2.2

Finding the effects of thermal drift on the output of an inverting op-amp circuit The inverting amplifier in Fig. 2.5(a) has $R_1 = 10$ kΩ, $R_F = 100$ kΩ, and $R_x = R_F \| R_1 = 9.091$ kΩ. The op-amp parameters are $V_{io} = \pm 6$ mV, $I_B = 500$ nA, $I_{io} = \pm 200$ nA, and PSRR = 150 μV/V. The thermal drifts are $D_v = 15$ μV/°C, $D_i = 0.5$ nA/°C, and $D_b = 0.5$ nA/°C at 25°C. The temperature is 55°C. The DC supply voltages change from $V_{CC} = 15$ V to 12 V and $-V_{EE} = -15$ V to -12 V. The input voltage is $v_S = 100$ mV (DC). Determine the output voltage v_O if **(a)** $R_x = R_F \| R_1 = 9.091$ kΩ and **(b)** $R_x = 0$.

SOLUTION

$R_1 = 10\ \text{k}\Omega$, $R_F = 100\ \text{k}\Omega$, $V_{io} = \pm 6\ \text{mV}$, $I_B = 500\ \text{nA}$, $I_{io} = \pm 200\ \text{nA}$, $D_v = 15\ \mu\text{V/°C}$, $D_i = 0.5\ \text{nA/°C}$, $D_b = 0.5\ \text{nA/°C}$, and $v_S = 100\ \text{mV}$. Then

$$\Delta T = 55 - 25 = 30°\text{C}$$

$$\Delta V_{DC} = (15 + 15) - (12 + 12) = 6\ \text{V}$$

$$\Delta V_{io} = D_v\,\Delta T + \text{PSRR}\,\Delta V_{DC} = 15 \times 10^{-6} \times 30 + 150 \times 10^{-6} \times 6 = 1.35\ \text{mV}$$

$$\Delta I_{io} = D_i\,\Delta T = 0.5 \times 10^{-9} \times 30 = 15\ \text{nA}$$

$$\Delta I_B = D_b\,\Delta T = 0.5 \times 10^{-9} \times 30 = 15\ \text{nA}$$

(a) With offset-minimizing resistance R_x, the total output voltage of the inverting amplifier is given by

$$v_O = -\frac{R_F}{R_1}v_S \pm \left(1 + \frac{R_F}{R_1}\right)(V_{io} + \Delta V_{io}) \pm R_F(I_{io} + \Delta I_{io}) \qquad (2.25)$$

$$= -\left(\frac{100\ \text{k}\Omega}{10\ \text{k}\Omega}\right) \times 100 \times 10^{-3} \pm \left(1 + \frac{100\ \text{k}\Omega}{10\ \text{k}\Omega}\right) \times (6 + 1.35) \times 10^{-3}$$

$$\pm\ 100 \times 10^3 \times (200 + 15) \times 10^{-9}$$

$$= -1000\ \text{mV} \pm 80.85\ \text{mV} \pm 21.5\ \text{mV}$$

$$= -1102.35\ \text{mV (min)} \quad \text{or} \quad -897.65\ \text{mV (max)}$$

(b) With $R_x = 0$, the total output voltage of the inverting amplifier is given by

$$v_O = -\frac{R_F}{R_1}v_S \pm \left(1 + \frac{R_F}{R_1}\right)(V_{io} + \Delta V_{io}) + R_F(I_B + \Delta I_B) \qquad (2.26)$$

$$= -\left(\frac{100\ \text{k}\Omega}{10\ \text{k}\Omega}\right) \times 100 \times 10^{-3} \pm \left(1 + \frac{100\ \text{k}\Omega}{10\ \text{k}\Omega}\right) \times (6 + 1.35) \times 10^{-3}$$

$$+\ 100 \times 10^3 \times (500 + 15) \times 10^{-9}$$

$$= -1000\ \text{mV} \pm 80.85\ \text{mV} + 51.5\ \text{mV}$$

$$= -1029.4\ \text{mV (min)} \quad \text{or} \quad -867.65\ \text{mV (max)}$$

2.3.6 Determining the Thermal Voltage Drift

Because of mismatches, an output voltage will exist even when the external input is zero. The differential voltage that must be applied to the input terminals of an amplifier to drive the output to zero is called the *input offset voltage* V_{io}. This input offset voltage is a function of temperature. The rate of change of the input offset voltage V_{io} per unit change in temperature is known as the *thermal voltage drift*, and it is expressed as

$$D_v = \frac{\Delta V_{io}}{\Delta T} \quad \text{(in V/°C)} \qquad (2.27)$$

BJT Amplifiers

The difference in the B-E voltages will cause an offset voltage. Using Eqs. (1.129) and (1.130), we can express this offset voltage as

$$v_{BE2} - v_{BE1} = V_T\left[\ln\frac{i_{C2}}{I_{S2}} - \ln\frac{i_{C1}}{I_{S1}}\right] = V_T \ln\left(\frac{i_{C2}}{I_{S2}} \times \frac{I_{S1}}{i_{C1}}\right) \tag{2.28}$$

Since the mismatch in the collector current is generally small (i.e., $i_{C1} \approx i_{C2}$), we can find the input offset voltage V_{io} from

$$V_{io} = v_{BE2} - v_{BE1} = V_T \ln\left(\frac{I_{S1}}{I_{S2}}\right) \tag{2.29}$$

Thus, the saturation current I_S is the prime contributing factor to V_{io} in a BJT amplifier and is proportional to the transistor base width W_B. Its value can vary from one transistor to another. Assuming that $W_{B1} = W_B$ and $W_{B2} = W_B + \Delta W_B$ are the base widths of transistors Q_1 and Q_2, respectively, Eq. (2.29) becomes

$$V_{io} = V_T \ln\left(\frac{W_B}{W_B + \Delta W_B}\right) = V_T \ln\left(\frac{1}{1 + \Delta W_B/W_B}\right) \tag{2.30}$$

For $W_B >> \Delta W_B$, which is usually the case, Eq. (2.30) can be approximated by

$$V_{io} = V_T\left(\frac{\Delta W_B}{W_B}\right) \tag{2.31}$$

The beta of a transistor is inversely proportional to the base width W_B. Therefore, any change in W_B will cause an almost equal change in beta; that is, $\Delta\beta_F/\beta_F = \Delta W_B/W_B$. Since the change in the base width is generally within 10%, the offset voltage can be found approximately from

$$V_{io} = 0.1V_T \tag{2.32}$$

For BJT input devices, this value is typically 2 mV to 5 mV (for compound devices) and can often be nullified with an external potentiometer.

The ratio $\Delta W_B/W_B$ is relatively independent of temperature. Since $V_T = kT/q$, $dV_T/dT = k/q = V_T/T$. Thus, we can find the thermal drift from Eq. (2.31) as follows:

$$D_v = \frac{dV_{io}}{dT} = \frac{V_T}{T}\left(\frac{\Delta W_B}{W_B}\right) = \frac{V_{io}}{T} \tag{2.33}$$

For example, if $V_{io} = 2.6$ mV and $T = 25°C = 273 + 25 = 298$ K,

$$D_v = \frac{2.6 \text{ mV}}{298} = 8.72\ \mu\text{V/K}$$

CMOS Amplifiers

From Eq. (1.54), the transconductance of a MOSFET is given by

$$g_m = 2K_p(V_{GS} - V_t) = \frac{2I_D}{V_{GS} - V_t} \tag{2.34}$$

If ΔI_D is the difference between the drain currents of MOSFETs, an input offset voltage $\Delta V_{GS} = V_{OS}$ must be applied between the two gates to cancel the difference. Any increase in ΔV_{GS} will cause an increase in ΔI_D. However, any differential increase ΔV_t in the threshold voltages will cause a decrease in drain current. That is, Eq. (2.19) gives

$$\Delta I_D = \frac{g_m}{2}(\Delta V_{GS} - \Delta V_t) = \frac{g_m}{2}(V_{io} - \Delta V_t)$$

which gives the input offset voltage V_{io} as

$$V_{io} = \Delta V_t + \frac{\Delta I_D}{I_D}(V_{GS} - V_t) \tag{2.35}$$

The ratio $\Delta I_D/I_D$ will depend on any change in the W/L ratio of the MOSFET. Thus, Eq. (2.35) can be written in terms of the W/L ratio:

$$V_{io} = \Delta V_t + \frac{\Delta(W/L)}{W/L}(V_{GS} - V_t) \tag{2.36}$$

In CMOS amplifiers, the value of ΔV_t alone, which is absent in BJTs, can be as high as 2 mV—a magnitude as large as V_{io} for BJTs. The value of $(V_{GS} - V_t)$ is usually much greater than V_T. Thus, the offset voltage of CMOS amplifiers will generally be considerably larger than that of BJT amplifiers. To have a low value of V_{io}, MOSFETs should be operated with a low value of $(V_{GS} - V_t)$.

For MOSFET amplifiers, thermal drift cannot be correlated with the offset voltage. The thermal drift of $\Delta V_t/dT$ can be quite significant.

2.3.7 Offset Voltage Adjustment

We saw in the previous section that an op-amp can have an output voltage V_{oo} without any external input signal. Op-amps are generally compensated internally; they have built-in offset adjustment terminals, as shown in Fig. 2.10. The output voltage can be adjusted to zero by an offset null potentiometer. The

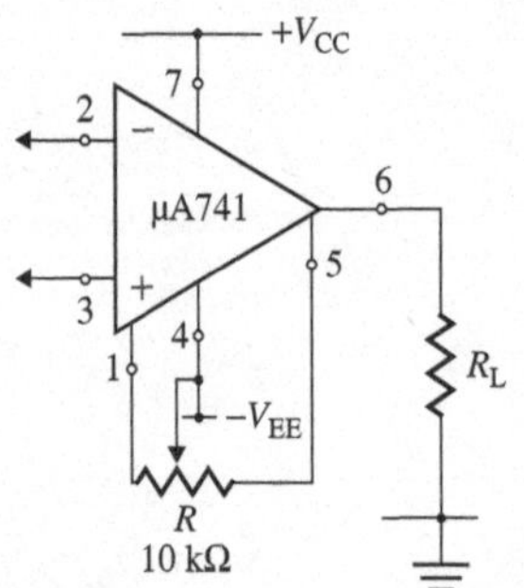

FIGURE 2.10 Op-amp with compensating terminals

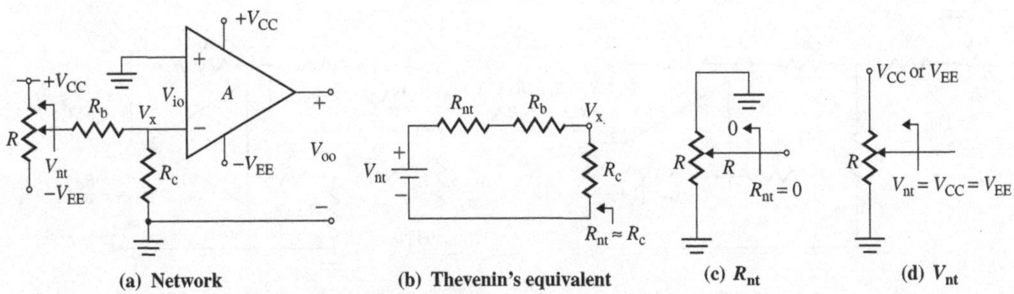

FIGURE 2.11 Offset-compensating network (external connection)

recommended value of the potentiometer is normally quoted in the data sheet; it is 10 kΩ for the μA741 series. By varying the potentiometer, the output offset voltage can be adjusted to zero within a certain input offset voltage adjustment range (±15 mV for the μA741 op-amp).

It is possible to compensate for the offset voltage by injecting a small voltage into the (+) terminal or the (−) terminal of an op-amp. An offset-compensating network is shown in Fig. 2.11(a). The potentiometer R is varied to produce an input offset voltage V_{io}, which should be just adequate to negate the output offset voltage. Thevenin's equivalent circuit of the network is shown in Fig. 2.11(b); the equivalent resistance of the compensating network is $R_{nt} = 0$, shown in Fig. 2.11(c), and the equivalent voltage is $V_{nt} \simeq V_{CC} = V_{EE}$, shown in Fig. 2.11(d). From the voltage division rule, the voltage V_x, which should be equal to V_{io}, is given by assuming $R_o >> R_{nt}$:

$$V_x \approx V_{io} = \frac{R_c V_{nt}}{R_{nt} + R_b + R_c} \tag{2.37}$$

The values for the network should be such that they do not alter the normal operation of the amplifier. That is, $R_b > R_{nt}$ so that R_b does not affect V_{nt} significantly; $R_b >> R_c$ so that the op-amp biasing current flows mostly through R_c (usually ≤100 Ω). The following relations are recommended:

$$R_b \geq 10R_{nt} \qquad [\text{where } R_{nt(max)} = (R/2) \parallel (R/2) = R/4]$$

$$R_b \geq 1000R_c$$

For $R_b > R_{nt} > R_c$, Eq. (2.37) can be approximated by

$$V_{io} \approx \frac{R_c V_{nt}}{R_b} = \frac{R_c V_{CC}(= V_{EE})}{R_b} \tag{2.38}$$

The compensating network can be used in inverting, noninverting, and differential amplifiers, as shown in Fig. 2.12. For a voltage follower, Fig. 2.12(a) can be modified by making $R_1 = 0$ and $R_F = \infty$ and then connecting one side of R_c to the output side instead of connecting to the ground.

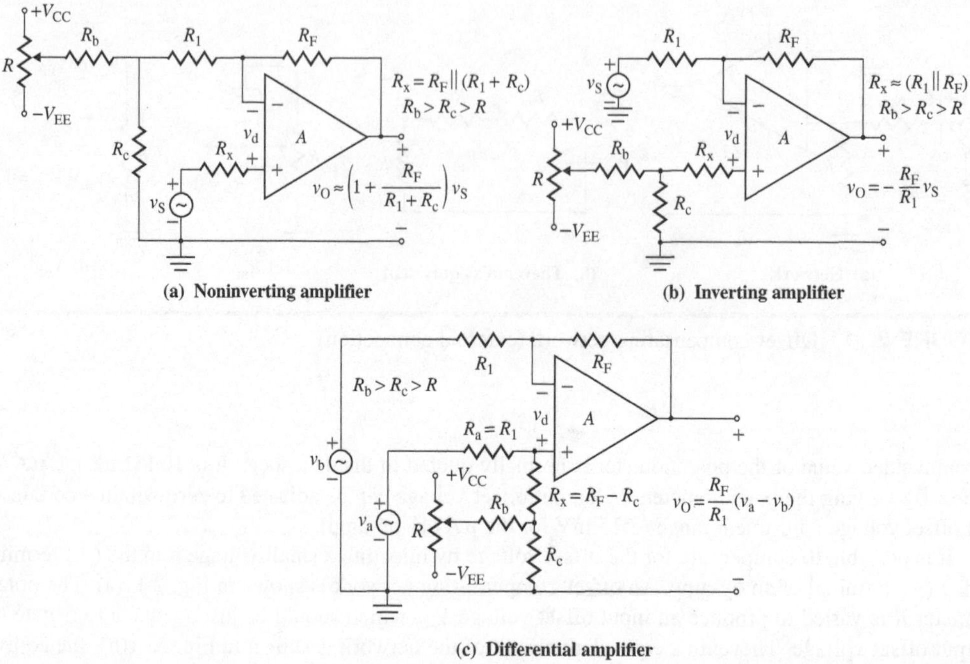

FIGURE 2.12 Op-amp amplifiers with offset-compensating network

EXAMPLE 2.3

Designing an offset-compensating network The noninverting amplifier in Fig. 2.12(a) has $R_1 = 10\ \text{k}\Omega$, $R_F = 100\ \text{k}\Omega$, and $R_x = R_F \| R_1 = 9.091\ \text{k}\Omega$. Design the offset-compensating network. The op-amp parameters are $V_{io} = 6$ mV, $I_B = 500$ nA, $I_{io} = 200$ nA, and PSRR = 150 μV/V. The DC supply voltages are $V_{CC} = 15$ V and $-V_{EE} = -15$ V.

SOLUTION

Since the offset due to the biasing current I_B is minimized by the resistance R_x, the output offset voltage will be contributed mostly by V_{io}. For $V_{io} = 6$ mV and $V_{nt} = V_{CC} = 15$ V, Eq. (2.38) gives 6 mV $= 15R_c/R_b$.

Letting $R_c = 10\ \Omega$, we get

$$R_b = \frac{15R_c}{6\ \text{mV}} = 25\ \text{k}\Omega$$

Letting $R_b = 10R_{nt(max)} = 10(R/4)$, we get

$$R = \frac{4R_b}{10} = 4 \times \frac{25\ \text{k}\Omega}{10} = 10\ \text{k}\Omega \quad \text{(potentiometer)}$$

The network will change the voltage gain from

$$1 + \frac{R_F}{R_1} = 1 + \frac{100\ \text{k}\Omega}{10\ \text{k}\Omega} = 11$$

to $$1 + \frac{R_F}{R_1 + R_c} = 1 + \frac{100\ \text{k}\Omega}{10{,}010\ \Omega} = 10.99$$

causing a 0.09% error.

2.3.8 Common-Mode Rejection Ratio

The CMRR, which is defined as the ratio of the differential voltage gain to the common-mode voltage gain, may also be defined as the change in input offset voltage per unit change in common-mode voltage. Let $v_{ic} = 0$ and apply v_{id} to drive the output voltage to zero. v_{id} should thus be equal to the input offset voltage V_{OS}. If we keep v_{id} constant and increase v_{ic} by an amount Δv_{ic}, the output voltage will change by an amount

$$\Delta v_O = A_c \Delta v_{ic} \tag{2.39}$$

To drive the output voltage to zero, we have to change v_{id} by an amount Δv_{id}, where

$$\Delta v_{id} = \frac{\Delta v_O}{A_d} = \frac{A_c \Delta v_{ic}}{A_d} = \Delta V_{OS} \tag{2.40}$$

Equation (2.40) indicates that any change in v_{ic} causes a corresponding change in V_{OS}. From Eq. (2.40), we get

$$\text{CMRR} = \frac{A_d}{A_c} = \left.\frac{\Delta v_{ic}}{\Delta v_{id}}\right|_{v_o=0} = \frac{\Delta v_{ic}}{\Delta V_{OS}} = \left.\frac{\delta v_{ic}}{\delta V_{OS}}\right|_{v_o=0} \tag{2.41}$$

which shows that the input offset voltage is dependent on the CMRR and the common-mode signal. For CMRR $= 10^5$ (or 100 dB) and $\Delta v_{ic} = 15$ V, the change in the input offset voltage will be $\Delta V_{OS} = 150\ \mu$V.

2.3.9 Input Resistance

The input resistance for a MOSFET input stage is very high, in the range of $10^9\ \Omega$ to $10^{12}\ \Omega$. For a BJT input stage, however, the input resistance is typically in the range of 100 kΩ to 1 MΩ. Usually, the voltage gain is large enough that this input resistance has little effect on the circuit performance. In some differential amplifiers, a compound transistor configuration known as a *Darlington pair* is used to give a much higher input resistance and a much lower input biasing current than a single transistor would provide. A Darlington pair is shown in Fig. 2.13. The effective B-E voltage of the equivalent transistor Q_T in Fig. 2.13(b) is

$$V_{BE} = V_{BE1} + V_{BE2} = V_T \ln\left(\frac{I_{C1}}{I_{S1}}\right) + V_T \ln\left(\frac{I_{C2}}{I_{S2}}\right)$$

$$= V_T \ln\left(\frac{I_{C1} I_{C2}}{I_{S1} I_{S2}}\right) \tag{2.42}$$

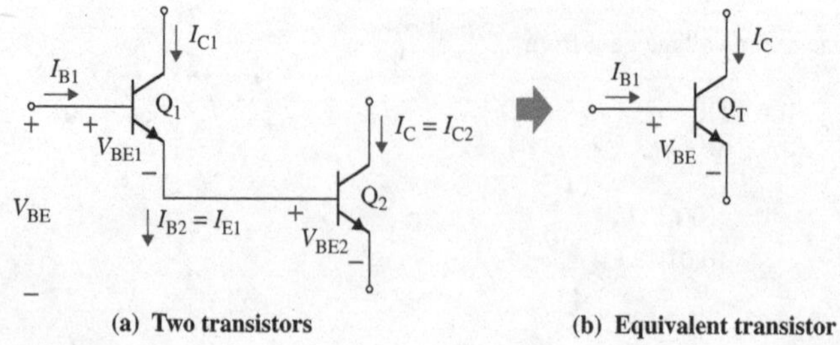

FIGURE 2.13 Darlington pair

Since $I_C = I_{C2} = I_{C1}\beta_F\beta_F/(1+\beta_F) \approx \beta_F I_{C1}$, Eq. (2.42) becomes

$$V_{BE} = V_T \ln\left(\frac{I_C^2}{\beta_F I_{S1} I_{S2}}\right) = 2V_T \ln\left(\frac{I_C}{\beta_F I_{S1} I_{S2}}\right) \tag{2.43}$$

Solving for I_C, we get

$$I_C = \sqrt{\beta_F I_{S1} I_{S2}}\left(\frac{V_{BE}}{2V_T}\right)$$

$$= I_S \exp\left(\frac{V_{BE}}{V'_T}\right) \tag{2.44}$$

where $I_S = \sqrt{\beta_F I_{S1} I_{S2}}$ = effective saturation current and $V'_T = 2V_T$ = effective thermal voltage.

The collector current I_C of equivalent transistor Q_T can be related to I_{B1} by

$$I_C = I_{C2} = \beta_F I_{B2} = (1+\beta_F)I_{C1} = \beta_F(1+\beta_F)I_{B1} \simeq \beta_F^2 I_{B1} \tag{2.45}$$

Thus, the effective input resistance of the compound pair is given by

$$r'_\pi = \frac{V'_T}{I_{B1}} \simeq \beta_F^2 \frac{2V_T}{I_C} \tag{2.46}$$

which will be $2\beta_F$ times greater than that for a single device. For a single equivalent transistor, $r'_\pi = 2\beta_F r_\pi$. Thus, if $I_C = 200\ \mu A$, $\beta_F = 100$, and $V_T = 26$ mV,

$$r_\pi = \frac{\beta_F V_T}{I_C} = 100 \times \frac{26\ \text{mV}}{200\ \mu\text{A}} = 13\ \text{k}\Omega$$

for a single transistor and

$$r'_\pi = 100^2 \times 2 \times \frac{26\ \text{mV}}{200\ \mu\text{A}} = 2.6\ \text{M}\Omega$$

for a Darlington pair. The input offset voltage V_{OS}, however, will increase (generally $\sqrt{2}$ times) as a result of the increase in the effective thermal voltage.

2.3.10 Output Resistance

The output stage is usually an emitter follower in class AB operation, and hence it gives a low output resistance, on the order of 40 Ω to 100 Ω. This resistance is low enough that it does not strongly affect performance. If $I_{C(out)}$ is the collector-biasing current of the output stage, then r_π of the output transistor becomes $r_\pi = \beta_F V_T / I_{C(out)}$, which, if converted to the emitter terminal by a dividing factor of $(1 + \beta_f)$, will give approximately the output resistance of the op-amp. That is,

$$R_{out} = \frac{\beta_F V_T}{I_{C(out)}(1 + \beta_F)} \simeq \frac{V_T}{I_{C(out)}} \tag{2.47}$$

The collector-biasing current of the output stage (which is in milliamperes) is generally much greater than that of the differential stage (which is in microamperes). For example, if $\beta_F = 100$, $V_T = 26$ mV, and $I_{C(out)} = 1$ mA, $R_{out} = 26$ mV/1 mA = 26 Ω.

2.3.11 Frequency Response

Because of the parasitic capacitances and the minority carrier charge storage in the devices within the op-amps, the voltage gain decreases at high frequency. The frequency at which the open-loop voltage gain falls to unity is defined as the *unity-gain bandwidth*, and it is in the range of 1 MHz to 20 MHz. The differential stage in Fig. 2.1 can be represented by a voltage-controlled current source, as shown in Fig. 2.14(a). C_1 is the effective capacitance due to the output capacitance of the differential stage and the input capacitance of the second stage. R_1 is the effective resistance due to the output resistance of the differential stage and the input resistance of the second stage. G_{m1} is the transconductance of the differential stage.

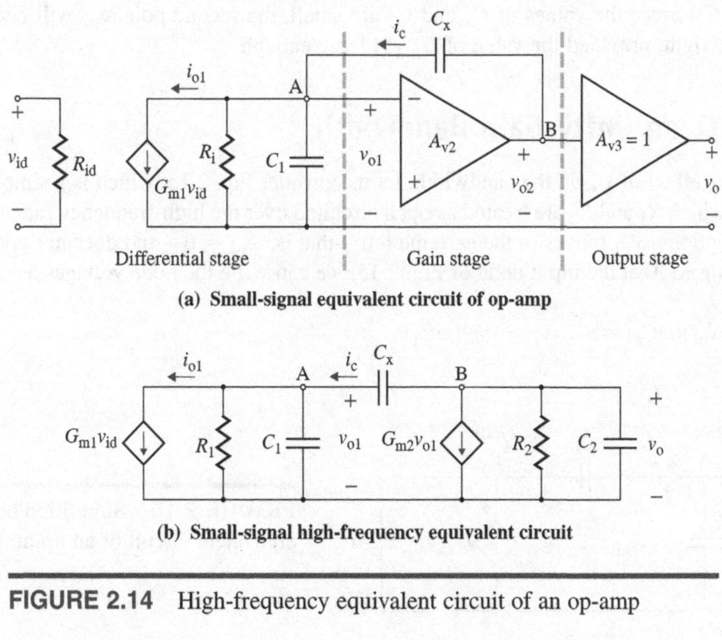

FIGURE 2.14 High-frequency equivalent circuit of an op-amp

The second stage generally has a common-emitter or common-source configuration with an active load for a large voltage gain. It can also be represented by another voltage-controlled current source, as shown in Fig. 2.14(b). G_{m2} is the transconductance of the second stage. C_2 and R_2 are the effective capacitance and the output resistance of the second stage, respectively. A_{v2} is the voltage gain of the second stage and is negative—that is, $A_{v2} = -G_{m2}R_2$.

The gain of the output stage is generally unity: $A_{v3} = 1$. Since the gain of the second stage is negative, the capacitance C_x will exhibit Miller's effect and split the poles. Also, C_x will influence both the frequency response and the slew rate considerably. From Eqs. (H.8) and (H.9) in Appendix H, we can find the new poles:

$$\omega_{p1} \approx \frac{1}{G_{m2}C_xR_1R_2} \tag{2.48}$$

$$\omega_{p2} \approx \frac{G_{m2}C_x}{C_1C_2 + C_x(C_1 + C_2)} \tag{2.49}$$

If $C_x >> C_1$ and C_2, Eq. (2.49) can be approximated by

$$\omega_{p2} \approx \frac{G_{m2}C_x}{C_x(C_1 + C_2)} = \frac{G_{m2}}{C_1 + C_2} \tag{2.50}$$

The first pole is due to the Miller capacitance

$$C_M = C_x(1 + G_{m2}R_2) \simeq C_xG_{m2}R_2 \quad (\text{for } G_{m2}R_2 >> 1)$$

which is much larger than C_1. To make ω_{p1} act as the unity-gain frequency ω_u, we can select the appropriate value of C_x. Since the values of C_1 and C_2 are small, the second pole ω_{p2} will become very large and move to the right, provided the value of G_{m2} is large enough.

Effects of C_x on Unity-Gain Bandwidth

To determine the effect of C_x on the bandwidth, let us consider Fig. 2.15, which is a simplified version of Fig. 2.14(b) in which R_1 and R_2 are treated as open circuited over the high-frequency range of interest. Under these assumptions, ω_{p1} moves to the extreme left—that is, $\omega_{p1} \approx 0$—and does not give the unity-gain bandwidth. Using KCL at the input node of Fig. 2.15, we can write the node voltages as

$$(v_o - v_{o1})(\omega C_x) = G_{m1}v_{id} + v_{o1}(\omega C_1) \tag{2.51}$$

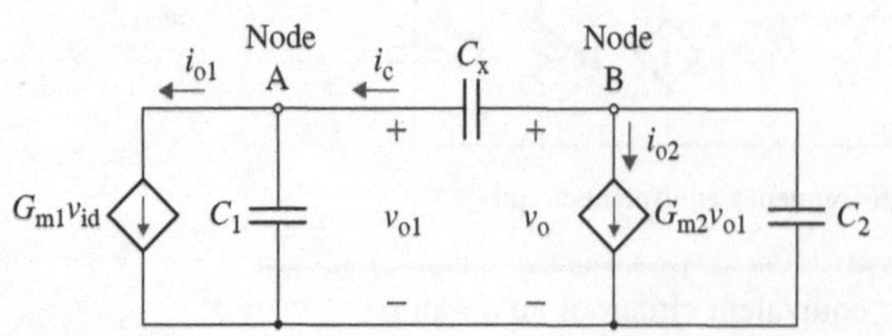

FIGURE 2.15 Simplified high-frequency equivalent circuit of an op-amp

where $G_{m1} = I_Q/2V_T$ is the transconductance of the differential stage. I_Q is the DC biasing current of the differential amplifier shown in Fig. 2.1.

Substituting $v_o = -A_{v2}v_{o1}$ (i.e., $v_{o1} = -v_o/A_{v2}$) into Eq. (2.51) and simplifying, we get the overall voltage gain A_o:

$$A_o(\omega) = \frac{v_o}{v_{id}} = \frac{G_{m1}}{(1 + 1/A_{v2})(\omega C_x) + (\omega C_1)/A_{v2}} \tag{2.52}$$

$$= \frac{G_{m1}/(\omega C_x)}{(1 + 1/A_{v2}) + (C_1/C_x)/A_{v2}} \tag{2.53}$$

where A_{v2} is the voltage gain of the second stage. For $A_{v2} >> 1$, which is usually the case, Eq. (2.53) can be simplified to

$$A_o(\omega) = \frac{G_{m1}}{\omega C_x} \tag{2.54}$$

At the unity-gain frequency ω_u, $|A(\omega)| = 1$. Thus, ω_u is given by

$$\omega_u = \frac{G_{m1}}{C_x} \tag{2.55}$$

which gives the corresponding frequency (i.e., the unity-gain bandwidth) f_u as

$$f_u = \frac{G_{m1}}{2\pi C_x} \tag{2.56}$$

$$= \frac{I_Q}{4\pi V_T C_x} \tag{2.57}$$

Thus, f_u is directly proportional to the biasing current I_Q of the differential stage and inversely proportional to the compensating capacitance C_x. A specific value of C_x is purposely added to the circuit, either on the chip for compensated op-amps or as an external capacitor, to set the desired value of f_u. For example, if $I_Q = 20\ \mu A$, $V_T = 26$ mV, and $C_x = 50$ pF, we get $f_u = 1.22$ MHz.

Effects of C_x on Zeros

The capacitance C_x also has a Miller's effect on the output side of the second stage, given by

$$C_N = C_x\left(1 + \frac{1}{G_{m2}R_2}\right) \approx C_x \quad (\text{for } G_{m2}R_2 >> 1)$$

and introduces a right-half-plane zero on the transfer function. The zero ω_z can be found from Fig. 2.16 by making the output voltage $v_o \approx 0$, as shown in Fig. 2.16(a). We get

$$G_{m2}v_{o1} = (v_{o1} - v_o)(\omega C_x) = v_{o1}(\omega C_x)$$

which gives the zero frequency ω_z as

$$\omega_z = \frac{G_{m2}}{C_x} \tag{2.58}$$

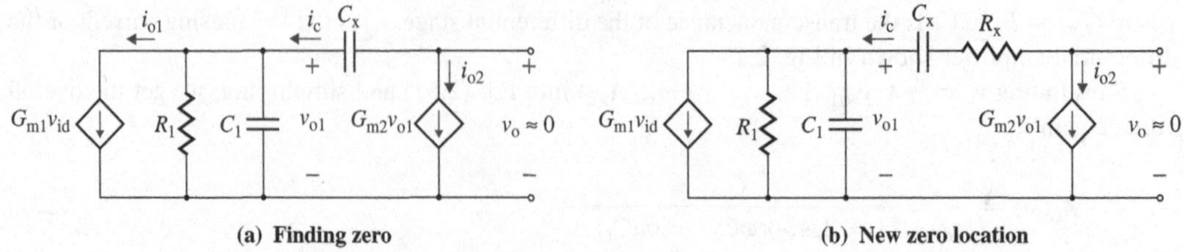

FIGURE 2.16 Equivalent circuit for determining zeros

If G_{m2} is large, which is usually the case for BJT amplifiers, then the zero will be at a very high frequency. However, if G_{m2} of the second stage is of the same magnitude as G_{m1} of the first stage, which is generally the case for CMOS amplifiers, the zero frequency will be close to the unity-gain frequency ω_u.

Since a zero introduces a phase shift, the phase margin of the amplifier will be decreased, affecting the amplifier stability. However, it is possible to remedy the stability problem by adding a resistance R_x in series with C_x, as shown in Fig. 2.16(b). The combination of R_x and C_x can move the zero and make ω_z a very large frequency. The zero frequency ω_z can be found by making the output voltage $v_o \simeq 0$; that is,

$$G_{m2}v_{o1} = \frac{(v_{o1} - v_o)}{R_x + 1/(j\omega C_x)}$$

which gives the new zero frequency ω_z as

$$\omega_z = \frac{1}{C_x(1/G_{m2} - R_x)} \tag{2.59}$$

Thus, as R_x approaches $1/G_{m2}$, the zero frequency ω_z tends to approach infinity. It is important to note that, by making $R_x > 1/G_{m2}$, we can locate the zero frequency at the negative real axis, which will increase the phase margin.

According to Eq. (2.50), the second pole ω_{p2} may be close to the unity-gain frequency ω_u for a low value of G_{m2}. As a result, ω_{p2} may introduce an appreciable phase shift and thus decrease the phase margin. This problem can be resolved by increasing the value of C_x in order to split the poles further.

2.3.12 Slew Rate

Op-amps are limited by the *slew rate* (SR), which specifies the maximum rate at which the output voltage can change without introducing any significant amount of distortion. That is, SR = $(dv_o/dt)_{max}$. Like the unity-gain frequency, the SR depends on the capacitance C_x. Figure 2.15 illustrates that the current i_c through the capacitor is related to the output voltage v_o by

$$i_c = C_x\frac{d}{dt}(v_o - v_{o1}) \tag{2.60}$$

$$= C_x\frac{d}{dt}\left(v_o + \frac{v_o}{A_{v2}}\right)$$

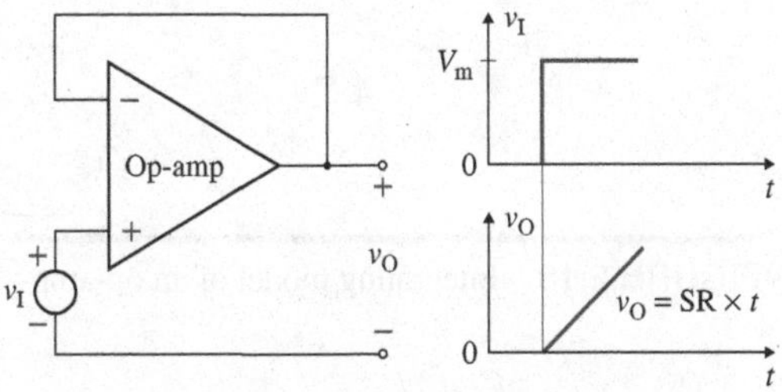

(a) Unity-gain follower (b) Waveforms

FIGURE 2.17 Unity-gain follower

(since $v_{o1} = -v_o/A_{v2}$), which gives dv_o/dt as

$$\frac{dv_o}{dt} = \frac{i_c}{C_x(1 + 1/A_{v2})} \tag{2.61}$$

Let us assume that $A_{v2} >> 1$—that is, v_{o1} tending to zero, $v_{o1} \approx 0$—and that i_c can be approximated by the output current of the differential amplifier (i.e., $i_c = i_{o1} = G_{m1}v_{id}$). Then Eq. (2.61) can be approximated by

$$\frac{dv_o}{dt} \simeq \frac{i_c}{C_x} = \frac{i_{o1}}{C_x} \tag{2.62}$$

Consider the unity-gain follower shown in Fig. 2.17(a). If a step input $v_I = V_m$ (say, 10 V) is applied to it, in zero time the output will not change, as shown in Fig. 2.17(b). Thus, V_m will appear as the differential voltage between the two input terminals, and the differential stage in Fig. 2.1 will be overdriven. Transistors Q_1 and Q_3 in Fig. 2.1 will carry the whole biasing current I_Q, and transistor Q_2 will be cut off. Q_4, however, is the mirror of Q_3, and it will produce a current of I_Q, which will flow into the second stage.

The biasing current limits the maximum value of the output current i_{o1} of the differential stage to the value I_Q in one direction and the value $-I_Q$ in the other direction; that is, $i_{o1(max)} = \pm I_Q$. Thus, the SR corresponding to $i_{o1(max)}$ is given by

$$\text{SR} = \left.\frac{dv_o}{dt}\right|_{max} = \frac{i_{o1(max)}}{C_x} = \pm\frac{I_Q}{C_x} \tag{2.63}$$

Like the unity-gain frequency f_u, the slew rate is directly proportional to I_Q and inversely proportional to C_x. If the gain of the second stage is large, then the op-amp behaves as an integrator, as shown in Fig. 2.18.

Relation between SR and f_u

Substituting $I_Q = 4\pi V_T C_x f_u$ from Eq. (2.57) into Eq. (2.63), we can relate the positive SR (known simply as SR) to f_u; that is,

$$\text{SR} = \frac{4\pi V_T C_x f_u}{C_x} = 4\pi V_T f_u \tag{2.64}$$

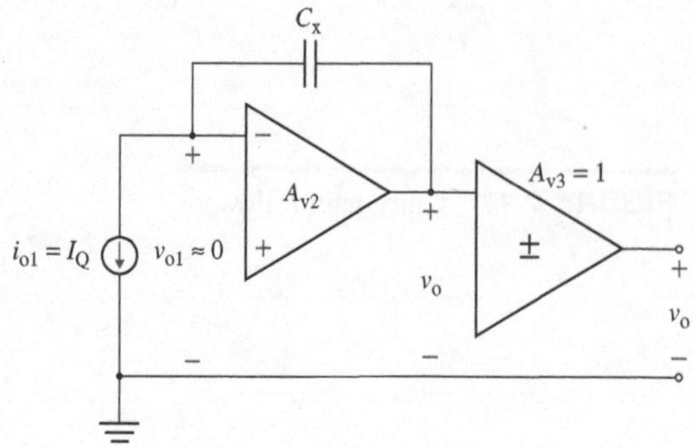

FIGURE 2.18 Integrating model of an op-amp

or

$$f_u = \frac{SR}{4\pi V_T} \tag{2.65}$$

Thus, there is a direct relation between the SR and the unity-gain frequency of an op-amp. For example, if $I_Q = 20\ \mu A$, $V_T = 26$ mV, and $C_x = 50$ pF, then Eq. (2.54) gives $f_u = 1.22$ MHz and Eq. (2.63) gives

$$SR = \frac{I_Q}{C_x} = \frac{20\ \mu A}{50\ pF} = 0.4\ V/\mu s$$

KEY POINTS OF SECTION 2.3

- The CMRR of an op-amp is very large (typically 10^5), and the output voltage due to a common-mode signal is negligible.
- The voltage gain of an op-amp decreases with frequency; however, the gain–bandwidth product remains constant. That is, if the gain decreases, the bandwidth increases.
- The slew rate SR of an op-amp limits the maximum input frequency at which the op-amp can amplify a signal without significant distortion. For minimum distortion, the SR of the input signal should be less than that of the op-amp.
- An imperfect op-amp produces an output offset voltage caused by parameters such as V_{io}, I_{io}, I_B, PSRR, and thermal drift. A resistor is usually connected at the (+) terminal to minimize offset due to DC-biasing currents.
- A compensating network (either internal or external) may be connected to negate the offset voltage.
- The input offset voltage of MOSFET amplifiers is considerably higher than that of BJT amplifiers. This difference is due to mismatches in base widths for BJTs, in I_{DSS} and V_p for MOSFETs, and in V_t and W/L for MOSFETs.
- A Darlington BJT pair is commonly used to provide high input resistance, low input base current, and high current gain.
- The unity-gain frequency f_u depends directly on the biasing current I_Q of the differential stage and inversely on the compensation capacitance C_x. The slew rate SR is directly proportional to f_u.
- If the transconductance of the gain stage is on the same order as that of the first stage, the zero frequency will be close to the unity-gain frequency and the phase margin will be reduced. A resistor R_x is usually connected in series with C_x to move the zero frequency to infinity.

2.4 CMOS Op-Amps

The internal structure of operational amplifiers has, in general, three stages: a differential stage, a gain stage, and an output stage. The gain and output stages are often combined into a single stage. This type of amplifier has a very high input resistance, and the voltage gain is generally less than that of other types of comparable amplifiers. In this section, we illustrate three CMOS amplifiers: one basic type and two commercial types.

2.4.1 Basic CMOS Op-Amp

A basic CMOS amplifier with a differential stage and a gain stage is shown in Fig. 2.19 [4, 5]. PMOS transistors M_6, M_7, and M_8 are used to generate biasing current sources. NMOS transistors M_1, M_2, M_3, and M_4 form the differential stage. NMOS M_1 operates as a common-source amplifier. C_x is used for frequency compensation by introducing a dominant pole. The main task in the design of CMOS amplifiers is to determine the width-to-length (W/ L) ratios of the NMOS and PMOS. Usually, W/L ratios of M_3, M_4, and M_8 are half of those for other MOSFETs. The differential pair is the "heart" of op-amps and can be formed by cascoding MOSFETs to yield high gain and low CMRR (see Sec. 1.2.1).

Using Eq. (1.61), the voltage gain of the differential stage is given by

$$A_1 = -(g_{m2} + g_{m4})(r_{o2} \parallel r_{o4})$$

and the voltage gain of the gain (second) stage is given by

$$A_2 = -g_{m5}(r_{o5} \parallel r_{o6})$$

Therefore, the overall low-frequency voltage gain of the amplifier is

$$A_o = A_1 A_2 = (g_{m2} + g_{m4})(r_{o2} \parallel r_{o4}) g_{m5}(r_{o5} \parallel r_{o6}) \tag{2.66}$$

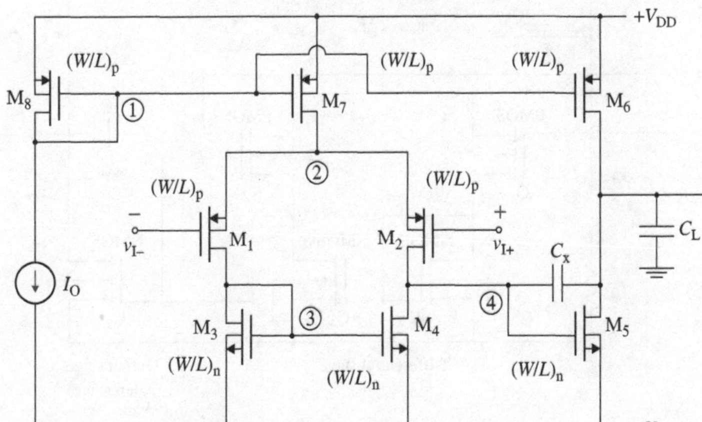

FIGURE 2.19 Basic CMOS amplifier

TABLE 2.1 Parameters of op-amp MC14573

Parameter	Minimum	Typical	Maximum	Units
DC supply voltages V_{DD}			±18	V
Input offset voltage V_{OS}		160	600	μV
Thermal drift D_v		1		μV/°C
Input biasing current I_B			1.0	nA
Input offset current I_{OS}			200	pA
Differential input resistance R_{id}		10^{12}		Ω
Common-mode input resistance R_{ic}		10^{12}		Ω
Output resistance R_o		50		Ω
Input capacitance C_i		1.0		pF
Open-loop voltage gain A_o		90		dB
Common-mode rejection ratio (CMRR)	75	95		dB
Unity-gain bandwidth f_u		70		MHz
Slew rate (SR)		2.5		V/μs
Power supply rejection ratio (PSRR)	75	97		dB

2.4.2 CMOS Op-Amp MC14573

The simplified schematic of op-amp MC14573 is shown in Fig. 2.20. Its internal structure can be divided into a differential CMOS stage, a gain stage, and a biasing circuit. This op-amp has a very high input resistance, but the gain is lower than that of other op-amps, as expected. Some parameters of the MC14573 op-amp are listed in Table 2.1.

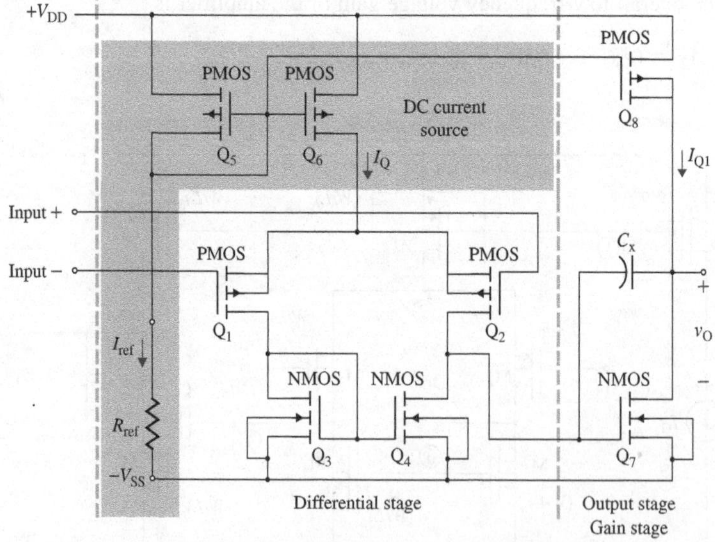

FIGURE 2.20 Schematic for op-amp MC14573 (Copyright of Motorola. Used by permission.)

Differential Stage

The input stage is a PMOS differential amplifier consisting of transistors Q_1 and Q_2. Its load is a current mirror consisting of NMOS transistors Q_3 and Q_4.

Gain Stage

The gain stage is also the output stage. Q_7 is operated as a common-source amplifier with Q_8 as the active load. C_x is the compensation capacitor connected to Q_7.

DC Biasing

The differential stage is biased by transistors Q_5 and Q_6 as a current source. An external resistance R_{ref} sets the DC biasing current I_Q, which can be found approximately from

$$I_Q = I_{ref} = \frac{V_{DD} + V_{SS} - V_{SG}}{R_{ref}} \tag{2.67}$$

$$= K_p(V_{GS} - V_t)^2$$

The value of V_{GS5} can be solved for known values of V_{DD}, V_{SS}, V_t, and K_p. Transistor Q_8 serves as the load to the output stage. Since the gate voltages of Q_5, Q_6, and Q_8 are the same, the biasing drain current Q_8 equals $I_{Q1} = I_Q$.

EXAMPLE 2.4

D **Analyzing the CMOS op-amp MC14573** The CMOS amplifier in Fig. 2.20 is operated at a biasing current of $I_Q = 40\ \mu A$. The parameters of the MOSFETs are $K_x = 10\ \mu A/V^2$, $|V_{M(NMOS)}| = V_{M(PMOS)} = 70$ V, $V_t = 0.5$ V, and $W/L = 160\ \mu m/10\ \mu m$, except for Q_7, for which $W/L = 320\ \mu m/10\ \mu m$. Assume $V_{DD} = -V_{SS} = 5$ V.

(a) Find V_{GS}, g_m, and r_o for all MOSFETs.

(b) Find the low-frequency voltage gain of the amplifier A_{vo}.

(c) Find the value of the external resistance R_{ref}.

(d) Find the value of compensation capacitance C_x that gives a unity-gain bandwidth of 1 MHz and the corresponding slew rate.

(e) Find the value of resistance R_x to be connected in series with C_x in order to move the zero frequency to infinity.

(f) Find the common-mode input voltage range.

(g) Find the output voltage range.

SOLUTION

(a) K_p, V_{GS}, g_m, and r_o are calculated using the following equations:

$$K_p = \frac{K_x W}{L} \tag{2.68}$$

TABLE 2.2 Calculated values for Example 2.4

	Q_1	Q_2	Q_3	Q_4	Q_5	Q_6	Q_7	Q_8
W/L (in μm/μm)	160/10	160/10	160/10	160/10	160/10	160/10	320/10	160/10
K_p (in μA/V^2)	160	160	160	160	160	160	320	160
I_D (in μA)	20	20	20	20	40	40	40	40
V_{GS} (in V)	0.854	0.854	0.854	0.854	1	1	0.854	1
g_m (in μA/V)	113.1	113.1	113.1	113.1	160	160	226.3	160
r_o (in MΩ)	3.5	3.5	3.5	3.5	1.75	1.75	1.75	1.75

$$V_{GS} = V_t + \sqrt{\frac{I_D}{K_p}} \tag{2.69}$$

$$g_m = 2K_p(V_{GS} - V_t) = 2\sqrt{K_p I_D} \tag{2.70}$$

$$r_o = \frac{|V_M|}{I_D} \tag{2.71}$$

Their values are shown in Table 2.2.

(b) The voltage gain of the first differential stage with the active load is given by

$$A_{v1} = -g_{m4}(r_{o2} \| r_{o4}) \tag{2.72}$$

$$= -113.1\ \mu \times (3.5\ \text{M} \| 3.5\ \text{M}) = -198$$

Since the relationship for the voltage gain of a transistor with a drain resistance of R_D is $A_{V0} = g_m R_D$, the voltage gain of the second stage with the active load is given by

$$A_{v2} = -g_{m7}(r_{o7} \| r_{o8}) \tag{2.73}$$

$$= -226.3\ \mu \times (1.75\ \text{M} \| 1.75\ \text{M}) = -198$$

Thus, the overall voltage gain is $A_{vo} = A_{v1}A_{v2} = 198 \times 198 = 39{,}204$ (or 91.87 dB).

(c) From Eq. (2.67), we get the value of the external resistance R_{ref} as

$$R_{ref} = \frac{V_{DD} - V_{SS} - V_{GS5}}{I_Q} = \frac{(5 + 5 - 1)\ \text{V}}{40\ \mu\text{A}} = 225\ \text{k}\Omega$$

(d) For $f_u = 1$ MHz, Eq. (2.56) gives the value of compensation capacitance C_x as

$$C_x = \frac{G_{m1}}{2\pi f_u} = \frac{G_{m2}}{2\pi f_u} = \frac{113.1\ \mu\text{A/V}}{2\pi \times 1\ \text{MHz}} = 18\ \text{pF}$$

From Eq. (2.63), we get the slew rate as

$$\text{SR} = \frac{I_Q}{C_x} = \frac{40\ \mu\text{A}}{18\ \text{pF}} = 2.22\ \text{V}/\mu\text{s}$$

(e) From Eq. (2.59), the value of resistance R_x for $\omega_z = 0$ is

$$R_x = \frac{1}{G_{m2}} = \frac{1}{g_{m7}} = \frac{1}{226.3\ \mu\text{A/V}} = 4.42\ \text{k}\Omega$$

(f) Transistor Q_6 will leave saturation when the common-mode input voltage becomes

$$v_{ic(max)} = V_{DD} - |V_{GS6}| + |V_t| - |V_{GS1}| = 5 - 1 + 0.5 - 0.854 = 3.646 \text{ V}$$

Transistors Q_1 and Q_2 will leave saturation when the common-mode input voltage falls below the voltage at the drain of Q_1 by $|V_t|$—that is, when

$$v_{ic(min)} = -V_{SS} + |V_{GS3}| - |V_t| = -5 + 1 - 0.5 = -4.5 \text{ V}$$

Thus, the common-mode input voltage range is −4.5 V to 3.646 V.

(g) Transistor Q_8 will leave saturation when the output voltage becomes

$$v_{o(max)} = V_{DD} - |V_{GS8}| + |V_t| = 5 - 1 + 0.5 = 4.5 \text{ V}$$

Transistor Q_7 will leave saturation when the output voltage becomes

$$v_{o(min)} = -V_{SS} + |V_{GS7}| - |V_t| = -5 + 0.854 - 0.5 = -4.646 \text{ V}$$

Thus, the output voltage range is −4.646 V to 4.5 V.

2.4.3 CMOS Op-Amp TLC1078

The simplified schematic of the TLC1078 op-amp is shown in Fig. 2.21. Its structure is similar to that of op-amp MC14573 in Fig. 2.20. The internal structure can be divided into a differential CMOS stage, a gain stage, an output stage, and a biasing circuit. Some parameters of the TLC1078 op-amp are listed in Table 2.3.

TABLE 2.3 Parameters of op-amp TLC1078

Parameter	Minimum	Typical	Maximum	Units
DC supply voltages V_{DD}			18	V
Input offset voltage V_{OS}		180	600	μV
Thermal drift D_v		1		μV/°C
Input biasing current I_B		0.7		pA
Input offset current I_{OS}		0.1		pA
Differential input resistance R_{id}		1.5		TΩ
Common-mode input resistance R_{ic}		1.5		TΩ
Output resistance R_o		50		Ω
Input capacitance C_i		1.0		pF
Open-loop voltage gain A_o	110	120		dB
Common-mode rejection ratio (CMRR)	75	97		dB
Unity-gain bandwidth f_u		110		MHz
Slew rate (SR)		47		V/ms
Power supply rejection ratio (PSRR)	75	97		dB

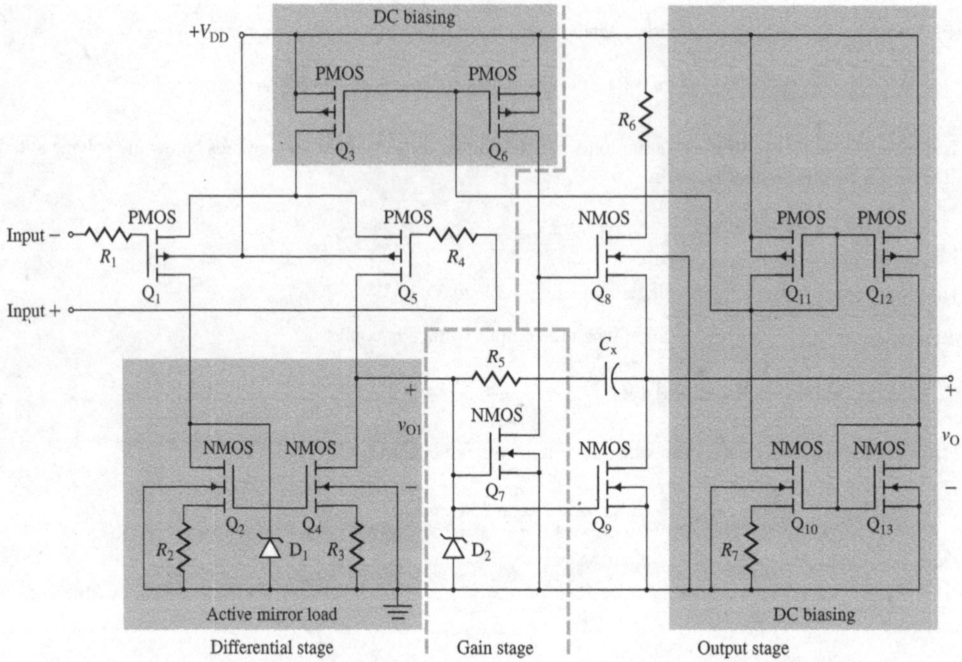

FIGURE 2.21 Schematic for op-amp TLC1078 (Reprinted by permission of Texas Instruments)

Differential Stage

The input stage is a PMOS differential amplifier consisting of transistors Q_1 and Q_5. Its load is a current mirror consisting of NMOS transistors Q_2 and Q_4, which have sources R_2 and R_3 to give high output resistances.

Gain Stage

The gain stage uses NMOS Q_7 in a common-source configuration with PMOS Q_6 as the current source active load. The combination of C_x and R_5 is a pole-zero circuit, which can reduce the lowest pole or break frequency to a low value to ensure stability. Also, it can produce a zero to cancel out one of the second poles of the amplifier's open-loop frequency response.

Output Stage

The output stage consists of a pair of NMOS transistors that operate in the class AB push-pull mode. Transistor Q_8 is the source follower, sourcing current to the load during the interval when the signal at the output of Q_7 goes up above the quiescent value. Transistor Q_9 acts as the common-source amplifier, sinking current from the load during the interval when the signal at the output of the differential stage goes down below the quiescent value.

A decreasing signal at the output of the differential stage will be amplified by Q_7 with a phase shift of 180° and then will appear through the source follower of Q_8 as an increasing signal to the load. However, an increasing signal will be amplified by the common-source amplifier of Q_9 with a phase shift of 180° and then will appear as a decreasing signal to the load. Thus, the voltages applied to the gates of Q_8 and Q_9 will be phase shifted by 180°. The voltage gain through the two paths will be the same, however.

DC Biasing

The differential amplifier is biased by transistors Q_3 and Q_6 as a current source whose gate voltage is set by the voltage divider circuit consisting of transistors Q_{10} through Q_{13}. All transistors are identical, except for Q_{10} and Q_{13}, whose channel widths differ from that of the others.

EXAMPLE 2.5

D **Analyzing the CMOS op-amp TLC1078** The CMOS amplifier in Fig. 2.21 is operated at a biasing current of $I_Q = 40\ \mu A$. The parameters of the MOSFETs are $K_x = 10\ \mu A/V^2$, $|V_{M(NMOS)}| = V_{M(PMOS)} = 70$ V, $V_t = 0.5$ V, and $W/L = 160\ \mu m/10\ \mu m$, except for Q_{10}, for which $W/L = 40\ \mu m/10\ \mu m$. Find the value of the resistance R_7. Assume $V_{DD} = V_{SS} = 5$ V.

SOLUTION

We have

$$K_p = \frac{K_x W}{L} = 10\ \mu \times \frac{160}{10} = 160\ \mu A/V^2 \qquad \text{(for all MOSFETs expect } Q_{10})$$

$$K_{p10} = \frac{K_x W}{L} = 10\ \mu \times \frac{40}{10} = 40\ \mu A/V^2 \qquad \text{(for } Q_{10})$$

For $I_Q = 40\ \mu A$ and $K_p = 160\ \mu A/V^2$, Eq. (2.69) gives

$$V_{GS} = V_t + \sqrt{I_D/K_p} = 0.5 + \sqrt{40/160} = 1 \text{ V}$$

Thus,

$$V_{GS3} = V_{GS6} = V_{GS11} = V_{GS12} = V_{GS13} = 1 \text{ V}$$

Also,

$$I_{D3} = I_{D6} = I_{D11} = I_{D12} = I_{D10} = I_{D13} = I_Q$$

The drain current I_{D13} of Q_{13} is given by

$$I_{D13} = K_{p13}(V_{GS13} - V_t)^2 \tag{2.74}$$

Since $V_{GS10} = V_{GS13} - I_Q R_7$, the drain current I_{D10} of Q_{10} is given by

$$I_{D10} = K_{p10}(V_{GS10} - V_t)^2 = K_{p10}(V_{GS13} - I_Q R_7 - V_t)^2 \tag{2.75}$$

The V_{GS} values of Q_{10} and Q_{13} are different, but their drain currents are equal; that is,

$$I_{D13} = I_{D10} \quad \text{and} \quad K_{p13}(V_{GS13} - V_t)^2 = K_{p10}(V_{GS13} - I_Q R_7 - V_t)^2$$

which relates R_7 to I_Q as follows:

$$R_7 = \frac{V_{GS13} - V_t}{I_Q}\left[\left(\frac{K_{p13}}{K_{p10}}\right)^{1/2} - 1\right] \tag{2.76}$$

Since K_p is proportional to the ratio W/L, Eq. (2.76) can be expressed as

$$R_7 = \frac{V_{GS13} - V_t}{I_Q}\left[\left(\frac{W_{13}}{W_{10}}\right)^{1/2} - 1\right] \tag{2.77}$$

which, for $I_Q = 40\ \mu A$, $W_{10} = 40\ \mu m$, $W_{13} = 160\ \mu m$, $V_{GS13} = 1$ V, and $V_t = 0.5$ V, gives $R_7 = 12.5\ k\Omega$. Thus, the ratio $\sqrt{W_{13}/W_{10}}$ sets the value of resistance R_7 or I_Q.

KEY POINTS OF SECTION 2.4

- CMOS op-amps can have either two or three stages. The gain and output stages are often combined into a single stage. This type of amplifier has a very high input resistance, and the voltage gain is generally less than that of other comparable types of amplifiers.
- Since MOSFETs have lower transconductance than other amplifiers, the gain stage often requires zero-pole compensation (see Fig. 2.21).

2.5 BJT Op-Amps

Like MOSFET op-amps, BJT amplifiers have three stages: a differential stage, a gain stage, and an output stage. They have the disadvantages of a lower input resistance and a higher input biasing current. However, BJTs provide higher voltage gain [6–8]. In this section, we consider two popular op-amps.

2.5.1 BJT Op-Amp LM124

The LM124 op-amp is designed to operate from a power supply of ±16 V, but it can operate from a single power supply as low as 5 V. The simplified schematic is shown in Fig. 2.22. Some parameters of the LM124 op-amp are listed in Table 2.4.

Differential Stage

The differential input stage consists of bipolar transistors Q_1 through Q_4. They are connected in Darlington pairs to give high input resistances and low input biasing currents. These BJTs drive the current mirror load consisting of transistors Q_8 and Q_9 for a high differential gain.

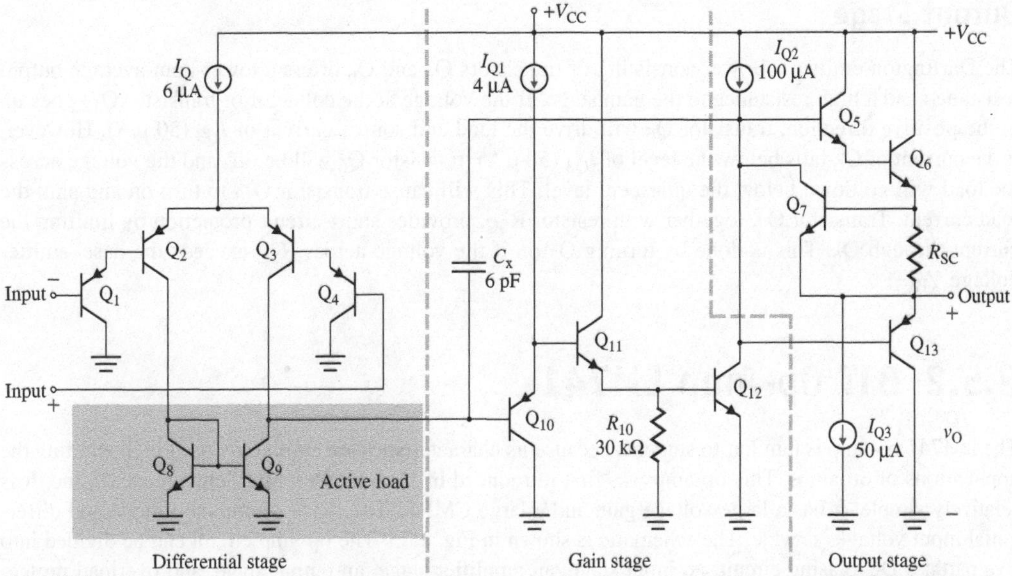

FIGURE 2.22 Simplified schematic for op-amp LM124 (Courtesy of National Semiconductor, Inc.)

Gain Stage

The gain stage is a common-emitter amplifier consisting of *pnp* transistor Q_{10} (with an active load), which is followed by a Darlington pair consisting of transistors Q_{11} and Q_{12}. The beta of a *pnp* transistor is generally low. The Darlington pair offers a high load to transistor Q_{10}. This arrangement ensures a high voltage gain (100 dB). The number of poles in the op-amp increases with the number of transistor stages. C_x is used for feedback compensation.

TABLE 2.4 Parameters of op-amp LM124

Parameter	Minimum	Typical	Maximum	Units
DC supply voltages V_S			±16	V
Input offset voltage V_{OS}		1.0	2.0	mV
Thermal drift D_v		100	200	µV/°C
Input biasing current I_B		20	50	nA
Input offset current I_{OS}		2.0	10	nA
Differential input resistance R_{id}		2		MΩ
Common-mode input resistance R_{ic}		2		MΩ
Output resistance R_o		75		Ω
Input capacitance C_i		4.0		pF
Open-loop voltage gain A_o	50	100		V/mV
Common-mode rejection ratio (CMRR)	70	85		dB
Unity-gain bandwidth f_u		1.0		MHz
Slew rate (SR)		3.0		V/µs
Power supply rejection ratio (PSRR)	650	100		dB

Output Stage

The Darlington emitter follower consisting of transistors Q_5 and Q_6 offers a lower-than-average output resistance and a high resistance to the gain stage. If the voltage at the collector of transistor Q_{12} goes up in the positive direction, transistor Q_6 will drive the load and source current of I_{Q3} (50 μA). However, if the current of Q_6 falls below the level of I_{Q3} (50 μA), transistor Q_6 will be off, and the voltage across the load will go down below the quiescent level. This will cause transistor Q_{13} to turn on and sink the load current. Transistor Q_7, together with resistor R_{SC}, provides short-circuit protection by limiting the current through Q_6. This is done by turning Q_7 on if the voltage across R_{SC} exceeds the base–emitter voltage V_{BE7}.

2.5.2 BJT Op-Amp LM741

The LM741 op-amp is familiar to students because its characteristics are commonly used in illustrating the applications of op-amps. This op-amp was first introduced in 1966 by Fairchild Semiconductor, Inc. It is relatively simple. It has a large voltage gain and a large CMRR. The range of common-mode and differential input voltages is wide. The schematic is shown in Fig. 2.23. The op-amp circuit can be divided into five parts: a DC-biasing circuit, an input stage, an amplifier stage, an output stage, and overload protection. Some parameters of the LM741 op-amp are listed in Table 2.5.

Differential Stage

The differential input stage consists of bipolar transistors Q_1 through Q_4. They form common-emitter and common-base configurations (see Fig. 1.42) for higher bandwidth and gain. This stage is biased by the current source consisting of Q_8, and it drives the current mirror active load consisting of Q_5, Q_6, and Q_7. The biasing feedback loop formed by Q_8 and Q_9 stabilizes the biasing currents in each of the input transistors at approximately one-half of the collector current of Q_{10}.

TABLE 2.5 Parameters of op-amp LM741

Parameter	Minimum	Typical	Maximum	Units
DC supply voltages V_{CC}			±22	V
Input offset voltage V_{OS}		1.0	5.0	mV
Thermal drift D_v		15		μV/°C
Input biasing current I_B		80	550	nA
Input offset current I_{OS}		20	200	nA
Differential input resistance R_{id}	0.3	2		MΩ
Common-mode input resistance R_{ic}		2		MΩ
Output resistance R_o		75		Ω
Input capacitance C_i		4.0		pF
Open-loop voltage gain A_o	50	200		V/mV
Common-mode rejection ratio (CMRR)	70	90		dB
Unity-gain bandwidth f_u		1.0		MHz
Slew rate (SR)		0.5		V/μs
Power supply rejection ratio (PSRR)	70	90		dB

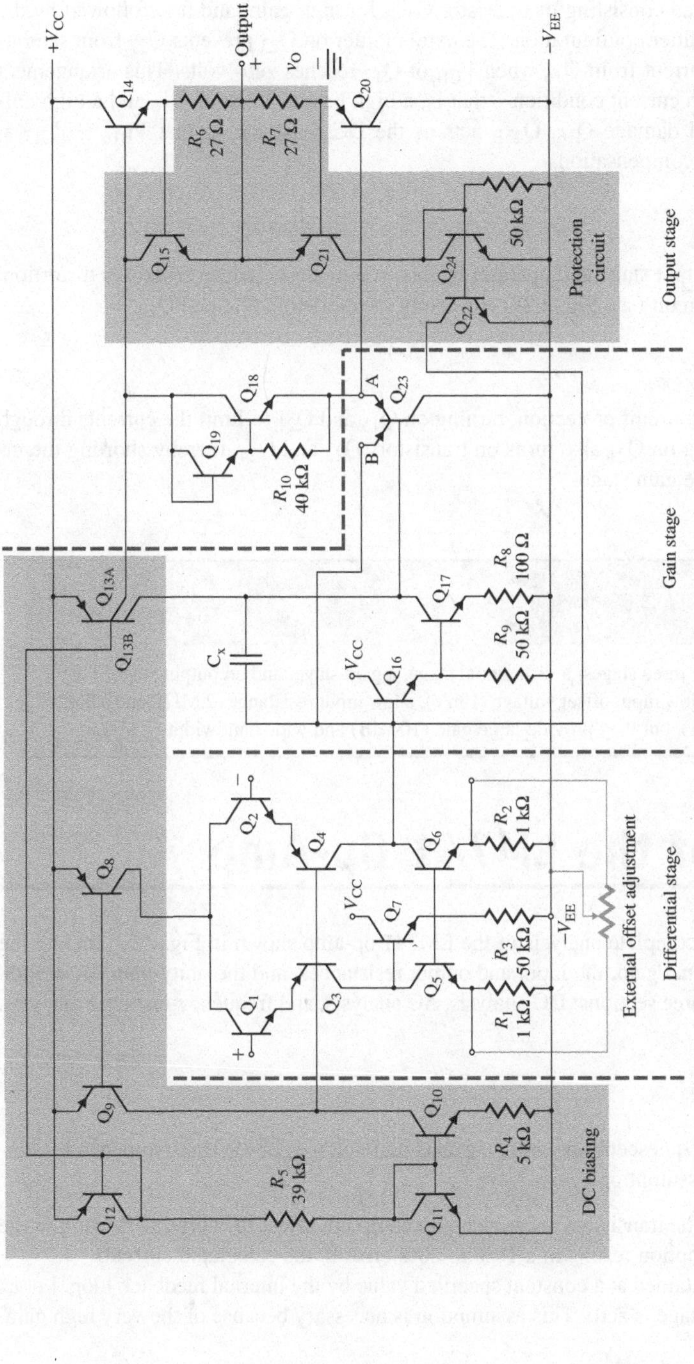

FIGURE 2.23 Schematic for op-amp LM741 (Courtesy of National Semiconductor, Inc.)

Gain Stage

The gain stage consists of a common-emitter Darlington pair amplifier made up of transistors Q_{16} and Q_{17}. Transistor Q_{17} drives an active load consisting of transistor Q_{13B} for high gain, and it is followed by the *pnp* transistor Q_{23} in common-emitter configuration. The extra emitter on Q_{23} prevents Q_{17} from saturating by diverting the base drive current from Q_{16} when V_{CB} of Q_{17} reaches zero volts. This arrangement eliminates the possibility of a high current condition—that is, a high base voltage of Q_{16} and a high collector voltage of Q_7—that could damage Q_{16}. Q_{23B} acts as the DC feedback so that $V_{B16} - V_{C17} = V_{EB23B}$. C_x is used for frequency compensation.

Output Stage

The output stage is a push-pull output stage and operates in class AB mode to reduce crossover distortion. It is biased by a V_{BE} multiplier circuit (see Fig. 2.24) consisting of transistors Q_{18} and Q_{19}.

Protection Circuitry

Resistors R_6 and R_7 provide short-circuit protection, turning on Q_{15} and Q_{21} to limit the currents through Q_{14} and Q_{20}, respectively. Turning on Q_{21} also turns on transistors Q_{22} and Q_{24}, thereby shorting the input signal to the base of Q_{16} in the gain stage.

KEY POINT OF SECTION 2.5

- BJT amplifiers usually have three stages: a differential stage, a gain stage, and an output stage. They have the disadvantages of a low input offset voltage (1 mV), a low input resistance (2 MΩ), and a high input biasing current (60 nA), but they provide large gain (100 dB) and wide bandwidth (1 MHz).

2.6 Analysis of the LM741 Op-Amp

As an example, let us carry out a complete analysis of the LM741 op-amp shown in Fig. 2.23, finding the DC biasing currents, the small-signal gain, the input and output resistances, and the unity-gain bandwidth. The analysis can be divided into three sections: DC analysis, AC analysis, and frequency-response analysis.

2.6.1 DC Analysis

The DC analysis to determine the quiescent operating currents and voltages of the transistors can be simplified by making the following assumptions:

1. The output resistances of the transistors are very high and do not affect the currents flowing in the circuit. Usually, this assumption results in a 10% to 20% error in the calculated currents.
2. The output voltage is maintained at a constant specified value by the internal feedback loop. Let us assume that the output voltage is zero. This assumption is necessary because of the very high gain,

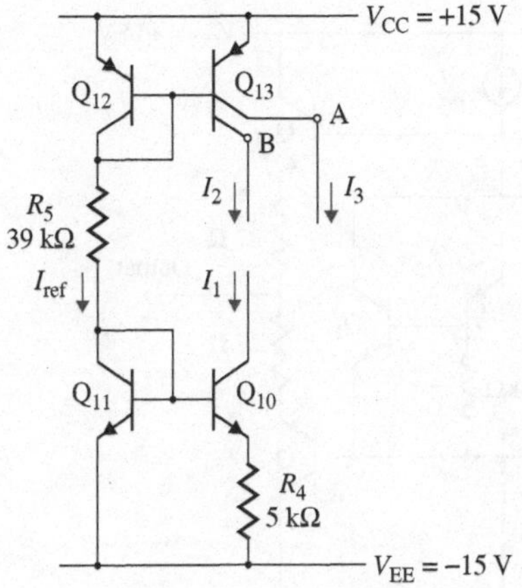

FIGURE 2.24 Biasing circuit for the LM741 op-amp

typically 10^5, which results in a very low input voltage. If the output voltage is calculated with two input terminals grounded, any change in the beta or output resistances could cause a large change in the output voltage, and the transistors could be operating in the saturation region rather than in the active region as expected.

3. The *npn* transistors have large betas; let us assume $\beta_{F(npn)} = 250$.
4. The betas of *pnp* transistors are much lower than those of *npn* transistors; let us assume $\beta_{F(pnp)} = 50$. Assume $V_T = 26$ mV for all transistors.

Biasing Circuit

The currents in the biasing current sources of Q_{10} and Q_{13AB} can be calculated from Fig. 2.24. Let us assume that all transistors are operating in the forward-active region, the base currents are negligible, V_T = 26 mV, and V_{BE} = 0.7 V. The reference current can be calculated as follows:

$$I_{ref} = \frac{V_{CC} - V_{EE} - V_{BE11} - V_{EB12}}{R_5} \tag{2.78}$$

$$= \frac{(15 + 15 - 2 \times 0.7)\text{ V}}{39\text{ k}\Omega} = 0.73\text{ mA}$$

The transistors Q_{10} and Q_{11} form a Widlar current source. Using Eq. (1.98), we can find the output current I_1 from

$$V_T \ln\left(\frac{I_{ref}}{I_1}\right) = R_4 I_1 \tag{2.79}$$

$$V_T \ln\left(\frac{0.73\text{ mA}}{I_1}\right) = 5\text{ k}\Omega \times I_1$$

which, by trial-and-error solution, gives I_1 = 19 μA.

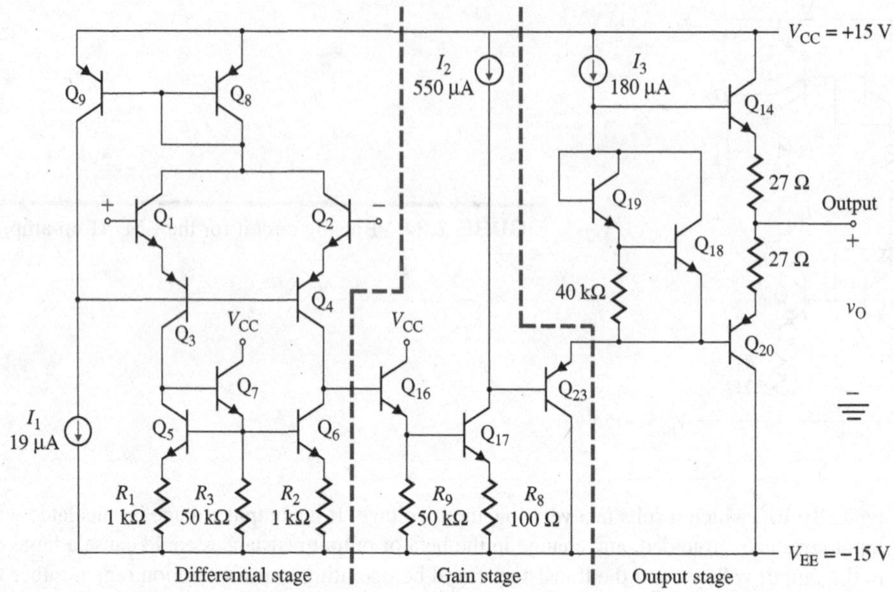

FIGURE 2.25 The LM741 op-amp with biasing current sources

Transistor Q_{13} is a multicollector lateral *pnp* device. Thus, currents I_2 and I_3 are three-fourths and one-fourth of the reference current I_{ref}, respectively:

$$I_2 = \left(\frac{3}{4}\right) \times 0.73 \text{ mA} = 0.55 \text{ mA}$$

$$I_3 = \left(\frac{1}{4}\right) \times 0.73 \text{ mA} = 0.18 \text{ mA}$$

If we replace the biasing circuit in Fig. 2.24 by the equivalent current sources of I_1, I_2, and I_3, the circuit in Fig. 2.23 can be simplified to that in Fig. 2.25.

Input Stage

The input stage provides a high input resistance for differential and common-mode input signals, the differential-to-single output, and some voltage gain. The input stage of the LM741 with biasing current sources is shown in Fig. 2.26. Since the *npn* transistors have large values of beta, the base currents are negligible compared to the collector currents. For identical transistors, $I_{C9} = I_{C8}$.

The current I_T, which is the sum of I_{C8} plus the base currents of Q_8 and Q_9, can be found from

$$I_T = I_{C8} + I_{B8} + I_{B9} = I_{C9}\left(1 + \frac{2}{\beta_{F(pnp)}}\right) \tag{2.80}$$

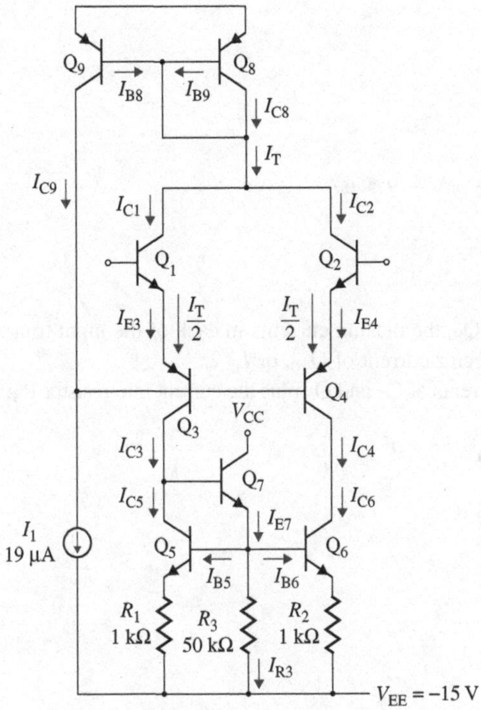

FIGURE 2.26 Input stage of the LM741 op-amp

where $\beta_{F(pnp)}$ is the current gain for *pnp* transistors. Neglecting the base currents of Q_1 and Q_2, we find that the emitter currents of Q_3 and Q_4 are

$$I_{E3} = I_{E4} = \frac{I_T}{2} = \frac{I_{C9}}{2}\left(1 + \frac{2}{\beta_{F(pnp)}}\right) \tag{2.81}$$

The sum of the base currents of Q_3 and Q_4 and the collector current of Q_9 must be equal to the biasing current of $I_1 = 19\ \mu A$. Thus,

$$I_1 = I_{C9} + \frac{I_{E3}}{1 + \beta_{F(pnp)}} + \frac{I_{E4}}{1 + \beta_{F(pnp)}} = I_{C9}\left[1 + \frac{1 + 2/\beta_{F(pnp)}}{1 + \beta_{F(pnp)}}\right] \tag{2.82}$$

which can be simplified to $I_1 = I_{C9}[(2 + \beta_{F(pnp)})/(1 + \beta_{F(pnp)})] \simeq I_{C9}$. Substituting I_{C9} from Eq. (2.82) into Eq. (2.80) yields

$$I_T \simeq I_1\left(1 + \frac{2}{\beta_{F(pnp)}}\right) \tag{2.83}$$

$$\simeq (19\ \mu A)\left(1 + \frac{2}{50}\right) = 19.38\ \mu A$$

Then

$$I_{C1} = I_{C2} = \frac{19.38\ \mu\text{A}}{2} = 9.68\ \mu\text{A}$$

$$I_{C3} = I_{C4} = \frac{\beta_{F(pnp)}}{1 + \beta_{F(pnp)}} I_{E3} = \frac{50}{1 + 50} \times 9.69\ \mu\text{A} = 9.5\ \mu\text{A}$$

$$I_{C5} = I_{C6} \approx I_{C3} = 9.5\ \mu\text{A}$$

Because of the biasing feedback loop formed by Q_8 and Q_9, the biasing currents in each of the input transistors Q_1 and Q_2 are approximately one-half of the collector current of Q_{10}, or $I_1/2$.

The emitter current I_{E7} of Q_7 is the sum of the base currents of Q_5 and Q_6 plus the current into resistor R_3:

$$I_{E7} = I_{B5} + I_{B6} + I_{R3} \approx I_{R3}$$

Assuming $I_{E5} \approx I_{C5}$ and $I_{E6} \approx I_{C6}$, the voltage across R_3 is

$$V_{R3} = V_{BE5} + R_1 I_{E5} = V_{BE6} + R_2 I_{E6}$$

$$= V_T \ln\left(\frac{I_{C5}}{I_S}\right) + R_1 I_{E5} \tag{2.84}$$

$$= 26\ \text{mV} \times \ln\left(\frac{9.5\ \mu\text{A}}{10^{-14}\ \text{A}}\right) + 1\ \text{k}\Omega \times 9.5\ \mu\text{A} = 537.5\ \text{mV} + 9.5\ \text{mV} = 547\ \text{mV}$$

The collector current of Q_7 is

$$I_{C7} \approx I_{E7} = \frac{V_{R3}}{R_3} = \frac{547\ \text{mV}}{50\ \text{k}\Omega} = 10.9\ \mu\text{A}$$

Gain Stage

The gain stage provides a high voltage gain with a very high input and output resistance. The amplifier stage is shown in Fig. 2.27. Since it is assumed that the output voltage of the amplifier is zero, the base current of Q_{23} is zero and the collector current of Q_{17} is $I_{C17} = I_2 = 550\ \mu\text{A}$. The voltage at the base of Q_{17}

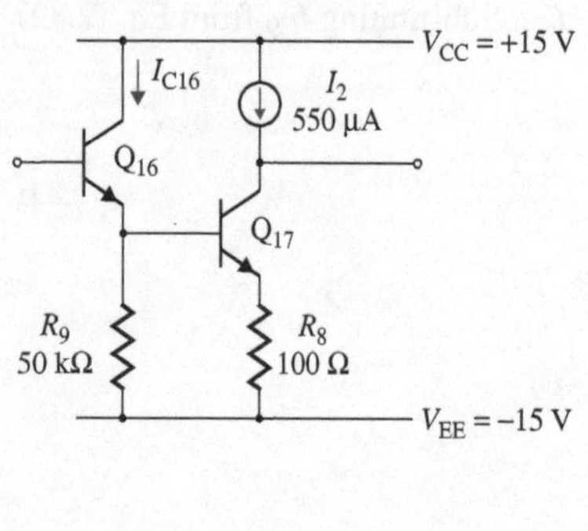

FIGURE 2.27 Gain stage

is equal to the base–emitter voltage of Q_{17} plus the voltage drop across resistor R_8. Assuming $I_{E17} = I_{C17}$ and $I_S = 10^{-14}$ A, the voltage at the base of Q_{17} with respect to $-V_{EE}$ is

$$V_{B17} = V_{BE17} + R_8 I_{E17}$$

$$= V_T \ln\left(\frac{I_{C17}}{I_S}\right) + R_8 I_{E17} \tag{2.85}$$

$$= V_T \ln\left(\frac{550\ \mu A}{10^{-14}\ A}\right) + 100\ \Omega \times 550\ \mu A = 643\ mV + 55\ mV = 698\ mV$$

The current through R_9 is

$$I_{R9} = \frac{V_{B17}}{R_9} = \frac{698\ mV}{50\ k\Omega} = 13.96\ \mu A$$

The base current of Q_{B17} is

$$I_{B17} = \frac{I_{C17}}{\beta_{F(npn)}} = \frac{550\ \mu A}{250} = 2.2\ \mu A$$

The collector current of Q_{16} is

$$I_{C16} \approx I_{E16} = I_{B17} + I_{R9} = 2.2\ \mu A + 13.94\ \mu A = 16.14\ \mu A$$

Output Stage

The output stage supplies a high load current and offers a low output resistance. This stage is shown in Fig. 2.28. Assuming that the base currents are negligible, the collector current of Q_{23} is $I_{C23} = I_3 = 180\ \mu A$. Further assuming that the circuit is connected with feedback in such a way that the output

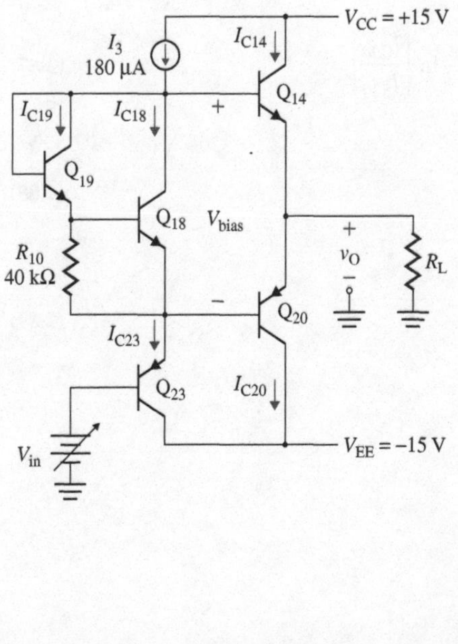

FIGURE 2.28 Output stage

voltage is driven to zero and the output current is also zero, I_{C14} and I_{C20} are approximately equal in magnitude. Thus,

$$I_{C14} = I_{C20} \tag{2.86}$$

Let us estimate that $V_{BE18} = 0.6$ V and the collector current of Q_{19} is given approximately by

$$I_{C19} \approx \frac{V_{BE18}}{R_{10}} = \frac{0.6}{40\ \text{k}\Omega} = 15\ \mu\text{A}$$

Then

$$I_{C18} = I_{C23} - I_{C19} = 180\ \mu\text{A} - 15\ \mu\text{A} = 165\ \mu\text{A}$$

Now we can calculate a more accurate value for V_{BE18}:

$$V_{BE18} = V_T \ln\left(\frac{I_{C18}}{I_S}\right) = (26\ \text{mV}) \ln\left(\frac{165\ \mu\text{A}}{10^{-14}\ \text{A}}\right) = 611.7\ \text{mV}$$

The collector current of Q_{19} is

$$I_{C19} = \frac{I_{C18}}{\beta_{F(npn)}} + \frac{V_{BE18}}{R_{10}} = \frac{165\ \mu\text{A}}{250} + \frac{611.7\ \text{mV}}{40\ \text{k}\Omega} = 16\ \mu\text{A}$$

and

$$I_{C18} = I_3 - I_{C19} = 180\ \mu\text{V} - 16\ \mu\text{V} = 164\ \mu\text{A}$$

which is very close to the original estimate of 165 μA. If there had been a significant difference between this value and the original estimate, we would have used this value to find new values of V_{BE18}, I_{C19}, and I_{C18}, continuing the iterations until the desired value was found.

Using KVL around the loop formed by Q_{14}, Q_{20}, Q_{18}, and Q_{19} in Fig. 2.28, we get

$$V_{BE18} + V_{BE19} = V_{BE14} + |V_{BE20}|$$

which can be written as

$$V_T \ln\left(\frac{I_{C18}}{I_{S18}}\right) + V_T \ln\left(\frac{I_{C19}}{I_{S19}}\right) = V_T \ln\left(\frac{I_{C14}}{I_{S14}}\right) + V_T \ln\left|\frac{I_{C20}}{I_{S20}}\right| \tag{2.87}$$

For an output voltage of $v_O = 0$ and $\beta_{F(npn)} >> 1$,

$$|I_{C14}| = |I_{C20}| \tag{2.88}$$

and Eq. (2.87) can be simplified to

$$\frac{I_{C18} I_{C19}}{I_{S18} I_{S19}} = \frac{I_{C14}^2}{I_{S14} I_{S20}} \tag{2.89}$$

from which we get

$$I_{C14} = I_{C20} = \sqrt{I_{C18} I_{C19}} \sqrt{\frac{I_{S14} I_{S20}}{I_{S18} I_{S19}}} \tag{2.90}$$

Thus, the collector currents depend on the I_S values, which depend on the physical geometry of the transistors. Q_{14} and Q_{20} are normally designed to carry much larger current than other transistors. The specific geometries used by different manufacturers may be different, but the I_S value of Q_{14} and Q_{20} is typically three times that of Q_{18} and Q_{19}. Thus,

$$I_{S14} = I_{S20} \approx 3I_{S18} = 3I_{S19} \tag{2.91}$$

Substituting I_{S14} from Eq. (2.91) into Eq. (2.90) yields

$$\begin{aligned} I_{C14} = I_{C20} &= 3\sqrt{I_{C18}I_{C19}} \\ &= 3 \times \sqrt{(164\ \mu\text{A})(16\ \mu\text{A})} = 153.7\ \mu\text{A} \end{aligned} \tag{2.92}$$

Overload Protection

Transistor Q_{15} (in Fig. 2.23) turns on only when the voltage across R_6 exceeds 550 mV at an output-sourcing current of 550 mV/R_6 = 550 mV/27 = 20 mA. When Q_{15} turns on, it limits the current to the base of Q_{14} and the output current cannot increase further. Thus, Q_{15} provides short-circuit protection, preventing damage to the op-amp due to excess current flow and power dissipation. These problems can occur if the output is shorted to a negative power supply. Similarly, transistors Q_{21}, Q_{22}, and Q_{24} protect transistor Q_{20} in the case of sinking current. When Q_{21} turns on, it protects Q_{20} by limiting the current to the base of Q_{20}, thereby turning on Q_{22} and Q_{24}.

If the inverting terminal were overdriven such that its voltage became more positive than that of the noninverting terminal, Q_1 would turn off. As a result, Q_6 would also be off, and the current into the base of Q_{16} would be I_T = 19.4 μA. This current would be amplified by the beta of Q_{16}, which could be as high as 1000, giving a collector current of $I_{C16} = 1000 \times 19.4\ \mu\text{A} \approx 19.4$ mA that would flow into the base of Q_{17}. Q_{17} would thus become saturated. Saturation would result in a power dissipation in Q_{16} of

$$I_{C16}(V_{CC} + V_{EE}) = 19.4\ \text{mA} \times (15 + 15)\ \text{V} \approx 580\ \text{mW}$$

The extra emitter on Q_{23} prevents Q_{17} from developing a high current condition that could damage Q_{16}.

2.6.2 Small-Signal AC Analysis

Small-signal analysis is performed to determine the input resistance, the output resistance, the transconductance, and the voltage gain. The op-amp circuit may be broken up into three stages: the input stage, the gain stage, and the output stage. Since the op-amp may be considered as three cascaded stages, we will represent the input and gain stages by their transconductance equivalents to simplify the analysis to determine the overall gain.

Notice from Eq. (2.41) that the CMRR is the change in the common-mode voltage per unit change in the input offset voltage ΔV_{OS}, which can be determined from Eq. (2.40) if A_d is known. Thus, we need to determine only A_d. The small-signal parameters of the transistors are

$$r_{\pi 16} = \frac{\beta_{F16}V_T}{I_{C16}} = \frac{250 \times 26\ \text{mV}}{16.1\ \mu\text{A}} = 403.7\ \text{k}\Omega$$

$$r_{\pi 17} = \frac{\beta_{F17} V_T}{I_{C17}} = \frac{250 \times 26\text{ mV}}{550\ \mu\text{A}} = 11.82\text{ k}\Omega$$

$$r_{\pi 23} = \frac{\beta_{F23} V_T}{I_{C23}} = \frac{50 \times 26\text{ mV}}{180\ \mu\text{A}} = 7.2\text{ k}\Omega$$

$$g_{m3} = g_{m4} = g_{m5} = g_{m6} = \frac{I_{C3}}{V_T} = \frac{I_{C4}}{V_T} = \frac{I_{C6}}{V_T} = \frac{9.5\ \mu\text{A}}{26\text{ mV}} = 365.4\ \mu\text{A/V}$$

$$g_{m13A} = g_{m23} = \frac{I_{C13A}}{V_T} = \frac{180\ \mu\text{A}}{26\text{ mV}} = 6.92\text{ mA/V}$$

$$g_{m13B} = g_{m17} = \frac{I_{C17}}{V_T} = \frac{550\ \mu\text{A}}{26\text{ mV}} = 21.15\text{ mA/V}$$

The output resistance r_o is given by

$$r_o = \frac{V_A}{I_C} = \frac{V_A}{V_T g_m} = \frac{1}{\eta_n g_m} \tag{2.93}$$

where $\eta_n = V_T/V_A$ is a transistor constant. Assuming $\eta_n = 5 \times 10^{-4}$ for *pnp* transistors, Eq. (2.93) gives

$$r_{o4} = \frac{1}{\eta_n g_{m4}} = \frac{1}{5 \times 10^{-4} \times 365.4\ \mu\text{A/V}} = 5.47\text{ M}\Omega$$

$$r_{o13A} = r_{o23} = \frac{1}{\eta_n g_{m13A}} = \frac{1}{5 \times 10^{-4} \times 6.92\text{ mA/V}} = 289\text{ k}\Omega$$

$$r_{o13B} = \frac{1}{\eta_n g_{m13B}} = \frac{1}{5 \times 10^{-4} \times 21.15\text{ mA/V}} = 94.56\text{ k}\Omega$$

Assuming $\eta_n = 2 \times 10^{-4}$ for *npn* transistors, Eq. (2.93) gives

$$r_{o6} = \frac{1}{\eta_n g_{m6}} = \frac{1}{2 \times 10^{-4} \times 365.4\ \mu\text{A/V}} = 13.68\text{ M}\Omega$$

$$r_{o17} = \frac{1}{\eta_n g_{m17}} = \frac{1}{2 \times 10^{-4} \times 21.15\text{ mA/V}} = 236.4\text{ k}\Omega$$

Input Stage

Let us assume that $v_{ic} = 0$ and only $v_{id}/2$ is applied. The AC equivalent circuit for the differential-mode signal is shown in Fig. 2.29(a). The input voltage to transistor Q_1 is $+v_{id}/2$ and that to transistor Q_2 is $-v_{id}/2$. Assuming that the transistors are identical and the circuit is balanced, an increase in voltage at the base junction of Q_3 and Q_4 due to $+v_{id}/2$ will be compensated by an equal decrease in voltage due to $-v_{id}/2$. The voltage at the base terminals of *pnp* transistors Q_3 and Q_4 will not vary at all. As a result, the bases of Q_3 and Q_4 are effectively at ground potential.

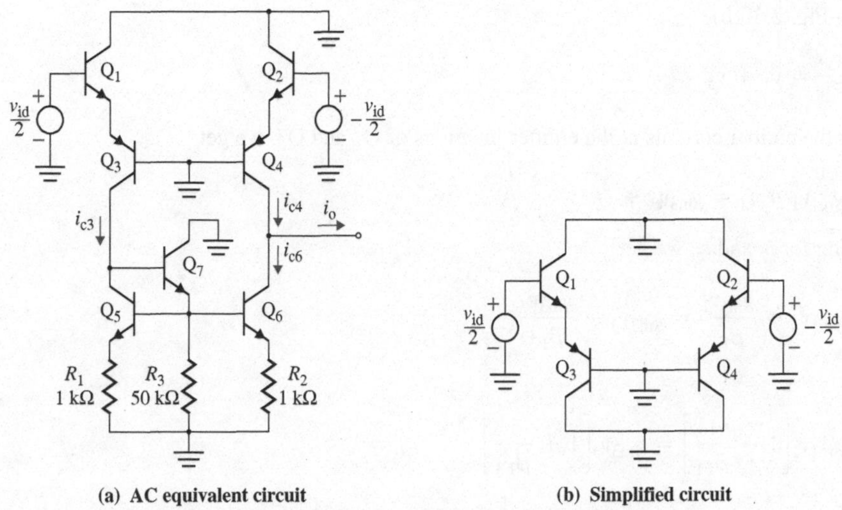

FIGURE 2.29 AC equivalent of input stage with differential input

The transconductance of the input stage can be found by shorting the output to the ground and calculating the resulting current. Since $i_{c3} = i_{c5}$ and i_{c6} is the current mirror of i_{c5}, the current in the active load circuit (i_{c6}) will be equal in magnitude to the collector current of Q_3 (i_{c3}); that is,

$$i_{c6} = i_{c3} \tag{2.94}$$

Thus, the output current under short-circuit conditions is

$$i_o = i_{c4} - i_{c6} = i_{c4} - i_{c3} \tag{2.95}$$

Under these conditions, Fig. 2.29(a) can be reduced to Fig. 2.29(b), which has two identical sides. The half-circuit AC equivalent is shown in Fig. 2.30(a), and its small-signal equivalent circuit is shown in Fig. 2.30(b), which can be simplified to Fig. 2.30(c).

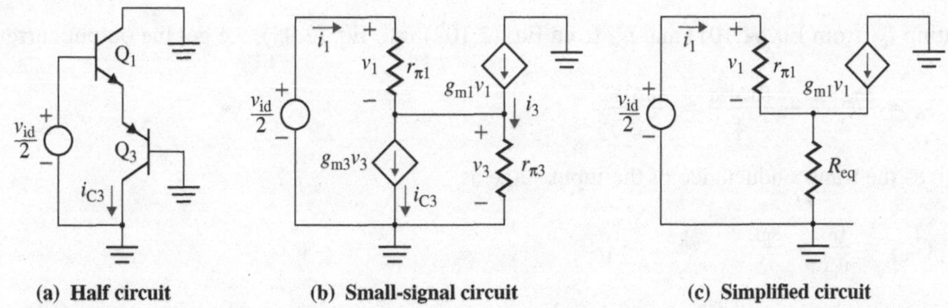

FIGURE 2.30 Half-circuit AC equivalent for differential input

From Fig. 2.30(b),

$$\frac{v_{id}}{2} = v_1 + v_3 \tag{2.96}$$

Summing the emitter currents at the emitter junctions of Q_1 and Q_3, we get

$$g_{m1}v_1 + i_1 = g_{m3}v_3 + i_3 \tag{2.97}$$

Substituting for i_1 and i_3, we get

$$g_{m1}v_1 + \frac{g_{m1}v_1}{\beta_{F1}} = g_{m3}v_3 + \frac{g_{m3}v_3}{\beta_{F3}}$$

or

$$g_{m1}v_1\left[1 + \frac{1}{\beta_{F1}}\right] = g_{m3}v_3\left[1 + \frac{1}{\beta_{F3}}\right] \tag{2.98}$$

where β_{F1} and β_{F3} are the small-signal current gains of transistors Q_1 and Q_3. Since $|I_{C1}| = |I_{C3}|$, $g_{m1} \approx g_{m3}$. Assuming that $\beta_{F1} >> 1$ and $\beta_{F3} >> 1$, Eq. (2.98) is reduced to

$$v_1 = v_3 \tag{2.99}$$

and Eq. (2.96) becomes

$$\frac{v_{id}}{2} = v_1 + v_3 = v_3 + v_3 = 2v_3$$

or

$$v_3 = \frac{v_{id}}{4} \tag{2.100}$$

Using Eq. (2.100), we can find the collector current of Q_3:

$$i_{c3} = g_{m3}v_3 = \frac{g_{m3}v_{id}}{4} \tag{2.101}$$

From the symmetry of the circuit in Fig. 2.29(a), we get

$$i_{c4} = -i_{c3} = \frac{g_{m3}v_{id}}{4} \tag{2.102}$$

Substituting i_{c3} from Eq. (2.101) and i_{c4} from Eq. (2.102) into Eq. (2.95), we get the output current as

$$i_o = \frac{g_{m3}v_{id}}{4} + \frac{g_{m3}v_{id}}{4} = \frac{g_{m3}v_{id}}{2} \tag{2.103}$$

which gives the transconductance of the input stage as

$$G_{m1} = \frac{i_o}{v_{id}} = \frac{g_{m3}}{2} \tag{2.104}$$

$$= \frac{I_{C3}}{2V_T} = \frac{9.5\ \mu A}{2 \times 26\ mV} = \frac{1}{5.47\ k\Omega} = 182.7\ \mu A/V$$

Q_3 can be replaced by the equivalent resistance R_{eq} seen looking from the emitter of Q_3, as shown in Fig. 2.30(c). R_{eq} can be found from

$$R_{eq} = \frac{v_3}{g_{m3}v_3 + v_3/r_{\pi 3}} = \frac{1}{g_{m3} + 1/r_{\pi 3}} = \frac{1}{g_{m3}(1 + 1/\beta_{F3})} \tag{2.105}$$

$$\approx \frac{1}{g_{m3}} = \frac{r_{\pi 3}}{\beta_{F3}} = \frac{r_{\pi 1}}{\beta_{F1}} \tag{2.106}$$

Using the relationship for the input resistance at the base of a transistor $R_x = r_\pi + (1 + \beta_F)\,R_{RE}$, we can relate v_{id} to i_1 as follows:

$$\frac{v_{id}/2}{i_1} = r_{\pi 1} + R_{eq}(1 + \beta_{F1}) \tag{2.107}$$

Substituting $R_{eq} = r_{\pi 1}/\beta_{F1}$ from Eq. (2.106) into Eq. (2.107) gives the input resistance as

$$R_{id} = \frac{v_{id}}{i_1} = 2\left[r_{\pi 1} + \frac{r_{\pi 1}}{\beta_{F1}}(1 + \beta_{F1})\right] \tag{2.108}$$

$$\approx 4r_{\pi 1}$$

$$= 4\frac{\beta_{F1}V_T}{I_{C1}} = 4 \times \frac{250 \times 26\text{ mV}}{9.69\ \mu\text{A}} = 2.68\text{ M}\Omega$$

Notice from Eq. (2.108) that the input resistance is four times the input resistance of the input transistors. Also note that when v_{id} changes, the output voltage changes and produces feedback to the input through the output resistance of Q_4. As a result, the input resistances seen from the two input terminals will not be exactly the same. This effect is neglected in the derivation of R_{id}.

The output resistance R_{o1} of the differential stage can be determined by setting the input voltage to zero and applying a test voltage v_x, as shown in Fig. 2.31(a). The analysis can be simplified by making the following assumptions:

1. Thevenin's equivalent resistance R_{th6} at the base of Q_6 is very small compared to r_π of Q_6, so the base of Q_6 is grounded. In reality, the voltage at this point is very small, and this point may be considered a virtual ground without greatly affecting the results. That is, $v_{B6} \approx 0$. Figure 2.31(a) can be viewed as shown in Fig. 2.31(b).
2. The output resistance of Q_2 is large and does not affect the results.

Figure 2.31(b) can be further simplified to the half circuit shown in Fig. 2.31(c). If Q_2 is replaced by $1/g_{m2}$, Fig. 2.31(c) can be represented by the equivalent circuit in Fig. 2.32(a). If R_{Q4} and R_{Q6} are the effective resistances seen from the collectors of Q_4 and Q_6, respectively, Fig. 2.32(a) can be represented by Fig. 2.32(b). Using Eq. (1.113), we get

$$R_{Q4} = r_{o4}\left[\frac{1 + g_{m4}(1/g_{m2})}{1 + g_{m4}/\beta_{F4}g_{m2}}\right] \tag{2.109}$$

$$\approx 2r_{o4} \quad (\text{for } g_{m4} = g_{m2} \text{ and } \beta_{F4} >> 1)$$

$$= 2 \times 5.47\text{ M}\Omega = 10.94\text{ M}\Omega$$

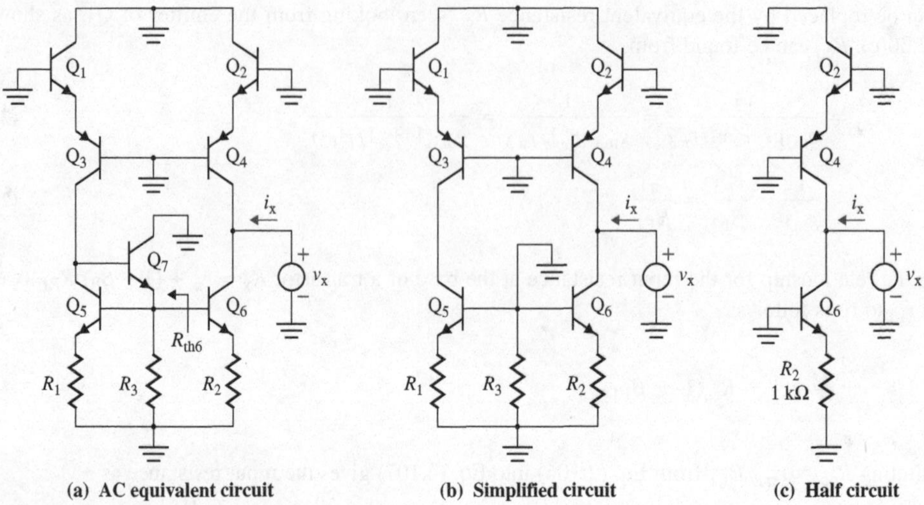

FIGURE 2.31 Test circuit for calculation of R_{o1}

$$R_{Q6} = r_{o6}\left[\frac{1 + g_{m6}R_2}{1 + g_{m6}R_2/\beta_{F6}}\right] \tag{2.110}$$

$$= r_{o6}\left[\frac{1 + 365.4\ \mu A/V \times 1\ k\Omega}{1 + 365.4\ \mu A/V \times 1\ k\Omega/250}\right] = 1.36r_{o6}$$

$$= 1.36 \times 13.68\ M\Omega = 18.65\ M\Omega$$

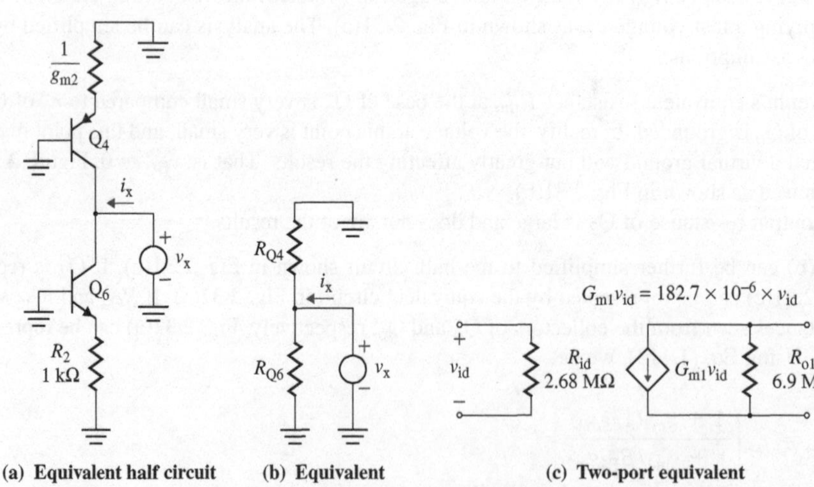

FIGURE 2.32 Two-port equivalent of the input stage

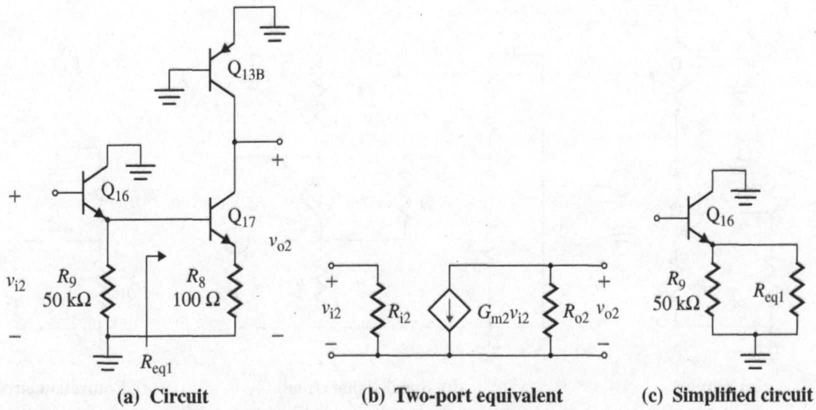

FIGURE 2.33 Small-signal AC equivalent circuit for the gain stage

The output resistance of the input stage is

$$R_{o1} = R_{Q4} \| R_{Q6} = 10.94 \text{ M}\Omega \| 18.65 \text{ M}\Omega = 6.9 \text{ M}\Omega$$

The two-port equivalent circuit of the input stage is shown in Fig. 2.32(c).

Gain Stage

The AC equivalent for the gain stage, shown in Fig. 2.33(a), can be represented by the equivalent circuit shown in Fig. 2.33(b). If R_{eq1} is Thevenin's equivalent resistance seen looking into the base of transistor Q_{17}, Fig. 2.33(a) can be reduced to the form shown in Fig. 2.33(c). From the relationship for the input resistance at the base of a transistor $R_x = r_\pi + (1 + \beta_F)R_{RE}$, we can find the input resistance of a common-emitter amplifier with a resistance in the emitter as follows:

$$R_{eq1} = r_{\pi 17} + (1 + \beta_{F17})R_8 \tag{2.111}$$

$$= 11.82 \text{ k}\Omega + (1 + 250) \times 100 \ \Omega = 36.9 \text{ k}\Omega$$

The input resistance of the gain stage is

$$R_{i2} = r_{\pi 16} + (1 + \beta_{F16})(R_{eq1} \| R_9) \tag{2.112}$$

$$= 403.7 \text{ k}\Omega + (1 + 250)(36.9 \text{ k}\Omega \| 50 \text{ k}\Omega) = 5.73 \text{ M}\Omega$$

Assuming that the voltage gain of the emitter follower Q_{16} is unity, the transconductance G_{m2} of this stage is that of the common-emitter amplifier of Q_{17} with resistance in the emitter; that is, the transconductance of the gain stage is

$$G_{m2} = \frac{g_{m17}}{1 + g_{m17}R_8} \tag{2.113}$$

$$= \frac{21.15 \text{ mA/V}}{1 + 21.15 \text{ mA/V} \times 100 \ \Omega} = \frac{1}{14 \ \Omega} = 6.79 \text{ mA/V}$$

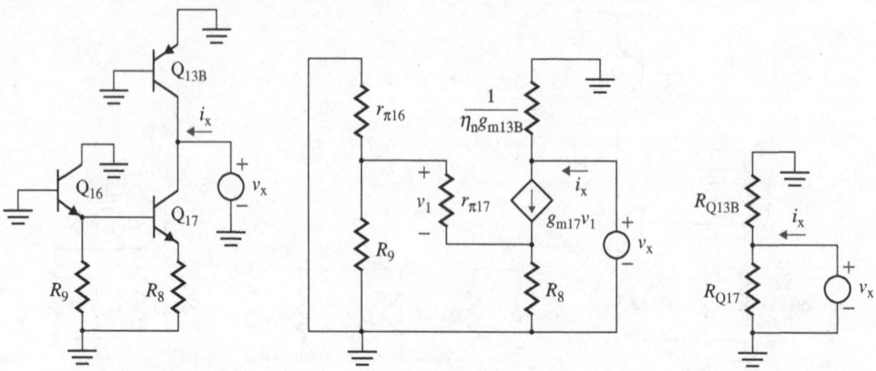

(a) Circuit (b) Small-signal circuit (c) Equivalent circuit

FIGURE 2.34 AC equivalent circuit for calculation of R_{o2}

The circuit for determining the output resistance is shown in Fig. 2.34(a). The small-signal AC equivalent is shown in Fig. 2.34(b). If R_{Q13B} and R_{Q17} are the effective resistances seen from the collectors of Q_{13B} and Q_{17}, respectively, Fig. 2.34(b) can be represented by the equivalent circuit in Fig. 2.34(c) and

$$R_{Q13B} = r_{o13B} = \frac{1}{\eta_n g_{m4}} = 94.56\ \text{k}\Omega$$

Using Eq. (1.113), we get

$$R_{Q17} \approx r_{o17}\left[\frac{1 + g_{m17}R_8}{1 + g_{m17}R_8/\beta_{F17}}\right] \tag{2.114}$$

$$\approx r_{o17}\left[\frac{1 + 21.15\ \text{mA/V} \times 100\ \Omega}{1 + 21.15\ \text{mA/V} \times 100\ \Omega/250}\right] = 3.09 r_{o17}$$

$$= 3.09 \times 236.4\ \text{k}\Omega = 730.5\ \text{k}\Omega$$

The output resistance of the gain stage is

$$R_{o2} = R_{Q13B} \| R_{Q17} = 94.56\ \text{k}\Omega \| 730.5\ \text{k}\Omega = 83.72\ \text{k}\Omega$$

Output Stage

Since V_{CE} of transistor Q_{18} is the sum of the V_{BE} voltages of Q_{18} and Q_{19}, transistors Q_{18} and Q_{19} can be replaced by two diodes, as shown in Fig. 2.35(a). The output could be either sourcing or sinking, depending on the output voltage and the load. As a result, the input and resistances of this stage greatly depend on the particular values of output voltage and current. We will make the following assumptions:

1. The output is sourcing—that is, the output current is flowing out of the output stage.
2. The load current is $i_L = 2$ mA at $R_L = 5$ kΩ.
3. Transistor Q_{14} is in the active region.
4. Transistor Q_{20} is conducting a very small amount of current and is considered to be off.

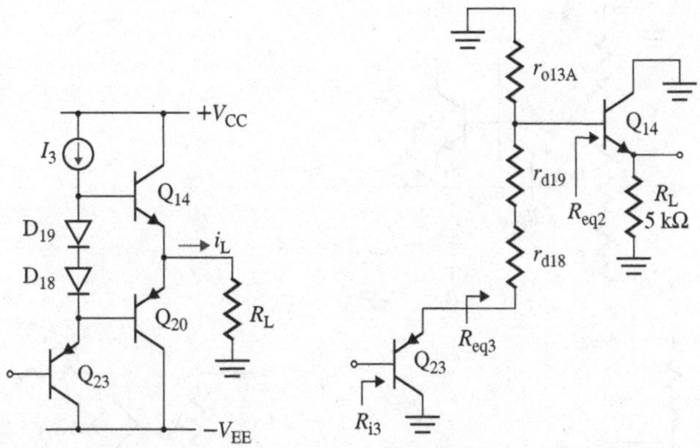

(a) AC equivalent circuit **(b) Small-signal equivalent circuit**

FIGURE 2.35 AC equivalent circuit for the output stage

The small-signal AC equivalent of Fig. 2.35(a) is shown in Fig. 2.35(b). Notice that the circuit consists of two emitter followers in series and that the voltage gain is approximately unity. Thus, the voltage gain of the output stage is

$$A_3 \approx 1 \tag{2.115}$$

Since Q_{14} carries the load current i_L and a no-load DC-biasing collector current of Q_{14} of 153.7 μA,

$$I_{C14} \simeq I_{E4} = 2\text{ mA} + 153.7\ \mu\text{A} = 2.15\text{ mA}$$

Thus,

$$r_{\pi 14} = \frac{\beta_{F14} V_T}{I_{C14}} = \frac{250 \times 26\text{ mV}}{2.15\text{ mA}} = 3.02\text{ k}\Omega$$

Since Q_{23} and diodes D_{18} and D_{19} operate approximately at a current of $I_3 \simeq 180$ μA,

$$r_{d18} = r_{d19} = \frac{V_T}{I_{d18}} = \frac{26\text{ mV}}{180\ \mu\text{A}} = 144\ \Omega$$

If R_{eq2} is the resistance seen looking at the base of Q_{14}, we can use the relationship for the input resistance at the base of a transistor $R_x = r_\pi + (1 + \beta_F)R_{RE}$ to find R_{eq2}:

$$\begin{aligned} R_{eq2} &= r_{\pi 14} + (1 + \beta_{F14})R_L \qquad (2.116) \\ &= 3.02\text{ k}\Omega + (1 + 250) \times 5\text{ k}\Omega = 1258\text{ k}\Omega \end{aligned}$$

Thevenin's equivalent resistance seen looking from the emitter of Q_{23} can be found from

$$\begin{aligned} R_{eq3} &= r_{d18} + r_{d19} + (r_{o13A} \parallel R_{eq2}) \qquad (2.117) \\ &= 144\ \Omega + 144\ \Omega + (289\text{ k}\Omega \parallel 1258\text{ k}\Omega) = 235.3\text{ k}\Omega \end{aligned}$$

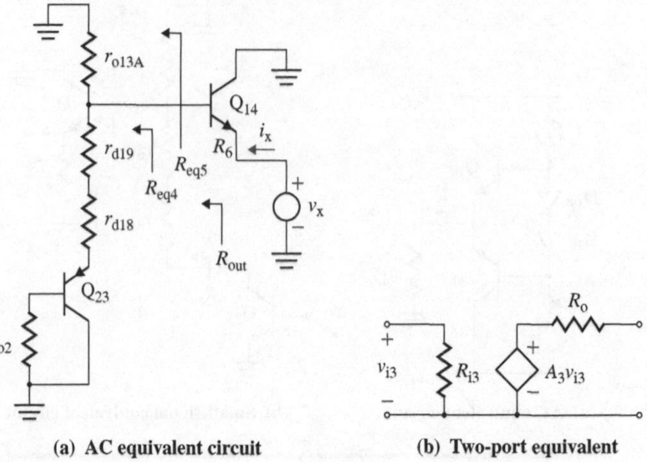

FIGURE 2.36 AC equivalent circuit for calculation of R_o

Using the relationship for the input resistance at the base of a transistor $R_x = r_\pi + (1 + \beta_F)R_{RE}$, we can find the input resistance of the stage:

$$R_{i3} = r_{\pi 23} + (1 + \beta_{F23})R_{eq3} \tag{2.118}$$

$$= 7.2\ \text{k}\Omega + (1 + 50) \times 235.3\ \text{k}\Omega = 12\ \text{M}\Omega$$

Notice that the input resistance of the output stage (12 MΩ) is much larger than the output resistance of the preceding stage (83.72 kΩ). As a result, the overall gain of the amplifier is contributed by the differential and gain stages, the overall gain is not affected by the variations in external load resistance.

The output resistance of the output stage can be determined from the AC equivalent circuit shown in Fig. 2.36(a), which includes the output resistance R_{o2} of the preceding stage. Converting R_{o2} at the base of Q_{23} to its emitter gives the equivalent resistance seen looking from the base of Q_{14} toward Q_{23}:

$$R_{eq4} = r_{d18} + r_{d19} + \frac{R_{o2} + r_{\pi 23}}{1 + \beta_{F23}} \tag{2.119}$$

$$= 144 + 144 + \frac{83.72\ \text{k}\Omega + 7.2\ \text{k}\Omega}{1 + 50} = 2.07\ \text{k}\Omega$$

The resistance seen looking from the base of Q_{14} to the left is

$$R_{eq5} = r_{o13A} \parallel R_{eq4} \tag{2.120}$$

$$= 289\ \text{k}\Omega \parallel 2.07\ \text{k}\Omega = 2.06\ \text{k}\Omega$$

The resistance seen looking into the output terminal is

$$R_{out} = \frac{R_{eq5} + r_{\pi 14}}{1 + \beta_{F14}} \tag{2.121}$$

$$= \frac{2.06\ \text{k}\Omega + 3.02\ \text{k}\Omega}{1 + 250} = 20.2\ \Omega$$

The current-limiting resistance R_6 must be added to R_{out} to find the actual output resistance of the op-amp:

$$\begin{aligned} R_o &= R_{out} + R_6 \\ &= 20.2 + 27 = 47.2\ \Omega \end{aligned} \tag{2.122}$$

The two-port equivalent circuit of the output stage is shown in Fig. 2.36(b).

2.6.3 Frequency-Response Analysis

For $G_{m1} = 182.7\ \mu A/V$ and $C_x = 30$ pF, Eq. (2.56) gives the unity-gain frequency (or bandwidth) as

$$f_u = \frac{G_{m1}}{2\pi C_x} = \frac{182.7\ \mu A/V}{2\pi \times 30\ pF} = 969.3\ kHz$$

From Eq. (2.64), the slew rate is

$$SR = 4\pi V_T f_u = 4\pi \times 26\ mV \times 969.3\ kHz = 0.317\ V/\mu s$$

The manufacturer-specified values are $f_u = 1$ MHz and SR = 0.5 V/μs. The discrepancy between the calculated values and the manufacturer-specified values is caused by the fact that Eq. (2.64) gives an approximate value of SR and does not take into account the input and output resistances of preceding and subsequent amplifier stages.

2.6.4 Small-Signal Equivalent Circuit

The small-signal equivalent of the complete circuit is shown in Fig. 2.37. The voltage gain is

$$\begin{aligned} A_v &= G_{m1}(R_{o1} \parallel R_{i2})\ G_{m2}(R_{o2} \parallel R_{i3}) \\ &= 182.7\ \mu A/V \times (6.9\ M\Omega \parallel 5.73\ M\Omega) \times 6.79\ mA/V \times (83.72\ k\Omega \parallel 12.0\ M\Omega) \\ &= 571.9 \times 564.5 = 323{,}000 \end{aligned} \tag{2.123}$$

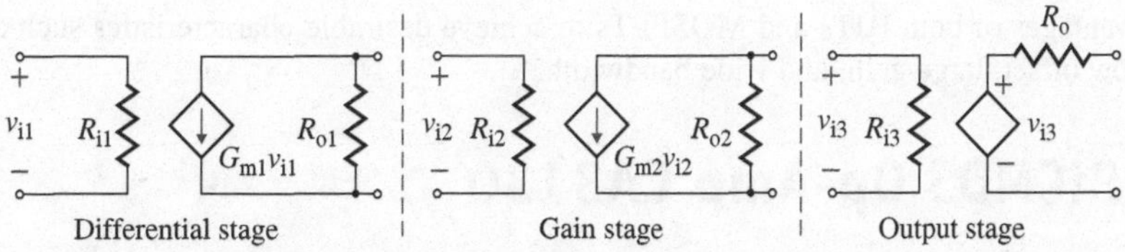

FIGURE 2.37 Small-signal equivalent circuit for the LM741 op-amp

where G_{m1} = the transconductance of the first (differential) stage
G_{m2} = the transconductance of the second (gain) stage
R_{o1} = the output resistance of the first (differential) stage
R_{o2} = the output resistance of the second (gain) stage
R_{i3} = the input resistance of the third (output) stage

Input and output resistances are

$$R_{id} = 2.68 \text{ M}\Omega \quad \text{and} \quad R_o = 47.2 \ \Omega$$

▶ NOTES

1. The input stage and the gain stage contribute about the same gain: 571.9 and 564.5, respectively. The gain stage loads the input stage and reduces its gain by about half. This loading makes the op-amp beta dependent, causing the gain to vary with temperature and fabrication-process tolerances.
2. The output stage does not significantly load the gain stage, and the voltage gain is almost independent of the external load resistance.
3. Although the results of the analysis are approximate, they give insight into the operation and performance of the op-amp.
4. The output resistances of transistors were neglected in the DC analysis. If these were taken into consideration, the biasing currents would change, thereby changing the small-signal output resistance and the results.
5. The variation in the transistor beta was neglected. But the transistor current gain falls at low collector current levels because of recombination in the emitter–base space charge layer.

KEY POINTS OF SECTION 2.6

- This section illustrated a three-part analysis of a complete op-amp circuit, the LM741: DC analysis to find the DC biasing currents, small-signal analysis to find the voltage gain, and analysis of the frequency response to find the unity-gain bandwidth for all stages of the op-amp.
- The values obtained from hand calculations correspond well to the values specified in the manufacturer's data sheet. That is, A_v = 323,000 (compared to 200,000), R_{id} = 2.68 MΩ (compared to 2 MΩ), R_o = 47.2 Ω (compared to 75 Ω), f_u = 969.3 kHz (compared to 1 MHz), and SR = 0.317 V/μs (compared to 0.5 V/μs).

2.7 BiCMOS Op-Amps

BiCMOS op-amps contain both CMOS and BJT transistors on the same chip. This arrangement incorporates the advantages of both BJTs and MOSFETs to achieve desirable characteristics such as high input resistance, low offset, large gain, and wide bandwidth.

2.7.1 BiCMOS Op-Amp CA3130

The simplified schematic of the BiCMOS op-amp CA3130 is shown in Fig. 2.38. The internal structure can be divided into three stages: a differential MOS stage, a gain stage, and an output stage. Some parameters of the CA3130 op-amp are listed in Table 2.6.

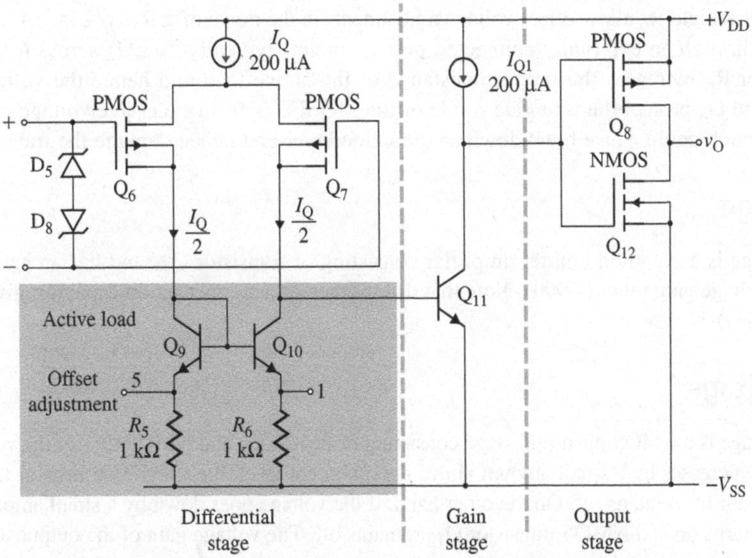

FIGURE 2.38 Simplified schematic for op-amp RCA-CA3130 (Courtesy of Harris Corporation, Semiconductor Sector)

Differential Stage

The input stage is a PMOS differential amplifier consisting of transistors Q_6 and Q_7. This stage is biased by a current source of I_Q, and it drives a current mirror load consisting of BJT transistors Q_9 and Q_{10}. Although a BJT active load offers high resistance, the voltage gain is only about 5 because of the relatively low transconductance of MOSFETs.

TABLE 2.6 Parameters of op-amp CA3130

Parameter	Minimum	Typical	Maximum	Units
DC supply voltages V_{DD}			±18	V
Input offset voltage V_{OS}		8.0	15.0	mV
Thermal drift D_v		15	30	μV/°C
Input biasing current I_B		5	50	pA
Input offset current I_{OS}		0.5	30	pA
Differential input resistance R_{id}		1.5		TΩ
Common-mode input resistance R_{ic}		1.5		TΩ
Output resistance R_o		75		Ω
Input capacitance C_i		4.3		pF
Open-loop voltage gain A_o	50	320		V/mV
Common-mode rejection ratio (CMRR)	70	90		dB
Unity-gain bandwidth f_u		15		MHz
Slew rate (SR)		10		V/μs
Power supply rejection ratio (PSRR)	70	90		dB

Resistors R_5 and R_6 allow offset voltage adjustment (in the range of $\pm R_5 I_Q/2 = \pm 1\ k\Omega \times 100\ \mu A = \pm 100\ mV$) through an externally connected potentiometer (typically 10 kΩ) across terminals 1 and 5. Also, resistor R_6 increases the output resistance of the active load and hence the voltage gain. Zener diodes D_5 and D_8 protect the thin gate-oxide of the MOSFETs from excessive voltage spikes and static discharge, which could cause breakdown of the oxide layer and hence damage the transistors.

Gain Stage

The gain stage is a common-emitter amplifier consisting of transistor Q_{11} and has an active current load for a large voltage gain (about 6000). Note that the absence of a compensation capacitor gives a high bandwidth (15 MHz).

Output Stage

The output stage is a CMOS push-pull stage consisting of PMOS Q_8 and NMOS Q_{12}. If the voltage at the collector of Q_{11} increases by a small amount above the quiescent level, then NMOS transistor Q_{12} turns on and PMOS transistor Q_8 remains off. On the other hand, if the voltage goes down by a small amount, then PMOS transistor Q_8 turns on and NMOS transistor Q_{12} remains off. The voltage gain of the output stage is about 30.

2.7.2 BiCMOS Op-Amp CA3140

The simplified schematic of the BiCMOS op-amp CA3140 is shown in Fig. 2.39. Its internal structure is similar to that of op-amp CA3130 in Fig. 2.38, except for the output stage and the addition of a compensation capacitor C_x in the second stage. The voltage gain of the differential stage is about 10. The offset

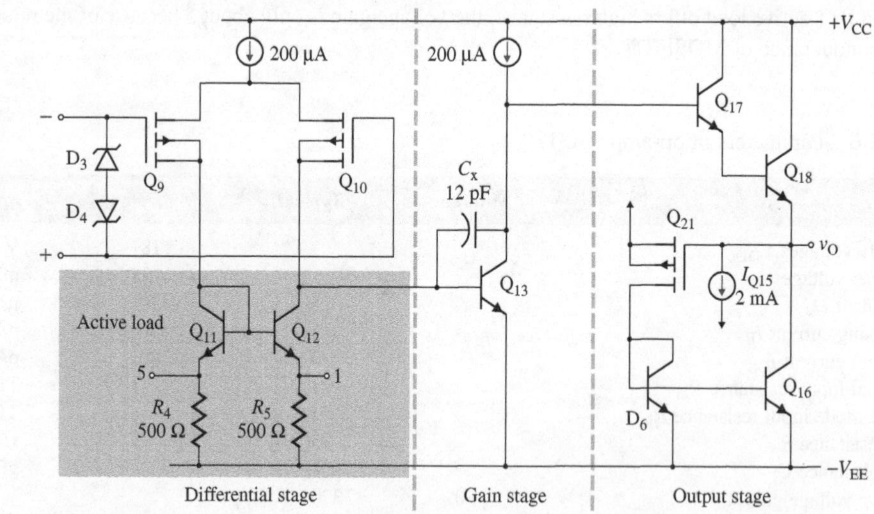

FIGURE 2.39 Simplified schematic for op-amp RCA-CA3140 (Courtesy of Harris Corporation, Semiconductor Sector)

TABLE 2.7 Parameters of op-amp CA3140

Parameter	Minimum	Typical	Maximum	Units
DC supply voltages V_{DD}			±18	V
Input offset voltage V_{OS}		8.0	15.0	mV
Thermal drift D_v		10	30	μV/°C
Input biasing current I_B		10	50	pA
Input offset current I_{OS}		0.5	30	pA
Differential input resistance R_{id}		1.5		TΩ
Common-mode input resistance R_{ic}		1.5		TΩ
Output resistance R_o		50		Ω
Input capacitance C_i		1		pF
Open-loop voltage gain A_o	20	100		V/mV
Common-mode rejection ratio (CMRR)	70	90		dB
Unity-gain bandwidth f_u		4.5		MHz
Slew rate (SR)		9		V/μs
Power supply rejection ratio (PSRR)	76	80		dB

voltage adjustment is in the range of $\pm R_5 I_Q/2 = \pm 500\ \Omega \times 100\ \mu A = \pm 50$ mV. C_x provides feedback stability, but it reduces the bandwidth. The Darlington *npn* emitter follower in the output stage increases the effective load resistance seen by transistor Q_{13} in the gain stage and hence increases the voltage gain to about 10. The voltage gain of the gain stage is about 10,000. Some parameters of the CA3140 op-amp are listed in Table 2.7.

Output Stage

The Darlington emitter follower offers a lower-than-average output resistance and a high resistance to the gain stage. If the voltage at the collector of transistor Q_{13} goes up in the positive direction above the quiescent value, transistor Q_{18} will drive the load and source current of I_{Q15} (2 mA). However, if the current of Q_{18} falls below the level of I_{Q15} (2 mA), the voltage across the load will go down below the quiescent level. Transistor Q_{18} will always remain on. This will cause MOSFET Q_{21} to turn on. Since the collector current of Q_{16} will be the mirror of the drain current of MOSFET Q_{21}, transistor Q_{16} will sink the load current.

2.7.3 BiCMOS Op-Amp LH0022

The simplified schematic of the LH0022 op-amp is shown in Fig. 2.40, which is a modification of the original circuit by replacing junction field-effect transistors Q_1 and Q_2 with depletion-type NMOS. The LH0022 is designed to operate from a ±15-V power supply. It is capable of a peak output voltage swing of about 12 V into a 1-kΩ load over the entire frequency range of 1 MHz. The circuit can be divided into a differential stage, a gain stage, an output stage, overload protection, and a DC-biasing circuit. Some parameters of the LH0022 are listed in Table 2.8.

Differential Stage

The differential input stage consists of Q_1 through Q_4. Transistors Q_1 and Q_2 are depletion NMOSs that operate in a common-drain (or source-follower) configuration. These NMOSs drive transistors Q_3 and Q_4,

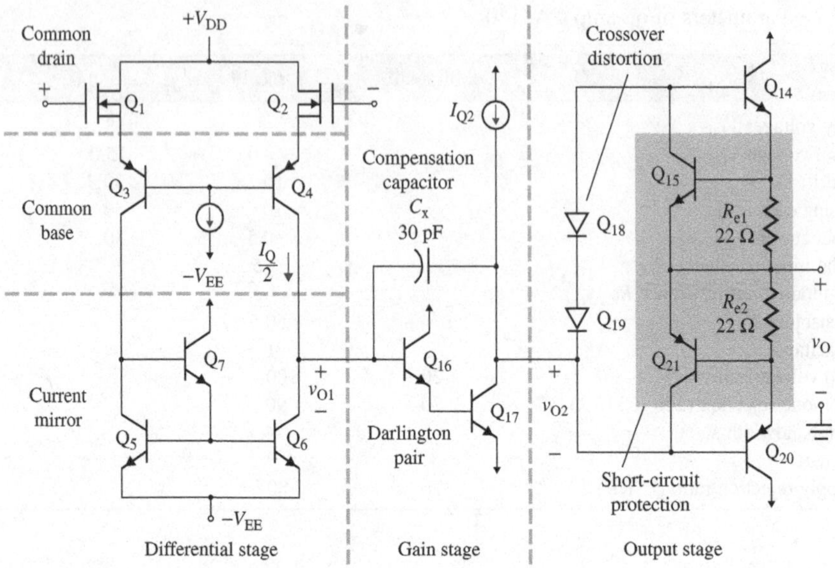

FIGURE 2.40 Simplified schematic for op-amp LH0022 (Courtesy of National Semiconductor, Inc.)

which operate in a common-base configuration. The current mirror, consisting of transistors Q_5, Q_6, and Q_7, acts as the active load.

The combination of common-drain and common-base configurations provides a very high input resistance and a very low input biasing current from the MOSFETs while giving a large voltage gain from the BJTs. Also, the common-drain configuration performs the function of shifting voltage levels

TABLE 2.8 Parameters of op-amp LH0022

Parameter	Minimum	Typical	Maximum	Units
DC supply voltages V_{CC}, V_{EE}			±15	V
Input offset voltage V_{OS}		6.0	20	mV
Thermal drift D_v		10		μV/°C
Input biasing current I_B		15	50	pA
Input offset current I_{OS}		2	10	pA
Differential input resistance R_{id}		10^{12}		Ω
Common-mode input resistance R_{ic}		10^{12}		Ω
Output resistance R_o		75		Ω
Input capacitance C_i		4.0		pF
Open-loop voltage gain A_o	25	100		V/mV
Common-mode rejection ratio (CMRR)	70	80		dB
Unity-gain bandwidth f_u		1.0		MHz
Slew rate (SR)	1.0	3.0		V/μs
Power supply rejection ratio (PSRR)	70	80		dB

toward the negative supply voltage $-V_{EE}$ so that the output signal can be shifted upward in the positive direction in the following stages. Note that the DC level of the output voltage of this stage will be approximately

$$-V_{EE} + V_{BE16} + V_{BE17} \equiv -V_{EE} + 1.2 \text{ V}$$

Gain Stage

The gain stage has the common-collector and common-emitter configurations forming a Darlington pair made up of Q_{16} and Q_{17}. This arrangement offers a high load resistance to the differential stage and hence provides a large voltage gain. Capacitor C_x is the pole-şplitting compensating capacitor connected in shunt-shunt feedback, and it controls the unity-gain bandwidth and the slew rate.

Output Stage

The output stage is a complementary class AB push-pull circuit consisting of transistors Q_{14} and Q_{20}. Diode-connected transistors Q_{18} and Q_{19} provide the biasing voltage for class AB operation in order to reduce crossover distortion.

Protection Circuitry

Transistors Q_{15} and Q_{21} act as the current booster. Resistors R_{e1} and R_{e2} provide short-circuit protection by turning on Q_{15} and Q_{21} so as to limit the currents through Q_{14} and Q_{20}, respectively.

Biasing Circuitry

The complete schematic, including the biasing circuitry, is shown in Fig. 2.41, which has an extra protection circuit (consisting of transistors Q_{23} and Q_{24}) to short the input terminal of Q_{16} (in the gain stage) to ground through Q_{23}. Transistors Q_{10} and Q_{11} form a Widlar current mirror, which biases transistors Q_3 and Q_4 of the common-base configuration. The current source consisting of transistors Q_{12} and Q_{13} biases the gain and output stages. The reference current I_{ref} can be found from

$$I_{ref} = \frac{V_{CC} + V_{EE} - V_{BE11} - V_{EB12}}{R_5} \tag{2.124}$$

Transistor Q_{13} is a multicollector lateral *pnp* device. Its geometry is shown in Fig. 2.42(a), and its symbol is shown in Fig. 2.42(b). The collector ring has been split into two parts—one part faces on three-fourths of the emitter periphery and collects the holes injected from that periphery, and the second one faces on one-fourth of the emitter periphery and collects the holes from that periphery. The structure is analogous to two *pnp* transistors whose base–emitter junctions are connected in parallel, one with an I_S that is one-fourth that of a standard *pnp* transistor and the other with an I_S that is three-fourths that of a standard *pnp* transistor. This electrical equivalence is shown in Fig. 2.42(c). Thus, the currents I_{Q2} and I_{Q3} are three-fourths and one-fourth of the reference current I_{ref}, respectively. That is,

$$I_{Q2} = \frac{3I_{ref}}{4} \quad \text{and} \quad I_{Q3} = \frac{I_{ref}}{4}$$

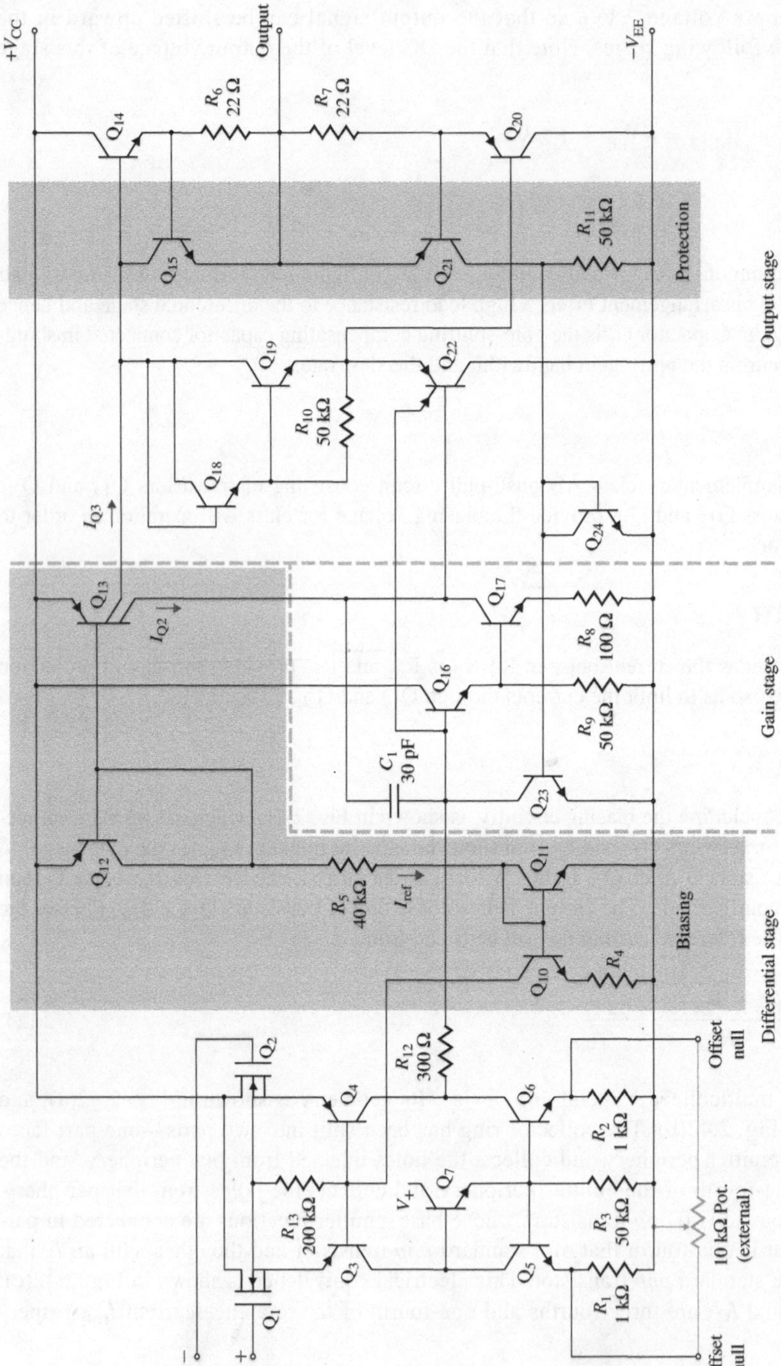

FIGURE 2.41 Schematic for op-amp LH0022 (Courtesy of National Semiconductor, Inc.)

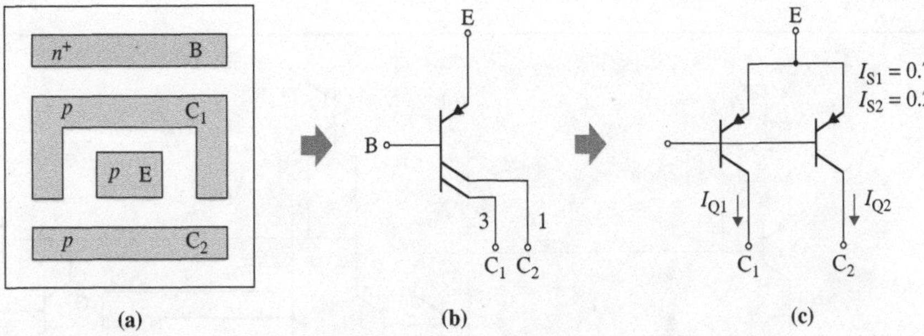

FIGURE 2.42 Electrical equivalent for multicollector lateral *pnp* transistor

2.7.4 BiCMOS Op-Amp LF411

The simplified schematic of the LF411 op-amp is shown in Fig. 2.43, which is a modification of the original circuit by replacing junction field-effect transistors J_1 and J_2 with depletion-type PMOS. Its internal structure is similar to that of the LH0022 op-amp. The differential input stage consists of *p*-channel depletion transistors M_1 and M_2 with a current mirror load. This stage has low offset, low drift, and a high unity-gain bandwidth. The complete schematic appears in Fig. 2.44, which shows in detail the biasing circuitry. There is a transistor Q_3 in the drain of M_2, but none in that of M_1. Depletion NMOS M_2 operates in a common-source configuration, whereas MOSFET M_1 operates in a common-drain configuration. This type of arrangement gives a lower voltage gain but a wider bandwidth.

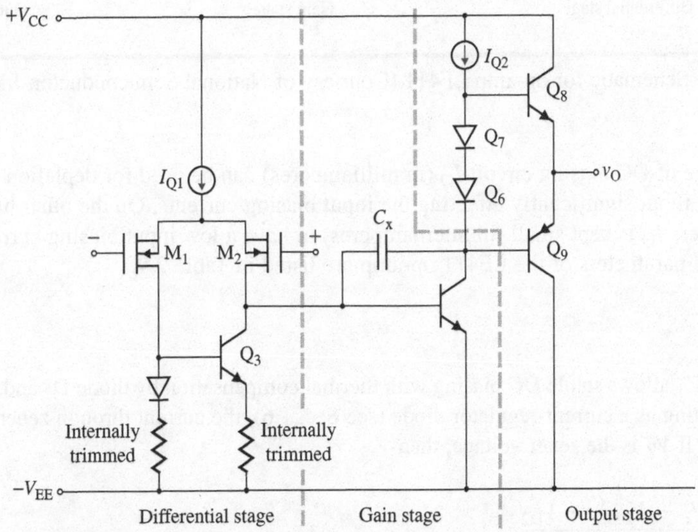

FIGURE 2.43 Simplified schematic for op-amp LF411 (Courtesy of National Semiconductor, Inc.)

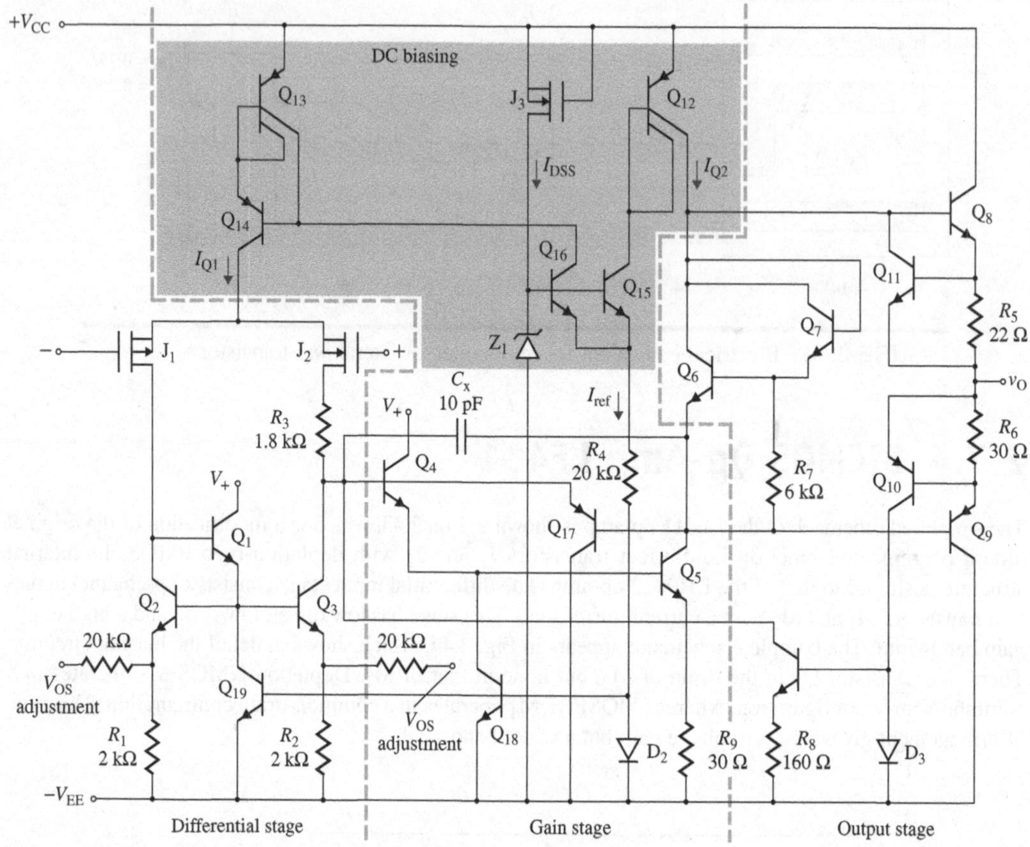

FIGURE 2.44 Schematic for op-amp LF411 (Courtesy of National Semiconductor, Inc.)

A large value of DC biasing circuit I_Q (in milliamperes) can be used for depletion NMOS differential amplifiers without significantly affecting the input biasing currents. On the other hand, for BJT differential amplifiers I_Q is kept small (in microamperes) to have a low input biasing current or high input resistance. Some parameters of the LF411 op-amp are listed in Table 2.9.

DC Biasing

The zener diode Z_1 allows stable DC biasing with thermal compensation by diode D_2 and resistor R_4. With MOSFET M_3 acting as a current-regulator diode (see Sec. 1.6), the current through zener diode Z_1 is kept constant at I_{DSS}. If V_Z is the zener voltage, then

$$I_{ref} = \frac{V_Z - V_{BE16} - V_{D2}}{R_4} \tag{2.125}$$

which in turn sets the DC biasing currents I_{Q1} and I_{Q2}.

TABLE 2.9 Parameters of op-amp LF411

Parameter	Minimum	Typical	Maximum	Units
DC supply voltages V_{CC}, V_{EE}			±18	V
Input offset voltage V_{OS}		0.8	2.0	mV
Thermal drift D_v		7	20	μV/°C
Input biasing current I_B		50	200	pA
Input offset current I_{OS}		25	100	pA
Differential input resistance R_{id}		10^{12}		Ω
Common-mode input resistance R_{ic}		10^{12}		Ω
Output resistance R_o		75		Ω
Input capacitance C_i		4.0		pF
Open-loop voltage gain A_o	25	200		V/mV
Common-mode rejection ratio (CMRR)	70	100		dB
Unity-gain bandwidth f_u	2.7	4.0		MHz
Slew rate (SR)	8.0	15		V/μs
Power supply rejection ratio (PSRR)	70	100		dB

Thermal Protection

Transistor Q_{19} is normally off. If the temperature rises, V_{BE16} will fall because of the negative temperature coefficient of Q_{16}, and the zener voltage V_Z will rise because of the positive temperature coefficient of zener diode Z_Z. As a result, the voltage at the emitter of Q_{16} will rise; thus, the voltage at the anode of D_2 will also rise. If the temperature rise is adequate, Q_{18} and Q_{19} will turn on and reduce the gain and the output voltage of the amplifier.

2.7.5 BiCMOS Op-Amp LH0062

The simplified schematic of the LH0062 op-amp is shown in Fig. 2.45, which is a modification of the original circuit by replacing junction field-effect transistors Q_1 and Q_2 with depletion-type NMOS. The internal structure is similar to that of op-amp LH0022.

Differential Stage

The differential input stage consists of *n*-channel transistors Q_1 and Q_2 with an active load. This is a normal source-coupled differential pair with a difference output, which gives a lower gain but a wider bandwidth. The input resistance is very high for the MOSFET input stage. The DC biasing current I_Q is set by the current source consisting of transistor Q_3 and zener diode D_1. Thus,

$$I_Q \equiv \frac{V_Z - V_{BE3}}{R_5} \tag{2.126}$$

$$= \frac{1.2\text{ V} - 0.6\text{ V}}{600\ \Omega} = 1\text{ mA}$$

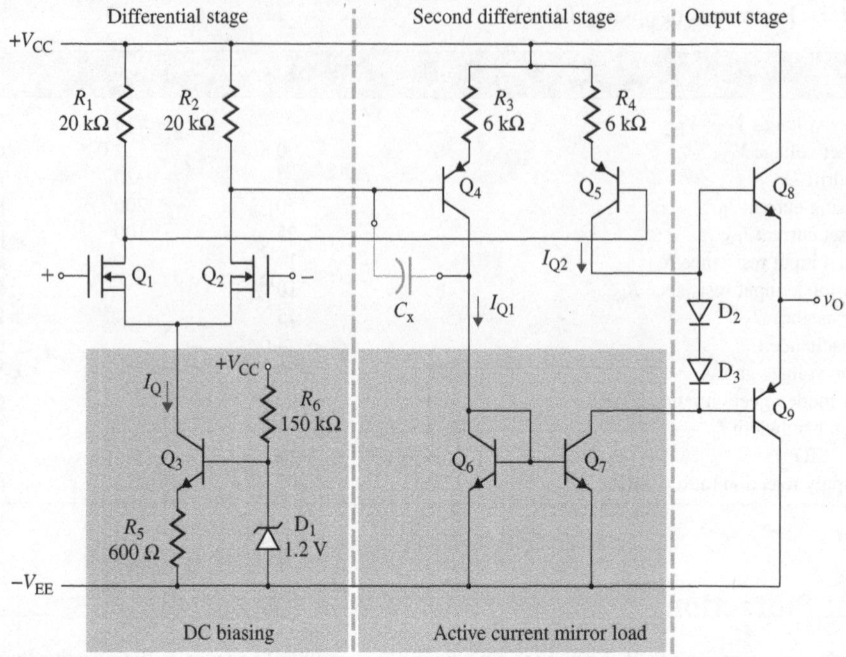

FIGURE 2.45 Simplified schematic for op-amp LH0062 (Courtesy of National Semiconductor, Inc.)

Gain Stage

The gain stage is a common-emitter coupled pair, which drives a current mirror load consisting of transistors Q_6 and Q_7. The single-ended output gives a larger voltage gain. The biasing current I_{Q1} can be found from

$$I_{Q1}R_3 + V_{EB4} = \frac{R_2 I_Q}{2}$$

That is,

$$I_{Q1} = \frac{R_2 I_Q/2 - V_{EB4}}{R_3} \tag{2.127}$$

$$= \frac{20\text{ k}\Omega \times 0.5\text{ mA} - 0.6\text{ V}}{6\text{ k}\Omega} = 1.6\text{ mA}$$

The collector current of Q_7 is the mirror of that of Q_6; that is, I_{Q2} is the mirror of I_{Q1}, and $I_{Q2} = I_{Q1} = 1.6$ mA, which is also the biasing current for the output stage. The high gain and wider bandwidth are obtained by cascading two differential stages. The slew rate is also high because of the higher DC biasing current. Some parameters of the LH0062 op-amp are listed in Table 2.10.

TABLE 2.10 Parameters of op-amp LH0062

Parameter	Minimum	Typical	Maximum	Units
DC supply voltages V_{CC}, V_{EE}			±18	V
Input offset voltage V_{OS}		10	15	mV
Thermal drift D_v		10	35	μV/°C
Input biasing current I_B		10	65	pA
Input offset current I_{OS}		1	5	pA
Differential input resistance R_{id}		10^{12}		Ω
Common-mode input resistance R_{ic}		10^{12}		Ω
Output resistance R_o		75		Ω
Input capacitance C_i		4.0		pF
Open-loop voltage gain A_o	25	160		V/mV
Common-mode rejection ratio (CMRR)	70	90		dB
Unity-gain bandwidth f_u		15		MHz
Slew rate (SR)	50	75		V/μs
Power supply rejection ratio (PSRR)	70	90		dB

2.7.6 BiCMOS Op-Amp LH0032

The LH0032 is an ultrafast op-amp. Its schematic is shown in Fig. 2.46, which is a modification of the original circuit by replacing junction field-effect transistors Q_1 and Q_2 with depletion-type NMOS. The internal structure is similar to that of op-amp LH0062, except that the second differential stage uses common-base

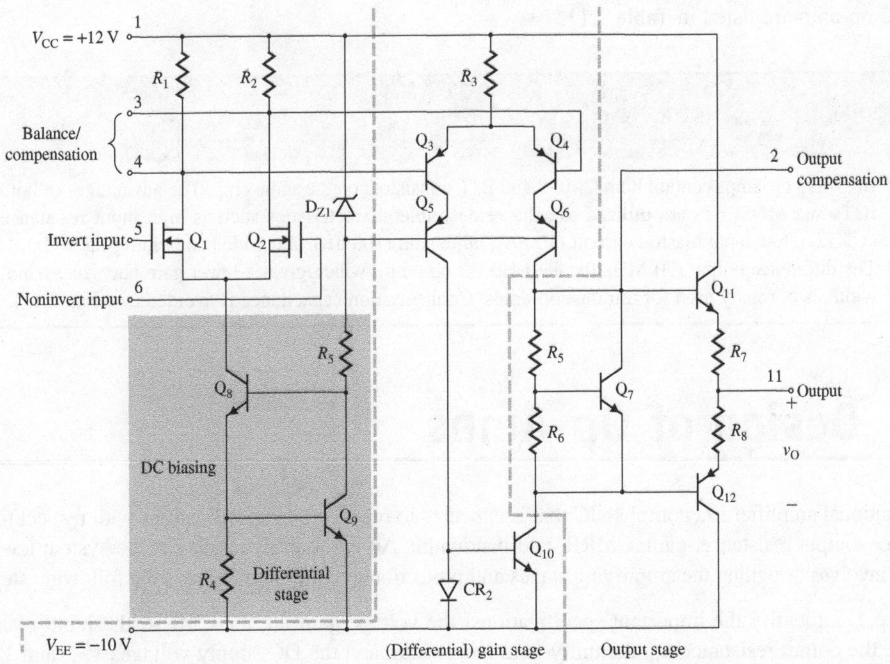

FIGURE 2.46 Schematic for op-amp LH0032 (Courtesy of National Semiconductor, Inc.)

TABLE 2.11 Parameters of op-amp LH0032

Parameter	Minimum	Typical	Maximum	Units
DC supply voltages V_S			±18	V
Input offset voltage V_{OS}		1.0	2.0	mV
Thermal drift D_v		15	30	μV/°C
Input biasing current I_B		50	150	pA
Input offset current I_{OS}		10	30	pA
Differential input resistance R_{id}		10^{12}		Ω
Common-mode input resistance R_{ic}		10^{12}		Ω
Output resistance R_o		75		Ω
Input capacitance C_i		4.0		pF
Open-loop voltage gain A_o	60	70		dB
Common-mode rejection ratio (CMRR)	50	60		dB
Unity-gain bandwidth f_u		70		MHz
Slew rate (SR)	350	500		V/μs
Power supply rejection ratio (PSRR)	50	60		dB

cascode BJTs for a higher voltage gain. The zener diode D_{Z1} gives a stable reference voltage for the biasing circuit consisting of transistors Q_8 and Q_9. In general, a differential stage with a difference output (i.e., the first differential stage) gives a lower voltage gain but a wider bandwidth. A differential pair with a single-ended output (i.e., the second differential stage) gives a higher voltage gain but a lower bandwidth. However, the common-base cascode connection gives a wider bandwidth.

The LH0032 op-amp has very high gain, a very large slew rate, and a very large bandwidth. Note that there is no compensation capacitor. As a result, the bandwidth is widened. Some parameters of the LH0032 op-amp are listed in Table 2.11.

KEY POINTS OF SECTION 2.7

- BiCMOS op-amps contain both CMOS and BJT transistors on the same chip. The advantages of both BJTs and MOSFETs are utilized to achieve desirable characteristics such as high input resistance (1.5 TΩ), low input biasing current (10 pA), large gain (100 dB), and wide bandwidth (4.5 MHz).
- The difference output (70 MHz for the LH0032 op-amp), which gives a lower gain but wider bandwidth, is normally used for ultrafast op-amps. Compensation capacitance is avoided (see Fig. 2.11).

2.8 Design of Op-Amps

An operational amplifier is a complete IC that is expected to meet certain specifications with respect to input resistance, output resistance, gain, CMRR, and bandwidth. An op-amp is designed at the system level. The process involves designing the amplifying stages and protection circuitry and requires the following steps [9]:

Step 1. Identify the important specifications: the voltage gain A_d, the CMRR, the input resistance R_{id}, the output resistance R_o, the unity-gain bandwidth, and the DC supply voltages V_{CC} and V_{EE} (or V_{DD} and V_{SS}).

Step 2. Select the type of op-amp (BJT, CMOS, or BiCMOS).

Step 3. Choose the circuit configurations and stages.

Step 4. Determine the biasing current requirement I_Q for the desired specifications.

Step 5. Choose the type of current source (BJT or MOSFET) and determine its component ratings.

Step 6. Choose the active load with a current mirror (BJT or MOSFET) in order to obtain the desired voltage gain and determine its component ratings.

Step 7. Choose the type of gain stage (BJT or MOSFET) and determine its component ratings.

Step 8. Choose the type of output stage (BJT or MOSFET) and determine its component ratings.

Step 9. Choose the type of frequency compensation (pole or pole-zero circuit) and determine its component ratings.

Step 10. Determine the voltage, current, and power ratings of active and passive components.

Step 11. Analyze and evaluate the complete differential amplifier to see that it meets the desired specifications.

Step 12. Use PSpice/SPICE to simulate and verify your design. If it would be implemented with discrete devices, use standard values of components, with tolerances of, say, 5%.

Summary

The open-loop voltage gain of an op-amp does not remain constant, but rather decreases with frequency. The frequency response of an internally compensated op-amp has the characteristic of a single time-constant network. The slew rate of the input signal should be less than that of the op-amp in order to avoid distortion of the output voltage. The common-mode rejection ratio is a measure of the ability of an op-amp to reject common-mode signals; the ratio should be as high as possible.

The characteristics of practical op-amps differ from those of ideal ones. The output of an op-amp is affected by parameters such as input offset voltage, input offset current, input biasing current, thermal drift, the power supply rejection ratio, and the input frequency. The effect of input biasing currents can be minimized by adding an offset-minimizing resistor.

The internal structure of an op-amp usually consists of differential, gain, and output stages. There are many possible combinations of op-amp stages, depending on whether the amplifier is a bipolar, MOSFET, or CMOS op-amp. In general, a MOSFET op-amp gives a very high input resistance but less gain, whereas a bipolar op-amp gives a higher gain but lower input resistance. A Darlington BJT pair is commonly used for high current gain and input resistance. A difference output of a differential stage gives less gain but yields wider bandwidth.

Op-amps exhibit offset voltages and currents, which can be minimized by appropriate internal design. The input offset voltage is dependent on the CMRR and the common-mode signal. A pole or pole-zero compensation circuit is commonly used for frequency compensation, but the compensation comes at the expense of unity-gain bandwidth.

The small-signal voltage gain and the input and output resistances of the LM741 op-amp can be determined for various load conditions. Since the voltage gain is very high, the amplifier offers offset voltages and currents because of variations in the transistor parameters and circuit resistances. Although the analysis required to derive equations describing the characteristics of current sources and differential amplifiers can be simplified with some assumptions, computer-aided analysis is generally required to evaluate the actual performance of an op-amp at the final design stage. The analysis of other op-amps would be similar to the analysis of the LM741 op-amp in this chapter.

References

1. A. B. Grebene, *Bipolar and MOS Analog Integrated Circuit Design*. New York: Wiley, 2003.
2. R. A. Gayakwad, *Op-Amps and Linear Integrated Circuits*. Englewood Cliffs, NJ: Prentice Hall, 1993.
3. W. D. Stanley, *Operational Amplifiers with Linear Integrated Circuits*. Upper Saddle River, NJ: Prentice Hall, 2002.
4. D. G. Ong, *Modern MOS Technology*. New York: McGraw-Hill, 1984.
5. P. R. Gray, P. J. Hurst, S. H. Lewis, and R. G. Meyer, *Analysis and Design of Analog Integrated Circuits*. New York: Wiley, 2001.
6. S. Soclof, *Design and Applications of Analog Integrated Circuits*. Englewood Cliffs, NJ: Prentice Hall, 1991.
7. R. F. Coughlin and F. F. Driscoll, *Operational Amplifiers and Linear Integrated Circuits*. Upper Saddle River, NJ: Prentice Hall, 2001.
8. R. C. Jaeger and T. Blalock, *Microelectronic Circuit Design*. New York: McGraw-Hill, 2008.
9. F. C. Franco, *Design with Operational Amplifiers and Analog Integrated Circuits*. New York: McGraw-Hill, 2002.

Review Questions

1. What are the main stages of an op-amp?
2. What is the input biasing current of an op-amp?
3. What is the input offset current of an op-amp, and what factors influence its value?
4. What is the input offset voltage of an op-amp, and what factors influence its value?
5. What is the CMRR of an op-amp?
6. What factors influence the unity-gain bandwidth of an op-amp?
7. What factors influence the slew rate of an op-amp?
8. What is the relation between slew rate and unity-gain bandwidth?
9. What is the typical value of the input resistance of an op-amp?
10. What is the typical value of the output resistance of an op-amp?
11. Ideally, what should be the differential voltage gain of an op-amp?
12. Ideally, what should be the common-mode voltage gain of an op-amp?
13. What is the unity-gain bandwidth of an op-amp?
14. What is the effect of rise time on the frequency response of an op-amp?
15. What is a slew rate?
16. What is the slew rate of a step input voltage?
17. What is the slew rate of a sinusoidal input voltage?
18. What is the feedback factor?
19. What is the typical break frequency of an op-amp?
20. What is the effect of input offset voltage on the output of inverting and noninverting amplifiers?
21. What is the effect of input biasing currents on the output of inverting and noninverting amplifiers?

22. What is the effect of input offset current on the output of inverting and noninverting amplifiers?
23. What is the common method for minimizing the effect of input biasing currents?
24. What is the cause of thermal drift?
25. What is the effect of thermal drift?
26. What is the effect of output offset voltage on an integrator?
27. What is the PSRR?
28. What is pole-zero compensation of an op-amp?
29. What circuit configurations are used for ultrafast op-amps?
30. What are the advantages and disadvantages of MOSFET op-amps?
31. What are the advantages and disadvantages of bipolar op-amps?
32. What are the advantages and disadvantages of CMOS op-amps?
33. What are the advantages and disadvantages of BiCMOS op-amps?
34. What is the function of short-circuit protection in op-amps?
35. What are the typical values of input resistance, output resistance, and voltage gain for the LM741 op-amp?

Problems

The symbol **D** indicates that a problem is a design problem. The symbol **P** indicates that you can check the solution to a problem using PSpice/SPICE or Electronics Workbench.

2.3 *Parameters and Characteristics of Practical Op-Amps*

2.1 The input voltages of an op-amp are $v_1 = 120\ \mu V$ and $v_2 = 80\ \mu V$. The op-amp parameters are CMRR = 90 dB and $A_d = A_o = 2 \times 10^5$. Determine **(a)** the differential voltage v_d, **(b)** the common-mode voltage v_c, **(c)** the magnitude of the common-mode gain A_c, and **(d)** the output voltage v_O.

2.2 The rise time of an op-amp is $t_r = 0.3\ \mu s$. What is the maximum frequency limit of the op-amp?

2.3 The inverting amplifier in Fig. 2.9(a) has $R_1 = 15\ k\Omega$ and $R_F = 50\ k\Omega$. The input offset voltage is $V_{io} = \pm 6$ mV at 25°C. Determine the output offset voltage V_{oo}.

2.4 The input biasing current I_B for the amplifier in Fig. 2.3 is $I_B = 500$ nA (DC) at 25°C. If $R_1 = 15\ k\Omega$ and $R_F = 50\ k\Omega$, determine **(a)** the output offset voltage due to input biasing current I_B and **(b)** the offset-minimizing resistance R_x.

2.5 The maximum input offset current of the amplifier in Fig. 2.4 is $I_{io} = \pm 200$ nA at 25°C. If $R_1 = 15\ k\Omega$ and $R_F = 50\ k\Omega$, determine the output offset voltage due to the input offset current.

2.6 The inverting amplifier in Fig. 2.5(a) has $R_1 = 15\ k\Omega$ and $R_F = 50\ k\Omega$. The op-amp has $V_{io} = \pm 6$ mV, $I_B = 500$ nA, and $I_{io} = \pm 300$ nA at 25°C. Determine the total output offset voltage v_{of} if **(a)** R_x ($= R_F \| R_1$) = 11.54 kΩ and **(b)** $R_x = 0$. Assume $v_S = 0$.

2.7 The noninverting amplifier in Fig. 2.5(b) has $R_1 = 15\ k\Omega$ and $R_F = 150\ k\Omega$. The op-amp has $V_{io} = \pm 6$ mV, $I_B = 500$ nA, and $I_{io} = \pm 300$ nA at 25°C. Determine **(a)** the total output offset voltage if $R_x = R_F \| R_1 = 11.54\ k\Omega$ and **(b)** the output offset voltage v_{of} if $R_x = 0$. Assume $v_S = 0$.

2.8 The integrator in Fig. 2.7 has $R_1 = 10\ k\Omega$, $R_x = 10\ k\Omega$, $C_1 = 0.1\ \mu F$, $V_{CC} = 15$ V, $-V_{EE} = -15$ V, and maximum saturation voltage = ±11 V. The op-amp has input offset voltage $V_{io} = 6$ mV, input biasing current $I_B = 500$ nA, and input offset current $I_{io} = 300$ nA at 25°C.

a. Determine the time required for the op-amp output offset voltage to reach the saturation limit of ±14 V.

b. Repeat part (a) if $R_x = 0$.

2.9 The inverting amplifier in Fig. 2.5(a) has $R_1 = 15$ kΩ, $R_F = 50$ kΩ, and $R_x = R_F \| R_1 = 11.54$ kΩ. The op-amp has $V_{io} = \pm 6$ mV, $I_B = 500$ nA, and $I_{io} = \pm 300$ nA. The thermal drifts are $D_v = 15$ μV/°C, $D_i = 0.5$ nA/°C, and $D_b = 0.5$ nA/°C at 25°C. The temperature is 55°C. Determine **(a)** the output offset voltage due to drifts V_{od} and **(b)** the total output voltage v_{of} if the input voltage is $v_S = 150$ mV (DC).

2.10 The noninverting amplifier in Fig. 2.5(b) has $R_1 = 15$ kΩ, $R_F = 50$ kΩ, and $R_x = R_F \| R_1 = 11.54$ kΩ. The op-amp has $V_{io} = \pm 6$ mV, $I_B = 500$ nA, and $I_{io} = \pm 300$ nA. The thermal drifts are $D_v = 15$ μV/°C, $D_i = 0.5$ nA/°C, and $D_b = 0.5$ nA/°C at 25°C. The temperature is 55°C. Determine **(a)** the output offset voltage due to drifts V_{od} and **(b)** the total output voltage v_o if the input voltage is $v_S = 150$ mV (DC).

2.11 The inverting amplifier in Fig. 2.5(a) has $R_1 = 15$ kΩ and $R_F = 150$ kΩ. The supply voltages change from ±12 V to ±10 V, and PSRR = 150 μV/V at 25°C. Determine **(a)** the input offset voltage V_{io} due to changes in the supply voltages and **(b)** the corresponding output offset voltage V_{oo}.

2.12 The noninverting amplifier in Fig. 2.5(b) has $R_1 = 15$ kΩ and $R_F = 50$ kΩ. The supply voltages change from ±12 V to ±10 V, and PSRR = 150 μV/V at 25°C. Determine **(a)** the input offset voltage V_{io} due to changes in the supply voltages and **(b)** the corresponding output offset voltage V_{oo}.

2.13 The inverting amplifier in Fig. 2.5(a) has $R_1 = 10$ kΩ, $R_F = 100$ kΩ, and $R_x = R_F \| R_1 = 9.091$ kΩ. The op-amp parameters are $V_{io} = 0.8$ mV, $I_B = 200$ nA, $I_{io} = 100$ nA, and PSRR = 150 μV/V. The drifts are $D_v = 15$ μV/°C, $D_i = 0.5$ nA/°C, and $D_b = 0.5$ nA/°C at 25°C. The temperature is 55°C. The DC supply voltages change from $V_{CC} = 12$ V to 10 V and $-V_{EE} = -12$ V to -10 V. The input voltage is $v_S = 100$ mV (DC). Determine the output voltage v_O if **(a)** $R_x = R_F \| R_1 = 9.091$ kΩ and **(b)** $R_x = 0$.

2.14 **D** The noninverting amplifier in Fig. 2.12(a) has $R_1 = 10$ kΩ, $R_F = 150$ kΩ, and $R_x = R_F \| R_1 = 9.091$ kΩ. Design the offset compensating network. The op-amp parameters are $V_{io} = 0.8$ mV, $I_B = 200$ nA, $I_{io} = 100$ nA, and PSRR = 150 μV/V. The DC supply voltages are $V_{CC} = 12$ V and $-V_{EE} = -12$ V.

2.15 **D** The inverting amplifier in Fig. 2.12(b) has $R_1 = 10$ kΩ, $R_F = 50$ kΩ, and $R_x = R_F \| R_1 = 8.33$ kΩ. Design the offset compensating network. The op-amp parameters are $V_{io} = 0.8$ mV, $I_B = 200$ nA, $I_{io} = 100$ nA, and PSRR = 150 μV/V. The DC supply voltages are $V_{CC} = 12$ V and $-V_{EE} = -12$ V.

2.16 **D** The differential amplifier in Fig. 2.12(c) has $R_a = R_1 = 12$ kΩ and $R_F = R_x = 24$ kΩ. Design the offset compensating network. The op-amp parameters are $V_{io} = 6$ mV, $I_B = 500$ nA, $I_{io} = 200$ nA, and PSRR = 150 μV/V. The DC supply voltages are $V_{CC} = 12$ V and $-V_{EE} = -12$ V.

2.17 The Darlington pair shown in Fig. 2.13 is biased in such a way that the collector-biasing current I_{C2} of Q_2 is 400 μA. The current gains of the two transistors are the same, $\beta_{F1} = \beta_{F2} = 80$, and the Early voltage is $V_A = 50$ V. Calculate **(a)** the effective input resistance r_π, **(b)** the effective transconductance g_m, **(c)** the effective current gain $\beta_{F(eff)}$, and **(d)** the effective output resistance r_o.

2.4 CMOS Op-Amps

2.18 The CMOS of the op-amp in Fig. 2.20 has $K_x = 10$ μA/V^2, $W/L = 160$ μm/10 μm, and $V_t = 0.5$ V. The biasing current is $I_Q = 40$ μA. If V_t changes by 2% and W/L by 1%, find the input offset voltage V_{OS}.

2.19 **D** The CMOS amplifier in Fig. 2.20 is operated at a biasing current of $I_Q = 50$ μA. The parameters of the MOSFETs are $K_x = 10$ μA/V^2, $|V_{M(NMOS)}| = V_{M(PMOS)} = 60$ V, $V_t = 1$ V, and $W/L = 80$ μm/10 μm, except for Q_7, for which $W/L = 160$ μm/10 μm. Assume $V_{DD} = V_{SS} = 5$ V.

a. Find V_{GS}, g_m, and r_o for all MOSFETs.

b. Find the low-frequency voltage gain of the amplifier A_{vo}.

c. Find the value of the external resistance R_{ref}.

d. Find the value of compensation capacitance C_x that gives a unity-gain bandwidth of 1 MHz and the corresponding slew rate.

e. Find the value of resistance R_x to be connected in series with C_x in order to move the zero frequency to infinity.

f. Find the common-mode input voltage range.
g. Find the output voltage range.

2.5 BJT Op-Amps

2.20 The BJTs for the op-amp shown in Fig. P2.20 have $\beta_{F(npn)} = 100$, $\beta_{F(pnp)} = 50$, $V_{A(pnp)} = V_{A(npn)} = 80$ V, $V_{BE} = 0.6$ V, and $I_S = 10^{-14}$ A. Find **(a)** the input biasing current I_B, **(b)** the input resistance R_i, **(c)** the unity-gain bandwidth f_u and the slew rate, and **(d)** the overall low-frequency voltage gain A_{vo}. Assume $V_{CC} = -V_{EE} = 15$ V.

FIGURE P2.20

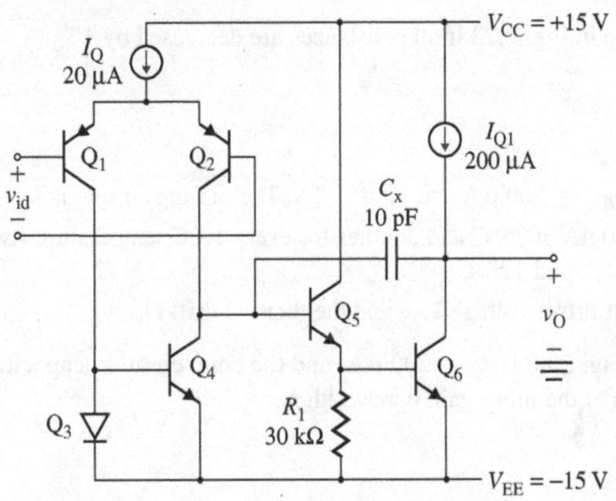

2.21 The BJTs for the op-amp shown in Fig. P2.21 have $\beta_{F(npn)} = 100$, $\beta_{F(pnp)} = 50$, $V_{A(pnp)} = V_{A(npn)} = 80$ V, $V_{BE} = 0.6$ V, and $I_S = 10^{-14}$ A. Find **(a)** the input biasing current I_B, **(b)** the input resistance R_i, and **(c)** the overall low-frequency voltage gain A_{vo}. Assume $V_{CC} = V_{EE} = 15$ V.

FIGURE P2.21

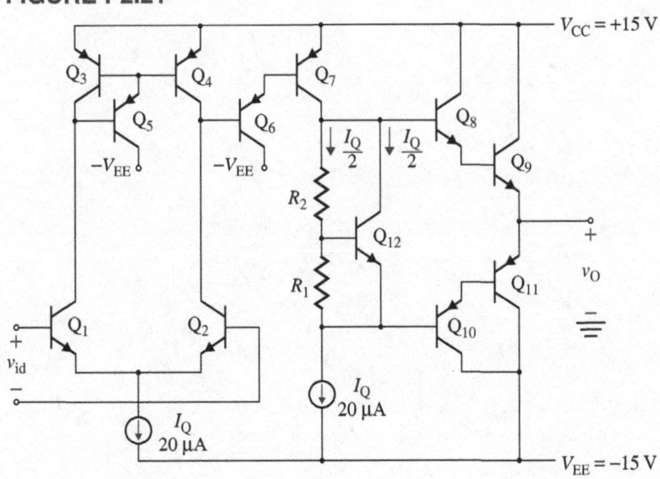

2.22 The BJTs for the LM124 op-amp shown in Fig. 2.22 have $\beta_{F(npn)} = 100$, $\beta_{F(pnp)} = 50$, $V_{A(pnp)} = V_{A(npn)} = 80$ V, $V_{BE} = 0.6$ V, and $I_S = 10^{-14}$ A. Assume $V_{CC} = 12$ V. Find **(a)** the input biasing current I_B, **(b)** the input resistance R_i, **(c)** the unity-gain bandwidth f_u and the slew rate, **(d)** the overall low-frequency voltage gain A_{vo}, and **(e)** the value of R_{SC} for a short-circuit current limit of 25 mA.

2.6 *Analysis of the LM741 Op-Amp*

2.23 Calculate the gain of the LM741 op-amp in Fig. 2.23 if R_g is reduced from 100 Ω to 0.

2.24 Calculate the gain of the LM741 op-amp in Fig. 2.23 if the current gain for *npn* transistors is changed from $\beta_F = 250$ to $1.1 \times 250 = 275.0$ and the saturation current is changed from $I_S = 10^{-14}$ A to 1.05×10^{-14} A.

2.25 Calculate the gain of the LM741 op-amp in Fig. 2.23 if all resistances are increased by 1%.

2.26 Calculate the gain of the LM741 op-amp in Fig. 2.23 if all resistances are decreased by 1%.

2.7 *BiCMOS Op-Amps*

2.27 The BiCMOS op-amp in Fig. 2.40 has $I_{DDS} = 500$ μA and $V_p = -4$ V. The biasing current is $I_Q = 200$ μA.

a. If the input biasing current is $I_B = 1.0$ nA at 25°C and doubles for every 10°C temperature rise, find the input biasing current I_B at 75°C, 100°C, and 125°C.

b. If I_{DDS} changes by 2%, find the input offset voltage V_{OS} and the thermal drift D_v.

2.28 The biasing current for the op-amp in Fig. 2.40 is $I_Q = 200$ μA, and the compensation capacitance is $C_x = 30$ pF. Find **(a)** the slew rate SR and **(b)** the unity-gain bandwidth f_u.

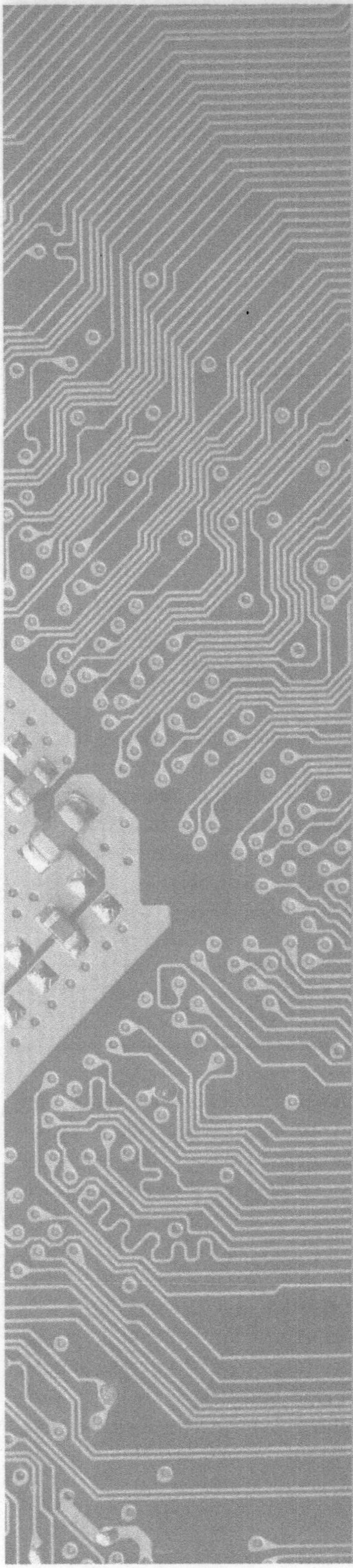

CHAPTER 3

INTRODUCTION TO DIGITAL ELECTRONICS

Learning Outcomes

After completing this chapter, students should be able to do the following:

- Describe the definition of logic states.
- List the performance parameters of logic gates.
- List the logic families and describe their internal circuitry.
- List the relative advantages and disadvantages of logic families.
- Design and analyze simple logic gates.

Symbols and Their Meanings

Symbol	Meaning
I_{IL}, I_{IH}	Input low and input high currents of a logic gate
I_{OL}, I_{OH}	Output low and output high currents of a logic gate
I_{Bn}, I_{Cn}, I_{En}	Base, collector, and emitter currents of an *n*th BJT
i_{Dn}, v_{GSn}	Drain current and gate–source voltage of an *n*th MOSFET
i_{DL}, v_{DSL}	Drain current and drain–source voltage of a load MOSFET
K_L, K_n	Constants of a load MOSFET and a driver NMOS of a logic gate

Symbol	**Meaning**
N	Fan-out of a logic gate
NM_L, NM_H	Low logic–noise margin and high logic–noise margin of a logic gate
P_{on}, P_{off}, P_{static}	On-state, off-state, and static low of a logic gate
R_{dN}, R_{dP}	Drain–source resistances of an NMOS and a PMOS
t_{pdo} or t_{pLH}	Propagation delay time from low to high
t_{pd1} or t_{pHL}	Propagation delay time from high to low
v_I, v_O	Input and output voltages of a logic gate
V_{tL}, V_{tD}	Threshold voltages of load and driver MOSFETs of a logic gate
V_{IL}, V_{IH}	Input low and input high voltages of a logic gate
V_{OL}, V_{OH}	Output low and output high voltages of a logic gate
V_{TW}, V_{LS}	Transition width and logic swing voltages of a logic gate

3.1 Introduction

If we observe the output (v-i) characteristic of a transistor closely, we notice that there are two distinct regions: a low-resistance region and a high-resistance region. In analog electronics, transistors are operated in the active region as amplifying devices, so they exhibit the characteristic of a high output resistance. However, with proper biasing conditions, transistors can also exhibit the characteristic of a low output resistance—that is, a low output voltage. In digital electronics, transistors are operated as on (low output) and off switches.

3.2 Logic States

Electronic circuits used to perform logic functions are known as *digital logic circuits* or *logic gates*. They use two binary variables, or states: 0 (low) and 1 (high). The binary states are normally represented by two distinct voltages: voltage V_H for logic 1 and voltage V_L for logic 0. If the logic 1 voltage is higher than the logic 0 voltage (e.g., $V_H = 5$ and $V_L = 0$), the circuit is said to use *positive logic*. If the logic 1 voltage is lower than the logic 0 voltage (e.g., $V_H = -5$ and $V_L = 0$), the circuit is said to use *negative logic*. In this chapter, we consider positive logic.

To accommodate variations in component tolerances, temperature, and noise in a logic circuit, two voltage ranges are usually used to define the two logic states. The ranges are illustrated in Fig. 3.1, in which V_{H1} is the lowest voltage that will always be recognized as logic 1 and V_{L2} is the highest voltage that will always be recognized as logic 0. If the voltage lies in the range V_{H2} to V_{H1}, the state of the digital circuit is interpreted as logic 1. If the voltage lies in the range V_{L2} to V_{L1}, the state of the digital circuit is interpreted as logic 0. The two voltage regions are separated by an *undefined*, or *excluded, region*. This is a forbidden band, and the signal voltage is not permitted to lie in this region. The difference $V_{H1} - V_{L2}$ is called the *transition region*.

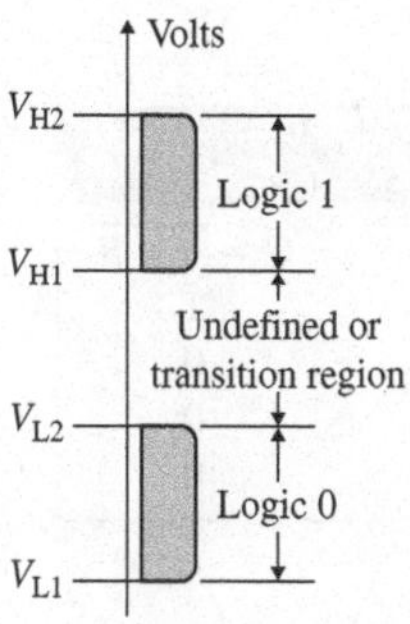

FIGURE 3.1 Voltage ranges and binary variables

KEY POINT OF SECTION 3.2

- The binary states of a logic circuit are normally represented by two distinct voltages: V_H for logic 1 and V_L for logic 0. V_H is positive for positive logic and negative for negative logic.

3.3 Logic Gates

The commonly used logic gates are NOT (inverter), AND, NAND, OR, and NOR. Consider a voltage-controlled switch with a resistance, as shown in Fig. 3.2(a), which is controlled by the input signal v_I applied between terminals 1 and 2. The output is taken across the switch between terminal 3 and the ground terminal 2. When v_I is low (around 0), the switch is open, and the output voltage v_O is high (equal to the DC supply voltage V_{CC}). When v_I is high enough to close the switch (generally above a specified threshold voltage), the switch closes, and the output voltage v_O is low (around 0). The input and output voltages are shown in Fig. 3.2(b). The output is the logical inversion of the input signal, and this circuit is known as an *inverter* or a *NOT gate*. Inverters are the building blocks of logic gates.

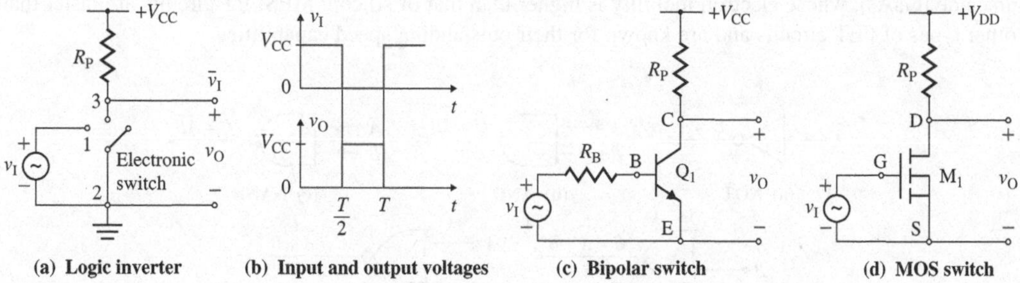

FIGURE 3.2 Inverter

TABLE 3.1 Truth table for NOT, AND, NAND, OR, and NOR

Input	NOT	Input	Input	AND	NAND	OR	NOR
A	$C=\overline{A}$	A	B	$C=AB$	$C=\overline{AB}$	$C=A+B$	$C=\overline{A+B}$
0	1	0	0	0	1	0	1
1	0	0	1	0	1	1	0
		1	0	0	1	1	0
		1	1	1	0	1	0

The switch can be realized by either a bipolar transistor, as shown in Fig. 3.2(c), or a MOS transistor, as shown in Fig. 3.2(d). Transistors M_1 and Q_1 are switched between two states: nonconducting (or off) and conducting (or on). R_P is called the *pull-up resistance* because the output is pulled up toward the positive supply voltage V_{CC} or V_{DD} when the switching transistor is off.

An inverter has only one input voltage. Logic gates usually combine one or more logic variable inputs to produce an output. All of the possible combinations of the input variables and the corresponding outputs are normally listed in a table called a *truth table*. The truth table for NOT, AND, NAND, OR, and NOR is shown in Table 3.1. If both inputs are high, the output becomes high in an AND gate and low in a NAND gate. If at least one input is high, the output becomes high in an OR gate and low in a NOR gate. The symbols for these logic gates are shown in Fig. 3.3.

Digital circuits are available exclusively in integrated circuits (ICs) and can be classified into families. Each member of a family is made with the same technology, has similar structure, and exhibits the same basic features. There are two MOS-logic families (NMOS, using only *n*-channel MOSFETs, and CMOS, using both *n*- and *p*-channel MOSFETs in a complementary configuration), and two BJT families, (a transistor-transistor logic [TTL] family and an emitter-coupled logic [ECL] family). Each of these families has unique advantages and disadvantages. The choice of a logic family is based on considerations such as logic functions, logic flexibility, speed, noise immunity, operating temperature range, power dissipation, and cost.

MOS transistor logic circuits have advantages over bipolar logic gates because MOSFETs are simpler to fabricate and occupy less space in integrated form than BJTs. A MOSFET can be connected to act as a resistive load to replace a diffused resistor (a resistor made by diffusion during the IC-manufacturing process) in bipolar integrated circuits. The packing density of MOSFETs is extremely high, and MOSFET circuits can be fabricated for LSI, VLSI, and ULSI applications. MOSFETs and metal–Schottky barrier FETs (MESFETs) can also be used in digital integrated circuits. MESFETs are usually fabricated of gallium arsenide (GaAs), whose electron mobility is higher than that of silicon. MESFET circuits are faster than other types of FET circuits and are known for their outstanding speed capabilities.

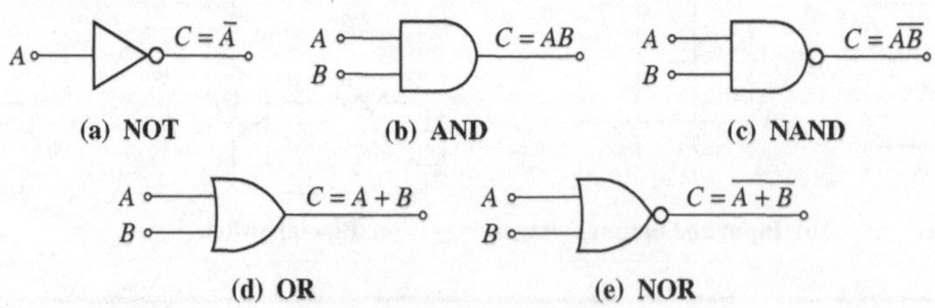

FIGURE 3.3 Symbols for logic gates

TABLE 3.2 Classification of digital integrated circuits

Degree of Integration	Number of Gates
Small-scale integration (SSI)	Fewer than 10
Medium-scale integration (MSI)	From 10 to 100
Large-scale integration (LSI)	From 100 to 1000
Very-large-scale integration (VLSI)	From 1000 to 10^5
Ultra-large-scale integration (ULSI)	More than 10^5

Depending on the complexity of the internal circuitry of the IC chip, digital IC packages can be classified by degree of integration, as shown in Table 3.2.

KEY POINT OF SECTION 3.3

- The commonly used logic gates are NOT (inverter), AND, NAND, OR, and NOR. Inverters are the building blocks of logic gates. There are many families of logic gates, among them NMOS, CMOS, TTL, and ECL gates.

3.4 Performance Parameters of Logic Gates

Figure 3.2(b) shows the characteristics of an ideal inverter [1–3]. However, the performance of actual inverters differs significantly from the ideal in the following ways:

1. The switch is not ideal; that is, when the switch is closed, it has a finite voltage drop rather than a short circuit.
2. The switch may not open and close instantaneously because of a time delay between the application of the input signal and the propagation of the desired output signal.
3. The input terminal of the inverter usually draws some current from the driving source.
4. The inverter usually drives a load or acts as a source to the next stage(s), and it also should be capable of supplying the driving current.

Manufacturers' data sheets on logic gates specify many performance parameters. As examples, typical data for TTL and CMOS devices are shown in Table 3.3, where t_{pd0} (also abbreviated as t_{pLH}) is the propagation delay time from low to high and t_{pd1} (also abbreviated as t_{pHL}) is the propagation delay time from high to low.

TABLE 3.3 Parameters of 54L/74L TTL and 54C/74C CMOS logic gates

Family	V_{CC}	V_{IL} max	I_{IL} max (mA)	V_{IH} min	I_{IH} 2.4 V (μA)	V_{OL} max	I_{OL} (μA)	V_{OH} min	I_{OH} (μA)	t_{pd0} TYP (ns)	t_{pd1} TYP (ns)	P_D/gate (μW)	P_D/gate 1 MHz, 50 pF (mW)
54L/74L	5	0.7	0.18	2.0	10	0.3	2000	2.4	100	31	35	1000	2.25
54C/74C	5	0.8	—	3.5	—	0.4	360	2.4	100	60	45	0.01	1.25
54C/74C	10	2.0	—	8.0	—	1.0	10	8.0	10	25	30	0.03	5

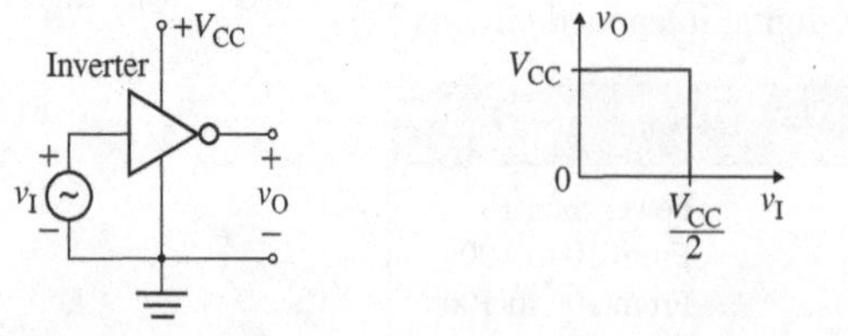

(a) Inverter connection (b) Transfer characteristic

FIGURE 3.4 Transfer characteristic of an ideal inverter

3.4.1 Voltage Transfer Characteristic

The *voltage transfer characteristic* (VTC) gives the relationship between the input voltage v_I and the output voltage v_O. An inverter with a single power supply V_{CC} and its VTC are shown in Figs. 3.4(a) and 3.4(b), respectively. The inverter has a threshold voltage of $V_{tI} = V_{CC}/2$; that is, the output is high (at $v_O = V_{CC}$) for $v_I < V_{CC}/2$ and low (at $v_O = 0$) for $v_I > V_{CC}/2$. The transition from low to high and vice versa is very sharp at $v_I = V_{CC}/2$. Thus, the incremental voltage gain is $dv_O/dv_I = 0$ for $v_I < V_{CC}/2$ and it is $dv_O/dv_I = \infty$ for $v_I = V_{CC}/2$. Therefore, the inverter exhibits a nonlinear characteristic. If it is connected with feedback, it will oscillate between the low and high states.

A practical inverter does not have a finite threshold voltage; rather, it goes through a *transition region* from the high state to the low state. The VTC of a practical inverter is shown in Fig. 3.5; the VTC has three distinct regions: the low-input region, $v_I < V_{IL}$; the transition region, $V_{IL} \le v_I \le V_{IH}$; and the high-input region, $v_I > V_{IH}$.

The VTC has two transitions, one at $v_I = V_{IL}$ and one at $v_I = V_{IH}$, with two corresponding output voltages, V_{OH} and V_{OL}. The transition voltages are defined as the points at which the slope of the VTC is -1 (that is, $dv_O/dv_I = -1$). The following variables are applicable to all logic circuits:

V_{OH} (high-level output voltage) is the *minimum* output voltage that will establish a high level (logic 1). Data sheets guarantee that the output voltage will exceed this level at all specified operating conditions.

V_{OL} (low-level output voltage) is the *maximum* output voltage that will establish a low level (logic 0). Data sheets guarantee that the output voltage will not exceed this level at all specified operating conditions.

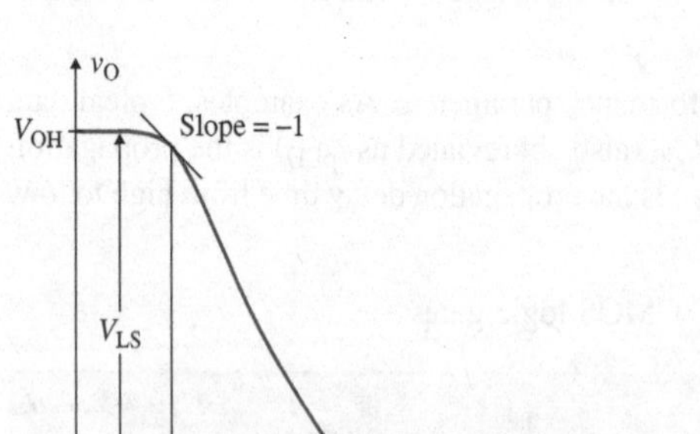

FIGURE 3.5 Transfer characteristic of a practical inverter

V_{IL} (low-level input voltage) is the *maximum* positive voltage that can be applied to an input terminal of a gate and still be recognized as logic low (0).

V_{IH} (high-level input voltage) is the *minimum* positive voltage that can be applied to an input terminal of a gate before the transition starts from the logic high (1) to logic low (0).

In the transition region, the output is undefined. The width of the transition region is a measure of ambiguity, and it is defined by

$$V_{TW} = V_{IH} - V_{IL} \tag{3.1}$$

A low value of V_{TW} is desirable to reduce ambiguity in the input logic state. *Logic swing* is also a measure of the ambiguity in the logic state, and it is defined by

$$V_{LS} = V_{OH} - V_{OL} \tag{3.2}$$

A high value of V_{LS} is desirable to reduce ambiguity and increase noise immunity. A gate for which $V_{OH} = 2.4$ V, $V_{OL} = 0.3$ V, $V_{IH} = 2$ V, and $V_{IL} = 0.7$ V has

$$V_{TW} = V_{IH} - V_{IL} = 2 - 0.7 = 1.3 \text{ V}$$

$$V_{LS} = V_{OH} - V_{OL} = 2.4 - 0.3 = 2.1 \text{ V}$$

3.4.2 Noise Margins

Noise generally is present in logic circuits, superimposed on input signals. Noise simply refers to extraneous signals, which may arise from inadequate regulation or decoupling of the power supply, electromagnetic radiation, inductive or capacitive coupling from other parts of the system, or line drops. One of the advantages of logic circuits is their tolerance for variations in the input signal, which results from the fact that the input signal is interpreted simply as high or low. *Noise immunity* is a measure of the tolerance for variations in the signal level, and it is that voltage that, applied to the input, will cause the output to change its state. Noise immunity is an important device characteristic. However, *noise margin* is of more use to the designer; it defines the amount of noise a system can tolerate under any circumstances and still maintain the integrity of the logic levels. That is, noise margin measures the ability of a gate to maintain its logic state under varying voltage levels.

In digital circuits, one gate usually drives another; that is, the output of the first gate is the input to the following gate. Thus, the gate whose output is high at V_{OH} will drive an identical gate whose high-level input voltage is V_{IH}. This concept is illustrated in Fig. 3.6(a). The difference $V_{OH} - V_{IH}$ represents a margin of safety at the logic high output. It is called the *logic 1*, or *high*, *noise margin* and is given by

$$NM_H = V_{OH} - V_{IH} \tag{3.3}$$

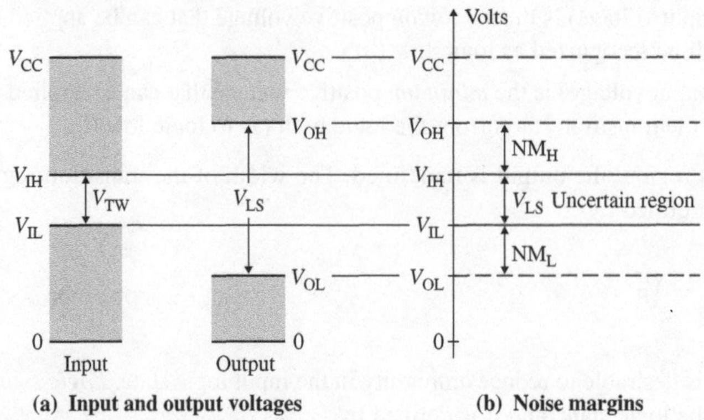

FIGURE 3.6 Noise margins

Similarly, the noise margin at the logic low output is called the *logic 0*, or *low*, *noise margin* and is denoted by

$$NM_L = V_{IL} - V_{OL} \tag{3.4}$$

Therefore, the noise margin is the difference between the guaranteed logic 1 (or 0) level output voltage and the guaranteed logic 1 (or 0) level input voltage. The *absolute noise margin* is the smaller of the two noise margins; that is,

$$NM = \min(NM_L, NM_H) \tag{3.5}$$

The noise margins are shown in Fig. 3.6(b). A gate with a high logic swing V_{LS} and a small transition width V_{TW} will have good noise immunity. There will be no uncertain or undefined region if $V_{IL} = V_{IH} = (V_{OH} + V_{OL})/2$, and the noise margins will be equal: $NM_L = NM_H$. However, this will cause the transition region to be eliminated with an abrupt switch in the ideal VTC, as shown in Fig. 3.4(b). For example, a gate for which $V_{OH} = 2.5$ V, $V_{OL} = 0.3$ V, $V_{IH} = 2$ V, and $V_{IL} = 0.7$ V has

$$NM_H = V_{OH} - V_{IH} = 2.5 - 2 = 0.5 \text{ V}$$

$$NM_L = V_{IL} - V_{OL} = 0.7 - 0.3 = 0.4 \text{ V}$$

$$NM = \min(NM_L, NM_H) = \min(0.4, 0.5) = 0.4 \text{ V}$$

3.4.3 Fan-Out and Fan-In

A gate draws an input current from the input signal and also delivers current to the load gate(s). Note that the current flowing out of a terminal will have a negative value. The gate acts as the load for the input signal and as the driver for the load gates. Just as there are four voltages, there are four currents associated with a gate and its load, and they are defined as follows:

I_{OH} (high-level output current) is the current flowing into the output terminal when it is in the high state (logic 1).

I_{OL} (low-level output current) is the current flowing into the output terminal when it is in the low state (logic 0).

I_{IL} (low-level input current) is the current flowing into the input terminal when a specified low-level voltage (logic 0) is applied to the input.

I_{IH} (high-level input current) is the current flowing into the input terminal when a specified high-level voltage (logic 1) is applied to the input.

A gate must be capable of accepting more than one input. The number of independent input nodes is known as the *fan-in*. The output of one gate must be capable of driving more than one input of subsequent gates. The number N of inputs that can be driven by a gate is known as the *fan-out*. Fan-out is illustrated in Fig. 3.7. More precisely, *fan-out* is defined as the maximum number of load gates of similar design that can be connected to the output of a logic gate (i.e., the driver gate) without changing its logic state. Since logic gates draw a different amount of current in the logic low and logic high states, the fan-out is the smaller of the numbers for the logic low and logic high states. That is, fan-out N is equal to either I_{OL}/I_{IL} or I_{OH}/I_{IH}, whichever gives the lower natural number. N is given by

$$N \equiv \min\left(\frac{I_{OL}}{I_{IL}}, \frac{I_{OH}}{I_{IH}}\right) \tag{3.6}$$

For example, if $I_{OL} = 2$ mA, $I_{IL} = 0.1$ mA, $I_{OH} = 100$ μA, and $I_{IH} = 10$ μA,

$$N \equiv \min\left[\left(\frac{2\text{ mA}}{0.1\text{ mA}}\right), \left(\frac{100\ \mu\text{A}}{10\ \mu\text{A}}\right)\right] = \min(20,10) = 10$$

The fan-out of a logic gate is usually quoted in terms of the number of inverters the gate can drive. This helps in comparing gates and determining the loading effects of the gate in a digital circuit with multiple gates of the same family.

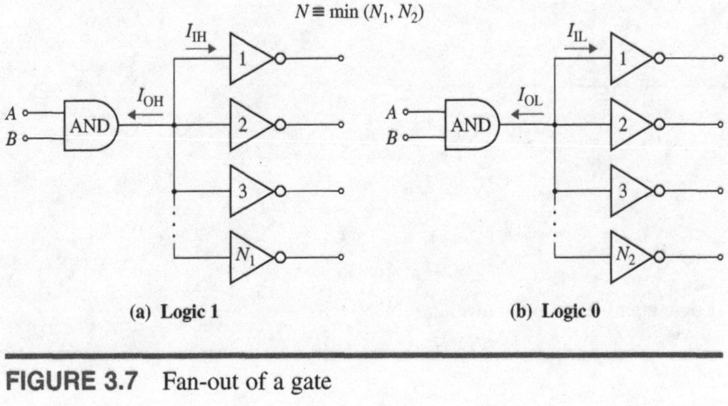

FIGURE 3.7 Fan-out of a gate

EXAMPLE 3.1

D **Designing a simple inverter** An inverter, as shown in Fig. 3.8, drives identical inverters and has $V_{OL} = 0.3$ V, $I_{OL} = 2$ mA, $V_{OH} = 2.4$ V, $I_{OH} = 100$ μA, and $V_{CC} = 5$ V. The input currents drawn by each load inverter are $I_{IH} = 0.18$ mA (at logic high) and $I_{IL} = 10$ μA (at logic low).

(a) If there are five load inverters, determine the value of pull-up resistance R_P that will ensure a logic 1 output of $V_{OH} = 2.4$ V.

(b) If $R_P = 4$ kΩ, find the fan-out N.

SOLUTION

(a) At high output, all load inverters are connected to the driving inverter, and each draws 0.18 mA. Because of the resistance R_P, the output voltage will be lower than V_{CC}. Thus,

$$v_O = V_{CC} - R_P(NI_{IH} + I_{OH})$$

To ensure that $v_O \geq V_{OH} = 2.4$ V, we calculate the value of R_P (for $N = 5$) as

$$R_P \leq \frac{V_{CC} - V_{OH}}{NI_{IH} + I_{OH}} = \frac{(5 - 2.4)\text{ V}}{5 \times 0.18\text{ mA} + 100\text{ μA}} = 2.6\text{ kΩ}$$

At low output, each load inverter draws $I_{IL} = 10$ μA. Thus,

$$V_{CC} = V_{OL} + R_P(I_{OL} + NI_{IL})$$

To ensure that $R_P \geq I_{OL} = 2$ mA, we calculate the value of R_P as

$$R_P \geq \frac{V_{CC} - V_{OL}}{I_{OL} + NI_{IL}} = \frac{(5 - 0.3)\text{ V}}{2\text{ mA} + 5 \times 10\text{ μA}} = 2.29\text{ kΩ}$$

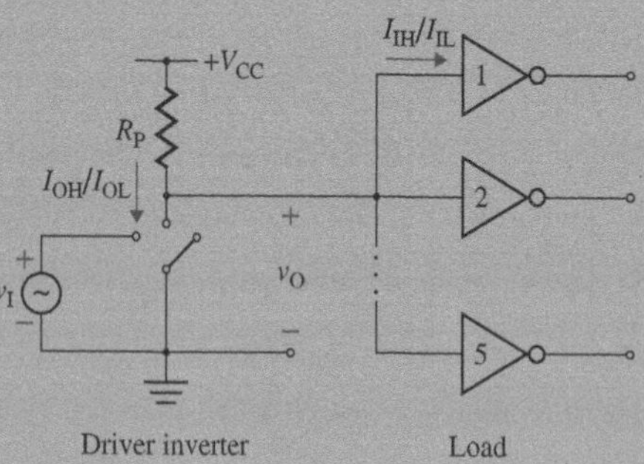

FIGURE 3.8 Logic high fan-out of an inverter

which depends mostly on the value of I_{OL} because $I_{IL} << I_{OL}$. Therefore, the value of R_P should be in the range $2.29\ \text{k}\Omega \leq R_P \leq 2.6\ \text{k}\Omega$. Let us choose $R_P = 2.5\ \text{k}\Omega$.

(b) For $R_P = 4\ \text{k}\Omega$, we get the value of N (for logic high) as

$$N = \frac{V_{CC} - V_{OH}}{R_P I_{IH}} - \frac{I_{OH}}{I_{IH}} = \frac{(5 - 2.4)\ \text{V}}{4\ \text{k}\Omega \times 0.18\ \text{mA}} - \frac{0.1\ \text{mA}}{0.18\ \text{mA}} = 3.06$$

Since fractional loads are not possible, the fan-out is $N = 3$. If we choose $N = 4$, v_O will be less than $V_{OH} = 2.4$ V, and the inverter will be in the transition (or logic 0) region.

3.4.4 Propagation Delay

A switching device such as a bipolar transistor exhibits junction capacitances. As a result, the output of an inverter may not respond instantaneously to the input signal. In addition, the load gate(s) offers a certain amount of capacitance C_L to the driving inverter, as shown Fig. 3.9(a). C_L is the equivalent input capacitance of the load gate(s), including any capacitance due to the wiring connection. Therefore, the input and output responses of the inverter will exhibit finite rise time t_r and fall time t_f, as shown in Fig. 3.9(b). The *rise time* t_r is the time required for the waveform to rise from 10% to 90% of its final (high) value. Similarly, the *fall time* t_f is the time required for the waveform to decrease from 90% to 10% of its final (low) value.

The speed of operation of a gate depends on how fast a change in the input propagates through it and causes a change at the output. There will be a delay time between the input and output waveform; this time is commonly known as the *propagation delay time* t_{pd}. The propagation delay time is defined as the time between when the input pulse waveform is at 50% of its logic high value and when the corresponding output pulse waveform is at 50% of its logic high value. For two edges (i.e., falling and rising), there are two delay times, denoted as t_{pd1} (or t_{pHL}) for the high-to-low logic and t_{pd0} (or t_{pLH}) for the low-to-high logic. The average of t_{pd1} and t_{pd0} is the average propagation time t_{pd}:

$$t_{pd} = \frac{t_{pd1} + t_{pd0}}{2} \tag{3.7}$$

This value is commonly used as a figure of merit to compare the performance of different logic families. For hand calculations of the propagation delays, the input voltage can be assumed to be ideal, as shown in Fig. 3.9(c). Typical values of t_{pd} range from 0.5 ns to 10 ns. Cycle time t_{cyc}, another parameter used to compare the performance of logic families, is the time between identical points of successive cycles in a signal waveform. The clock frequency f_{clk}, which is the reciprocal of the cycle time, is more often used. Practical digital systems are usually designed to operate with a cycle time 20 to 50 times the propagation time of a single gate.

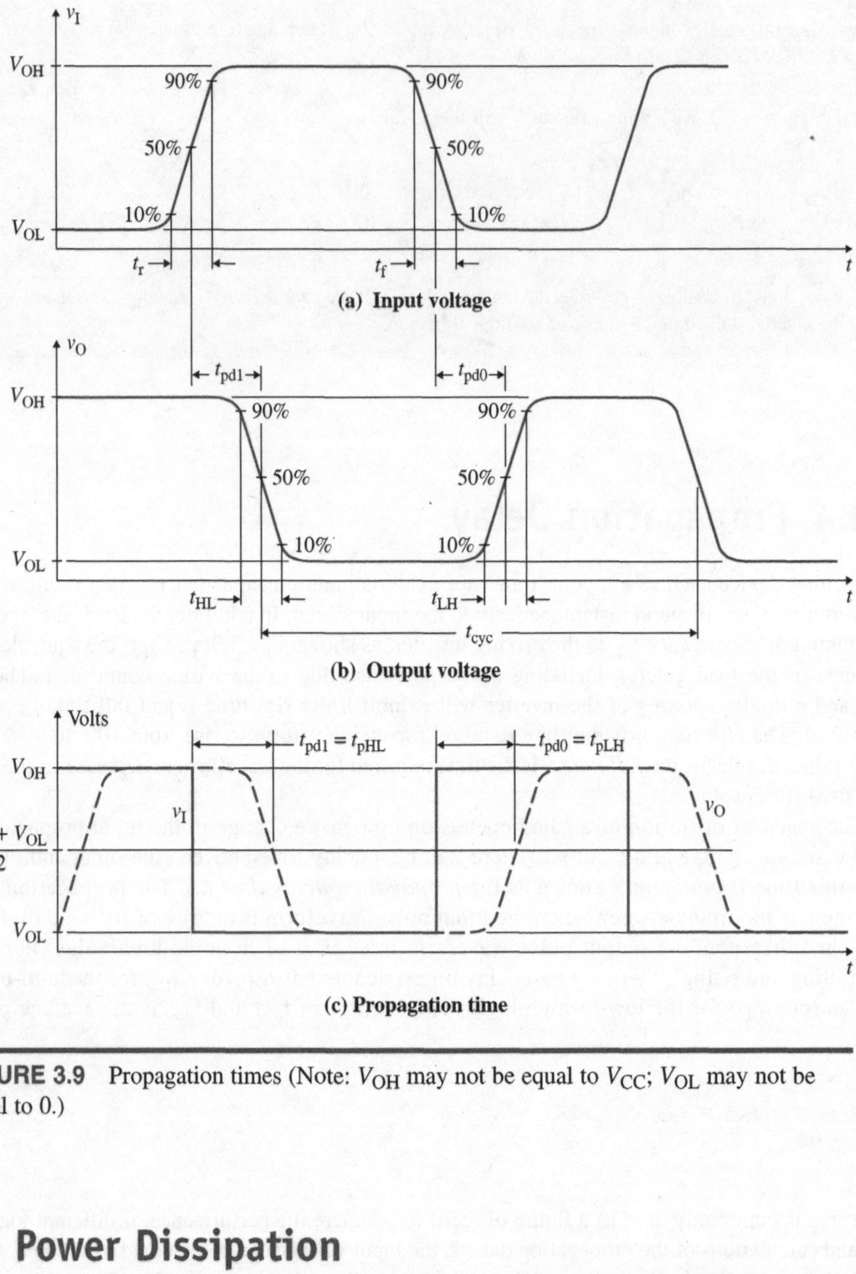

FIGURE 3.9 Propagation times (Note: V_{OH} may not be equal to V_{CC}; V_{OL} may not be equal to 0.)

3.4.5 Power Dissipation

The amount of power (P_D) consumed by a digital circuit is also an important parameter. Knowing this parameter enables the designer to determine the amount of current that will be drawn from the power supply. The power dissipation has static and dynamic components. As an example, consider the inverter in Fig. 3.10(a), where C_L is the load capacitance (normally the input capacitance of another gate or the wiring capacitance or internal capacitance of the switching device itself).

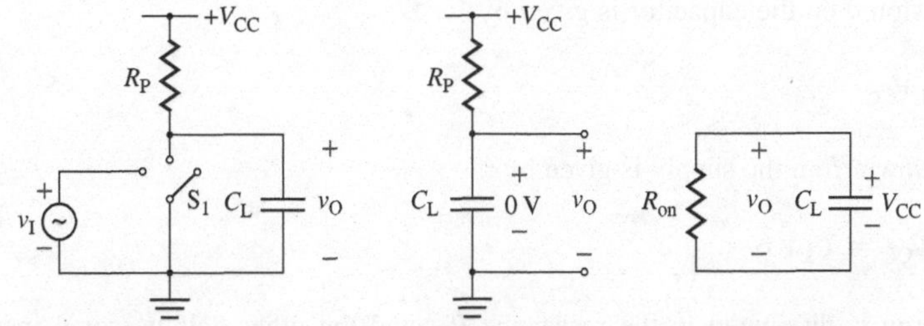

(a) Inverter (b) Charging (c) Discharging

FIGURE 3.10 Charging and discharging of capacitor C_L

Static Power

The power consumed by an inverter depends on its logic state. When the switch S_1 is closed at logic 0, the current is drawn from the supply. The power delivered by the supply is called the *static*, or *quiescent*, *power*. Thus, the static power at logic 0 is given by

$$P_{on} = \frac{V_{CC}^2}{R_P + R_{on}} \tag{3.8}$$

where R_{on} is the on-state switch resistance, whose value is usually low, and $R_{on} << R_P$. When the switch is open at logic 1, a small leakage current I_{leak} flows through the switch. The static power at logic 1 is given by

$$P_{off} = \frac{V_{CC}^2}{R_P + R_{off}} \tag{3.9}$$

where R_{off} is the off-state switch resistance, whose value is usually very large, and $R_{off} >> R_P$. Thus, for the idealized inverter, $P_{on} \approx V_{CC}^2/R_P$ and $P_{off} \approx 0$. The static power is usually expressed as an average value. Assuming that a gate, on average, spends half the time in each state, the *average static power dissipation* becomes

$$P_{static} = \frac{1}{2}(P_{on} + P_{off}) \tag{3.10}$$

$$\approx \frac{V_{CC}^2}{2R_P} \quad \left(\text{for } P_{on} \approx \frac{V_{CC}^2}{R_P} \text{ and } P_{off} \approx 0\right) \tag{3.11}$$

Dynamic Power

An inverter also consumes power each time it changes state. Let us assume that at $t = 0^-$, the input is high and the switch is closed. Thus, the capacitor C_L is discharged, and it has no charge, as shown in Fig. 3.10(a). When the switch is opened at $t = 0^+$, the capacitor will charge exponentially to the supply voltage V_{CC} (approximately) through R_P, as shown in Fig. 3.10(b). The charging current will also flow through R_P, and thus power will be dissipated in R_P.

The charge stored on the capacitor is given by

$$Q = C_L V_{CC} \tag{3.12}$$

and the energy drawn from the supply is given by

$$E = QV_{CC} = C_L V_{CC}^2 \tag{3.13}$$

Half of this energy is dissipated in the resistance R_P, and the other half is stored in the capacitor as $(1/2)C_L V_{CC}^2$. The next time the input becomes high, the switch is closed, as shown in Fig. 3.10(c), and the capacitor discharges through the switch resistance R_{on}. That is, the energy stored in the capacitor is dissipated as heat in R_{on}. Therefore, every time the capacitor C_L is charged or discharged, an amount of energy must be provided by the power supply. The energy per cycle is $C_L V_{CC}^2$. Since energy per unit time is the power, the *dynamic power dissipation* is given by

$$P_{dynamic} = f_{clk} C_L V_{CC}^2 \tag{3.14}$$

where f_{clk} is the clock frequency of the inverter in hertz. Therefore, the total power that must be supplied by the power supply is given by

$$P_D = P_{static} + P_{dynamic} \tag{3.15}$$

Since P_D is dependent on the clock frequency, power dissipation can be a severe problem in digital circuits with frequencies over 100 MHz.

EXAMPLE 3.2

Finding the delay times and power dissipation of an inverter The inverter shown in Fig. 3.10(a) has $V_{OL} = 0.3$ V, $V_{OH} = 2.4$ V, $V_{CC} = 5$ V, $R_{on} = 500\ \Omega$, $R_{off} \approx \infty$, $R_P = 2.6$ kΩ, $C_L = 5$ pF, and $f_{clk} = 10$ MHz.

(a) Find the delay times for the output voltage to rise from 0.3 V to 2.4 V (t_{pd0}) and to fall from 5 V to 0.3 V (t_{pd1}).

(b) Find the power dissipation P_D.

SOLUTION

(a) When the switch is open, the capacitor C_L charges exponentially from 0.3 V to 5 V. The output voltage $v_O(t)$ can be expressed in the general form

$$v_O(t) = v_O(t = \infty) + [v_O(t = 0) - v_O(t = \infty)]e^{-t/\tau} \tag{3.16}$$

For $v_O(t = 0) = 0.3$ V and $v_O(t = \infty) = 5$ V, Eq. (3.16) becomes

$$v_O(t) = 5 - 4.7e^{-t/\tau}$$

which, for $v_O(t) = 2.4$ V and $\tau = R_P C_L = 2.6$ k$\Omega \times 5$ pF $= 13$ ns, gives the delay time $t_{pd0} = -\tau \ln [(5 - 2.4)/(5 - 0.3)] = 7.7$ ns. When the switch is closed, the capacitor C_L discharges exponentially from 5 V to 0.3 V. The output voltage $v_O(t)$ can be expressed in the general form

$$v_O(t) = v_O(t = 0)e^{-t/\tau}$$

For $v_O(t = 0) = 2.4$ V, the above equation becomes

$$v_O(t) = 2.4e^{-t/\tau}$$

which, for $v_O(t)$ to fall from 2.4 V to 0.3 V and $\tau = R_{on}C_L = 0.5$ k$\Omega \times 5$ pF $= 2.5$ ns, gives the delay time $t_{pd1} = -\tau \ln (0.3/2.4) = 5.2$ ns.

(b) Substituting the numerical values in Eq. (3.10) gives

$$P_{static} = \frac{V_{CC}^2}{2\,(R_P + R_{on})} = 4.03 \text{ mW}$$

From Eq. (3.14),

$$P_{dynamic} = f_{clk} C_L V_{CC}^2 = 10 \text{ MHz} \times 5 \text{ pF} \times 5^2 = 1.25 \text{ mW}$$

Thus, $P_D = 4.03 + 1.25 = 5.28$ mW.

3.4.6 Delay-Power Product

It is desirable to have both a low propagation delay time (high speed) and low power dissipation. However, these two requirements conflict with each other. For example, if the power dissipation is reduced by decreasing the supply current, the delay will increase. The *delay-power product* (DP) is the product of average propagation delay time (t_{pd}) and power dissipation (P_D):

$$\text{DP} = t_{pd} P_D \tag{3.17}$$

DP is a figure of merit for comparing logic gates. A small value of DP indicates that a circuit has a fast switching speed and dissipates very little power. In practice, a trade-off between power dissipation and switching speed is usually required. Typical values of DP range from 5 pJ to 50 pJ.

KEY POINT OF SECTION 3.4

- The performance specifications of a logic gate normally include a description of the voltage transfer characteristic, noise margins, fan-out, fan-in, propagation delay, power dissipation, and delay-power product.

3.5 NMOS Inverters

NMOS inverters are the building blocks of NMOS digital circuits [4–6]. They use only *n*-channel MOSFETs, which have low channel resistance because of the greater mobility of electrons in an *n*-channel. If all the MOSFETs are enhancement-type devices, the fabrication procedure becomes simpler. But combining an enhancement-type device (as the driver) and a depletion-type device (as the load) increases the switching speed. An enhancement-type transistor is used as the driver because (1) it will be off when the input voltage v_I is low and (2) its drain and gate voltages have the same polarity. This feature allows direct coupling between stages (i.e., one stage can be connected to the next one without any coupling capacitor). Although depletion-load inverters are generally used in integrated circuits for high-speed switching, we will analyze inverters with both enhancement and depletion MOSFETs.

3.5.1 NMOS Inverter with Enhancement Load

In ICs, an inverter uses a transistor as the load because it requires much less chip area (typically 50 times less than for a diffused resistor with the same value). The resistor in Fig. 3.2(a) can be replaced by a MOSFET.

Enhancement Load

Consider an enhancement-type *n*-channel MOSFET (NMOS) whose gate is connected to the drain terminal, as shown in Fig. 3.11(a). For $v_{GSL} = v_{DSL} \le V_{tL}$ (threshold voltage), the drain current will be zero; that is,

$$i_{DL} = 0$$

$$v_{DSL} = v_{GSL}$$

where L in the subscript after D, DS, and GS refers to the load MOSFET.

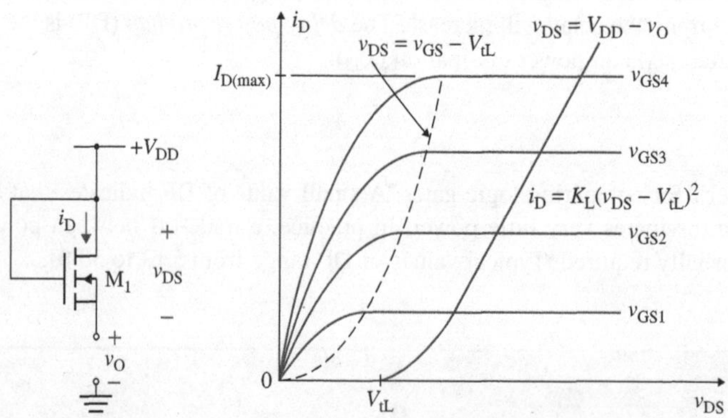

(a) Circuit **(b) *v-i* characteristics**

FIGURE 3.11 *n*-channel enhancement MOSFET as a load with $v_{GS} = v_{DS}$

For $v_{GSL} = v_{DSL} > V_{tL}$, a drain current will flow and the transistor will always operate in the saturation (pinch-down) mode under the following condition:

$$v_{DS} > (v_{GSL} - V_{tL}) = v_{DSL} - V_{tL} = V_{DS(sat)}$$

The corresponding drain current is given by

$$i_{DL} = K_L(v_{GSL} - V_{tL})^2 \tag{3.18}$$

where K_L is the MOS constant and V_{tL} is the threshold voltage of the load MOSFET. The output characteristics, which are shown in Fig. 3.11(b), indicate that the transistor will act as a nonlinear resistor if operated in the saturation region.

Static Characteristics

An NMOS inverter with an enhancement load is shown in Fig. 3.12(a). The substrates of the MOSFETs (M_D and M_L) are connected to the ground. The substrate in an IC logic gate is common to all devices, and therefore it is connected to the ground (or to the most negative potential). Consequently, there is a nonzero source-to-body voltage V_{SB} (which varies with logic levels) for the device. There is also a body-to-source capacitance, which influences both the frequency and the transient response of the device. However, this effect, called the *body effect*, is not present if the substrate (i.e., the body) is connected directly to the source. To simplify the analysis, we will assume that the substrates are connected to the source terminals so that $V_{SB} = 0$ for both M_D and M_L.

The output characteristics of the load and driver transistors are shown in Fig. 3.12(b). Before the application of high input, $v_I = V_{OL}$ and $v_O = V_{OH}$. The inverter is operating at point *A*. When v_I goes high (to V_{OH}), transistor M_D turns on, and the operating point jumps from *A* to *B* and then moves along the *i*-*v* curve of M_D until it finally reaches the quiescent point *Q* so that $v_O = V_{OL}$. When v_I goes from high to low (to V_{OL}), the operating point jumps from *Q* to *A* and then moves along the *i*-*v* curve of M_L until it finally reaches $v_O = V_{OH}$. Depending on the input voltage v_I, the circuit operation can be divided into three regions, as shown in Fig. 3.12(c): region I, region II, and region III.

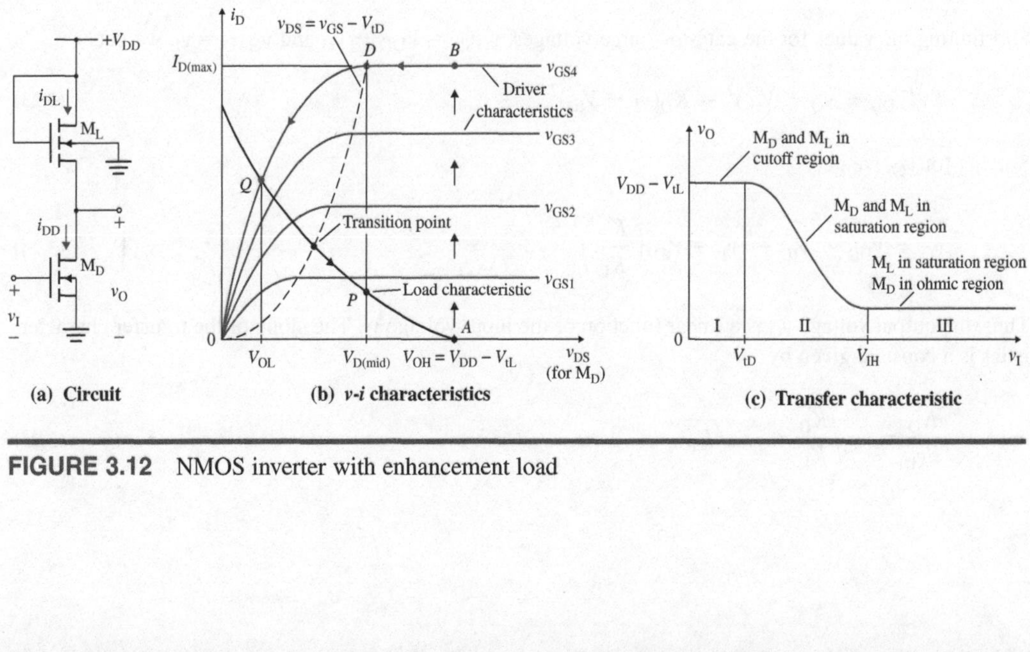

FIGURE 3.12 NMOS inverter with enhancement load

In region I, the input voltage $v_I \leq V_{tD}$, where V_{tD} is the threshold voltage of the drive transistor M_D. Thus, M_D is cut off, so the drain currents of the two transistors must be zero. For the load transistor M_L,

$$i_{DL} = 0 = K_L(v_{GSL} - V_{tL})^2 \tag{3.19}$$

where i_{DL} = drain current of the load transistor M_L
v_{GSL} = gate-to-source voltage of the load transistor M_L
V_{tL} = threshold voltage of the load transistor M_L
K_L = MOS constant parameter of the load transistor M_L

Using KVL, we can write

$$v_{GSL} = V_{DD} - v_O \tag{3.20}$$

Substituting v_{GSL} from Eq. (3.20) into Eq. (3.19) yields

$$0 = K_L(V_{DD} - v_O - V_{tL})^2$$

which gives the output voltage as

$$v_O = V_{OH} = V_{DD} - V_{tL} \tag{3.21}$$

Therefore, the high output voltage V_{OH} is less than V_{DD} by the amount of the threshold voltage of the load transistor M_L. The input voltage corresponding to $v_O = V_{OH}$ is

$$V_{IL} = V_{tD} \tag{3.22}$$

In region II, $v_I > V_{tD}$. As v_I becomes slightly greater than V_{tD}, M_D begins to conduct and operates in the saturation region. The drain currents of the two transistors will be equal; that is, $i_{DL} = i_{DD}$, which, for both transistors in saturation, gives

$$K_L(v_{GSL} - V_{tL})^2 = K_D(v_{GSD} - V_{tD})^2$$

▶ **NOTE** D in the subscripts after D, DS, and GS refers to the driver MOSFET.

Substituting the values for the gate-to-source voltages, $v_{GSL} = V_{DD} - v_O$ and $v_{GSD} = v_I$, we get

$$K_L(V_{DD} - v_O - V_{tL})^2 = K_D(v_I - V_{tD})^2 \tag{3.23}$$

Solving for v_O, we get

$$v_O = V_{DD} - V_{tL} - (v_I - V_{tD})\left(\frac{K_D}{K_L}\right)^{1/2} \tag{3.24}$$

Thus, the output voltage v_O is a linear function of the input voltage v_I. The slope of the transfer characteristics is a constant given by

$$\frac{dv_O}{dv_I} = -\sqrt{\frac{K_D}{K_L}} = -\sqrt{K_R} \tag{3.25}$$

where K_R is known as the *geometry ratio*. The slope changes abruptly from 0 to $-\sqrt{K_R}$ at $V_{IL} = V_{tD}$. The slope, which is inversely proportional to the transition width V_{TW}, can be increased and the transition region can be narrowed with a large value of K_D/K_L. Since the MOS constant parameters (K_D and K_L) are proportional to the ratio W/L, the slope can be rewritten as

$$\frac{dv_O}{dv_I} = -\left[\frac{(W/L)_D}{(W/L)_L}\right]^{1/2} \tag{3.26}$$

The voltage $V_{D(mid)} = v_I = v_O$ can be found from Eq. (3.23):

$$K_L(V_{DD} - V_{MD} - V_{tL})^2 = K_D(V_{MD} - V_{tD})^2$$

which can be solved for V_{MD} as follows:

$$V_{MD} = \frac{V_{DD} - V_{tL} + V_{tD}\sqrt{K_R}}{1 + \sqrt{K_R}} \tag{3.27}$$

If v_I is increased sufficiently, M_D will operate at the edge of the saturation region. If $v_{DSD(sat)}$ is the drain–source voltage at the transition point,

$$v_{DSD(sat)} = v_{GSD} - V_{tD} \tag{3.28}$$

Since $v_{DSD(sat)} = v_O$ and $v_{GSD} = v_I$, Eq. (3.28) can be written as

$$v_O = v_I - V_{tD} \tag{3.29}$$

Equating v_O in Eq. (3.24) to the output voltage in Eq. (3.29) gives the input voltage $V_{I(tran)}$ at the transition point between the saturation and nonsaturation (ohmic) regions; that is,

$$v_O = V_{I(tran)} - V_{tD} = V_{DD} - V_{tL} - (V_{I(tran)} - V_{tD})\sqrt{K_R}$$

which, solved for $V_{I(tran)}$, yields

$$V_{I(tran)} = \frac{V_{DD} - V_{tL} + V_{tD}(1 + \sqrt{K_R})}{1 + \sqrt{K_R}} \tag{3.30}$$

In region III, $v_I > V_{I(tran)}$. M_L and M_D operate in the saturation and nonsaturation (ohmic) regions, respectively. The drain current i_D is expressed in a general form in terms of external voltages V_{GS} and V_{DS} as given by

$$i_{DD} = K_D[2(v_{GSD} - V_{tD})v_{DSD} - v_{DSD}^2] \tag{3.31}$$

Since the two drain currents must be the same, we can find the relation between the input and the output voltages. Thus, from Eqs. (3.18) and (3.31), we get

$$K_L(v_{GSL} - V_{tL})^2 = K_D[2(v_{GSD} - V_{tD})v_{DSD} - v_{DSD}^2] \tag{3.32}$$

Substituting the values for the gate-to-source voltages, $v_{GSL} = V_{DD} - v_O$, $v_{GSD} = v_I$, and $v_{DSD} = v_O$, we get the relationship between the input and output voltages:

$$K_L(V_{DD} - v_O - V_{tL})^2 = K_D[2(v_I - V_{tD})v_O - v_O^2] \tag{3.33}$$

If we use $dv_O/dv_I = -1$, Eq. (3.33) gives

$$v_I = V_{tD} + 2v_O + \frac{v_O + V_{tL} - V_{DD}}{K_R} \tag{3.34}$$

Solving Eqs. (3.33) and (3.34) for the logic high input voltage $V_I = V_{IH}$ and the corresponding V_O, we get

$$v_{IH} = \frac{(V_{DD} - V_{tL})(2 + 1/K_R)}{\sqrt{1 + 3K_R}} + V_{tD} - \frac{V_{DD} - V_{tL}}{K_R} \tag{3.35}$$

$$V_O = \frac{V_{DD} - V_{tL}}{\sqrt{1 + 3K_R}} \tag{3.36}$$

▶ **NOTE** V_O in Eq. (3.36) is not V_{OL}, which is found by substituting $v_I = V_{OH} = V_{DD} - V_{tL}$ and $v_O = V_{OL}$ in Eq. (3.33) and then solving the quadratic equation for V_{OL}.

The output voltage and the drain currents depend on the input voltage v_I. The transfer characteristics are shown in Fig. 3.13 for various values of the ratio K_D/K_L. As the ratio K_D/K_L becomes larger, a steeper characteristic is obtained, which is highly desirable in digital circuits. At a larger value of K_D/K_L, the low output voltage V_{OL} becomes smaller. To guarantee turnoff of the succeeding stages, the low output voltage V_{OL} should be less than the threshold voltage of the driver transistors.

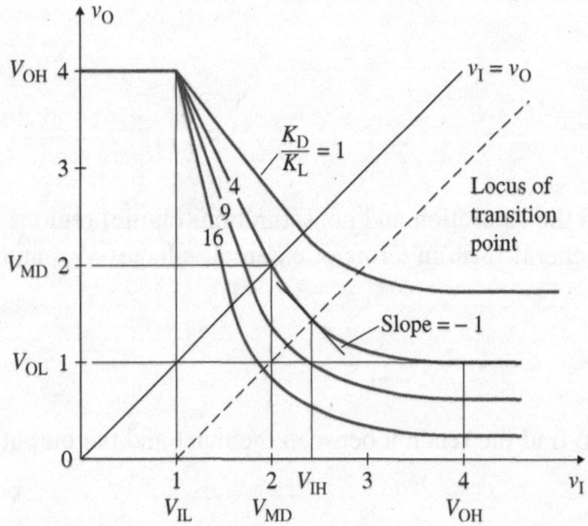

FIGURE 3.13 Transfer characteristics for various values of the ratio K_D/K_L

Body Effect

The previous equations were derived by neglecting the body effect in the load transistor M_L. In IC inverters, the body effect increases the threshold voltage. A higher value of V_{IL} causes V_{OH} to drop significantly, and the noise margin is decreased. In addition, a large geometry ratio K_R (large K_D and low K_L) is required to achieve a steep VTC. The threshold voltage V_t is related to V_{SB} [7] by

$$V_t = V_{t0} + \gamma |\sqrt{V_{SB} + 2\phi_f} - \sqrt{2\phi_f}| \tag{3.37}$$

where V_{t0} = *threshold voltage* at $V_{SB} = 0$, typically 1 V to 1.5 V
γ = *fabrication process constant*, typically 0.3 $\sqrt{V}$ to 1 $\sqrt{V}$
$2\phi_f$ = *equilibrium electrostatic potential* of the *p*-type body material, typically 0.6 V

Using Eq. (3.37), $V_{tD0} = V_{t0}$ for M_D, and $V_{SB} = v_O$ for M_L, we find that V_{tL} is given by

$$V_{tL} = V_{tD0} + \gamma |\sqrt{v_O + 2\phi_f} - \sqrt{2\phi_f}| \tag{3.38}$$

This relationship for $V_{tL}(v_O)$, which is a nonlinear function of v_O, can be used to find the transfer characteristic of the inverter. However, several iterations are required, and the process can be tedious. It is rarely necessary to obtain a detailed analysis for circuit design, and such a task is usually left for computer-aided simulation.

EXAMPLE 3.3

D **Designing an enhancement-load NMOS inverter** Design an enhancement-load NMOS inverter, as shown in Fig. 3.12(a), to obtain a noise margin of $NM_L \geq 0.8$ V. The threshold voltages are $V_{tL} = V_{tD} = 1$ V, and $K_L = 20\ \mu A/V^2$. The supply voltage is $V_{DD} = 5$ V. Assume load capacitance $C_L = 0.5$ pF and clock frequency $f_{clk} = 5$ MHz.

(a) Find the design parameter $K_R = K_D/K_L$, neglecting the body effect.

(b) Calculate NM_L if the body effect is included. Assume $\gamma = 0.5\sqrt{V}$ and $2\phi_f = 0.6$ V.

(c) Calculate the low-to-high propagation time t_{pLH}.

(d) Calculate the high-to-low propagation time t_{pHL}.

(e) Calculate the delay-power product (DP).

SOLUTION

$NM_L = V_{IL} - V_{OL} \geq 0.8$ V. Thus, for $V_{IL} = V_{tD} = 1$ V,

$$V_{OL} = V_{IL} - NM_L \leq 0.2 \text{ V}$$

Technically, $V_{OH} = V_{DD} - V_{tD} = 5 - 1 = 4$ V. However, to simplify the analysis for finding an expression for K_R, we will assume $V_{OH} = V_{IH} = V_{DD} = 5$ V. For $V_{IH} = 5$ V, M_L and M_D operate in the saturation and nonsaturation (ohmic) regions, respectively. Thus, for $v_I = V_{IH} = V_{OH}$ and $v_O = V_{OL}$, Eq. (3.33) yields

$$K_D[2(V_{OH} - V_{tD})V_{OL} - V_{OL}^2] = K_L(V_{DD} - V_{OL} - V_{tL})^2 \tag{3.39}$$

which gives the desired value of the geometry ratio K_R as

$$K_R = \frac{K_D}{K_L} = \frac{(V_{DD} - V_{OL} - V_{tL})^2}{2(V_{OH} - V_{tD})V_{OL} - V_{OL}^2} \tag{3.40}$$

(a) Without the body effect, $V_{tD} = 1$ V. For $V_{OL} = 0.2$ V and $V_{OH} = 5$ V, Eq. (3.40) gives $K_R = 9.26$. Since a higher value gives a lower value of V_{OL}, we choose $K_R = 10$. For this value of K_R, we can find V_{OL} by solving the quadratic equation in Eq. (3.39):

$$10[2(5 - 1)V_{OL} - V_{OL}^2] = (5 - V_{OL} - 1)^2$$

which gives $V_{OL} = 0.19$ V or 7.81 V (not physically meaningful). Thus, $V_{OL} = 0.19$ V, which is less than 0.2 V.

(b) With the body effect of $V_O = 0.2$ V, Eq. (3.38) gives

$$V_{tL} = V_{tD} + \gamma|\sqrt{v_O + 2\phi_f} - \sqrt{2\phi_f}| \tag{3.41}$$

$$= 1 + 0.5|\sqrt{0.2 + 0.6} - \sqrt{0.6}| = 1.06 \text{ V}$$

With this value of $V_{tL} = 1.06$ V, Eq. (3.39) gives

$$10[2(5 - 1)V_{OL} - V_{OL}^2] = (5 - 1.06 - V_{OL})^2$$

which gives $V_{OL} = 0.18$ V. Thus,

$$NM_L = V_{tL} - V_{OL} = 1 - 0.18 = 0.82 \text{ V}$$

which is better than the specified 0.8 V.

(c) The NMOS inverter with a load capacitor C_L is shown in Fig. 3.14(a). As v_I goes low to V_{OL}, M_D turns off immediately, and the capacitor C_L is charged up by the drain current i_{DL}. The operating point, shown in Fig. 3.14(b), moves along the load line of M_L from point Q of V_{OL} to midpoint P of $V_{O(mid)} = (V_{OH} + V_{OL})/2$. The charging of the capacitor is shown in Fig. 3.14(c). The output voltage is given by

$$C_L\frac{dv_O}{dt} = i_{DL} = K_L(V_{DD} - v_O - V_{tL})^2 \tag{3.42}$$

which can be integrated from V_{OL} to $V_{O(mid)}$ to give the low-to-high propagation time t_{pLH}, as shown in Fig. 3.14(d). To simplify the analysis, we will find the average value of the charging current I_{CLH} and then find an approximate value of t_{pLH}; that is,

$$t_{pLH} = \frac{C_L(V_{O(mid)} - V_{OL})}{I_{CLH}} \tag{3.43}$$

where I_{CLH} is given by

$$I_{CLH} = \frac{i_C(V_{OL}) + i_C(V_{O(mid)})}{2} \tag{3.44}$$

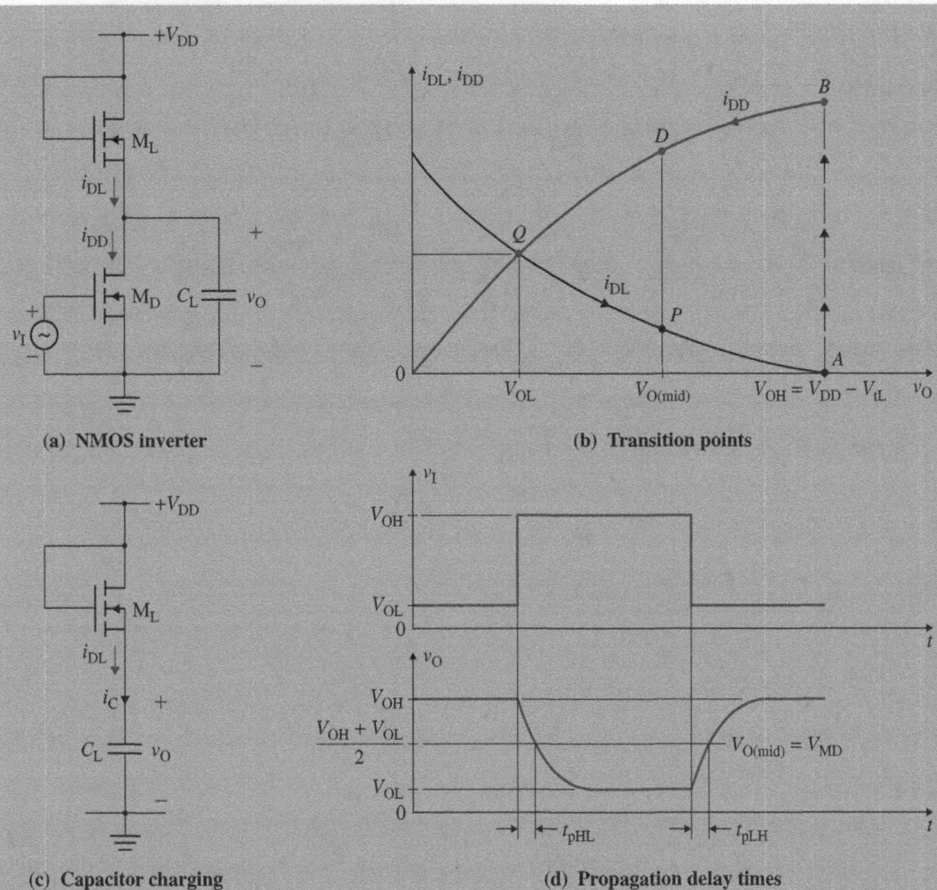

FIGURE 3.14 Switching speed of enhancement-load NMOS inverter

For $V_{OL} = 0.2$ V,

$$V_{OH} = V_{DD} - V_{tL} = 5 - 1 = 4 \text{ V}$$

$$V_{O(mid)} = \frac{V_{OH} + V_{OL}}{2} = 2.1 \text{ V}$$

Then $i_C(V_{OL}) = i_{DL}(V_{OL}) = K_L(V_{DD} - v_O - V_{tL})^2 = 20\ \mu \times (5 - 0.2 - 1)^2 = 288.8\ \mu\text{A}$

$$i_C(V_{O(mid)}) = K_L(V_{DD} - V_{O(mid)} - V_{tL})^2 = 20\ \mu \times (5 - 2.1 - 1)^2 = 72.2\ \mu\text{A}$$

$$I_{CLH} = \frac{288.8\ \mu + 72.2\ \mu}{2} = 180.5\ \mu\text{A}$$

Thus, $t_{pLH} = 0.5 \text{ pF} \times \dfrac{(2.1 - 0.2)\text{ V}}{180.5\ \mu\text{A}} = 5.26 \text{ ns}$

(d) As v_I goes high to V_{OH}, M_D turns on and the capacitor C_L discharges through M_D. The operating point moves along the load line of M_D from point *B* of V_{OH} to midpoint *D* of $V_{O(mid)} = (V_{OH} + V_{OL})/2$. The output voltage is given by

$$C_L \frac{dv_O}{dt} = i_{DD} - i_{DL} \tag{3.45}$$

$$\text{where} \quad i_{DD} = \begin{cases} K_D (V_{OH} - V_{tD})^2 & \text{for } (V_{OH} - V_{tD}) \geq v_O > V_{O(mid)} \\ K_D[2(V_{OH} - V_{tD})v_O - v_O^2] & \text{for } v_O \leq V_{O(mid)} \end{cases}$$

Using the average value of the capacitor discharging current i_{CHL}, we can find the approximate value of t_{pHL} from

$$t_{pHL} = \frac{C_L(V_{OH} - V_{O(mid)})}{I_{CHL}} \tag{3.46}$$

where i_{CHL} is given by

$$i_{CHL} = \frac{i_{DD}(V_{OH}) + i_{DD}(V_{O(mid)}) - i_{DL}(V_{O(mid)})}{2} \tag{3.47}$$

For $K_D = K_R K_L = 200\ \mu\text{A/V}^2$,

$$V_{OL} = 0.2\text{ V}$$

$$V_{OH} = V_{DD} - V_{tL} = 5 - 1 = 4\text{ V}$$

$$V_{O(mid)} = \frac{V_{OH} + V_{OL}}{2} = 2.1\text{ V}$$

Thus, $i_{DD}(V_{OH}) = K_D(V_{OH} - V_{tD})^2 = 200\ \mu \times (4 - 1)^2 = 1800\ \mu\text{A}$

$$i_{DD}(V_{O(mid)}) = K_D[2(V_{OH} - V_{tD})V_O - V_O^2] = 200\ \mu \times [2 \times (4 - 1) \times 2.1 - 2.1^2] = 1638\ \mu\text{A}$$

$$i_{DL}(V_{O(mid)}) = K_L(V_{DD} - V_O - V_{tL})^2 = 20\ \mu \times (5 - 2.1 - 1)^2 = 72.2\ \mu\text{A}$$

$$i_{CHL} = \frac{1800 + 1638 - 72.2}{2} = 1683\ \mu\text{A}$$

$$t_{pHL} = 0.5\text{ pF} \times \frac{(4 - 2.1)\text{ V}}{1683\ \mu\text{A}} = 0.56\text{ ns}$$

which is much shorter than $t_{pLH} = 5$ ns. The value of $i_{DL}(V_{O(mid)})$, which is much smaller than that of $i_{DD}(V_{OH})$, can often be neglected.

(e) The propagation time t_{pd}, which is the average of t_{pLH} and t_{pHL}, is

$$t_{pd} = \frac{t_{pLH} + t_{pHL}}{2} = \frac{5.26\text{ ns} + 0.56\text{ ns}}{2} = 2.91\text{ ns}$$

Then $P_{static} = \frac{V_{DD}\, i_D(V_{DL})}{2} = \frac{V_{DD}\, K_L(V_{DD} - V_{OL} - V_{tL})^2}{2}$ (3.48)

$$= \frac{5 \times 20\ \mu \times (5 - 0.2 - 1)^2}{2} = 722\ \mu\text{W}$$

and $P_{dynamic} = f_{clk} C_L V_{DD}^2 = 5\ \text{MHz} \times 0.5\ \text{pF} \times 5^2 = 62.5\ \mu\text{W}$

so $P_D = P_{static} + P_{dynamic} = 722\ \mu + 62.5\ \mu = 784.5\ \mu\text{W}$

Therefore, the delay-power product is

$$\text{DP} = P_D \times t_{pd} = 784.5\ \mu\text{W} \times 2.91\ \text{ns} = 2.28\ \text{pJ}$$

3.5.2 NMOS Inverter with Depletion Load

The load in an inverter can be a depletion-type MOSFET. Consider an *n*-channel depletion MOSFET whose gate is connected to the source terminal, as shown in Fig. 3.15(a). With this connection, $v_{GS} = 0$; the output characteristic is shown in Fig. 3.15(b), which indicates that v_{DS} must be zero to obtain a zero drain current.

Static Characteristics

An NMOS inverter with a depletion-load transistor M_L is shown in Fig. 3.16(a), in which the substrates of the MOSFETs are connected to the ground. To simplify the analysis, we will assume that the substrates are connected to the source terminals so that $V_{SB} = 0$ for both M_D and M_L. The output characteristics of the load and driver transistors are shown in Fig. 3.16(b). If $v_I = V_{OL}$, then $v_O = V_{OH}$. When v_I goes high (to V_{OH}), transistor M_D turns on, and the operating point jumps from *A* to *B* and then moves along the *i*-*v* curve of M_D until it finally reaches the quiescent point *Q* so that $v_O = V_{OL}$. When v_I goes from high to low (to V_{OL}), the operating point jumps from *Q* to *A* and then moves along the *i*-*v* curve of M_L until it finally reaches $v_O = V_{OH}$. Depending on the input voltage v_I, the VTC can be divided into four regions, as shown in Fig. 3.16(c).

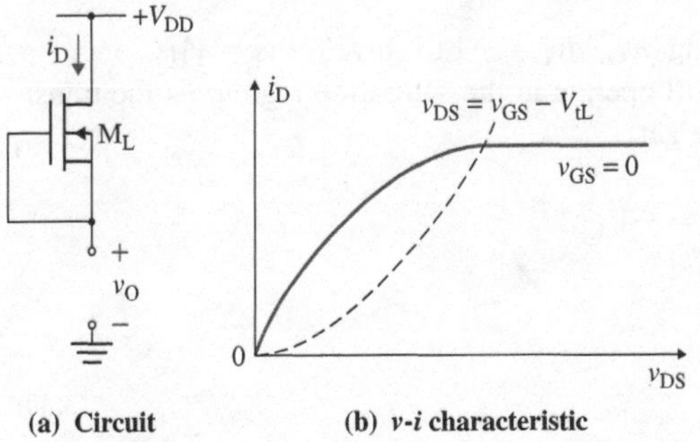

FIGURE 3.15 *n*-channel depletion-type MOSFET with $v_{GS} = 0$

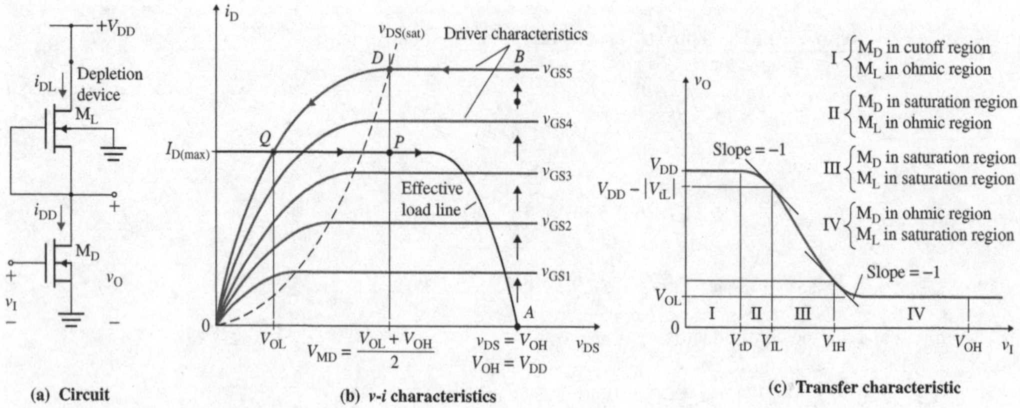

FIGURE 3.16 NMOS inverter with depletion load

In region I, the input voltage $v_I \le V_{tD}$, where V_{tD} is the threshold voltage of the drive transistor M_D. M_D is cut off, so the drain currents of the two transistors must be zero. M_L operates in the nonsaturation (ohmic) region. The drain current of load transistor M_L is $i_{DL} = 0$, which, for $v_{DSL} = 0$, gives the output voltage as

$$v_O = V_{OH} = V_{DD} - v_{DSL} = V_{DD}$$

Thus, V_{OH} is not reduced by V_{tL}, as it is in the case of an enhancement-load inverter.

In region II, $v_I > V_{tD}$. As v_I becomes slightly greater than V_{tD}, M_D begins to conduct and operates in the saturation region. M_L is still in the nonsaturation region. The two drain currents must be equal; that is, $i_{DL} = i_{DD}$, which, for M_L in the nonsaturation region and M_D in the saturation region, gives

$$K_L[2(v_{GSL} - V_{tL})v_{DSL} - v_{DSL}^2] = K_D(v_{GSD} - V_{tD})^2$$

Substituting the values for $v_{GSL} = 0$, $v_{DSL} = V_{DD} - v_O$, and $v_{GSD} = v_I$, we get the relationship between the input and output voltages as

$$K_L[-2V_{tL}(V_{DD} - v_O) - (V_{DD} - v_O)^2] = K_D(v_I - V_{tD})^2 \tag{3.49}$$

V_{IL} can be found by differentiating Eq. (3.49) and setting $dv_O/dv_I = -1$ to solve for $v_I = V_{IL}$.

If v_I is increased sufficiently, both M_D and M_L will operate in the saturation region. At the transition point from nonsaturation to saturation for M_L, we get

$$v_{DSL} = V_{DD} - v_O = v_{GSL} - V_{tL} = 0 - V_{tL} = -V_{tL}$$

which gives v_O at the edge of the transition as

$$V_{O(\text{tran1})} = V_O = V_{DD} + V_{tL} \tag{3.50}$$

Substituting v_O from Eq. (3.50) into Eq. (3.49) gives the corresponding input voltage $V_I = V_{I(tran1)}$; that is,

$$K_L[-2V_{tL}(V_{DD} - V_{DD} - V_{tL}) - (V_{DD} - V_{DD} - V_{tL})^2] = K_D(v_I - V_{tD})^2$$

which, for $v_I > V_{tD}$, gives

$$V_{I(tran1)} = v_I = V_{tD} - V_{tL}\left(\frac{K_L}{K_D}\right)^{1/2} = V_{tD} - V_{tL}\sqrt{\frac{1}{K_R}} \tag{3.51}$$

▶ **NOTE** $V_{I(tran1)}$ can also be found from

$$K_L(v_{GSL} - V_{tL})^2 = K_D(v_{GSD} - V_{tD})^2$$

for $v_{GSL} = 0$ and $v_{GSD} = v_I = V_{I(tran1)}$.

In region III, $v_I = V_{I(tran1)}$. Both M_L and M_D operate in the saturation region. Since the two currents must be equal, $i_{DL} = i_{DD}$, or

$$K_L(v_{GSL} - V_{tL})^2 = K_D(v_{GSD} - V_{tD})^2$$

Substituting $v_{GSL} = 0$ and $v_{GSD} = v_I$, we get the input voltage when both transistors are in saturation:

$$v_I = V_{tD} - V_{tL}\left(\frac{K_L}{K_D}\right)^{1/2} = V_{tD} - V_{tL}\sqrt{\frac{1}{K_R}}$$

The corresponding value of drain current is

$$i_{DL} = i_{DD} = K_L(v_{GSL} - V_{tL})^2 = K_L(-V_{tL})^2 = K_L V_{tL}^2 \tag{3.52}$$

which indicates that the drain current is independent of v_I and remains constant. At this current level, M_L will continue to operate in the saturation region and M_D will be forced into nonsaturation. There will be a quick transition as M_D switches from the saturation to the nonsaturation region. The output voltage will change to

$$V_{O(tran2)} = v_O = v_{DSD} = v_{GSD} - V_{tD} = v_I - V_{tD} \tag{3.53}$$

The input voltage at this transition is the same as $V_{I(tran1)}$; that is,

$$V_{I(tran2)} = V_{I(tran1)} = V_{tD} - V_{tL}\left(\frac{K_L}{K_D}\right)^{1/2} \tag{3.54}$$

Note that at $v_I = V_{I(tran1)} = V_{I(tran2)}$, there will be two transitions: the first as M_D switches from saturation to nonsaturation at an output voltage of $v_O = V_{DD} + V_{tL}$ and then the second as M_L switches from nonsaturation to saturation at an output voltage of $v_O = v_I - V_{tD}$.

In region IV, $v_I > V_{I(tran1)} = V_{I(tran2)}$. The drive transistor M_D goes into nonsaturation, and M_L operates in the saturation region. Since the two currents must be equal, $i_{DL} = i_{DD}$, which, for M_L in saturation and M_D in nonsaturation, gives

$$K_L(v_{GSL} - V_{tL})^2 = K_D[2(v_{GSD} - V_{tD})v_{DSD} - v_{DSD}^2]$$

Substituting $v_{GSL} = 0$, $v_{DSD} = v_O$, and $v_{GSL} = v_I$, we get the relationship between the input and output voltages:

$$K_L(-V_{tL})^2 = K_D[2(v_I - V_{tD})v_O - v_O^2] \tag{3.55}$$

If we set $dv_O/dv_I = -1$, Eq. (3.55) yields

$$v_I = 2v_O + V_{tD}$$

which, after substitution into Eq. (3.55), gives

$$V_{IH} = \frac{2|V_{tL}|}{\sqrt{3K_R}} + V_{tD} \tag{3.56}$$

$$v_O = \frac{|V_{tL}|}{\sqrt{3K_R}} \tag{3.57}$$

The typical loci for the transition points for both the load and the drive transistors are shown in Fig. 3.17. The typical transfer characteristics are shown in Fig. 3.18 for several values of the ratio K_D/K_L. The depletion load provides a constant current once M_L goes into saturation.

Body Effect

With the body effect, V_{tL} varies with $v_{SB} = v_O$ by

$$V_{tL} = V_{tL0} + \gamma|\sqrt{v_O + 2\phi_f} - \sqrt{2\phi_f}| \tag{3.58}$$

where V_{tL0} is the threshold voltage at $v_{SB} = 0$. Since $|V_{tL}|$ decreases with falling v_O, the load current will decrease slightly and v_O will still drop rapidly with v_I.

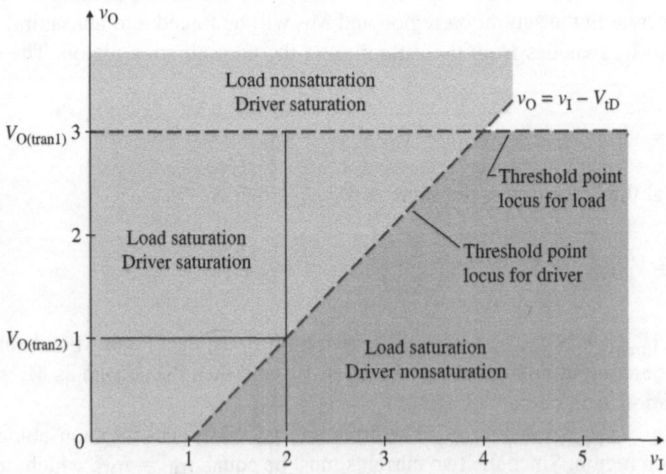

FIGURE 3.17 Transition loci for driver and load transistors

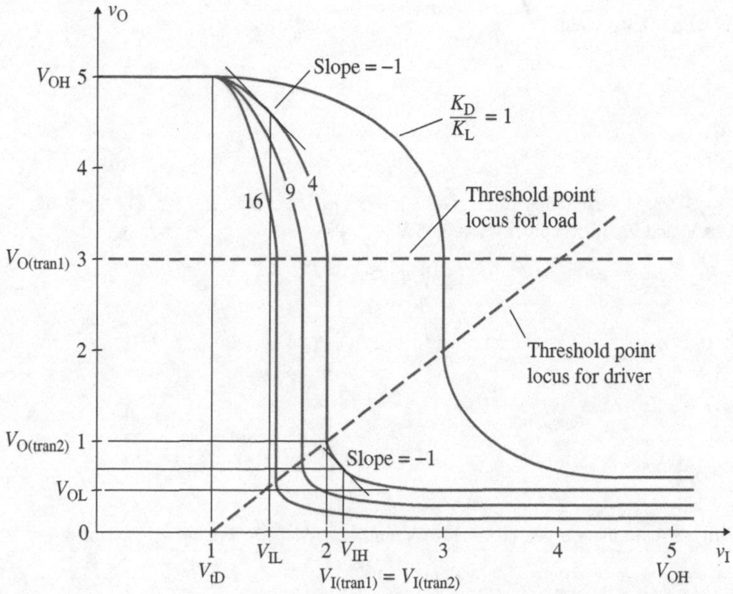

FIGURE 3.18 Transfer characteristics for various values of the ratio K_D / K_L

EXAMPLE 3.4

D **Designing a depletion-load NMOS inverter** Design a depletion-load NMOS inverter, as shown in Fig. 3.16(a), to obtain a noise margin of $NM_H \geq 2.7$ V. The threshold voltages are $V_{tL} = -2.5$ V, $V_{tD} = 1$ V, and $K_L = 20\ \mu A/V^2$. The supply voltage is $V_{DD} = 5$ V. Assume $V_{OH} = V_{DD} = 5$ V, load capacitance $C_L = 0.5$ pF, and frequency $f_{clk} = 5$ MHz.

(a) Find the design parameter $K_R = K_D/K_L$, neglecting the body effect.

(b) Calculate NM_L if the body effect is neglected.

(c) Calculate NM_H if the body effect is included. Assume $\gamma = 0.5\ \sqrt{V}$ and $2\phi_f = 0.6$ V.

(d) Calculate the low-to-high propagation time t_{pLH}.

(e) Calculate the high-to-low propagation time t_{pHL}.

(f) Calculate the delay-power product (DP).

(g) Use PSpice/SPICE to verify your results.

SOLUTION

(a) $NM_H = V_{OH} - V_{IH} \geq 2.7$ V. Thus, for $V_{OH} = V_{DD} = 5$ V,

$$V_{IH} = V_{OH} - NM_H \leq 2.3 \text{ V}$$

For $V_{IH} \leq 2.3$ V, Eq. (3.56) yields

$$K_R = \frac{(2V_{tL})^2}{3(V_{IH} - V_{tD})^2} \tag{3.59}$$

which gives $K_R = 4.93$. Since a higher value gives a lower V_{IH}, we choose $K_R = 5$. For this value of K_R, Eq. (3.56) gives $V_{IH} = 2.29$ V, and Eq. (3.57) gives $v_O = 0.65$ V.

(b) For $K_D = K_L \times K_R = 100\ \mu A/V^2$, Eq. (3.49) gives

$$20\ \mu \times [(-2) \times (-2.5) \times (5 - v_O) - (5 - v_O)^2] = 20\ \mu \times 5(v_I - 1)^2 \tag{3.60}$$

Setting $dv_O/dv_I = -1$, we get

$$v_I = 0.2v_O + 0.5$$

Solving these two equations, we get $V_{IL} = 1.46$ V and $v_O = 4.78$ V. For $v_I = V_{OH}$ and $v_O = V_{OL}$, Eq. (3.55) yields

$$K_D[2(V_{OH} - V_{tD})V_{OL} - V_{OL}^2] = K_L V_{tL}^2 \tag{3.61}$$

which, for $V_{OH} = 5$ V and $K_R = 5$, gives $V_{OL} = 0.16$ V. Thus,

$$NM_L = V_{IL} - V_{OL} = 1.46 - 0.16 = 1.3 \text{ V}$$

(c) With the body effect, at $v_O = 0.65$ V, Eq. (3.58) gives

$$V_{tL} = V_{tL0} + \gamma |\sqrt{v_O + 2\phi_f} - \sqrt{2\phi_f}| \tag{3.62}$$

$$= -2.5 + 0.5|\sqrt{0.65 + 0.6} - \sqrt{0.6}| = -2.33 \text{ V}$$

With this value, Eq. (3.56) gives $V_{IH} = 2.20$ V. Thus,

$$NM_H = V_{OH} - V_{IH} = 5 - 2.20 = 2.80 \text{ V}$$

which is better than the specified 2.7 V.

(d) The inverter with a load capacitor C_L is shown in Fig. 3.19(a). As v_I goes low to V_{OL}, M_D turns off immediately and the capacitor C_L is charged up by the drain current i_{DL}. The operating point, shown in Fig. 3.19(b), moves along the load line of M_L from point Q of V_{OL} to midpoint P of $V_{O(mid)} = (V_{OH} + V_{OL})/2$. The charging of the capacitor is shown in Fig. 3.19(c). The output voltage is given by

$$C_L \frac{dv_O}{dt} = i_{DL} = K_L V_{tL}^2 \tag{3.63}$$

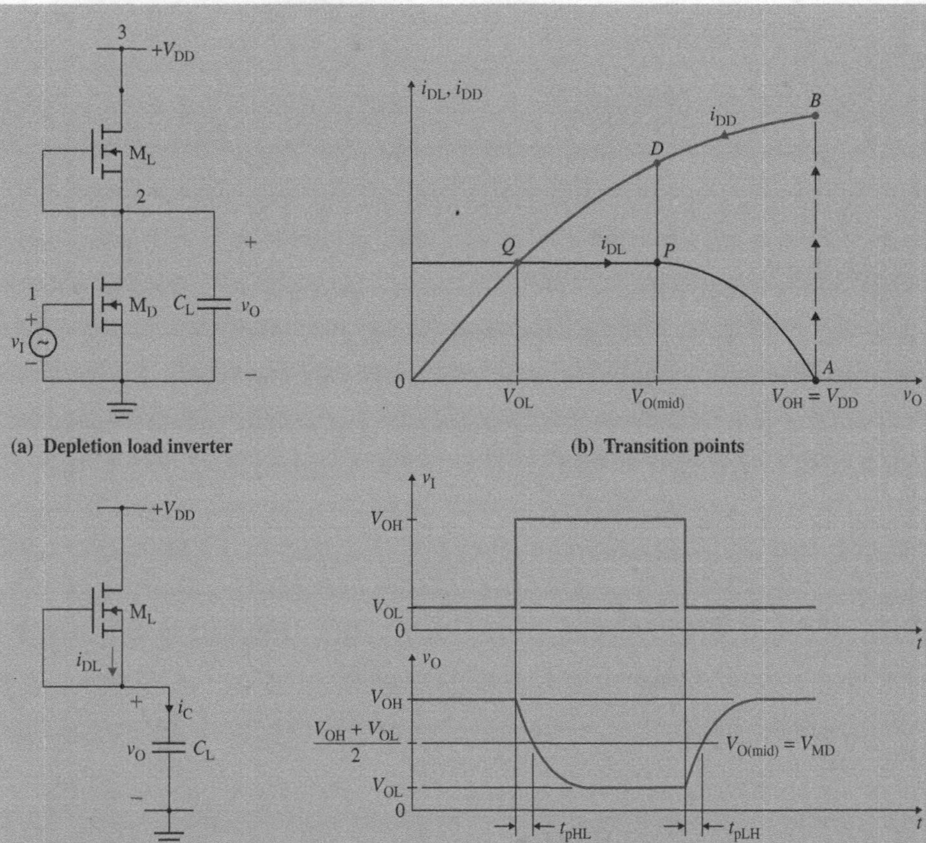

(a) **Depletion load inverter** (b) **Transition points** (c) **Capacitor charging** (d) **Propagation delay times**

FIGURE 3.19 Switching speed of depletion-load NMOS inverter

which, after integration from V_{OL} to $V_{O(mid)}$, gives t_{pLH} as

$$t_{pLH} = \frac{C_L(V_{OH} - V_{OL})}{2K_L V_{tL}^2} \tag{3.64}$$

$$= \frac{0.5 \text{ pF} \times (5 - 0.16)}{2 \times 20\ \mu \times 2.5^2} = 9.7 \text{ ns}$$

(e) As v_I goes high to V_{OH}, M_D turns on and the capacitor C_L discharges through M_D. The operating point moves along the load line of M_D from point B of V_{OH} to midpoint D of $V_{O(mid)}$. The output voltage is given by

$$C_L \frac{dv_O}{dt} = i_{DD} - i_{DL} = i_{DD} - K_L V_{tL}^2 \tag{3.65}$$

$V_{OL} = 0.16$ V, $V_{OH} = V_{DD} = 5$ V, and

$$V_{O(mid)} = \frac{V_{OH} + V_{OL}}{2} = 2.6 \text{ V}$$

Thus, $i_{DD}(V_{OH}) = K_D(V_{OH} - V_{tD})^2 = 100\ \mu \times (5 - 1)^2 = 1600\ \mu\text{A}$

$$i_{DD}(V_{O(mid)}) = K_D[2(V_{OH} - V_{tD})v_O - v_O^2] = 100\ \mu \times [2 \times (5 - 1) \times 2.6 - 2.6^2] = 1404\ \mu\text{A}$$

$$i_{DL}(V_{O(mid)}) = K_L V_{tL}^2 = 20\ \mu \times 2.5^2 = 125\ \mu\text{A}$$

$$i_{CHL} = \frac{1600\ \mu + 1404\ \mu - 125\ \mu}{2} = 1440\ \mu\text{A}$$

Using Eq. (3.46), we have

$$t_{pHL} = 0.5 \text{ pF} \times \frac{(5 - 2.6) \text{ V}}{1440\ \mu\text{A}} = 0.83 \text{ ns}$$

which is much shorter than $t_{pLH} = 9.7$ ns. Assuming that i_{DD} is much greater than i_{DL}, Eq. (3.65) can be simplified to

$$C_L \frac{dv_O}{dt} = i_{DD} = K_D(V_{DD} - V_{tD})^2$$

which, after integration between V_{OH} and $(V_{OH} + V_{OL})/2$, gives t_{pHL} as

$$t_{pHL} = \frac{C_L(V_{OH} - V_{OL})}{2K_D(V_{DD} - V_{tD})^2} \tag{3.66}$$

$$= \frac{0.5 \text{ pF} \times (5 - 0.16)}{2 \times 100\ \mu \times (5 - 1)^2} = 0.76 \text{ ns}$$

(close to 0.83 ns).

(f) The propagation time t_{pd}, which is the average of t_{pLH} and t_{pHL}, is

$$t_{pd} = \frac{t_{pLH} + t_{pHL}}{2} = \frac{9.7 \text{ ns} + 0.76 \text{ ns}}{2} = 5.2 \text{ ns}$$

The current at $v_O = V_{OL}$ becomes

$$i_{DL} = i_{DD} = K_L V_{tL}^2 = 125\ \mu\text{A}$$

$$P_{static} = \frac{5 \times 125\ \mu\text{A}}{2} = 312.5\ \mu\text{W}$$

Since $P_{dynamic} = f_{clk} C_L V_{DD}^2 = 5 \text{ MHz} \times 0.5 \text{ pF} \times 5^2 = 62.5\ \mu\text{W}$,

$$P_D = P_{static} + P_{dynamic} = 312.5\ \mu + 62.5\ \mu = 375\ \mu\text{W}$$

Therefore, the delay-power product is

$$\text{DP} = P_D \times t_{pd} = 375\ \mu\text{W} \times 5.2 \text{ ns} = 1.95 \text{ pJ}$$

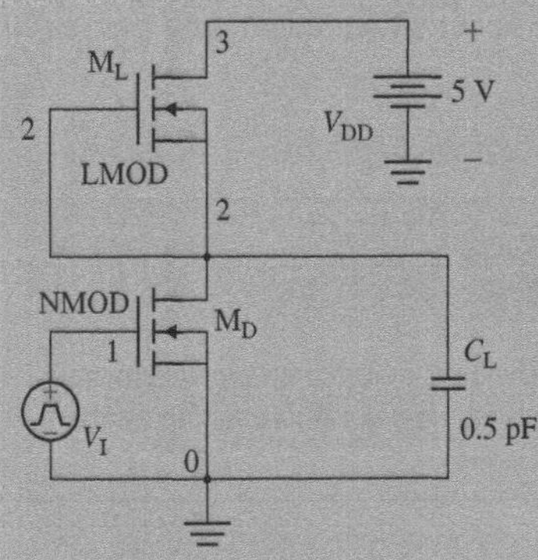

FIGURE 3.20 PSpice schematic for Example 3.4

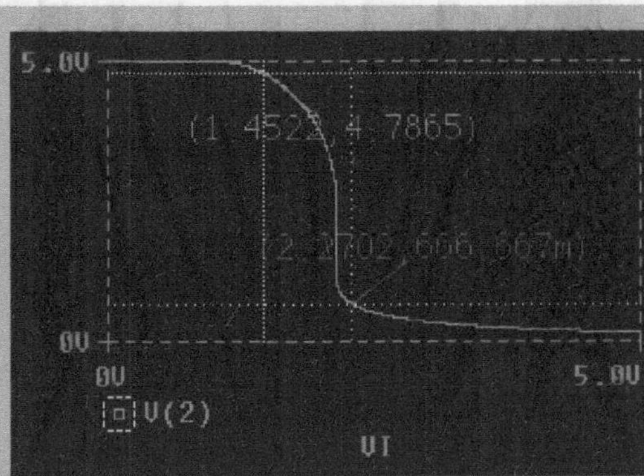

FIGURE 3.21 PSpice VTC plot for depletion-load NMOS (using transistor circuit model) for Example 3.4

(g) The PSpice schematic is shown in Fig. 3.20. The PSpice plot of the VTC is shown in Fig. 3.21, which gives $V_{IL} = 1.4522$ V (expected value is 1.46 V), $V_{IH} = 2.2702$ V (expected value is 2.29 V), $V_{OL} = 0.16$ V (expected value is 0.16 V) at $V_{OH} = 5$ V, $V_{I(tran1)} = 2.1172$ (expected value is $1 + 2.5\sqrt{1/5} = 2.118$ V), and $V_{O(tran1)} = 2.815$ V (expected value is $5 - 2.5 = 2.5$ V). The VTC values are very close to the expected values.

NOTE: We can find dv_O/dv_I $(=-1)$ points by plotting dv(2) in Probe.

3.5.3 Comparison of NMOS Inverters

The load lines for three basic types of inverters are shown in Fig. 3.22, superimposed on the drain characteristics of the driver transistor. The resistive load line exhibits linear characteristics, and its high output voltage is dependent on drain resistance R_D. An inverter with an enhancement load has a lower drain current for the same value of v_{DS}. An inverter with a depletion load shows a constant current over a wide range

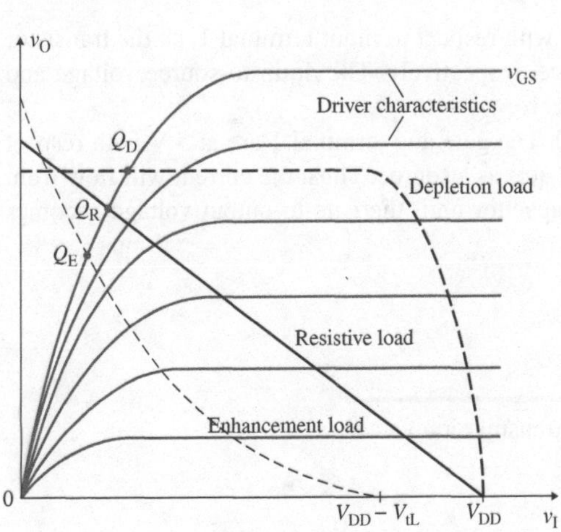

FIGURE 3.22 Load lines for three NMOS inverters

of v_{DS}; as a consequence of this characteristic, it will switch a capacitive load more rapidly than the other two inverters. Also, it exhibits a higher noise margin. Depletion-load inverters are commonly used in ICs for high-speed switching.

KEY POINT OF SECTION 3.5

- An NMOS inverter uses an enhancement MOSFET as the driver. The load is either an enhancement or a depletion MOSFET. Depletion-load inverters allow more rapid switching and thus are commonly used in integrated circuits.

3.6 NMOS Logic Circuits

In digital circuits, NMOS inverters are frequently used in transmission gates and NOR and NAND gates.

3.6.1 NMOS Transmission Gates

An NMOS transmission gate is basically an NMOS that is connected to an effective load capacitance C_L, as shown in Fig. 3.23. The substrate is connected to the most negative potential in the circuit rather than to the source terminal. The MOSFET may be assumed to be completely bilateral; that is, the drain and source terminals are identical.

To examine the operation of NMOS transmission gates, we will assume that the substrate is connected to zero (that is, $v_{SUB} = 0$) and consider the following cases based on the level of input voltage v_I:

In case 1, $v_G = 0$ and $v_I = 0$ or 5 V. The gate will not be positive with respect to either input terminal 1 or output terminal 2, and the transistor will always be cut off. Thus, the input and output terminals will be isolated from each other. The input voltage v_I may have any value without affecting the output voltage v_O.

In case 2, $v_G = 5$ V and $v_I = 0$. The gate is at 5 V with respect to input terminal 1, so the transistor is turned on. Terminals 1 and 2 act as drain and source, respectively. The drain-to-source voltage and drain current will be zero. The output voltage will be zero: $v_O = 0$.

In case 3, $v_G = 5$ V, $v_I = 5$ V, and initially $v_O = 0$. The gate and terminal 1 are at 5 V with respect to terminal 2. Terminal 1 acts as a drain, and terminal 2 acts as a source. Thus, the current will flow from terminal 1 to terminal 2, and it will charge the load capacitor until the gate-to-output voltage becomes

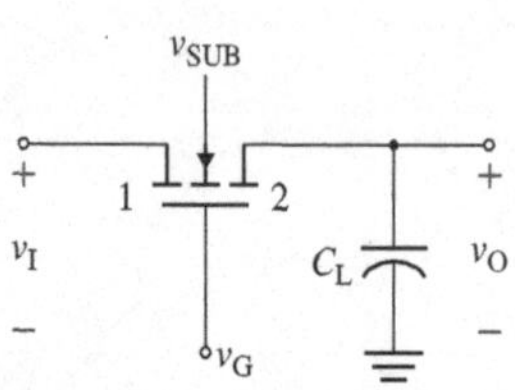

FIGURE 3.23 NMOS transmission gate

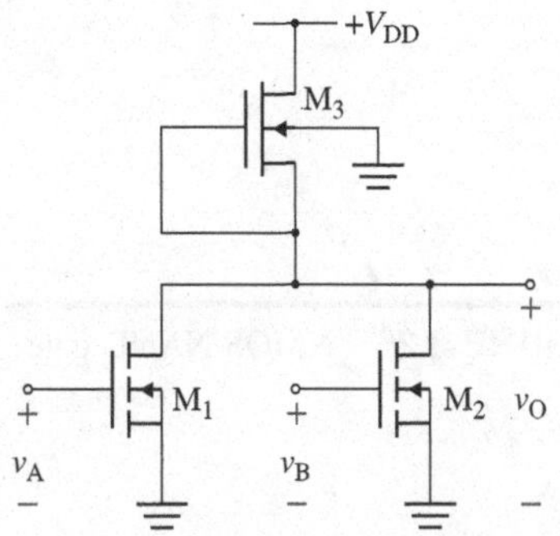

FIGURE 3.24 NMOS NOR gate

equal to the threshold voltage V_t of the transistor; that is, $v_G - v_O = V_t$. For a transistor with $V_t = 1$ V, $v_O = v_G - V_t = 5 - 1 = 4$ V.

In case 4, $v_G = 5$ V, $v_I = 0$, and initially $v_O = v_G - V_t = 4$ V. The gate is at 5 V, and terminal 2 is at 4 V with respect to terminal 1. Terminal 2 acts as a drain, and terminal 1 acts as a source. The transistor will be on. Thus, the current will flow from terminal 2 to terminal 1, and it will discharge the load capacitor; that is, $v_O = 0$.

In case 5, $v_G = 0$, $v_I = 0$ or 5 V, and initially $v_O = v_G - V_t = 4$ V. When v_G goes to zero, the situation is similar to that in case 1, and the transistor will be cut off. The input and output terminals are isolated.

Therefore, as long as v_G is high, the transistor transmits the input voltage to the output. If v_G is low, the transistor is turned off, and the input and output terminals are isolated. Depending on the levels of the gate and input voltages, the transmission gate can be operated as (1) an analog switch, (2) a sample-and-hold circuit that converts analog signals to their digital equivalents, or (3) a pass transistor for steering logic signals.

3.6.2 NMOS NOR Gates

An NMOS NOR logic gate is shown in Fig. 3.24. Two parallel NMOS transistors are connected in series with a depletion-type load. The transfer characteristic is similar to that of a depletion-load inverter. If both input voltages v_A and v_B are less than the threshold voltage of the driver transistors, $v_O = V_{OH}$ (for logic 1). If $v_A = V_{OH}$ (for logic 1), then M_1 will turn on, and the output voltage will drop to low output, $v_O = V_{OL}$. If both v_A and v_B are high (at V_{OH}), then both M_1 and M_2 will turn on, and the output voltage will be low (at V_{OL}). The exact value of the output voltage will depend on the transistor parameters, as discussed in Sec. 3.5, and it will vary slightly depending on whether one or both driver transistors are turned on. The logic function for an NMOS NOR gate is shown in Table 3.4.

TABLE 3.4 NMOS NOR gate logic function

v_A	v_B	v_O
V_{OL} (0)	V_{OL} (0)	V_{OH} (1)
V_{OL} (0)	V_{OH} (1)	V_{OL} (0)
V_{OH} (1)	V_{OL} (0)	V_{OL} (0)
V_{OH} (1)	V_{OH} (1)	V_{OL} (0)

TABLE 3.5 NMOS NAND gate logic function

v_A	v_B	v_O
V_{OL} (0)	V_{OL} (0)	V_{OH} (1)
V_{OL} (0)	V_{OH} (1)	V_{OH} (1)
V_{OH} (1)	V_{OL} (0)	V_{OH} (1)
V_{OH} (1)	V_{OH} (1)	V_{OL} (0)

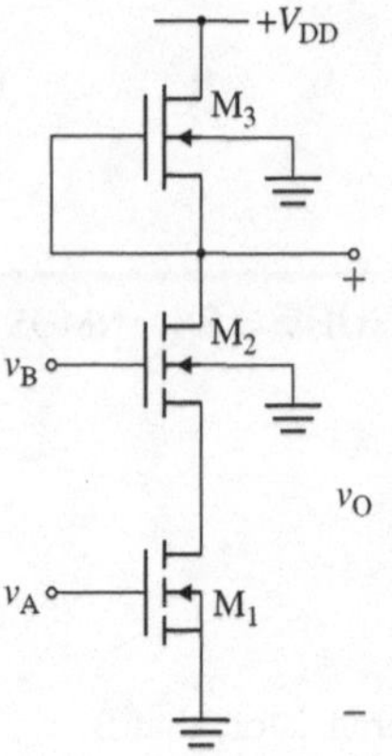

FIGURE 3.25 NMOS NAND gate

3.6.3 NMOS NAND Gates

An NMOS NAND gate is shown in Fig. 3.25. Two transistors are connected in series with a depletion-type load. If both input voltages v_A and v_B are less than the threshold voltage of the driver transistors, $v_O = V_{OH}$ (for logic 1). If $v_A = V_{OH}$ and v_B is still less than the threshold voltage V_{t2} of M_2, then M_2 is still cut off, and $v_O = V_{OH}$. If both v_A and v_B are high (at V_{OH}), then both M_1 and M_2 will turn on, and the output voltage will be low (at V_{OL}). The exact value of the output voltage will depend on the transistor parameters. The logic function for an NMOS NAND gate is shown in Table 3.5.

KEY POINT OF SECTION 3.6

- In digital circuits, NMOS inverters are frequently used in transmission gates and NOR and NAND gates.

3.7 CMOS Inverters

Complementary, or CMOS, circuits use both n-channel and p-channel enhancement-type MOSFETs in the same circuit. Because of their very low power consumption, CMOS circuits are commonly used in ICs. A CMOS inverter is shown in Fig. 3.26(a). Transistor M_P is a p-channel device, and transistor M_N is an n-channel device. Each substrate is tied to its source. The input signal is connected to both transistor gates, and the output terminal is common to both drain terminals. The load characteristics of two CMOS devices are shown in Fig. 3.26(b) for two extreme inputs: $v_I = 0$ and $v_I = V_{DD}$. For $v_I = 0$, M_N is off and its drain current is zero. M_P has the characteristic corresponding to $v_{GSP} = V_{DD}$. For $v_I = V_{DD}$, M_P is off and its drain current is zero. M_N has the characteristic corresponding to $v_{GSN} = V_{DD}$. Thus, their drain currents are zero at these inputs, and the current drawn from the supply is zero.

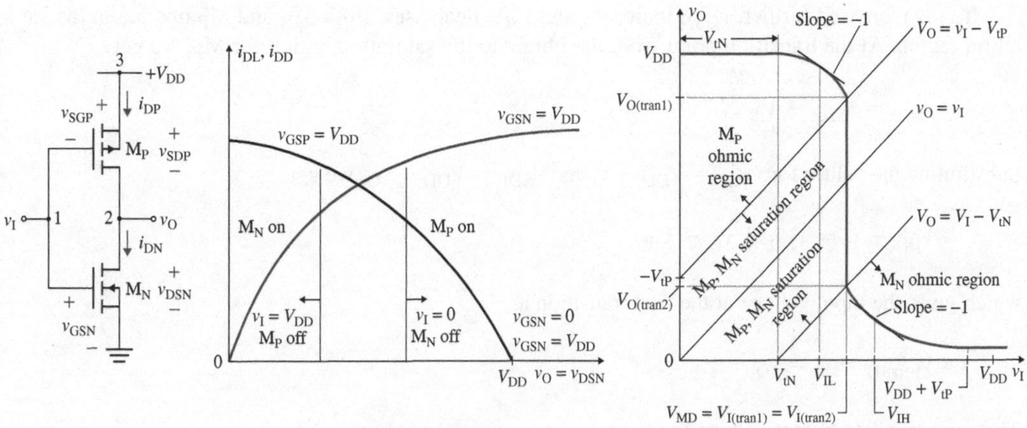

(a) Circuit **(b) Load characteristics** **(c) Transfer characteristic**

FIGURE 3.26 CMOS inverter

As the input voltage v_I is varied from zero to the maximum value V_{DD}, the output voltage v_O falls from V_{DD} to zero. The transfer characteristic is shown in Fig. 3.26(c). Depending on the input voltage v_I, the VTC can be divided into five regions.

In region I, $0 \le v_I < V_{tN}$, where V_{tN} is the threshold voltage of transistor M_N. Since $v_{GSN} = v_I < V_{tN}$, M_N remains off. At a low value of v_I, $V_{SGP} = V_{DD} - v_I$ is high and positive. M_P is turned on, and it is driven into the nonsaturation (ohmic) region. Since the channel resistance of the off-transistor M_N is very much greater than that of the on-transistor M_P and since M_N and M_P form a voltage divider, the output voltage is $v_O \approx V_{DD}$, as shown in Fig. 3.26(c). The various voltages are $v_{GSN} = v_I$, $v_{GSP} = -V_{DD} + v_I$, and $v_O = V_{DD}$.

In region II, $V_{tN} \le v_I \le V_{MD} = V_{I(tran1)}$ and $v_{GSP} = (-V_{DD} + v_I) < V_{tP}$, where V_{tP} is the threshold voltage of transistor M_P and $V_{I(tran1)}$ is defined by Eq. (3.68). As v_I becomes greater than or equal to V_{tN}, M_N conducts and operates in the saturation region, for which v_{GSN} is described by $V_{tN} < v_{GSN}$ $(= v_I)$ $< (v_{DSN} + V_{tN})$.

For $v_{GSP} < V_{tP}$, M_P remains in the ohmic region. With M_N in the saturation region and M_P in the ohmic region, the two drain currents must be equal; that is, $i_{DN} = i_{DP}$. Applying the expressions for the saturation and ohmic regions gives

$$K_n(v_{GSN} - V_{tN})^2 = K_p[2(v_{GSP} + V_{tP})v_{SDP} - v_{SDP}^2]$$

where K_n and K_p are the constants for n-type and p-type transistors, respectively. Substituting $v_{GSN} = v_I$, $v_{SGP} = V_{DD} - v_I$, and $v_{SDP} = V_{DD} - v_O$ into the above equation, we get the relationship between v_I and v_O as

$$K_n(v_I - V_{tN})^2 = K_p[2(V_{DD} - v_I + V_{tP})(V_{DD} - v_O) - (V_{DD} - v_O)^2] \tag{3.67}$$

V_{IL} can be found by differentiating Eq. (3.67) and setting $dv_O/dv_I = -1$ to solve for $v_I = V_{IL}$.

If v_I is increased further, v_{GSN} increases and v_{SGP} decreases. Both M_N and M_P operate in the saturation region. At the transition point from the ohmic to the saturation region for M_P, we get

$$v_{SGP} = v_{SDP} - V_{tP}$$

Substituting the values for $v_{SGP} = V_{DD} - v_I$ and $v_{SDP} = V_{DD} - v_O$, we get

$$V_{DD} - v_I = V_{DD} - v_O - V_{tP}$$

which gives the input voltage at the first transition as

$$V_{I(tran1)} = v_I = v_O + V_{tP}$$

The corresponding output voltage is

$$V_{O(tran1)} = v_O = v_I - V_{tP}$$

whose plot is a straight line that intercepts the output axis at $-V_{tP}$ (a positive quantity), as shown in Fig. 3.26(c). The intersection with the transfer characteristic gives $V_{I(tran1)}$ and $V_{O(tran1)}$. To solve for the value of $V_{I(tran1)}$ or $V_{O(tran1)}$, one of these quantities must be known.

In region III, $v_I = V_{I(tran1)}$. Both M_N and M_P operate in the saturation region. Since the two drain currents must be equal, $i_{DN} = i_{DP}$, or

$$K_n(v_{GSN} - V_{tN})^2 = K_p(v_{SGP} + V_{tP})^2$$

Substituting $v_{GSN} = v_I$ and $v_{SGP} = V_{DD} - v_I$, we get the input voltage at the transition of M_P from the ohmic to the saturation region:

$$V_{I(tran1)} = V_{MD} = \frac{V_{DD} + V_{tP} + V_{tN}\sqrt{K_n/K_p}}{1 + \sqrt{K_n/K_p}} = \frac{V_{DD} + V_{tP} + V_{tN}\sqrt{K_R}}{1 + \sqrt{K_R}} \tag{3.68}$$

which is independent of output voltage v_O. For identical transistors, $K_n = K_p$ and $V_{tN} = |V_{tP}|$, and Eq. (3.68) is reduced to

$$V_{I(tran1)} = V_{MD} = \frac{V_{DD}}{2} \tag{3.69}$$

which is desirable to maximize the noise immunity of the circuit. Once we determine the value of $V_{I(tran1)}$ from Eq. (3.69), we can find the output voltage at the edge of the transition of M_P from

$$V_{O(tran1)} = v_O = V_{I(tran1)} - V_{tP}$$

This segment ends when M_N enters the ohmic region, which is defined by

$$v_{GSN} = v_{DSN} + V_{tN}$$

Substituting $v_{GSN} = v_I$ and $v_{DSN} = v_O$ into the preceding equation yields

$$v_I = v_O + V_{tN}$$

which gives the input voltage at the transition of M_N from the saturation region to the ohmic region as

$$V_{I(tran2)} = v_I = v_O + V_{tN}$$

The corresponding output voltage is given by

$$v_O = v_I - V_{tN}$$

which intercepts the input axis at V_{tN}, as shown in Fig. 3.26(c). The intersection with the transfer characteristic gives $V_{I(tran2)}$ and $V_{O(tran2)}$. Since $V_{I(tran1)}$ is independent of v_O and $v_I = V_{I(tran1)} = V_{I(tran2)}$, the quantities $V_{O(tran1)}$ and $V_{O(tran2)}$ must be different. There will be two transitions—the first for M_P from the ohmic region to the saturation region at an output voltage of $v_O = v_I + V_{tP}$ and then the second for M_N from the saturation region to the ohmic region at an output voltage of $V_{O(tran2)} = V_{I(tran1)} - V_{tN}$.

In region IV, $V_{I(tran1)} = V_{I(tran2)} \leq v_I \leq (V_{DD} + V_{tP})$. M_N goes into the ohmic region, and M_P still operates in the saturation region. Since the two drain currents must be equal, $i_{DN} = i_{DP}$, which, for M_P in the saturation region and M_N in the ohmic region, gives

$$K_n[2(v_{GSN} - V_{tN})v_{DSN} - v_{DSN}^2] = K_p(v_{SGP} + V_{tP})^2$$

Substituting $v_{GSN} = v_I$, $v_{DSN} = v_O$, and $v_{SGP} = V_{DD} - v_I$, we can express the relationship between v_O and v_I as

$$K_n[2(v_I - V_{tN})v_O - v_O^2] = K_p(V_{DD} - v_I + V_{tP})^2 \tag{3.70}$$

V_{IH} can be found by differentiating Eq. (3.70) and setting $dv_O/dv_I = -1$ to solve for $v_I = V_{IH}$.

In region V, $v_I > (V_{DD} + V_{tP})$. M_P is in the cutoff region, and M_N is in the ohmic region. There will be virtually no current through the transistors, and the output voltage will be zero; that is,

$$i_{DN} = i_{DP} = 0$$
$$v_O = 0$$

EXAMPLE 3.5

D **Designing a CMOS inverter** Design a CMOS inverter, as shown in Fig. 3.26(a), to operate at a transition voltage $V_{MD} = 2.5$ V. The threshold voltages are $V_{tP} = -1$ V, $V_{tN} = 1$ V, and $K_p = 20\ \mu A/V^2$. The supply voltage is $V_{DD} = 5$ V. Assume $V_{IH} = V_{OH} = 5$ V, load capacitance $C_L = 0.5$ pF, and frequency $f_{clk} = 5$ MHz.

(a) Find the design parameter $K_R = K_n/K_p$, neglecting the body effect.

(b) Calculate NM_L and NM_H.

(c) Calculate the propagation delay t_{pd}.

(d) Calculate the delay-power product (DP).

(e) Use PSpice/SPICE to verify your results.

SOLUTION

(a) Solving Eq. (3.68) for K_R, we have

$$K_R = \frac{(W/L)_N}{(W/L)_P} = \left[\frac{V_{DD} + |V_{tP}| - V_M}{V_M - V_{tN}}\right]^2 \tag{3.71}$$

which, for $V_{MD} = 2.5$ V, gives $K_R = 1$. Thus, $K_n = K_p = 20\ \mu A/V^2$.

(b) $V_{OH} = V_{DD} = 5$ V. Then

$$V_{O(tran1)} = V_{MD} - V_{tP} = 2.5 + 1 = 3.5 \text{ V}$$

$$V_{O(tran2)} = V_{MD} - V_{tN} = 2.5 - 1 = 1.5 \text{ V}$$

Substituting numerical values of V_{DD}, V_{tN}, V_{tP}, K_n and K_p in Eq. (3.67), we get

$$v_I^2 = 14 - 8v_I + 2v_O + 2v_Iv_O - v_O^2$$

which, at $dv_O/dv_I = -1$, gives $v_O = v_I + 2.5$. Solving these two equations, we get $V_{IL} = v_I = 2.13$ V at $v_O = 4.63$ V.

Substituting numerical values of V_{DD}, V_{tN}, V_{tP}, K_n and K_p in Eq. (3.70), we get

$$16 - 8v_I + v_I^2 = 2v_Iv_O - 2v_O - v_O^2$$

which, at $dv_O/dv_I = -1$, gives $v_I = v_O + 2.5$. Solving these two equations, we get $V_{IH} = v_I = 2.88$ V at $v_O = 0.38$ V.

At $v_I = V_{DD} + V_{tP} = 5 - 1 = 4$, M_P turns off and $V_{OL} = v_O = 0$. Thus,

$$NM_L = V_{IL} - V_{OL} = 2.13 - 0 = 2.13 \text{ V}$$

$$NM_H = V_{OH} - V_{IH} = 5 - 2.88 = 2.12 \text{ V}$$

(c) The inverter with a load capacitor C_L has an arrangement similar to that shown in Fig. 3.19(a). As v_I goes low to V_{OL}, M_N turns off immediately, and capacitor C_L is charged up by the drain current i_{DP}. Since M_P is in saturation and $v_I = V_{OL} = 0$, the output voltage is related to the charging current by

$$C_L\frac{dv_O}{dt} = i_{DP} = K_p(V_{DD} - v_I + V_{tP})^2$$

which, after integration between V_{DD} and $V_{DD}/2$, gives t_{pLH} as

$$t_{pLH} = \frac{C_L V_{DD}}{2K_p(V_{DD} - |V_{tP}|)^2} \tag{3.72}$$

$$= \frac{0.5 \text{ pF} \times 5 \text{ V}}{2 \times 20\ \mu A/V^2 \times (5 \text{ V} - 1 \text{ V})^2} = 3.91 \text{ ns}$$

Because of the topology of the CMOS inverter, t_{pHL} will have the same value as t_{pLH}. Thus, the propagation time $t_{pd} \approx t_{pLH} = 3.91$ ns. In practice, a symmetric CMOS does not occupy the minimum chip area, and thus not all CMOS designs are symmetric.

(d) A CMOS inverter draws a negligible current (on the order of nanoamperes) from the power supply in both high and low states. Hence, the static power dissipation is almost zero: $P_{static} = 0$. This is a distinct advantage

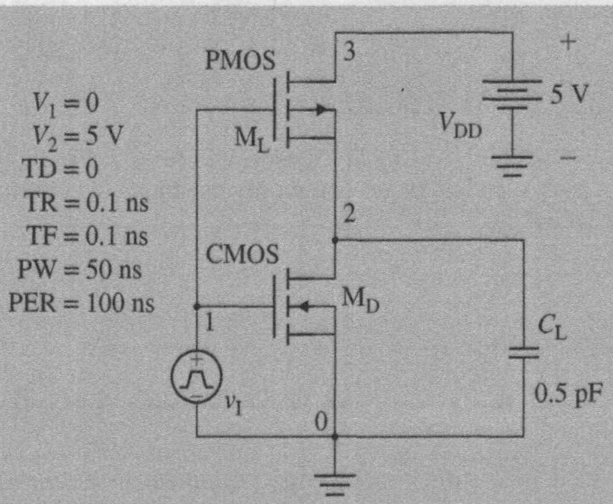

FIGURE 3.27 PSpice schematic for Example 3.5

for portable CMOS equipment because standby operation of the equipment will not discharge the battery. We have

$$P_{\text{dynamic}} = f_{\text{clk}} C_L V_{DD}^2 = 5 \text{ MHz} \times 0.5 \text{ pF} \times 5^2 = 62.5 \ \mu\text{W}$$

so $$P_D = P_{\text{static}} + P_{\text{dynamic}} = 0 + 62.5 \ \mu\text{W} = 62.5 \ \mu\text{W}$$

Therefore, the delay-power product is

$$DP = P_D \times t_{pd} = 62.5 \ \mu\text{W} \times 3.91 \text{ ns} = 0.244 \text{ pJ}$$

(e) The PSpice schematic is shown in Fig.3.27. The PSpice plot of the VTC is shown in Fig. 3.28(a), which gives $V_{IL} = 2.1245$ V (expected value is 2.13 V), $V_{IH} = 2.8755$ V (expected value is 2.88 V), $V_{OL} = 0$ (expected value is 0) at $V_{OH} = 5$ V, and $V_{I(\text{tran1})} = 2.5$ (expected value is 2.5 V). The VTC values are very close to the hand calculations, as expected. The transient response is shown in Fig. 3.28(b), which gives $t_{pHL} = 4.18$ ns and $t_{pLH} = 4.255$ ns (for $L = 50$ μm and $W = 100$ μm). The transient performance, however, will depend on the values of length (L) and width (W). A typical value of the transconductance parameter for the NMOS process is $\mu_n \epsilon / t_{ox} = 40 \ \mu\text{A/V}^2$. [Note: We can find $dv_O/dv_I = -1$ points by plotting dv(2) in Probe.]

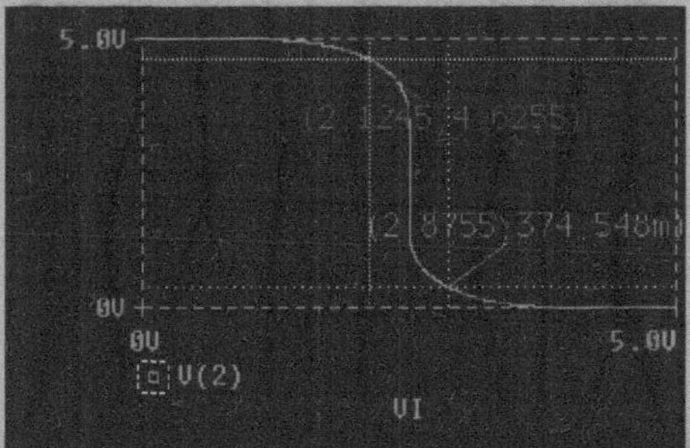

(a) VTC

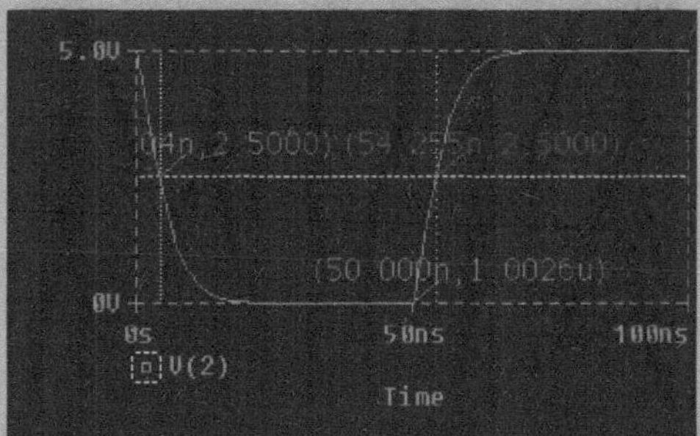

(b) Transient response

FIGURE 3.28 PSpice plots for CMOS inverter (using transistor circuit model) for Example 3.5

KEY POINT OF SECTION 3.7

- CMOS circuits use both *n*-channel and *p*-channel enhancement MOSFETs in the same circuit. Because they have very low power consumption and offer very high speed, they are commonly used in ICs.

3.8 CMOS Logic Circuits

Like NMOS inverters, CMOS inverters are frequently used in digital circuits in transmission gates and NOR and NAND gates.

3.8.1 CMOS Transmission Gates

A CMOS transmission gate consists of an NMOS and a PMOS that are connected in parallel and feed an effective load capacitance C_L. This arrangement is shown in Fig. 3.29. The substrate of the NMOS (M_N) is usually connected to the most negative potential (assumed to be the ground in Fig. 3.29), and the substrate of the PMOS (M_P) is connected to the most positive potential (usually the positive supply voltage V_{DD}) in the circuit rather than to the source terminal. MOSFETs may be assumed to be completely bilateral; that is, the drain and source terminals of each transistor are identical. The gate control voltage $\overline{v}_G = v_{cnt}$ of the PMOS is the complement of the gate voltage of the NMOS. Assuming that v_I operates between 0 and 5 V, $V_{DD} = 5$ V, $V_{tN} = 1$ V, and $V_{tP} = -1$ V, we will consider the following cases based on the level of input voltage v_I.

In case 1, $v_G = 0$, $v_{cnt} = 5$ V, $v_I = 0$ or 5 V, and initially $v_O = 0$ or 5 V. The gate of M_N is never positive with respect to either terminal, so M_N is always cut off. The gate of M_P is never negative with respect to either terminal, so M_N is always cut off. Thus, the input and output terminals are isolated from each other. The input voltage v_I may take on any value without affecting the output voltage v_O.

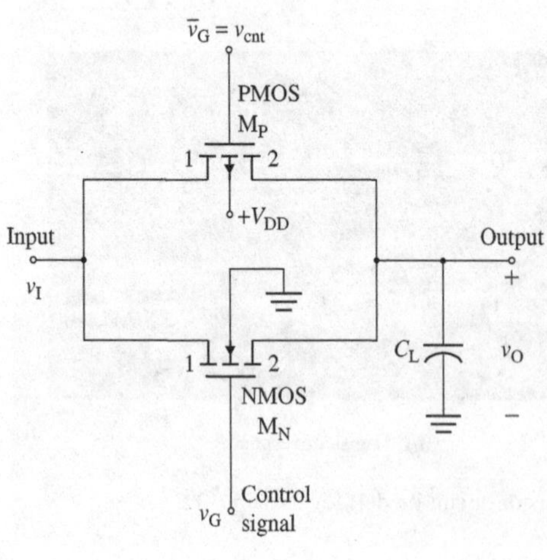

FIGURE 3.29 CMOS transmission gate

In case 2, $v_G = 5$ V, $v_{cnt} = 0$, and $v_I = 0$. The gate-to-terminal-1 voltage of M_N is 5 V, so M_N will be turned on. Terminals 1 and 2 of M_N act as the drain and the source, respectively. The drain-to-source voltage and drain current are zero. The output voltage is zero: $v_O = 0$. With $v_{cnt} = 0$ and $v_I = v_O = 0$, M_P will be cut off.

In case 3, $v_G = 5$ V, $v_{cnt} = 0$, $v_I = 5$ V, and initially $v_O = 0$. The gate and terminal 1 of M_N are at 5 V with respect to terminal 2 of M_N. Terminal 1 and terminal 2 of M_N act as the drain and the source, respectively. The drain current of M_N flows from terminal 1 to terminal 2 and charges the load capacitor until the gate-to-output voltage becomes equal to the threshold voltage V_{tN}; that is,

$$v_{GSN} = v_G - v_O = V_{tN}$$

For a transistor with $V_{tN} = 1$ V, $v_O = v_G - V_{tN} = 5 - 1 = 4$ V. When v_O rises to $v_G - V_{tN} = 5 - 1 = 4$ V, the gate-to-source voltage of M_N is equal to the threshold voltage V_{tN}, and M_N turns off. However, with $v_{cnt} = 0$ on the gate of M_P, M_P is turned on, with terminal 1 and terminal 2 acting as the drain and the source, respectively. After M_N is turned off, M_P is still turned on; the capacitor is charged all the way up to v_I, so the output voltage becomes $v_O = v_I$, at which point the drain-to-source voltage on M_P is zero and M_P is turned off. It is important to note that the output of a CMOS transmission gate is the full value of v_I.

In case 4, $v_G = 5$ V, $v_{cnt} = 0$, $v_I = 5$ V, and initially $v_O = v_I$. The gate and terminal 2 of M_N are at 5 V with respect to terminal 1 of M_N. Terminal 1 and terminal 2 of M_N act as the source and the drain, respectively. M_N is always turned on, and the drain current flows from terminal 2 to terminal 1, causing the load capacitor to be discharged completely to zero, so $v_O = 0$. M_P, whose terminal 2 acts as the source, always remains turned off.

In case 5, $v_G = 0$, $v_{cnt} = 5$ V, $v_I = 0$ or 5 V, and initially $v_O = v_I$. When v_G goes to zero, the situation is similar to case 1, and both transistors will be cut off. The input and output terminals are isolated.

The advantage of a CMOS transmission gate is that the output voltage v_O is always equal to v_I when the transmission gate is turned on, so $v_O = v_I$. The logic function can be described by

$$v_O = \begin{cases} v_I & \text{if } v_G \text{ is high (at logic 1)} \\ v_O & \text{if } v_G \text{ is low (at logic 0)} \end{cases}$$

With an NMOS transmission gate, the output voltage is reduced by the threshold voltage, so $v_O = v_I - V_{tN}$. The major disadvantage of the CMOS gate is that it requires both a gate voltage v_G and its complement for successful operation.

Propagation Delay

For $v_I = V_{DD}$ and $v_O = 0$ (initially), both M_N and M_P are in saturation. Capacitor C_L, which arises from the interconnecting area of the two MOSFETs and the load, charges to V_{DD}. Thus, the *charging time constant* τ_{LH} (for output low to high) can be found from the channel resistance of the complementary pair; that is,

$$\tau_{LH} = C_L(R_L \parallel R_{dN} \parallel R_{dP}) \tag{3.73}$$

where R_L is the load resistance. R_{dN}, which is the static resistance of the NMOS, is given by

$$R_{dN} = \frac{v_{DSN}}{i_{DN}} = \frac{V_{DD} - v_O}{K_n(V_{DD} - v_O - V_{tN})^2} \tag{3.74}$$

R_{dP}, which is the static resistance of the PMOS, is given by

$$R_{dP} = \frac{v_{SDP}}{i_{DP}} = \frac{V_{DD} - v_O}{K_p(V_{DD} - |V_{tP}|)^2} \tag{3.75}$$

The charging time constant τ_{LH} is usually higher than the discharging time. Assuming $t_{pLH} \approx t_{pHL}$ and $t_{pd} \approx t_{pLH}$, its value can be estimated from τ_{LH}. Thus, the propagation time t_{pd} to charge the capacitor C_L from 0 to $V_{DD}/2$ is given by

$$t_{pd} = \tau_{LH} \ln 2 = 0.69315\tau_{LH} \tag{3.76}$$

3.8.2 CMOS NOR and NAND Gates

A two-input NOR logic gate is shown in Fig. 3.30. Two parallel NMOSs are connected in series with two PMOSs. The substrates of the PMOSs are connected to the most positive potential V_{DD}; the substrates of the NMOSs are connected to the most negative potential—that is, the ground. The transfer characteristic is similar to that of a CMOS inverter. If both input voltages v_A and v_B are less than the threshold voltage (assuming $v_A = v_B = V_{OL}$), M_{N1} and M_{N2} are cut off. At the same time, the *p*-channel transistors M_{P1} and M_{P2} are turned on. Thus, the output voltage becomes high: $v_O = V_{OH}$ (at logic 1).

If $v_A = V_{OH}$, then M_{N1} is turned on, and M_{P1} is turned off. In this case, the output voltage drops to low: $v_O = V_{OL}$. If both v_A and v_B are equal to V_{OH}, then both M_{N1} and M_{N2} turn on, and the output voltage becomes low. For $v_A = v_B = V_{OL}$, both M_{N1} and M_{N2} are cut off, and the drain currents will be zero. The logic function is the same as that for an NMOS, shown in Table 3.4. If one or both of the logic inputs are at logic high, then at least one PMOS is cut off, and the drain current(s) is again zero. Therefore, the steady-state current is zero, and the power dissipation is essentially zero. Current flow and power dissipation occur only during the transition from one state to another.

A two-input NAND logic gate is shown in Fig. 3.31. Two NMOSs are connected in series, and two PMOSs are connected in parallel. The transfer characteristic is similar to that of a CMOS inverter. If both

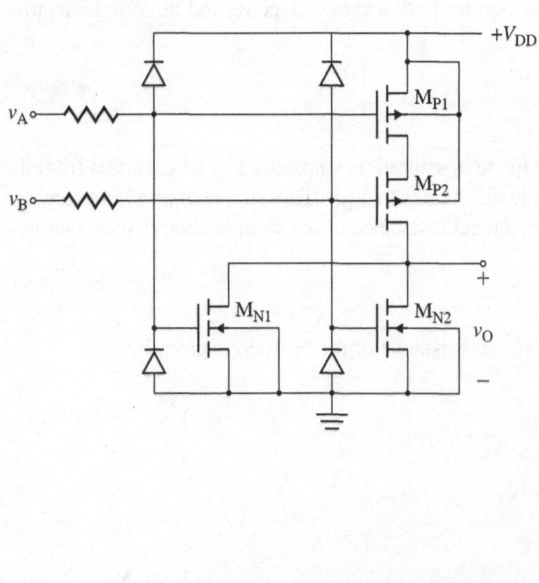

FIGURE 3.30 CMOS NOR gate

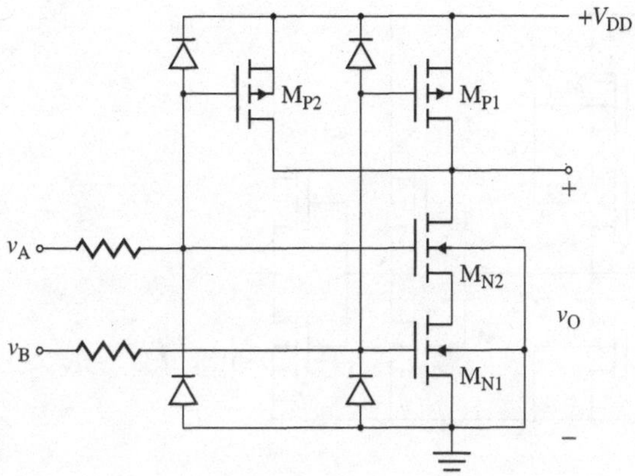

FIGURE 3.31 CMOS NAND gate

input voltages v_A and v_B are less than the threshold voltage (assuming $v_A = v_B = V_{OL}$ at logic 0), M_{N1} and M_{N2} are turned off, and M_{P1} and M_{P2} are turned on. In this case, the output voltage becomes high: $v_O = V_{OH}$ (at logic 1). If $v_A = V_{OH}$ and $v_B = V_{OL}$, then M_{N1} is turned on, M_{N2} is cut off, M_{P1} is off, and M_{P2} is turned on. Because of the high impedance from the drain to the source of M_{N2}, the output voltage becomes high: $v_O = V_{OH}$. If $v_A = v_B = V_{OH}$, then both M_{N1} and M_{N2} are turned on, and both M_{P1} and M_{P2} are turned off. In this case, the output voltage is low: $v_O = V_{OL}$. The logic function is the same as that for an NMOS, shown in Table 3.5.

CMOS gates have extremely high input resistance, on the order of hundreds of megohms, which is desirable for the driving circuitry. Unfortunately, they are susceptible to damage to the thin gate-oxide layer (typically 0.1 μm for a metal gate and less for a polysilicon gate), which has a breakdown voltage of 50 V to 100 V. Since the capacitance of a typical human body is 100 pF to 300 pF, a person walking across a waxed laboratory floor or brushing against a garment may generate static voltage in excess of 10 kV! Stray electrostatic discharge from a person handling the CMOS can easily release enough energy to cause permanent damage. A charged body can release tens of kilowatts of power over hundreds of nanoseconds. To protect a CMOS, clamping diodes, shown in Fig. 3.30, are built in to CMOS gates to limit any input voltages that fall outside the range from V_{SS} to V_{DD}. Also, any distributed input resistance (R_s), which is typically 1.5 kΩ for a metal gate and 250 Ω for a polysilicon gate, limits the transient gate current.

3.8.3 CMOS Families

The CMOS family of gates dates from the late 1960s. The 74Cxx logic families are second-generation CMOS circuits. These are polysilicon rather than metal gates, and the devices are smaller and faster than earlier designs. In addition, outputs are double buffered, as shown in Fig. 3.32. A 74Cxx gate has two cascaded inverters to isolate the logic from the output. As a result, the voltage gain is increased with a very steep VTC.

The 74HCxx series is the third generation of the CMOS family, which continues the trend toward smaller size and considerably less power, typically $P_D = 500$ μW and $t_{pd} = 10$ ns. The 74ACxx series is the fourth generation of the CMOS family, which is faster at the same power, typically $P_D = 500$ μW and $t_{pd} = 4$ ns.

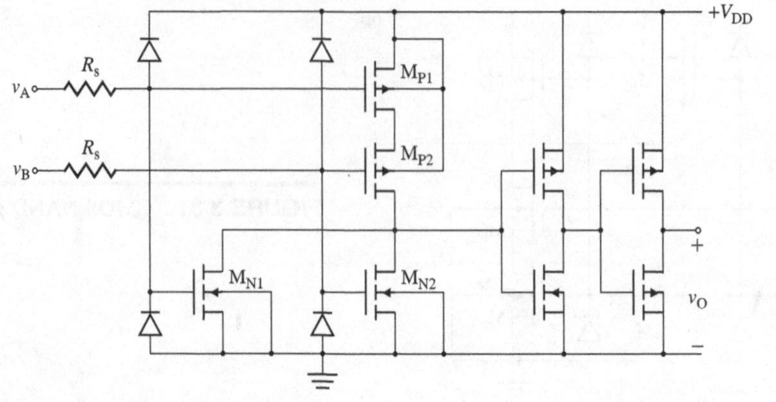

FIGURE 3.32 Buffered CMOS NOR gate

KEY POINT OF SECTION 3.8

- Like NMOS inverters, CMOS inverters are frequently used in digital circuits in transmission gates and NOR and NAND gates.

3.9 Comparison of CMOS and NMOS Gates

The major advantages and disadvantages of CMOS and NMOS gates are listed below:

1. The output voltage of a CMOS gate is the full input voltage v_I without the drop due to the threshold voltage, as in the case of NMOS gates.
2. A CMOS gate requires more transistors than an NMOS gate to perform similar logic functions.
3. CMOS gates consume very little power, thereby enabling very-large-scale integration (VLSI). NMOS gates consume more power than CMOS gates and have thermal limitations that make them less attractive for VLSI.
4. CMOS gates draw spikes of current during the transition from one state to another; the current peaks occur when both NMOS and PMOS transistors are in saturation.
5. CMOS gates occupy a larger area and have larger capacitances than NMOS gates.

3.10 BJT Inverters

BJT switches are the building blocks of bipolar logic circuits [8, 2, 5]. The earliest logic device using BJTs was the basic inverter, developed in the 1960s. This device was followed by the *resistor-transistor logic* (RTL), *diode-transistor logic* (DTL), and *transistor-transistor logic* (TTL) families of logic circuits. In RTL, DTL, and TTL families, the BJTs operate by switching between the saturation (on) and cutoff (off) regions of operation; hence, they are generally called *saturation-logic* families. Because their BJT saturation-logic circuits experience delay due to the storage time in saturated devices, RTL and DTL circuits are no longer used. In some BJTs (e.g., Schottky BJTs), the time delay required to bring a transistor out of saturation is avoided by preventing the BJT from becoming saturated. Although TTL logic circuits are being challenged by CMOS circuits, TTL technology has improved over the years and continues to be popular. The gate delay of a modern TTL circuit may be as low as 1.5 ns.

There are two other families of logic circuits: the emitter-coupled logic (ECL) circuit and the integrated-injection logic (I^2L) circuit. They are operated by switching a constant current between two parts of a circuit and avoiding transistor saturation. The gate delay of an ECL circuit can be less than 1 ns, and ECLs find applications in digital communication circuits and high-speed circuits in supercomputers. I^2L circuits, however, have lost ground to CMOS logic circuits in many applications.

3.10.1 Voltage Transfer Characteristics

A BJT with a collector (pull-up) resistance R_C is shown in Fig. 3.33(a). When the input v_I is low, such that the base-to-emitter voltage is less than the forward-bias cut-in voltage, $v_{BE} < V_{BE(\text{cut-in})}$, transistor Q_1 is off, and the output v_O is high; that is, $v_O = V_{OH} = V_{CC}$. The transistor operates at point *A*, as shown in Fig. 3.33(b). When v_I goes high to V_{OH}, the operating point jumps to point *B* and moves along the output characteristic (in the active region) toward the quiescent point *Q* (in the saturation region). The output voltage becomes low—that is, the collector–emitter saturation voltage, $V_{CE(sat)}$. The VTC is shown in Fig. 3.33(c).

The collector current at saturation is given by

$$I_{C(sat)} = \frac{V_{CC} - V_{CE(sat)}}{R_C} \tag{3.77}$$

and the corresponding base current at saturation is given by

$$I_{B(sat)} = \frac{I_{C(sat)}}{\beta_F} \quad (\text{at } V_{CE(sat)})$$

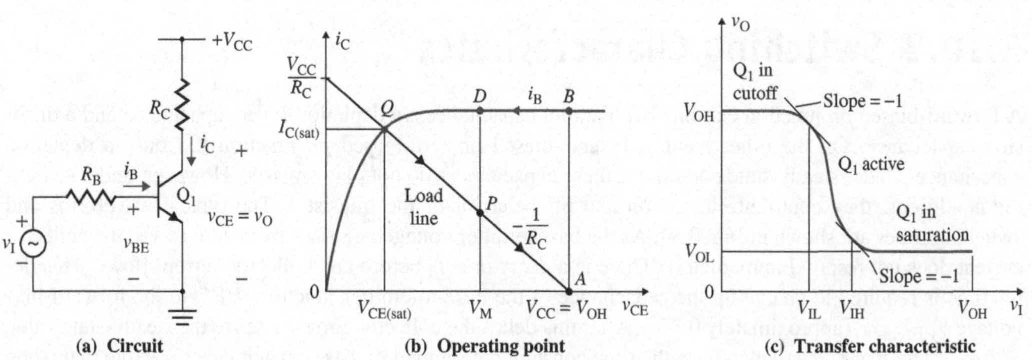

FIGURE 3.33 BJT inverter

where β_F is the forward current gain of the transistor. Normally, the circuit is designed with I_B higher than $I_{B(sat)}$ to ensure that Q_1 is driven into saturation. I_B can be found from

$$I_B = \frac{v_I - V_{BE(sat)}}{R_B}$$

The ratio of I_B to $I_{B(sat)}$ is called the *overdrive factor* k_{ODF}:

$$k_{ODF} = \frac{I_B}{I_{B(sat)}}$$

The ratio of $I_{C(sat)}$ to I_B is called the *forced* β_F, given by

$$\beta_{F(forced)} = \frac{I_{C(sat)}}{I_B} = \frac{I_{B(sat)}\beta_F}{I_B} = \frac{\beta_F}{k_{ODF}}$$

$V_{CE(sat)}$, which changes slightly with collector current $I_{C(sat)}$, can be found from

$$V_{CE(sat)} = V_T \ln \frac{I_B\beta_F + I_C\beta_F(1-\alpha_R)}{I_B\beta_F\alpha_R - I_C\alpha_R}$$

which, for a typical value of reverse current gain $\alpha_R = 0.1$, $I_C = I_{C(sat)}$, and $I_B = I_{B(sat)}$, becomes

$$V_{CE(sat)} = V_T \ln\left(\frac{10 + 9I_{C(sat)}/I_{B(sat)}}{1 - I_{C(sat)}/\beta_F I_{B(sat)}}\right) = V_T \ln\left(\frac{10 + 9\beta_{F(forced)}}{1 - \beta_{F(forced)}/\beta_F}\right) \quad \textbf{(3.78)}$$

(See Appendix D.) The value of $V_{CE(sat)}$ usually falls in the range of 0.1 V to 0.3 V. The base-to-emitter junction characteristic is similar to that of a diode, and $V_{BE(sat)}$ usually falls in the range of 0.65 V to 0.8 V.

3.10.2 Switching Characteristics

A forward-biased *pn* junction exhibits two parallel capacitances: a depletion-layer capacitance and a diffusion capacitance. On the other hand, a reverse-biased or zero-biased *pn* junction has only a depletion capacitance. Under steady-state conditions, these capacitances do not play any role. However, under switching conditions, they contribute to the on and off behavior of the transistor. The typical waveforms and switching times are shown in Fig. 3.34. As the base–emitter voltage v_{BE} rises from zero to V_B, the collector current does not respond immediately. There is a *delay time* t_d before any collector current flows. This delay time is required to charge up the capacitance of the base-to-emitter junction (BEJ) to the forward-bias voltage $V_{BE(cut\text{-}in)}$ (approximately 0.7 V). After this delay, the collector current rises to the steady-state value of $I_{C(sat)}$. The *rise time* t_r depends on the time constant determined by base capacitance C_B. Note that when the collector current i_C rises, the output voltage v_O falls. Thus, t_r for i_C in Fig. 3.34(b) corresponds to t_f for v_O in Fig. 3.34(c).

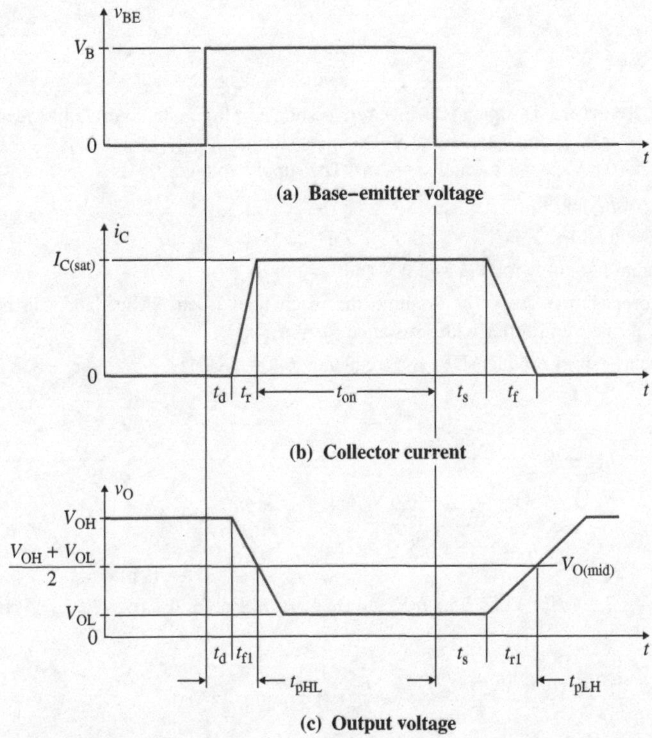

FIGURE 3.34 Switching times of bipolar transistors

The base current is normally more than that required to saturate the transistor. As a result, the excess minority carrier charge is stored in the base region. The higher the overdrive factor k_{ODF}, the greater will be the amount of extra charge stored in the base. This extra charge, which is called the *saturating charge*, is proportional to the excess base drive Δi_B given by

$$\Delta i_B = I_B - I_{B(sat)} = k_{ODF} I_{B(sat)} - I_{B(sat)} = I_{B(sat)}(k_{ODF} - 1)$$

The saturating charge is given by

$$Q_s = \tau_s \, \Delta i_B = \tau_s I_{B(sat)}(k_{ODF} - 1)$$

where τ_s is known as the *storage time constant* of the transistor. When v_{BE} falls to zero, the collector current does not change for a time t_s, called the *storage time*, which is the time required to remove the saturating charge from the base. Once the extra charge is removed, the BEJ capacitance discharges to zero. *Fall time* t_f depends on the time constant, which is determined by the capacitance of the reverse-biased BEJ.

The propagation time t_{pHL} consists of the delay time t_d and the time t_{f1} for the output to fall from V_{OH} to $(V_{OH} + V_{OL})/2$; that is, $t_{pHL} = t_d + t_{f1}$. The propagation time t_{pLH} consists of the storage time t_s and the time t_{r1} for the output to rise from V_{OL} to $(V_{OH} + V_{OL})/2$. That is, $t_{pLH} = t_s + t_{r1}$.

EXAMPLE 3.6

D

Designing a BJT inverter Design a BJT inverter, as shown in Fig. 3.35, to drive five identical inverters ($N = 5$) and to give $V_{OH} = 3.5$ V and $NM_L = 0.4$ V. The transistor has $I_{C(max)} = 5$ mA, $\beta_F = 100$ to 150, $V_{BE(cut\text{-}in)} = 0.6$ V, $V_{BE(sat)} = 0.8$ V, $t_d = 1$ ns, and $t_s = 2$ ns. The supply voltage V_{CC} is 5 V.

(a) Find the values of R_C and R_B.

(b) Calculate k_{ODF} and NM_H.

(c) Find the maximum fan-out N for $V_{IH} = 2.0$ V and $k_{ODF} = 1$.

(d) Calculate the propagation delay t_{pd}. Assume that each load is an NMOS and can be represented by a capacitance $C_B = 2$ pF in series with resistance $R_B = 1.8$ kΩ.

(e) Calculate the delay-power product (DP) at a frequency $f_{clk} = 5$ MHz.

SOLUTION

(a) $N = 5$, $V_{OH} = 3.5$ V, $V_{IL} = V_{BE(cut\text{-}in)} = 0.6$ V, and

$$V_{OL} = V_{IL} - NM_L = 0.6 - 0.4 = 0.2 \text{ V}$$

For $V_{CE(sat)} = V_{OL} = 0.2$ V (for $V_T = 25.8$ mV and $\beta_F = 100$), Eq. (3.78) gives $\beta_{F(forced)} = 71.8$. Using Eq. (3.77), we have

$$R_C \geq \frac{(V_{CC} - V_{CE(max)})}{I_{C(sat)}} = \frac{(5 - 0.2)\text{ V}}{5 \text{ mA}} = 960\ \Omega$$

We choose $R_C = 1$ kΩ. When the input is low, Q_1 is off and $v_O = V_{OH}$. The base current I_B of each load stage is

$$I_B = \frac{V_{OH} - V_{BE(sat)}}{R_B}$$

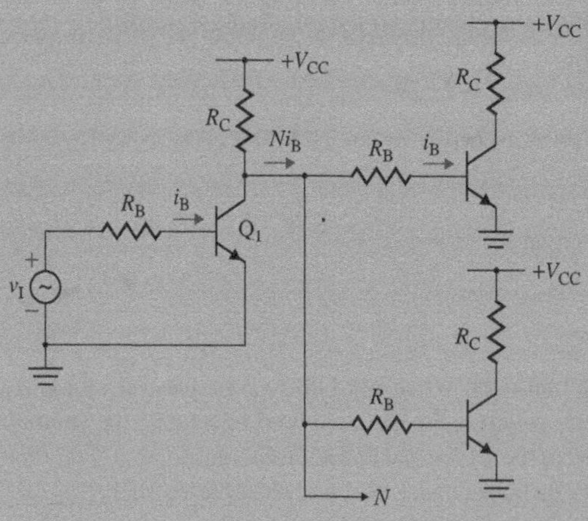

FIGURE 3.35 BJT inverter driving identical inverters

and $V_{OH} = V_{CC} - NI_BR_C$. Eliminating I_B and solving for V_{OH}, we get

$$V_{OH} = \frac{V_{CC} + NR_CV_{BE(sat)}/R_B}{1 + NR_C/R_B} = \frac{R_BV_{CC} + NR_CV_{BE(sat)}}{R_B + NR_C} \tag{3.79}$$

which, for $V_{OH} = 3.5$ V, $N = 5$, and $V_{BE(sat)} = 0.8$ V, gives $R_B = 9R_C = 9$ kΩ.

(b) We have

$$I_B = \frac{I_{C(sat)}}{\beta_{F(forced)}} = \frac{5 \text{ mA}}{71.8} = 70\ \mu\text{A}$$

$$I_{B(sat)} = \frac{I_{C(sat)}}{\beta_F} = \frac{5 \text{ mA}}{100} = 50\ \mu\text{A}$$

$$k_{ODF} = \frac{I_B}{I_{B(sat)}} = \frac{70}{50} = 1.40$$

$$V_{IH} = R_BI_B + V_{BE(sat)} \tag{3.80}$$

$$= 9 \text{ k}\Omega \times 70\ \mu\text{A} + 0.8 \text{ V} = 1.43 \text{ V}$$

Thus, $NM_H = V_{OH} - V_{IH} = 3.5 - 1.43 = 2.07$ V. This noise margin is a measure of the safety factor that allows the load transistors to remain saturated despite changes in supply voltage, temperature, and manufacturing tolerances.

(c) When Q_1 is off, the output voltage v_O becomes

$$v_O = V_{CC} - NI_BR_C = V_{CC} - Nk_{ODF}I_{B(sat)}\,R_C = V_{CC} - N\frac{k_{ODF}(V_{CC} - V_{CE(sat)})}{\beta_F}$$

For regeneration of logic levels at the load gates, $v_O \geq V_{IH}$, and we find the fan-out N as

$$N \leq \frac{\beta_F(V_{CC} - V_{IH})}{k_{ODF}(V_{CC} - V_{CE(sat)})} \tag{3.81}$$

$$\leq 100 \times \frac{5 - 1.4}{1.39 \times (5 - 0.2)} = 53.4$$

Hence, the maximum fan-out is $N = 53$.

(d) When Q_1 turns off, each load is represented by a capacitance C_B and a resistance R_B. The equivalent circuit is shown in Fig. 3.36(a). For identical loads, all branches are effectively in parallel as far as the output node is concerned, and Fig. 3.36(a) can be simplified to Fig. 3.36(b). That is, the time constant for low to high can be found from

$$\tau_{r1} \approx NC_B\left(R_C + \frac{R_B}{N}\right) \tag{3.82}$$

$$= 5 \times 2 \text{ pF} \times \left(1 \text{ k}\Omega + \frac{9 \text{ k}\Omega}{5}\right) = 28 \text{ ns}$$

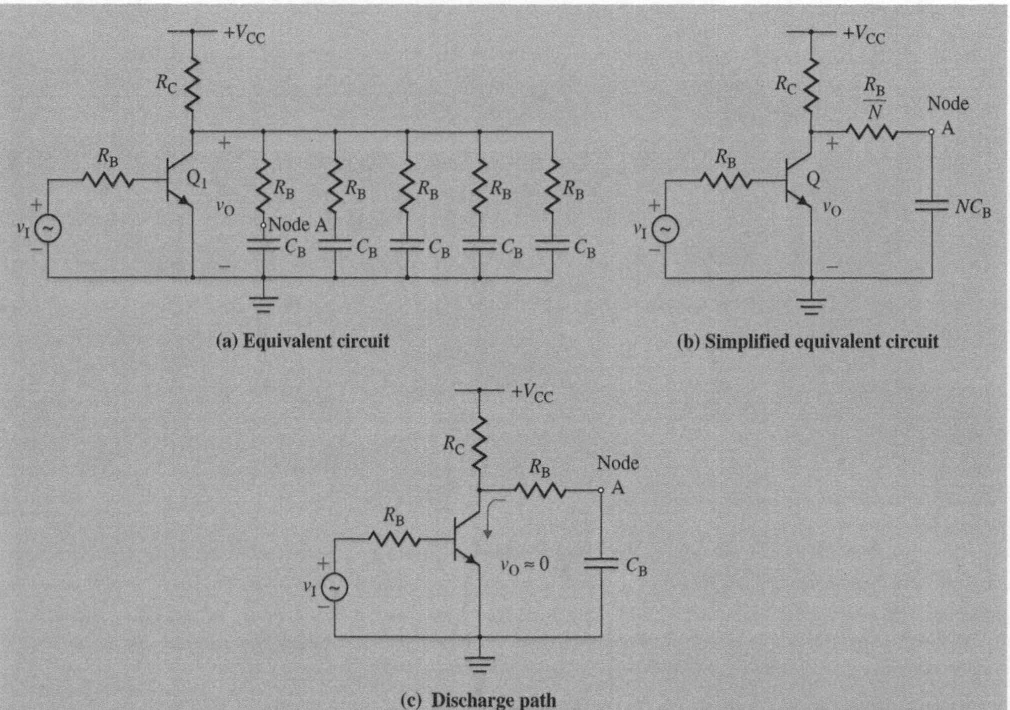

(a) Equivalent circuit

(b) Simplified equivalent circuit

(c) Discharge path

FIGURE 3.36 Equivalent circuits for Example 3.6

The output voltage v_O during charging can be found from

$$v_O = V_{CC} - \frac{R_C V_{CC}}{R_C + R_B/N} e^{-t/\tau_{r1}}$$

from which we can find the time required for v_O to rise from $V_{OL} = 0.2$ V to $V_{O(mid)}$, which is equal to $(V_{OH} + V_{OL})/2 = 1.85$ V; that is,

$$t_{r1} = t_2(\text{for } v_O = V_{O(mid)}) - t_1(\text{for } v_O = V_{OL}) = \tau_{r1} \ln \frac{V_{CC} - V_{OL}}{V_{CC} - V_{O(mid)}} \tag{3.83}$$

$$= 28 \text{ ns} \times \ln\left[\frac{(5 - 0.2)}{(5 - 1.85)}\right] = 11.79 \text{ ns}$$

Thus, $t_{pLH} = t_s + t_{r1} = 2 + 11.79 = 13.79$ ns.

When Q_1 is turned on to saturation, each capacitor discharges through the transistor. The equivalent circuit is shown in Fig. 3.36(c). The time constant for high to low can be found from

$$\tau_{f1} \approx C_B R_B \tag{3.84}$$

$$= 2 \text{ pF} \times 9 \text{ k}\Omega = 18 \text{ ns}$$

The output voltage v_O during discharging can be found from

$$v_O = V_{OH} e^{-t/\tau_{f1}}$$

from which we can find the time for v_O to fall from $V_{OH} = 3.5$ V to $(V_{OH} + V_{OL})/2 = 1.85$ V; that is,

$$t_{f1} = \tau_{f1} \ln\left(\frac{2V_{OH}}{V_{OH} + V_{OL}}\right) \tag{3.85}$$

$$= 18 \text{ ns} \times \ln\left(2 \times \frac{3.5}{3.5 + 0.2}\right) = 11.48 \text{ ns}$$

Thus, $t_{pHL} = t_d + t_{f1} = 1 + 11.48 = 12.48$ ns. The propagation time is

$$t_{pd} = \frac{t_{pHL} + t_{pLH}}{2} = 13.135 \text{ ns}$$

(e) The total power loss in the two junctions is found as follows:

$$P_{static} = V_{CE(sat)} I_{C(sat)} + V_{BE(sat)} I_B$$

$$= 0.2 \times 5 \text{ mA} + 0.8 \times 70\ \mu\text{A} = 1.056 \text{ mW}$$

$$P_{dynamic} = N f_{clk} C_B V_{OH}^2 = 5 \times 5 \text{ MHz} \times 2 \text{ pF} \times 3.5^2 = 0.61 \text{ mW}$$

$$P_D = P_{static} + P_{dynamic} = 1.67 \text{ mW}$$

Therefore, the delay-power product is

$$\text{DP} = P_D \times t_{pd} = 1.67 \text{ mW} \times 13.135 \text{ ns} = 21.9 \text{ pJ}$$

▶ **NOTE** The value of DP is much higher than that of a CMOS or an NMOS inverter.

KEY POINT OF SECTION 3.10

- A BJT is operated as a switching device. It is usually overdriven to ensure operation in the saturation region. This causes a reduction in the switching speed due to charge recovery. The higher the amount of overdrive, the greater will be the amount of extra charge stored in the base.

3.11 Transistor-Transistor Logic Gates

Equation (3.81) shows that, to achieve a high fan-out, β_F should be as high as possible. In 1965 the first TTL family was introduced specifically to increase switching speed without sacrificing fan-out or noise margin and without increasing power dissipation. TTL gates were leading DTL gates in sales by 1970 and dominated the digital IC market for more than 10 years. Many developments in the process and manufacturing technology as well as in circuit design techniques have led to several new generations (families) of TTL gates. Each generation has its advantages and disadvantages relative to the previous generations. As the technology has advanced, it has become economically feasible to obtain high performance by using

TABLE 3.6 Parameters of 54/74 series TTL logic gates

Parameter	74	74H	74L	74S	74LS	74AS	74ALS	74F
V_{IL}, in V	0.8	0.8	0.8	0.8	0.8	0.8	0.8	0.8
V_{IH}, in V	2.0	2.0	2.0	2.0	2.0	2.0	2.0	2.0
V_{OL}, in V	0.4	0.4	0.4	0.5	0.5	0.5	0.5	0.5
V_{OH}, in V	2.4	2.4	2.4	2.7	2.7	2.7	2.7	2.7
t_{pd}, in ns at $C_L = 50$ pF	10	6	30	3	10	1.5	4	2.5
P_D, in mW	10	25	1	20	2	8	1	5
PD, in pJ	100	150	30	60	20	12	4	12.5

more transistors per gate, and several families have emerged, among which are standard TTL gates, high-speed (H) TTL gates, low-power (L) TTL gates, Schottky (S) TTL gates, low-power Schottky (LS) gates, advanced Schottky (AS) gates, advanced low-power Schottky (ALS) gates, and Fairchild advanced Schottky (F or Fast) gates.

The performance parameters of TTL 54/74 families are summarized in Table 3.6. In early TTL families, the improvement in switching speed resulted in more power dissipation. The Schottky series, however, has both lower power dissipation and improved speed. Currently, TTL applications focus on low-power Schottky (LS) and advanced low-power Schottky (ALS) families of gates. The ALS family offers the highest speed (although it is surpassed by the ECL family and rivaled by the CMOS family). As examples, we will analyze the TTL, and ALS types.

3.11.1 Standard TTL Gates

The input stage of a TTL gate uses a multiemitter transistor Q_1, as shown in Fig. 3.37(a). In isoplanar ICs, multiemitters are normally fabricated within the same base region. Figure 3.37(b) shows a cross section of a three-emitter transistor. Q_1 can be viewed as three diodes from the base to the emitter and one from the base to the collector, as shown in Fig. 3.37(c).

A TTL NAND gate with a multiemitter transistor input is shown in Fig. 3.38. The combination of Q_4, D_1, and Q_3 is called the *totem-pole output stage*. The transistor Q_2 forms a phase splitter since the collector and emitter voltages are 180° out of phase. The circuit operation can be divided into two modes: mode 1 and mode 2.

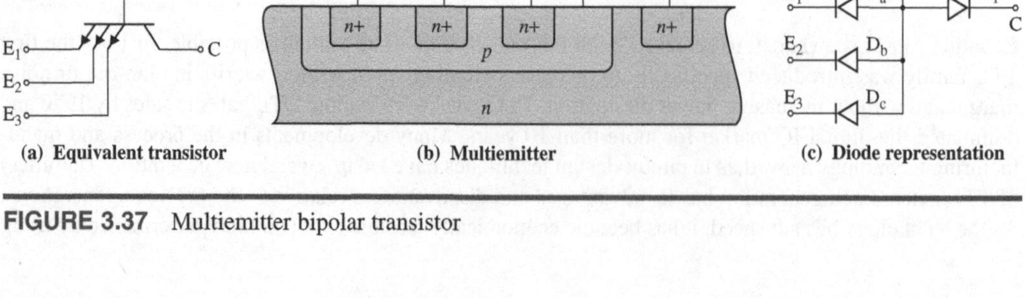

FIGURE 3.37 Multiemitter bipolar transistor

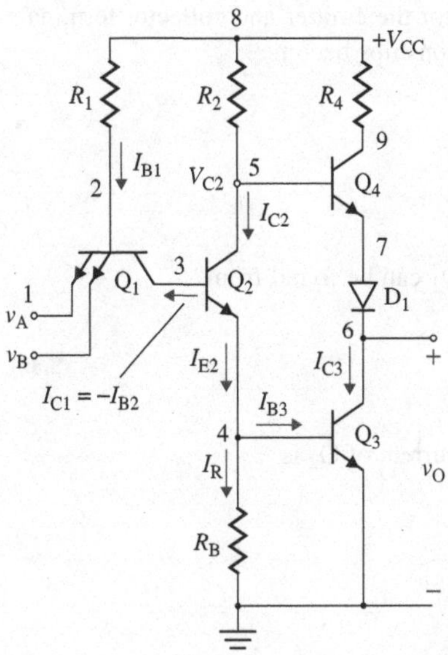

FIGURE 3.38 TTL NAND gate

During mode 1, the input voltages v_A and v_B are high, at V_{OH}; that is, $v_A = v_B = V_{OH}$. The TTL NAND gate is at low-state output voltage, as shown in Fig. 3.39, with N similar load gates. I_{C3} becomes the collector saturation current I_{CS} of Q_3. The two emitter junctions of Q_1 are reverse biased. The voltage v_{B1} at the base of Q_1 is large enough to forward-bias the base–collector junction of Q_1 and drive Q_2 into saturation. Since the base–collector junction is forward biased and the BEJ is reverse biased,

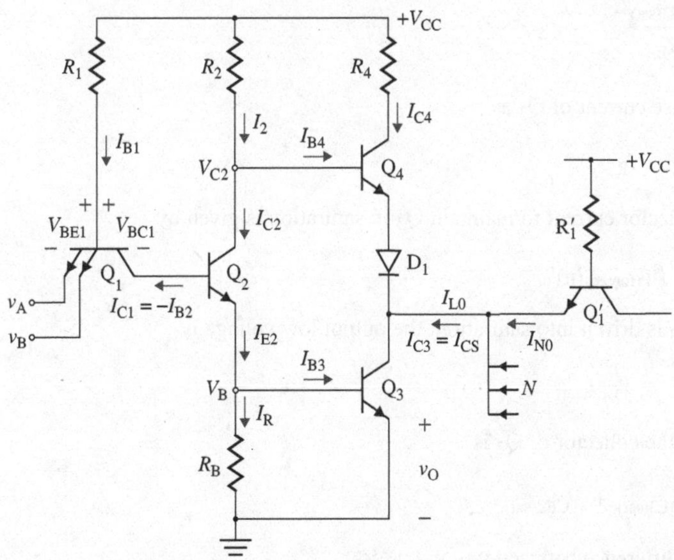

FIGURE 3.39 TTL gate in the output low state

Q_1 will be operating in its *inverse-active mode*; that is, the roles of the emitter and collector terminals are interchanged. Under this condition, the terminal current relationships become

$$I_{E1} = -\beta_R I_{B1}$$

$$I_{C1} = -(1 + \beta_R) I_{B1}$$

where β_R is the inverse-mode current gain of the transistor Q_1.

Assuming that Q_2 and Q_3 are saturated, the base current of Q_1 can be found from

$$I_{B1} = \frac{V_{CC} - V_{BE2(sat)} - V_{BE3(sat)} - V_{BC1}}{R_1} \tag{3.86}$$

where V_{BC1} is the base-to-collector voltage of Q_1. Thus, the base current of Q_2 is

$$I_{B2} = -I_{C1} = (1 + \beta_R) I_{B1}$$

Since the transistor Q_2 is driven into saturation, its collector current is

$$I_{C2} = \frac{V_{CC} - V_{BE3(sat)} - V_{CE2(sat)}}{R_2} \tag{3.87}$$

Thus, the emitter current of Q_2 is

$$I_{E2} = I_{B2} + I_{C2}$$

and the current through the recovery resistance R_B is

$$I_R = \frac{V_{BE3(sat)}}{R_B}$$

which gives the base current of Q_3 as

$$I_{B3} = I_{E2} - I_R$$

The maximum collector current to maintain Q_3 in saturation is given by

$$I_{C3(sat)} = \beta_{F(forced)} I_{B3}$$

Since transistor Q_3 is driven into saturation, the output low voltage is

$$v_O = V_{OL} = V_{CE3(sat)}$$

and the voltage at the collector of Q_2 is

$$V_{C2} = V_{BE3(sat)} + V_{CE2(sat)}$$

Thus, the voltage difference between v_{C2} and v_O is

$$v_{C2} - v_O = V_{BE3(sat)} + V_{CE2(sat)} - V_{CE3(sat)} \approx V_{BE3(sat)}$$

which is the voltage across the BEJ of Q_4 and D_1 and will not be sufficient to turn on both Q_4 and D_1. With Q_3 in saturation and the output voltage in its low state, Q_4 will be off. Thus, Q_3 will sink to the (low-state) load current I_{LO} given by

$$I_{LO} = N_L I_{NO}$$

where I_{NO} is the individual load current for each fan-out at low output (i.e., $I_{NO} = I_{IL}$) and is obtained by multiplying the base current of Q_1' by β_R; that is,

$$I_{NO} = \beta_R \left[\frac{V_{CC} - V_{CE3(sat)} - V_{BE1(sat)}}{R_1} \right] \tag{3.88}$$

Thus, the low-output fan-out is given by

$$N_L = \frac{I_{C3(sat)}}{I_{NO}}$$

During mode 2, at least one of the inputs v_A or v_B is low, at V_{OL}; that is, v_A (or v_B) = V_{OL}. The TTL NAND gate in the output high state is shown in Fig. 3.40. The BEJ is forward biased through R_1 and V_{CC}. The base current I_{B1} causes an emitter current through the particular emitter that is connected to the low input. Transistor action forces collector current into Q_1. But the collector current of Q_1, which equals the reverse-biased saturation current out of the base of Q_2, is usually much smaller than its base current. As a result, Q_1 will be in saturation. With v_A (or v_B) = V_{OL}, the base current of Q_1 becomes

$$I_{B1} = \frac{V_{CC} - V_{BE1(sat)}}{R_1}$$

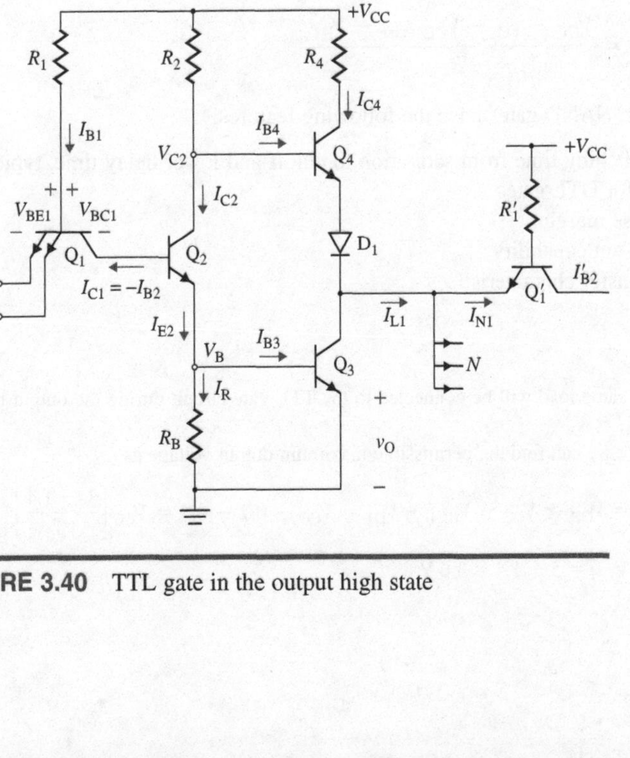

FIGURE 3.40 TTL gate in the output high state

Transistors Q_2 and Q_3 are cut off. Transistor Q_4 and diode D_1 supply the (high-state) load current I_{L1} given by

$$I_{L1} = N_H I_{N1}$$

where I_{N1} is the individual load current for each fan-out at high output (i.e., $I_{N1} = I_{IH}$). Since Q_1 is operating in the inverse-action mode, its base current must be multiplied by β_R to give $I_{N1} = \beta_R I_{B1}$; that is,

$$I_{N1} = \beta_R\left[\frac{V_{CC} - V_{BE2(sat)} - V_{BE3(sat)} - V_{BC1}}{R_1}\right] \tag{3.89}$$

With Q_4 in the active region, the base current of Q_4 is

$$I_{B4} = \frac{I_{L1}}{1 + \beta_F} = \frac{N_H I_{N1}}{1 + \beta_F}$$

Using KVL around the loop formed by V_{CC}, R_2, Q_4, and D_1, we get

$$\begin{aligned} V_{CC} &= R_2 I_{B4} + V_{BE4} + V_{D1} + v_O \\ &= \frac{R_2 N_H I_{N1}}{1 + \beta_F} + V_{BE4} + V_{D1} + v_O \end{aligned} \tag{3.90}$$

which gives the value of the high-output fan-out as

$$N_H = \frac{(1 + \beta_F)(V_{CC} - V_{BE4} - V_{D1} - v_O)}{R_2 I_{N1}} \tag{3.91}$$

If R_4 were not present, the collector current of Q_4 would be $\beta_F I_{B4}$, which would be highly undesirable. To limit the collector current of Q_4 to an acceptable value, R_4 is introduced. The maximum collector current of Q_4 is

$$I_{C4(max)} = \frac{V_{CC} - v_O - V_{CE4(sat)} - V_{D1}}{R_4} \tag{3.92}$$

In summary, TTL NAND gates have the following features:

1. Short switching time from saturation to cutoff and lower delay time, typically 10 ns as compared to 40 ns for DTL gates
2. High noise margins
3. High fan-out capability
4. Sharp transfer characteristic

▶ **NOTES**

1. In practice, the same load will be connected to the TTL gate circuit during the output high and output low states; that is, $N_L = N_H = N$.
2. From Eq. (3.90), we can find the permissible maximum output voltage as

$$v_{O(max)} = V_{OH} \approx V_{CC} - V_{BE4} - V_{D1} = V_{CC} - 0.7 - 0.7 = V_{CC}\left[\frac{1 - 1.4}{V_{CC}}\right]$$

For $V_{CC} = 5$ V, $V_{OH} = 5 - 1.4 = 3.6$ V.

EXAMPLE 3.7

D

Designing a TTL NAND gate

(a) Design the TTL NAND gate of the circuit in Fig. 3.38. It has two inputs and feeds four identical NAND gates. The output voltage at output high is $V_{OH} = 3.5$ V, and $I_{N1} = 72.5$ μA. Assume $V_{CC} = 5$ V, $V_D = 0.7$ V, $V_{CE(sat)} = 0.2$ V, $V_{BE} = 0.7$ V, $V_{BC} = 0.7$ V, $\beta_{F(forced)} = 10$, $\beta_R = 0.1$, $I_{C3(sat)} = 1$ mA, and $I_{C4(max)} = 1$ mA. Capacitance of each load is, $C_i = 5$ pF.

(b) Use PSpice/SPICE to check your design by plotting the transfer function. Determine NM_L and NM_H. Model parameters for diodes are

```
RS=4 TT=0.1NS
```

and for transistors are

```
BF=10 BR=0.1 TF=0.1NS TR=10NS VJC=0.85 VAF=50
```

SOLUTION

(a) The design can be carried out using the following steps:

Step 1. Using Eq. (3.89), calculate the value of R_1:

$$R_1 = \beta_R \frac{V_{CC} - 2V_{BE(sat)} - V_{BC}}{I_{N1}} = \frac{0.1 \times [5 - (2 \times 0.7) - 0.7]\ \text{V}}{72.5\ \mu\text{A}} = 4\ \text{k}\Omega$$

Step 2. Using Eq. (3.91) for $v_O = V_{OH}$, find the value of R_2 to meet the output requirement:

$$R_2 = \frac{(1 + \beta_{F(forced)})(V_{CC} - V_{BE4} - V_{D1} - V_{OH})}{NI_{N1}}$$

$$= \frac{(1 + 10) \times (5 - 0.7 - 0.7 - 3.5)\ \text{V}}{4 \times 72.5\ \mu\text{A}} = 3.79\ \text{k}\Omega$$

Step 3. Using Eq. (3.86), find the value of I_{B1} at output high:

$$I_{B1} = \frac{(5 - 0.7 - 0.7 - 0.7)\ \text{V}}{4\ \text{k}\Omega} = 725\ \mu\text{A}$$

Step 4. Using Eq. (3.88), calculate the low-state load current:

$$I_{N0} = \beta_R \left[\frac{V_{CC} - V_{CE3(sat)} - V_{BE1(sat)}}{R_1}\right] = 0.1\left[\frac{(5 - 0.2 - 0.7)\ \text{V}}{4\ \text{k}\Omega}\right] = 102.5\ \mu\text{A}$$

The total load current at low state for $N = 4$ is

$$I_{L0} = 4 \times 102.5\ \mu\text{A} = 410\ \mu\text{A}$$

Step 5. Calculate the high-state I_{B2}, I_{C2}, and I_{E2}. The high-state I_{B2} is

$$I_{B2} = -I_{C1} = (1 + \beta_R)I_{B1} = (1 + 0.1) \times 725\ \mu A = 798\ \mu A$$

The high-state I_{C2} is

$$I_{C2} = \frac{V_{CC} - V_{BE3(sat)} - V_{CE2(sat)}}{R_2} = \frac{(5 - 0.7 - 0.2)\ V}{3.79\ k\Omega} = 1082\ \mu A$$

The high-state I_{E2} is

$$I_{E2} = I_{B2} + I_{C2} = 798\ \mu A + 1082\ \mu A = 1.88\ mA$$

Step 6. Assume that most of I_{E2} flows through the base of Q_3 rather than through R_B. Let $I_R = 0.4I_{E2} = 0.4 \times 1.88\ m = 0.75\ mA$. Then calculate the value of R_B:

$$R_B = \frac{V_{BE3(sat)}}{I_R} = \frac{0.7\ V}{0.75\ mA} = 933\ \Omega$$

Choose $R_B = 1\ k\Omega$.

Step 7. Calculate the maximum permissible collector low-state current to drive Q_3 in saturation:

$$I_{B3} = I_{E2} - I_R = 1.88\ m - 0.75\ m = 1.13\ mA$$

$$I_{C3(max)} = \beta_F I_{B3} = 10 \times 1.13\ mA = 11.3\ mA$$

which is higher than $I_{LO} = 410\ \mu A$, from step 4. Thus, the design should be satisfactory. (Otherwise steps 1 through 7 should be repeated with a lower value of N, a higher value of $\beta_{F(sat)}$, or a lower value of I_{N1}.) The overdrive factor is

$$k_{ODF} = \frac{I_{C3(max)}}{I_{LO}} = \frac{11.3\ m}{410\ \mu} = 27.6$$

Step 8. Calculate the base current of Q_4 at output high:

$$I_{B4} = \frac{NI_{N1}}{1 + \beta_F} = \frac{4 \times 72.5\ \mu A}{1 + 10} = 26.4\ \mu A$$

Step 9. Using Eq. (3.92), calculate the value of R_4 for $v_O = V_{OH}$:

$$R_4 = \frac{V_{CC} - V_{OH} - V_{CE4(sat)} - V_{D1}}{I_{C4(max)}} = \frac{(5 - 3.5 - 0.2 - 0.7)\ V}{1\ mA} = 600\ \Omega$$

The power rating of R_4 is $P_{R4} = (1\ mA)^2 R_4 = (1\ mA)^2 \times 600 = 0.6\ mW$.

Step 10. Calculate the power rating of R_2 due to I_{C2} or I_{B2}, whichever has the higher value. Since $I_{C2} = 1082\ \mu A$ and $I_{B2} = 798\ \mu A$, use I_{C2}.

$$P_{R2} = (1.082\ mA)^2 R_2 = (1.082\ mA)^2 \times 3.79\ k\Omega = 4.44\ mW$$

Step 11. Calculate the power rating of R_1 due to I_{B1} at output high or I_{B1} at output low, whichever has the higher value. Since the base current of Q_1 at output high is

$$I_{B1} = \frac{V_{CC} - V_{BE2(sat)}}{R_1} = \frac{(5 - 0.7)\text{ V}}{4\text{ k}\Omega} = 1.075\text{ mA}$$

and I_{B1} at output low is 725 μA (from step 3), use I_{B1} at output high.

$$P_{R1} = (1.075\text{ mA})^2R_2 = (1.075\text{ mA})^2 \times 4\text{ k}\Omega = 4.6\text{ mW}$$

(b) Each transistor has junction capacitances that will affect the switching speed. To examine the propagation time, we will assume that a load resistance of

$$R_L = \frac{V_{OH}}{NI_{N1}} = \frac{3.5\text{ V}}{4 \times 72.5\ \mu\text{A}} = 12\text{ k}\Omega$$

and an equivalent capacitance of

$$C_L = 4 \times 5\text{ pF} = 20\text{ pF}$$

is connected to the output.

The PSpice schematic is shown in Fig. 3.41. The PSpice plots of the characteristics, which are shown in Fig. 3.42, give $V_{OH} = 3.53$ V (expected value is 3.5 V), $V_{IL} = 0.6$ V at $V_O = 3.49$ V, $V_{OL} = 42$ mV at $V_O = 5$ V, and $V_{IH} = 1.5$ V at $V_O = 62$ mV.

$$NM_L = V_{IL} - V_{OL} = 0.6 - 0.042\text{V} \approx 0.56\text{ V}$$

$$NM_H = V_{OH} - V_{IH} = 3.53 - 1.5 = 2.03\text{ V}$$

$t_{pHL} = 4.53$ ns, $t_{pLH} = 6.68$ ns, and $t_{pd} = 5.61$ ns.

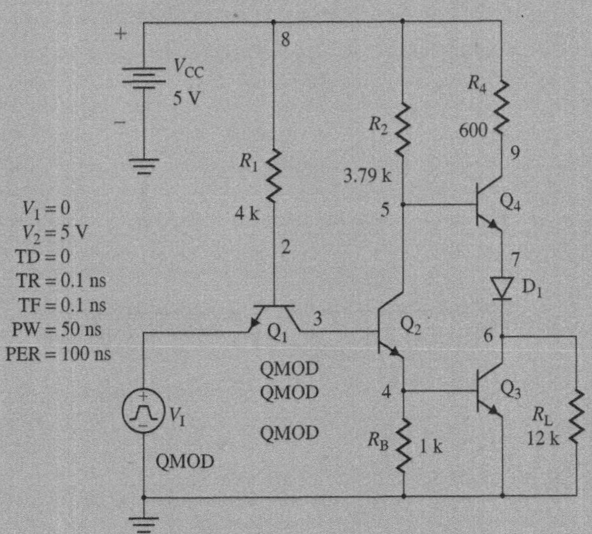

FIGURE 3.41 PSpice schematic for Example 3.7

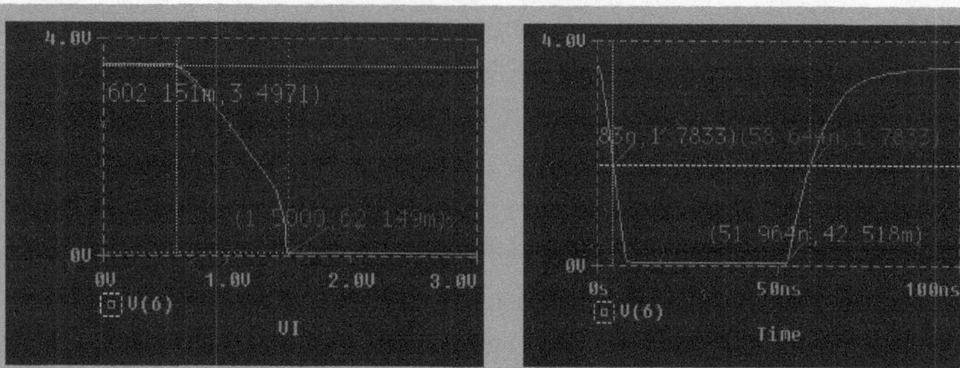

(a) VTC (b) Transient response

FIGURE 3.42 PSpice plots of TTL characteristics (using transistor circuit model) for Example 3.7

3.11.2 High-Speed TTL NAND Gates

Notice from Eq. (3.90) that the voltage drop across R_2 caused by the load current flow reduces the output high voltage v_O. The output voltage can be increased by replacing Q_4 in Fig. 3.38 by a Darlington pair, as shown in Fig. 3.43. Transistors Q_4 and Q_5 form the Darlington pair. The BEJ of Q_5 offers the voltage offset and serves the purpose of diode D_1 in Fig. 3.38. Resistor R_5 is included to aid the reverse-biased recovery current of Q_4. The operation and analysis of this gate are similar to those of the gate in Fig. 3.38.

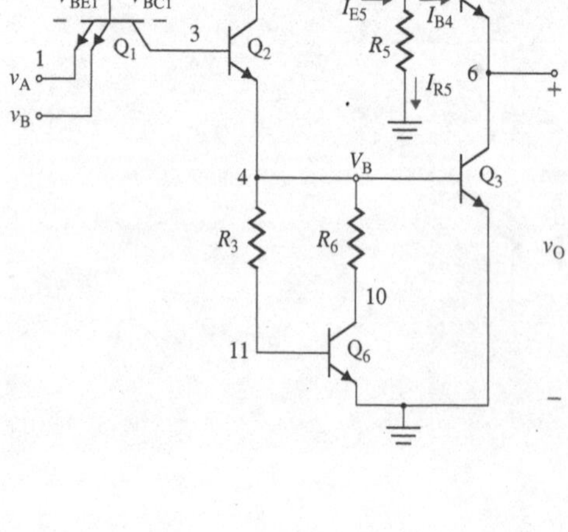

FIGURE 3.43 High-speed TTL NAND gate

The current through R_5 is

$$I_{R5} = \frac{V_{B4}}{R_5} = \frac{v_O + V_{BE4}}{R_5}$$

The emitter current of Q_5 is

$$I_{E5} = I_{B4} + I_{R5}$$

which gives its base current as

$$I_{B5} = \frac{I_{E5}}{1 + \beta_{F5}} = \frac{I_{B4} + I_{R5}}{1 + \beta_{F5}} = \frac{I_{E4}}{(1 + \beta_{F4})(1 + \beta_{F5})} + \frac{I_{R5}}{1 + \beta_{F5}}$$

Equation (3.90) can be modified to give the output voltage as

$$V_{CC} = R_2 I_{B5} + V_{BE4} + V_{BE5} + v_O$$

or $$v_O = V_{CC} - (R_2 I_{B5} + V_{BE4} + V_{BE5}) \tag{3.93}$$

Also, resistance R_B is replaced by an active base recovery circuit consisting of R_3, R_6, and Q_6. This recovery circuit reduces the amount of diverting current at the base of Q_3. As a result, the VTC becomes sharper, the noise margin is enhanced, and the delay time is reduced. The recovery circuit and its equivalent are shown in Fig. 3.44[(a) and (b)]. I_{BR} is the base recovery current of Q_3.

Assuming that Q_6 is close to saturation but not saturated, the collector current of Q_6 is given by

$$I_{C6} = \frac{I_{BR}}{1 + 1/\beta_{F6}}$$

which, near the edge of saturation, must be equal to

$$I_{C6} = \frac{V_{BE3} - V_{CE6(sat)}}{R_6}$$

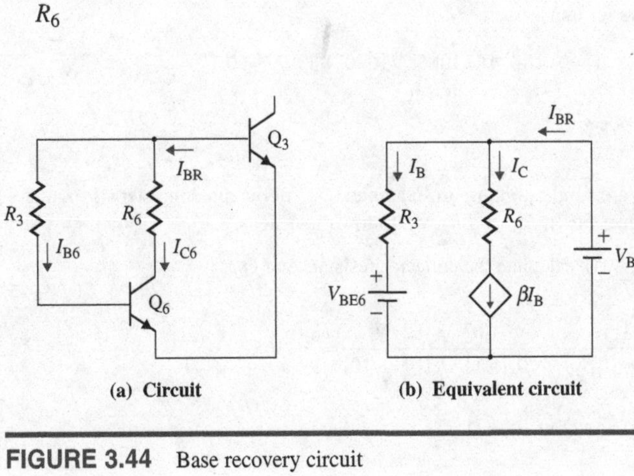

FIGURE 3.44 Base recovery circuit

Equating these two and assuming $\beta_F = \beta_{F(forced)}$, we can find the value of R_6 as

$$R_6 = \frac{V_{BE3} - V_{CE6(sat)}}{I_{BR}/(1 + 1/\beta_{F(forced)})} \tag{3.94}$$

Suppose that R_6 is chosen to yield the same value of I_{BR} as was obtained when a passive resistance R_B was used; that is, $I_{BR} = V_{BE3}/R_B$. Substituting into Eq. (3.94) yields

$$R_6 = \frac{R_B(V_{BE3} - V_{CE6(sat)})}{V_{BE3}/(1 + 1/\beta_{F(forced)})} \tag{3.95}$$

In summary, high-speed TTL gates have the following features:

1. Short propagation delay time
2. Low noise immunity
3. Sharp voltage transfer characteristic
4. High fan-out
5. Maximum output voltage limited to, typically, 3.6 V

EXAMPLE 3.8

Designing a TTL NAND gate

(a) Design the TTL NAND gate of the circuit in Fig. 3.43. It has two inputs and feeds four identical NAND gates. The output voltage at output high is $V_{OH} = 3.5$ V, and $I_{N1} = 72.5\ \mu$A. Assume $V_{CC} = 5$ V, $V_D = 0.7$ V, $V_{CE(sat)} = 0.2$ V, $V_{BE} = 0.7$ V, $\beta_{F(forced)} = 10$, $\beta_R = 0.1$, $I_{C3(sat)} = 1$ mA, $I_{C4(max)} = 1$ mA, and $I_{BR} = 750\ \mu$A. Capacitance of each load is, $C_i = 5$ pF.

(b) Use PSpice/SPICE to check your design by plotting the transfer characteristic. Determine NM_L and NM_H. Model parameters for transistors are

```
BF=10 BR=0.1 TF=0.1NS TR=10NS VJC=0.85 VAF=50
```

SOLUTION

(a) The design steps are similar to those in Example 3.7. After completing steps 1, 6, and 9 in Example 3.7 to determine the values of R_1, R_B, and R_4, we follow these steps.

Step 1. Using Eq. (3.94), calculate the collector resistance of Q_6:

$$R_6 = \frac{V_{BE3} - V_{CE6(sat)}}{I_{BR}/(1 + 1/\beta_F)} = \frac{(0.7 - 0.2)\text{ V}}{0.75\text{ mA}/(1 + 1/10)} = 733\ \Omega$$

Choose $R_3 = 2R_6 = 2 \times 733 = 1466\ \Omega$.

Step 2. Choose a value of I_{R5}: Let $I_{R5} = 1130\ \mu A$ (the same as the value for the TTL) ≈ 1.2 mA. For $v_O = V_{OH}$, find the value of R_5:

$$R_5 = \frac{V_{OH} + V_{BE4}}{I_{R5}} = \frac{(3.5 + 0.7)\text{ V}}{1.2\text{ mA}} = 3.5\text{ k}\Omega$$

Step 3. Calculate the base current of Q_4 at output high and the base current of Q_5:

$$I_{B4} = \frac{N_H I_{N1}}{1 + \beta_{F4}} = \frac{4 \times 72.5\ \mu\text{A}}{1 + 10} = 26.4\ \mu\text{A}$$

$$I_{B5} = \frac{I_{B4} + I_{R5}}{1 + \beta_{F5}} = \frac{26.4\ \mu\text{A} + 1.2\text{ mA}}{1 + 10} = 111.5\ \mu\text{A}$$

Step 4. Using Eq. (3.93) for $v_O = V_{OH}$, find the value of R_2 that meets the output requirement:

$$R_2 = \frac{V_{CC} - V_{BE4} - V_{BE5} - V_{OH}}{I_{B5}} = \frac{(5 - 0.7 - 0.7 - 3.5)\text{ V}}{111.5\ \mu\text{A}} = 897\ \Omega$$

Step 5. Using Eq. (3.89), calculate the value of R_1:

$$R_1 = \beta_R\left[\frac{V_{CC} - 2V_{BE(sat)} - V_{BC1}}{I_{N1}}\right] = 0.1\left[\frac{(5 - 2 \times 0.7 - 0.7)\text{ V}}{72.5\ \mu\text{A}}\right] = 4\text{ k}\Omega$$

Step 6. Using Eq. (3.86), find the value of I_{B1} at output low, and calculate the high-state I_{B2}:

$$I_{B1} = \frac{V_{CC} - V_{BE2(sat)} - V_{BE3(sat)} - V_{BC1}}{R_1} = \frac{(5 - 0.7 - 0.7 - 0.7)\text{ V}}{4\text{ k}\Omega} = 725\ \mu\text{A}$$

$$I_{B2} = -I_{C1} = (1 + \beta_R)I_{B1} = (1 + 0.1) \times 725\ \mu\text{A} = 798\ \mu\text{A}$$

Step 7. Calculate the high-state I_{C2}, I_{E2}, and I_{B3}. The high-state I_{C2} for low output is

$$I_{C2} = \frac{V_{CC} - V_{BE3(sat)} - V_{CE2(sat)}}{R_2} = \frac{(5 - 0.7 - 0.2)\text{ V}}{897} = 4.57\text{ mA}$$

In the high state,

$$I_{E2} = I_{B2} + I_{C2} = 798\ \mu\text{A} + 4.57\text{ mA} = 5.37\text{ mA}$$

For the base current,

$$I_{B3} = I_{E2} - I_R = 5.37\text{ mA} - 0.75\text{ mA} = 4.62\text{ mA}$$

Step 8. Calculate the maximum permissible collector low-state current to drive Q_3 in saturation:

$$I_{C3(max)} = \beta_{F(forced)} I_{B3} = 10 \times 4.62\text{ mA} = 46.2\text{ mA}$$

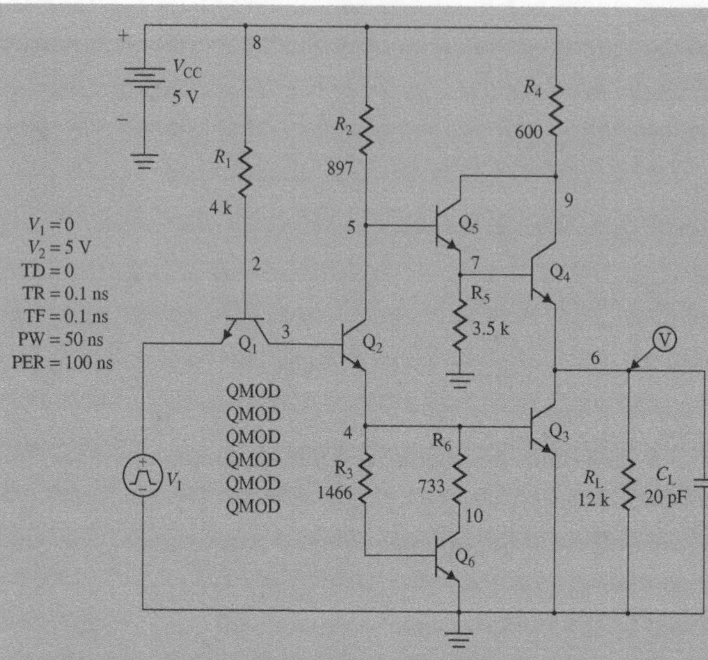

FIGURE 3.45 PSpice schematic for Example 3.8

which is higher than $I_{LO} = 410\ \mu A$, from step 4 of Example 3.7. Thus, the design should be satisfactory. (Otherwise, steps 1 through 8 should be repeated with a lower value of N, a higher value of $\beta_{F(sat)}$, or a lower value of I_{N1}.) The overdrive factor is

$$k_{ODF} = \frac{I_{C3(sat)}}{I_{LO}} = \frac{46.2\ mA}{410\ \mu A} = 113$$

which is high; that is, the fan-out can be much more than $N = 4$.

(b) To examine the propagation time, we will assume that a load resistance of

$$R_L = \frac{V_{OH}}{NI_{LO}} = \frac{3.5\ V}{4 \times 72.5\ \mu A} = 12\ k\Omega$$

and an equivalent capacitance of

$$C_L = 4 \times 5\ pF = 20\ pF$$

is connected to the output.

The PSpice schematic is shown in Fig. 3.45. The PSpice plots, which are shown in Fig. 3.46, give $V_{OH} = 3.3$ V (expected value is 3.5 V), $V_{IL} = 1.35$ V at $V_O = 3.3$ V, $V_{OL} = 62$ mV at $v_O = 3.3$ V, and $V_{IH} = 1.6$ V at $V_O = 62$ mV.

$$NM_L = V_{IL} - V_{OL} = (1.35 - 0.062)\ V \approx 1.29\ V$$

$$NM_H = V_{OH} - V_{IH} = 3.3 - 1.6 = 1.7V$$

$t_{pHL} = 1.93$ ns, $t_{pLH} = (57.57\ ns - 51.77\ ns) = 5.8$ ns, and $t_{pd} = (t_{pHL} + t_{pLH})/2 = 3.87$ ns.

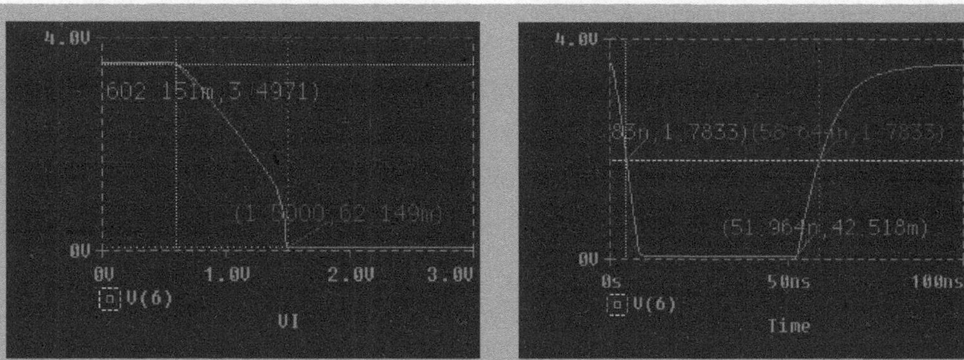

(a) VTC

(b) Transient response

FIGURE 3.46 PSpice plots for TTL characteristics (using transistor circuit model) for Example 3.8

3.11.3 Schottky TTL NAND Gates

In TTL gates, the transistors are driven into saturation. Since the delay time of a TTL gate is a strong function of the storage time of the saturated transistors, a nonsaturating logic gate has an advantage. A Schottky clamped transistor, which is prevented from being driven into saturation, can switch faster than a saturated transistor. Schottky clamped transistors are incorporated in many transistor logic gates.

A Schottky clamped transistor is basically a bipolar transistor with a built-in clamped Schottky diode, as shown in Fig. 3.47(a). Its symbol is shown in Fig. 3.47(b). The forward voltage drop of a Schottky diode is low, typically 0.3 V. When transistor Q_1 is in its active region of operation, the base–collector junction is reverse biased. The clamped diode is also reverse biased and does not affect the circuit operation. Q_1 behaves as a normal *npn* transistor. As transistor Q_1 goes into saturation, the base–collector junction becomes forward biased, and it is clamped to the 0.3-V Schottky diode voltage. The excess base current is shunted through the diode, and the transistor is prevented from going heavily into saturation.

Assuming that transistor Q_1 is clamped at the edge of saturation, $I_C = \beta I_B$. The diode current can be related to the input and base currents by

$$I_D = I_I - I_B = I_I - \frac{I_C}{\beta_F}$$

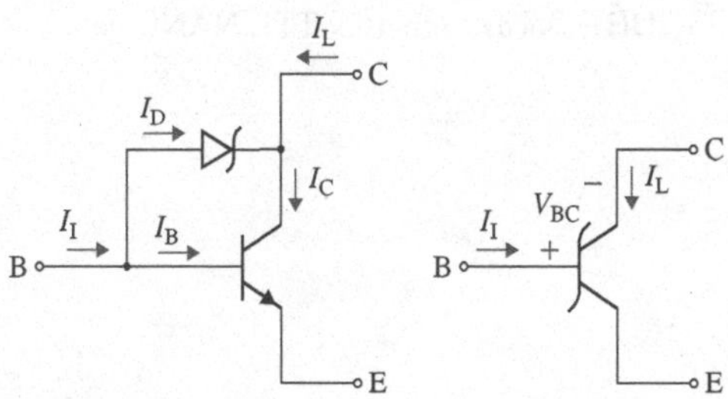

(a) Schottky clamped transistor

(b) Symbol

FIGURE 3.47 Schottky clamped transistor

Using KCL at the collector junction, we get

$$I_C = I_D + I_L = I_I - \frac{I_C}{\beta_F} + I_L$$

which yields

$$I_C = \frac{I_I + I_L}{1 + 1/\beta_F}$$

Thus, for an increased value of load current I_L, the value of I_C is increased and the value of diode current I_D is reduced. That is, the major part of the input current is diverted into the base of the transistor, keeping the transistor at the edge of saturation. For a small value of load current, the value of I_C becomes small, and a large part of the input current is shunted through the diode. The base and diode currents change with the load conditions, while the transistor remains at the edge of saturation. The Schottky barrier diode has no minority carrier charge storage, and the transistor is never fully saturated. Thus, the recovery is very quick.

In the Schottky TTL NAND gate shown in Fig. 3.48, all of the transistors except Q_4 are Schottky clamped transistors. This circuit is similar to the one in Fig. 3.43. The two Schottky diodes from the input terminals to the ground act as clamps, to suppress any ringing that might occur from voltage transients and clamp any negative undershoots at approximately −0.3 V.

The analysis of a Schottky TTL gate is similar to that of a standard TTL gate. When the output transistor Q_3 is on, $V_{BE} = 0.7$ V, the voltage drop across the Schottky diode is clamped to $V_{BC} = 0.3$ V, and

$$V_{CE} = V_{CB} + V_{BE} = -V_{BC} + V_{BE} = -0.3 + 0.7 = 0.4 \text{ V}$$

Thus, the output voltage of a Schottky gate in its output low state is slightly higher than the value of $V_{CE(sat)}$ for standard TTL gates. The output voltage in the output high state is essentially the same as that of the standard TTL gate.

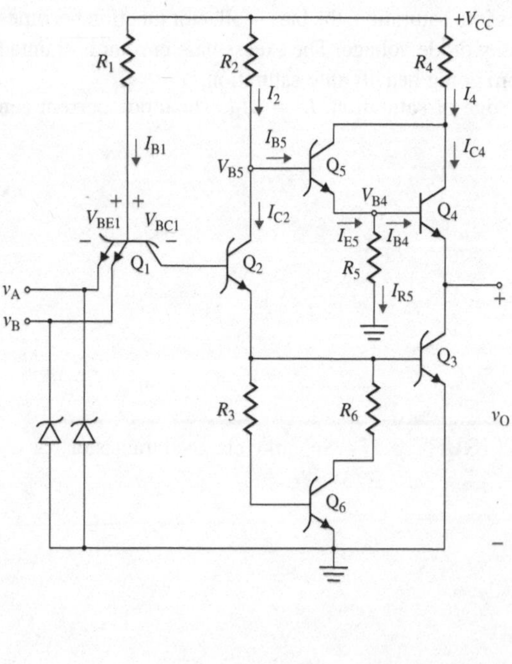

FIGURE 3.48 Schottky TTL NAND gate

In summary, Schottky TTL gates have the following features:

1. Minimal delay time. Because a Schottky clamped transistor operates at a low saturation, the delay time is approximately 2 ns to 5 ns, compared to 10 ns to 15 ns in standard and high-speed TTL gates; that is, the propagation delay time is reduced by a factor of 5 to 10.
2. Sharper voltage transfer characteristic than standard TTL gates
3. Low noise immunity
4. Slightly higher output voltage in the output low state than standard TTL gates, typically 0.4 V

The success of the Schottky gate led to the development of other Schottky gates: low-power Schottky (LS), advanced Schottky (AS), and advanced low-power Schottky (ALS) gates.

KEY POINT OF SECTION 3.11

- The TTL families have gone through many developmental stages, which have resulted in improved switching speed and lower power dissipation. Applications focus on the low-power Schottky (LS) and advanced low-power Schottky (ALS) families because they offer the highest speed.

3.12 Emitter-Coupled Logic OR/NOR Gates

Storage time is the dominant parameter affecting the delay time of transistors that are driven into saturation. Storage time can be minimized by preventing the transistors from operating at saturation. Schottky TTL gates minimize transistor saturation by clamping the base–collector junction at the edge of the saturation condition. This significantly reduces the storage time and delay time, but slight saturation of the Schottky transistors adds to the switching time. The transistors in ECL gates are never driven into saturation, and so the storage time is virtually zero. The first ECL nonsaturated logic gates were introduced in 1962 by Motorola under the family name MECL I. Since then, the MECL gates have progressed through several generations: MECL II, MECL III, MECL 10K, and MECL 10KH.

An ECL uses an *emitter-coupled pair* as a current switch circuit, as shown in Fig. 3.49(a). It consists of two identical transistors Q_1 and Q_2, two matched resistors $R_C = R_{C1} = R_{C2}$, and a current

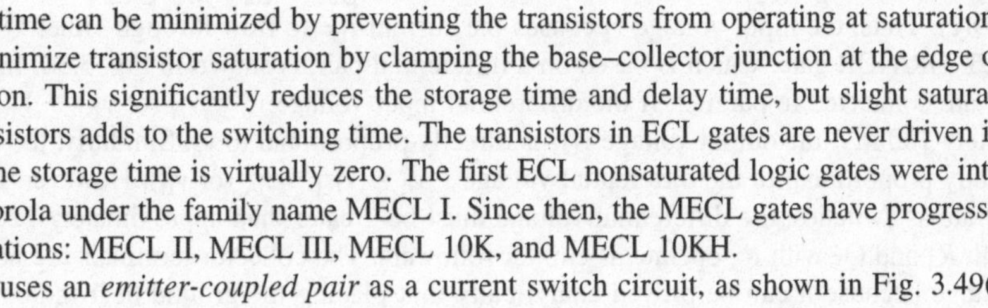

(a) Current switch circuit　**(b) Switch position for $v_I > V_{ref}$**　**(c) Switch position for $v_I < V_{ref}$**

FIGURE 3.49 Current switch circuit for ECL

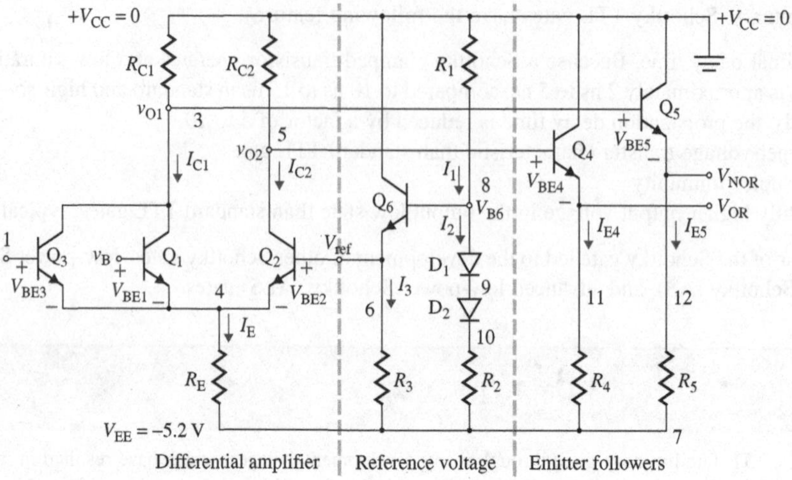

FIGURE 3.50 ECL OR/NOR gate

source I_{EE}. The input voltage v_I, which is applied to the base of Q_1, is compared to the reference voltage V_{ref}, which is applied to the base of Q_2. If v_I is greater than V_{ref} by a few hundred millivolts, the source current I_{EE} flows through Q_1. Thus, the output voltage becomes $v_{O1} = v_{C1} = -R_C i_{CI} = -R_C I_{EE}$, as shown in Fig. 3.49(b). On the other hand, if v_I is less than V_{ref} by a few hundred millivolts, the source current I_{EE} flows through Q_2. The output voltage becomes $v_{O2} = v_{C2} = -R_C i_{C2} = -R_C I_{EE}$, as shown in Fig. 3.49(c). Thus, the input voltage v_I causes the current I_{EE} to flow through either Q_1 or Q_2.

An ECL OR/NOR gate, which is based on a differential pair, is shown in Fig. 3.50. Input transistors Q_1 and Q_3 are connected in parallel. If the differential input voltage $v_d = v_A$ (or v_B) $- V_{ref}$ is more than approximately 100 mV, the output voltage v_{O1} is directly proportional to v_d. Similarly, the output voltage v_{O2} is directly proportional to the differential voltage $-v_d = V_{ref} - v_A$ (or v_B). However, for the transistors to operate as switches, the differential voltage must be greater than approximately 120 mV. Transistors Q_4 with R_4 and Q_5 with R_5 operate as emitter followers. The collector terminals are normally placed at zero voltage because it can be proved analytically that placing the ground near the collectors of the transistors results in less noise sensitivity. For this reason, the supply voltages are generally $V_{CC} = 0$ and $V_{EE} = -5.2$ V.

The reference circuit consists of resistors R_1, R_2, and R_3, diodes D_1 and D_2, and transistor Q_6. The diodes D_1 and D_2 provide temperature compensation for the BEJ of Q_6. Neglecting the base current of Q_6, we get

$$I_1 = I_2 = \frac{V_{CC} - 2V_D - V_{EE}}{R_1 + R_2}$$

The voltage at the base of transistor Q_6 is

$$V_{B6} = V_{CC} - I_1 R_1$$

which gives the reference voltage as

$$V_{ref} = V_{B6} - V_{BE6}$$

The emitter current of Q_6 is given by

$$I_{E6} \approx \frac{V_{ref} - V_{EE}}{R_3}$$

The circuit operation can be divided into two modes.

During mode 1, either input v_A or input v_B is at logic high (V_{OH}); that is, v_A (or v_B) = $V_{CC} - 0.7$ V. The OR logic is at the v_{O2} output, and the NOR logic is at the v_{O1} output. Transistor Q_2 is off. That is, $i_{C2} = 0$, and $v_{O2} = V_{CC}$. The output voltage V_{OR} is given by

$$V_{OR} = V_{O2} - 0.7 \text{ V} = V_{CC} - 0.7 \text{ V}$$

If $v_A = V_{OH} = V_{CC} - 0.7$ V, then Q_1 is on, and we get

$$V_E = v_A - V_{BE1}$$

which gives the emitter current I_E as

$$I_E = \frac{V_E - V_{EE}}{R_E}$$

Assuming that $i_{C1} \approx I_E$, the voltage v_{O1} can be found from

$$v_{O1} = V_{CC} - i_{C1}R_{C1}$$

and the output V_{NOR} becomes

$$V_{NOR} = v_{O1} - V_{BE5}$$

The current I_{E4}, which is the emitter current of Q_4, is

$$I_{E4} = \frac{V_{E4} - V_{EE}}{R_4} = \frac{V_{OR} - V_{EE}}{R_4}$$

and the emitter current of Q_5 is

$$I_{E5} = \frac{V_{E5} - V_{EE}}{R_5} = \frac{V_{NOR} - V_{EE}}{R_5}$$

During mode 2, inputs v_A and v_B are at logic low (V_{OL}). Transistors Q_1 and Q_3 are off; that is, $i_{C1} = 0$, and $v_{O1} = V_{CC}$. The output voltage V_{NOR} is given by

$$V_{NOR} = V_{OH} = v_{O1} - 0.7 \text{ V} = V_{CC} - 0.7 \text{ V}$$

Transistor Q_2 is on, and we get

$$V_E = V_{ref} - V_{BE2}$$

which gives the emitter current I_E as

$$I_E = \frac{V_E - V_{EE}}{R_E}$$

Assuming that $i_{C2} \approx I_E$, the voltage v_{O2} becomes

$$v_{O2} = V_{B4} = V_{CC} - i_{C2}R_{C2}$$

and the output V_{OR} is

$$V_{OR} = v_{O2} - V_{BE4}$$

The current I_{E4}, which is the emitter current of Q_4, is

$$I_{E4} = \frac{V_{E4} - V_{EE}}{R_4} = \frac{V_{OR} - V_{EE}}{R_4}$$

and the emitter current of Q_5 is

$$I_{E5} = \frac{V_{E5} - V_{EE}}{R_5} = \frac{V_{NOR} - V_{EE}}{R_5}$$

Defining logic high as

$$V_{OH} = V_{CC} - 0.7 = -0.7 \text{ V}$$

and substituting the numerical values, we get the logic low as

$$V_{OL} = v_{O1} - V_{BE5} = V_{CC} - i_{C1}R_{C1} - V_{BE5} = V_{NOR} = -1.63 \text{ V}$$

gives the input and output voltages shown in Table 3.7.

In summary, ECL OR/NOR gates have the following features:

1. No transistor saturation and negligible delay time
2. High power dissipation, typically 50 mW to 70 mW (compared to 2 mW to 10 mW for Schottky TTL circuits)
3. Availability of complementary outputs, which eliminates the need to include separate inverters to provide these complementary outputs
4. High fan-out, typically in the range of 50 to 100
5. Low noise margin, since the logic high and logic low output voltages are only approximately −0.7 V and −1.63 V, respectively
6. Sharp voltage transfer characteristic

TABLE 3.7 ECL/OR gate logic function

v_A	v_B	V_{OR}	V_{NOR}
V_{OL} (for 0)	V_{OL} (for 0)	V_{OL} (for 0)	V_{OH} (for 1)
V_{OH} (for 1)	V_{OL} (for 0)	V_{OH} (for 1)	V_{OL} (for 0)
V_{OL} (for 0)	V_{OH} (for 1)	V_{OH} (for 1)	V_{OL} (for 0)
V_{OH} (for 1)	V_{OH} (for 1)	V_{OH} (for 1)	V_{OL} (for 0)

EXAMPLE 3.9

D

Designing an ECL OR/NOR gate

(a) Design the ECL OR/NOR gate of the circuit in Fig. 3.50. It has two inputs. The desired collector currents are $I_{C1} = 3$ mA and $I_{C4} = I_{C5} = 3$ mA. At logic high, $V_{OR} = -0.7$ V, $V_{NOR} = -1.63$ V, and v_A (or v_B) $= V_{OR} = -0.7$ V. Assume $V_{BE} = 0.7$ V, $V_{EE} = -5.2$ V, $V_{BE} = 0.7$ V, and $\beta_F = 100$.

(b) Calculate the maximum fan-out with similar ECL gates if V_{OR} is allowed to fall from $V_{OR(max)} = -0.7$ V to $V_{OR(min)} = -0.75$ V.

(c) Use PSpice/SPICE to check your design by plotting the transfer function. Calculate NM_L and NM_H. Use PSpice/SPICE to find P_{static} if $v_A = v_B = -0.7$ V.

SOLUTION

(a) Assuming v_A (or v_B) $= V_{OR} = -0.7$ V, the steps to complete the design are as follows:

Step 1. Calculate V_{CC}, v_A, and V_E. The value of V_{CC} is

$$V_{CC} = V_{OR} + 0.7 = -0.7 + 0.7 = 0$$

The input voltage for logic high is

$$v_A = V_{OH} = V_{CC} - 0.7 = 0 - 0.7 = -0.7 \text{ V}$$

The emitter voltage is

$$V_E = v_A - V_{BE1} = -0.7 - 0.7 = -1.4 \text{ V}$$

Step 2. Calculate the emitter resistance R_E:

$$R_E = \frac{V_E - V_{EE}}{I_E \simeq I_C} = \frac{-1.4 \text{ V} - (-5.2 \text{ V})}{3 \text{ mA}} = \frac{(-1.4 + 5.2) \text{ V}}{3 \text{ mA}} = 1.27 \text{ k}\Omega$$

Step 3. Calculate the voltage v_{O1}, the collector resistance R_{C1}, R_4, and R_5:

$$v_{O1} = V_{NOR} + V_{BE5} = -1.63 + 0.7 = -0.93 \text{ V}$$

$$R_{C1} = \frac{V_{CC} - v_{O1}}{I_{C1}} = \frac{V_{CC} - v_{O1}}{I_E} = \frac{0 - (-0.93 \text{ V})}{3 \text{ mA}} = \frac{0.93 \text{ V}}{3 \text{ mA}} = 310 \ \Omega$$

$$R_4 = \frac{V_{OR} - V_{EE}}{I_{E4} \simeq I_{C3}} = \frac{-0.7 \text{ V} - (-5.2 \text{ V})}{3 \text{ mA}} = 1.5 \text{ k}\Omega$$

$$R_5 = \frac{V_{NOR} - V_{EE}}{I_{E5} \simeq I_{C3}} = \frac{-1.63 \text{ V} - (-5.2 \text{ V})}{3 \text{ mA}} = 1.19 \text{ k}\Omega$$

Step 4. Since the input voltages v_A and v_B are greater than V_{ref} when the circuit is in the logic high state and less than V_{ref} when it is in the logic low state, set V_{ref} at the midpoint between the logic low and logic high levels and calculate the voltage at the base of Q_6:

$$V_{ref} = \frac{V_{NOR} + V_{OR}}{2} = \frac{-1.63 - 0.7}{2} = -1.165 \text{ V}$$

$$V_{B6} = V_{ref} + V_{BE6} = -1.165 + 0.7 = -0.465 \text{ V}$$

Step 5. Calculate the values of R_1, R_2, and I_1. Using KVL, we can write

$$I_1(R_1 + R_2) = -V_{EE} - 2V_D$$

$$I_1R_1 = -V_{B6}$$

which gives

$$\frac{I_1(R_1 + R_2)}{I_1R_1} = \frac{-V_{EE} - 2V_D}{-V_{B6}}$$

or $$1 + \frac{R_2}{R_1} = \frac{-V_{EE} - 2V_D}{-V_{B6}} = \frac{-(-5.2) - 2 \times 0.7}{-(-0.465)} = 8.172$$

Choose $R_1 = 300\ \Omega$. Then

$$R_2 = (8.172 - 1)R_1 = (8.172 - 1) \times 300 = 2.15 \text{ k}\Omega$$

Calculate the value of I_1 as

$$I_1 = I_2 = -\frac{V_{B6}}{R_1} = -\frac{-0.465}{300} = 1.55 \text{ mA}$$

Step 6. Calculate the value of R_3. For good temperature compensation, the current through the emitter of Q_6 should be the same as the current through diodes D_1 and D_2; that is,

$$I_3 = I_2 = \frac{V_{ref} - V_{BE}}{R_3} = \frac{-1.165 - (-5.2)}{R_3} = 1.55 \text{ mA}$$

or $R_3 = 2.6\ \text{k}\Omega$.

Step 7. Choose R_{C2} to be slightly more than R_{C1} (approximately 3% more for resistors of 1% tolerance)—that is, 2% more than the tolerance value. Thus,

$$R_{C2} = 1.03 \times R_{C1} = 1.03 \times 310 = 319.3\ \Omega$$

Let $R_{C2} = 320\ \Omega$.

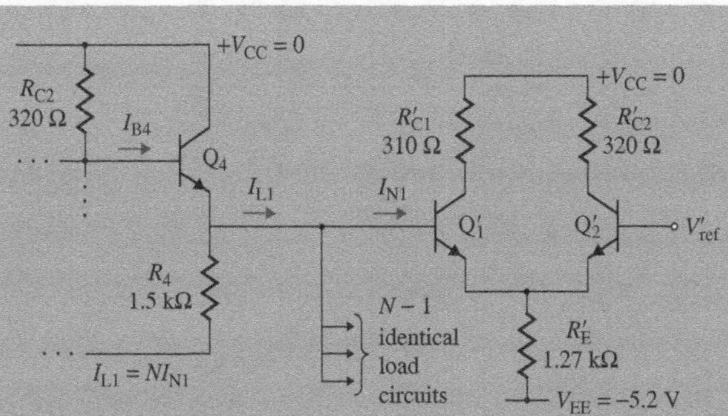

FIGURE 3.51 ECL gate driver and ECL load gate circuits

(b) The fan-out can be determined from Fig. 3.51, which shows the emitter-follower output stage of an ECL circuit driving a difference amplifier input stage of an ECL load. This circuit is shown for V_{OR} at the logic high level. The load transistor Q_1 is on, and the load emitter current is given by

$$I_E = \frac{V_{OR(max)} - V_{BE} - V_{EE}}{R_E}$$

$$= \frac{-0.7 \text{ V} - 0.7 \text{ V} - (-5.2 \text{ V})}{1.27 \text{ k}\Omega} = 2.99 \text{ mA}$$

The input base current for each fan-out is given by

$$I_{N1} = \frac{I_E}{1 + \beta_F} = \frac{2.99 \text{ mA}}{1 + 100} = 29.6 \ \mu\text{A}$$

Thus, the total load current is

$$I_{L1} = NI_{N1} = N \times 29.6 \ \mu\text{A}$$

The emitter current of Q_4 is

$$I_{E4} = \frac{V_{OR(min)} - V_{EE}}{R_4} = \frac{-0.75 \text{ V} - (-5.2 \text{ V})}{1.5 \text{ k}\Omega} = 2.97 \text{ mA}$$

The base current I_{B4} required to supply the load current I_L and the current I_{E4} is

$$I_{B4} = \frac{I_{E4} + I_L}{1 + \beta_F} = \frac{V_{CC} - V_{BE4} - V_{OR(min)}}{R_{C2}}$$

which gives

$$\frac{2.97 \text{ mA} + (N \times 29.6 \ \mu\text{A})}{1 + 100} = \frac{0 - 0.7 - (-0.75)}{320}$$

or $N = 432$.

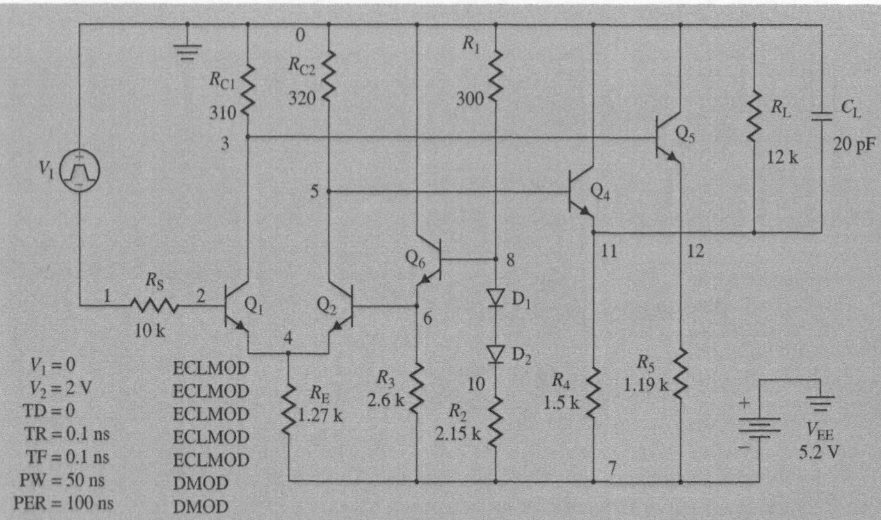

FIGURE 3.52 PSpice schematic for Example 3.9

(c) The PSpice schematic is shown in Fig. 3.52. The PSpice plots shown in Fig. 3.53 give $V_{OH} = -0.81$ V (expected value is -0.7 V), $V_{IL} = 0.95$ V at $v_O = -0.855$ V, $V_{OL} = -1.59$ V at $V_I = 2$ V, and $V_{IH} = 1.33$ V at $V_O = -1.55$ V. Note that V(11) and V(12) are the voltages at the OR and NOR terminals, respectively.

$$NM_L = V_{IL} - |V_{OL}| = 0.95 - 1.59 = -0.64 \text{ V}$$

$$NM_H = |V_{OH}| - V_{IH} = 0.81 - 1.33 = -0.52 \text{ V}$$

$t_{pHL} = 1.05$ ns, $t_{pLH} = 1.22$ ns, and $t_{pd} = 1.13$ ns. The PSpice output file gives $P_{static} = 62$ mW.

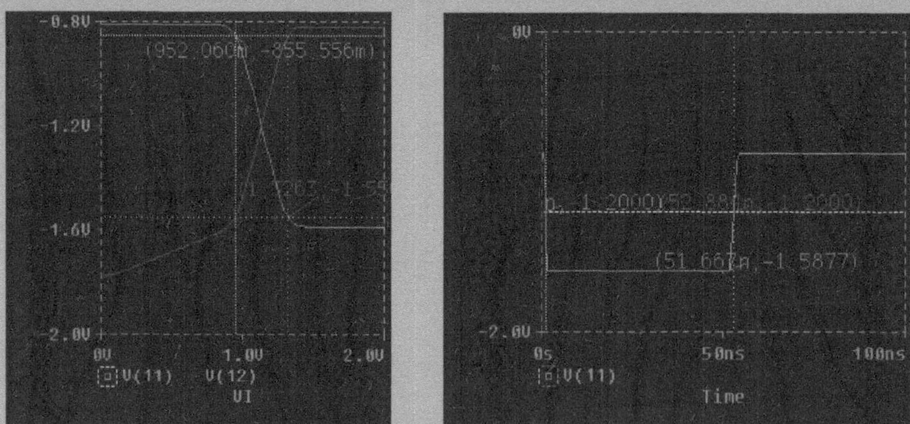

(a) VTC

(b) Transient response

FIGURE 3.53 PSpice plots for ECL characteristics (using transistor circuit model) for Example 3.9

KEY POINT OF SECTION 3.12

- The transistors in ECL gates are never driven into saturation, and thus the storage time is virtually zero. Since their introduction in 1962 by Motorola, the MECL gates have progressed through several generations: MECL I, MECL II, MECL III, MECL 10K, and MECL 10KH.

3.13 BiCMOS Inverters

The CMOS inverter is a low-power, compact inverter that exhibits a high input resistance. Its bipolar circuits, however, have low propagation delays. A BiCMOS combines a CMOS with a BJT output buffer and thus incorporates the best features of both technologies. A BiCMOS inverter is shown in Fig. 3.54. The BJT totem-pole output stage provides the high current capability to charge the load capacitance C_L rapidly while maintaining the low-power advantage of the CMOS. In practice, R_1 and R_2 are polysilicon resistors or MOS resistors. To examine the charging and discharging of C_L, we will divide the operation into two modes.

During mode 1, the input voltage v_I is low (at V_{OL}). The NMOS M_N is off. Q_1 is also off because its base is effectively grounded through R_1. The PMOS M_P is on, with $v_{GSP} = -V_{DD}$, and it provides the base current to Q_2. Thus, the load capacitance C_L is charged up through Q_2 (in the active region) to approximately $V_{DD} - V_{BE2}$. At $v_O = V_{DD} - V_{BE2}$, Q_2 is cut off, and C_L then continues to charge toward V_{DD} through M_P and R_2. Thus, V_{BE2} decreases to approximately 0, and $v_O \approx V_{OH} = V_{DD}$. The base of Q_2 also discharges through R_2.

During mode 2, the input voltage v_I goes high (to V_{OH}). Both M_P and Q_2 are off. M_N is turned on. Capacitor C_L discharges through R_2 and M_N. As a result of the current flow through R_1, Q_1 is turned on, and the collector current of Q_1 causes C_L to discharge rapidly until V_{BE1} drops below cut-in. After that, Q_1 turns off, and C_L continues to discharge through R_2, M_N, and R_1 to approximately zero. That is, $V_{OL} \approx 0$. The base of Q_1 also discharges through R_1.

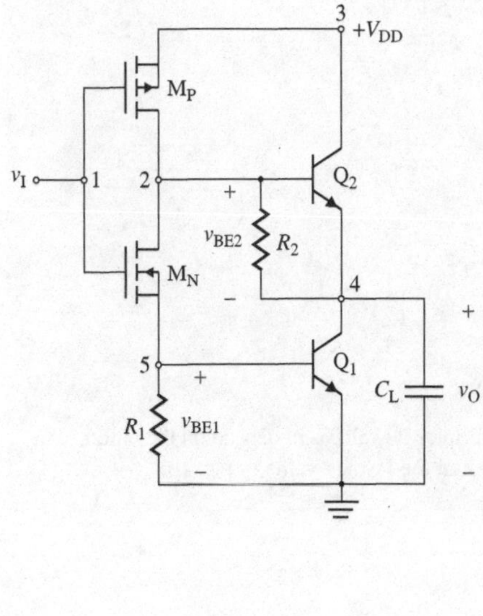

FIGURE 3.54 BiCMOS inverter

3.13.1 Propagation Delay

The propagation delay time of a BiCMOS is reduced because of the increased charging and discharging currents available from the BJTs. When the input switches from logic high to logic low, M_P supplies the charging current I_B for Q_2. Because of Q_1, the minimum value of v_I is V_{BE}. Thus, $V_{SGP} = V_{DD} - V_{BE}$, and I_B is given by

$$I_B = K_p(V_{DD} - V_{BE} - |V_{tP}|)^2$$

The charging current supplied by Q_2 in the active mode is

$$I_E = \beta_F I_B$$

Thus, the time t_{pLH} required to charge C_L from V_{OL} to $(V_{OL} + V_{OH})/2 \approx V_{DD}/2$ (neglecting the junction capacitances of the transistors) is given by

$$t_{pLH} \approx \frac{(V_{OL} + V_{OH})C_L}{2I_E} = \frac{V_{DD}C_L}{2\beta_F K_p(V_{DD} - V_{BE} - |V_{tP}|)^2} \tag{3.96}$$

If the input switches from logic low to logic high, M_N will provide the discharging path of C_L. Thus, the time t_{pHL} to discharge C_L from $V_{OH} = (V_{OL} + V_{OH})/2 \approx V_{DD}/2$ to V_{OL} is given by

$$t_{pHL} = \frac{(V_{OL} + V_{OH})C_L}{2I_E} = \frac{V_{DD}C_L}{2\beta_F K_n(V_{DD} - V_{BE} - V_{tN})^2} \tag{3.97}$$

When the BJTs are identical and the threshold voltages of the MOSFETs have the same magnitude, $t_{pLH} = t_{pHL}$ and $t_{pd} = t_{pLH}$.

EXAMPLE 3.10

Finding the propagation delays of a BiCMOS inverter The BiCMOS inverter shown in Fig. 3.54 has $V_{DD} = 5$ V, $V_{BE} = 0.7$ V, $R_1 = 1$ kΩ, $R_2 = 4$ kΩ, $C_L = 0.5$ pF, and $\beta_F = 100$. The threshold voltages are $V_{tP} = -1$ V, $V_{tN} = 1$ V, and $K_p = 20$ μA/V^2. Assume $V_{OL} = 0$ and $V_{OH} = 5$ V.

(a) Calculate t_{pLH}, t_{pHL}, and t_{pd}.

(b) Use PSpice/SPICE to plot the VTC and transient response.

SOLUTION

(a) Using Eq. (3.96), we have

$$t_{pLH} = \frac{5\text{ V} \times 0.5\text{ pF}}{2 \times 100 \times 20\ \mu\text{A/V}^2 \times (5\text{ V} - 0.7\text{ V} - 1\text{ V})^2} = 0.06\text{ ns}$$

$$t_{pd} = t_{pHL} = t_{pLH} = 0.06\text{ ns}$$

Although t_{pLH} is independent of the L/W ratio of the MOSFETs, in PSpice its value will depend on the ratio L/W. This causes the difference between the calculated value in part (a) and the PSpice value in part (b).

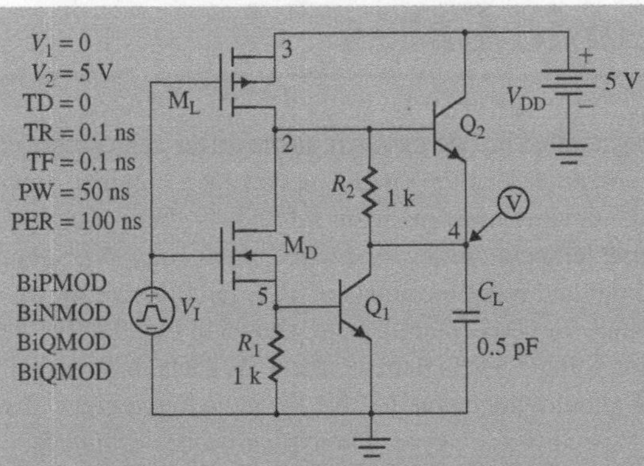

FIGURE 3.55 PSpice schematic for Example 3.10

(b) The PSpice schematic is shown in Fig. 3.55. The PSpice plot of the VTC, shown in Fig. 3.56(a), gives V_{IL} = 2.125 V, V_{IH} = 2.875 V, V_{OL} = 0 at V_{OH} = 5 V, and $V_{I(tran1)}$ = 2.5 V (expected value is 2.5 V). The transient response is shown in Fig. 3.56(b), which gives t_{pHL} = 2.93 ns, t_{pLH} = 0.65 ns, and t_{pd} = 1.79 ns, which is lower than t_{pd} = 3.91 ns for the CMOS in Example 3.5.

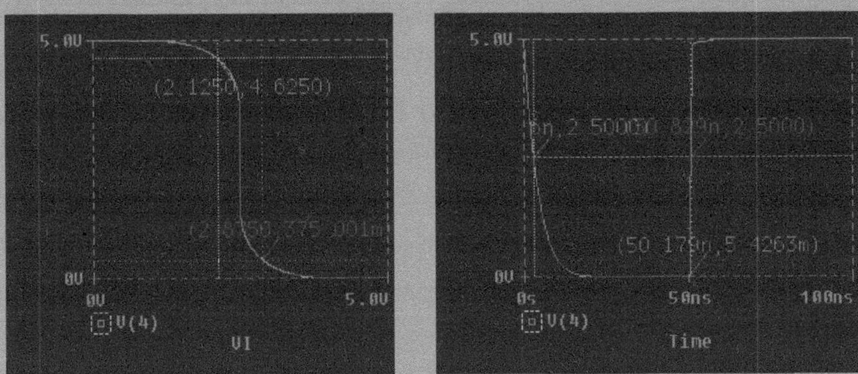

(a) VTC (b) Transient response

FIGURE 3.56 PSpice plots for BiCMOS inverter (using transistor circuit model) for Example 3.10

NOTE: Figure 3.56 illustrates the very sharp rise and fall times that occur while the BJTs are conducting. The output signal is nearly V_{DD}, and the noise margins for this circuit are high, both being more than 2 V.

KEY POINT OF SECTION 3.13

- A BiCMOS combines a CMOS with a BJT output buffer and thus incorporates the best features of both technologies.

3.14 Interfacing of Logic Gates

In practice, the load of a logic gate is another logic gate; that is, one gate acts as the driver and another one as the load. The interfacing of different logic gates requires that the circuits operate at a common supply voltage and have logic-level compatibility. Also, the devices must maintain safe power dissipation levels and good noise immunity over the required operating temperatures.

The voltage characteristics required at the output and input terminals of TTL families are shown in Fig. 3.57(a) for $V_{CC} = 5$ V; the voltage characteristics for CMOS families are shown in Fig. 3.57(b). The CMOS gates are designed to switch states at higher voltage levels than are the TTL gates. In interfacing one type of logic gate with another, attention should be given to the logic swing, output drive capability, DC input current, noise immunity, and speed of each type. The typical values of interfacing parameters such as V_{OL}, V_{OH}, I_{OL}, I_{OH}, V_{IL}, V_{IH}, I_{IL}, and I_{IH} are shown in Table 3.3 for CMOS and TTL devices.

3.14.1 TTL Driving CMOS

When a TTL device is used to drive a CMOS, the output drive capability of the driving device and the switching levels and input currents of the driven devices are important considerations. The input currents for a CMOS are very small in both the 1 and the 0 state, typically $I_{IL} = I_{IH} = 10$ pA, and the thresholds for a CMOS are typically $V_{IL(max)} = 1.5$ V and $V_{IH(min)} = 3.5$ V. Thus, to obtain some noise immunity, the output of the TTL driver must be no more than $V_{OL} = 1.5$ V at logic 0 and no less than $V_{OH} = 3.5$ V at logic 1. Depending on whether the device is in the 0 or the 1 state, the driver will be sinking or sourcing current.

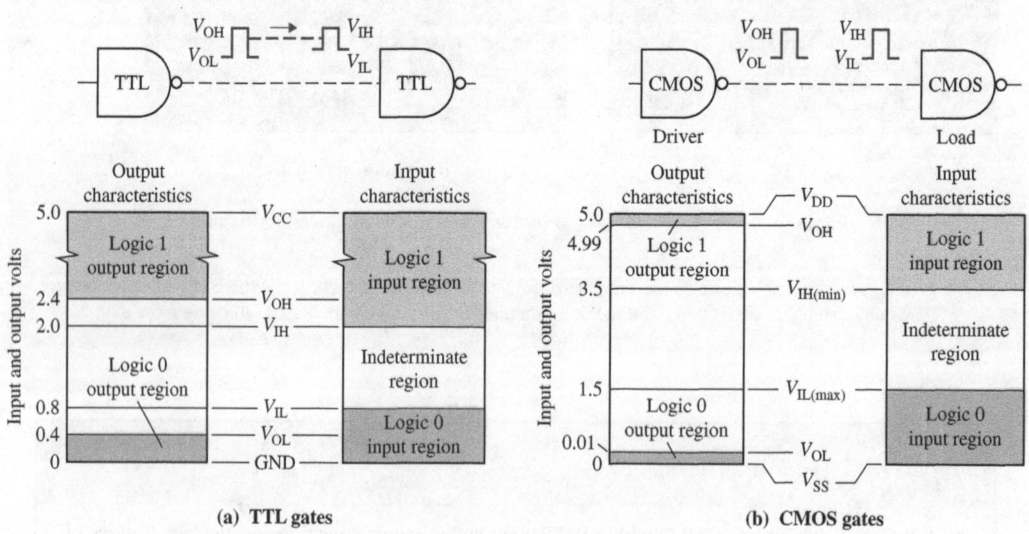

FIGURE 3.57 Input and output characteristics of logic gates

Current Sinking

When the output of the TTL device is in the low state (at V_{OL}), the collector of Q_1, shown in Fig. 3.58(a), is essentially at ground potential. Transistor Q_1 must go into saturation to ensure a stable 0 level, typically 0.3 V. To attain this voltage level, there should be a high impedance path from the output to the power supply V_{CC}. Since the CMOS has extremely high input impedances, typically 10^{11} Ω, it will not contribute any significant current through Q_1. Thus, the current-sinking capability is not a problem. The voltage level is not a problem either, as CMOS devices have high noise immunity, usually greater than 1 V.

Current Sourcing

When Q_1, shown in Fig. 3.58(b), is off (at logic 1 output V_{OH}), there is a current flow from the V_{CC} terminal of the driver, through a pull-up resistor R_P, into the input stages of the load; that is, the driving gate acts as the current source for the load. The total load current (NI_{IH}) should not reduce the output level below V_{IH} required by the CMOS load. A driver with a built-in pull-up resistor R_P, as shown in Fig. 3.58(b), presents no problem in the interface with the CMOS. However, a driver with an open collector requires an external pull-up resistor R_P, as shown in Fig. 3.58(c).

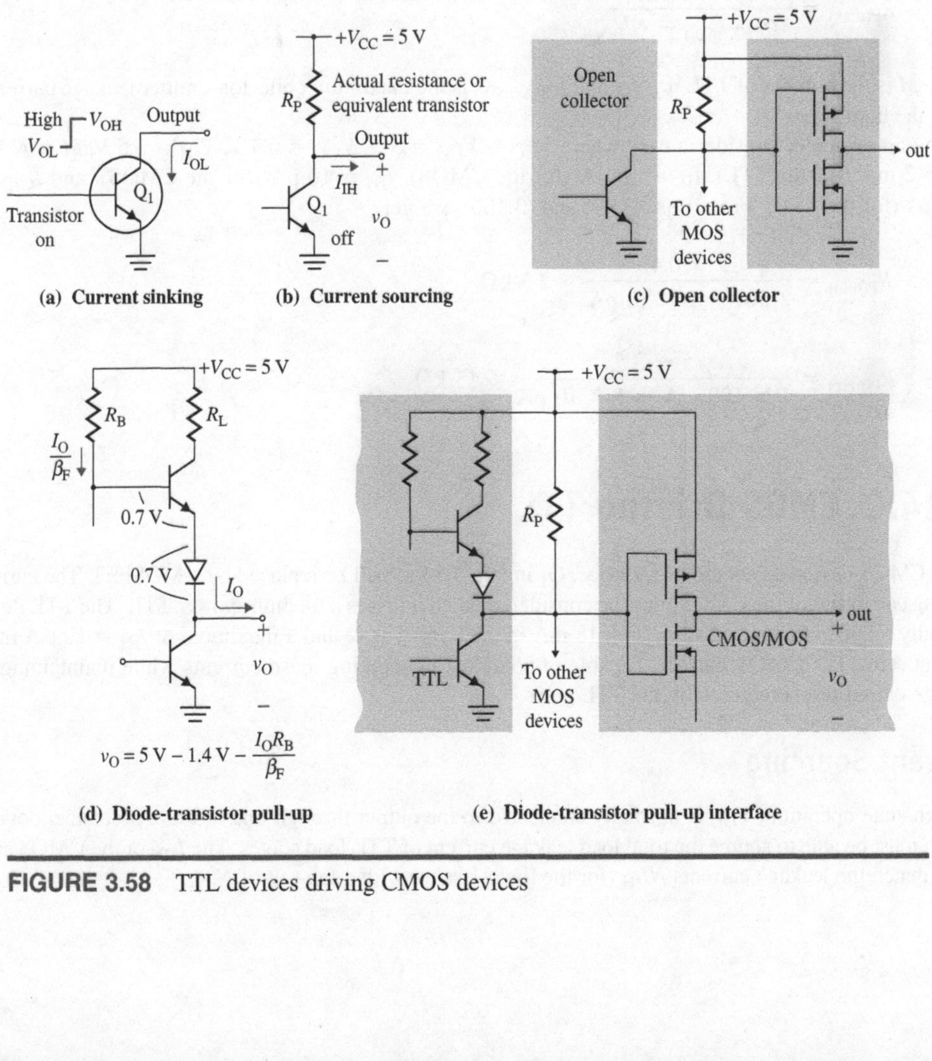

FIGURE 3.58 TTL devices driving CMOS devices

Consider the diode-transistor arrangement shown in Fig. 3.58(d), which will reduce the output voltage v_O given by

$$v_O = V_{CC} - V_{BE} - V_D - \frac{i_O R_B}{\beta_F} \tag{3.98}$$

Depending on the load current i_O, the minimum output level of $V_{OH} = 2.4$ V may not ensure an acceptable state of $V_{IH(min)} = 3.5$ V for the CMOS device, and this could cause a problem in the logic 1 state. However, the minimum level of $V_{OH} = 2.4$ V for a TTL gate is often specified at a given load current ($I_{OH} = 200\ \mu$A). Since the CMOS draws very little current, an output level of $V_{OH} = 3.5$ V will be available, but there is no noise immunity. Therefore, a pull-up resistor R_P should be added from the output terminal of the driver, as shown in Fig. 3.58(e). The minimum value of R_P can be found from

$$R_{P(min)} = \frac{V_{DD} - V_{OL(max)}}{I_{OL} - NI_{IL}} \tag{3.99}$$

where N is the number of CMOS load gates. The maximum value of the external pull-up resistor can be found from

$$R_{P(max)} = \frac{V_{CC} - V_{IH(min)}}{MI_{CEX(max)} - NI_{IH}} \tag{3.100}$$

where M is the number of TTL drivers and $I_{CEX(max)}$ is the maximum collector–emitter leakage current of Q_1 in the high state.

As an example, consider a case where $V_{CC} = V_{DD} = 5$ V, $V_{OL} = 0.4$ V, $V_{IH} = 3.5$ V, $M = N = 1$, $I_{OL} = 2$ mA (for the TTL), $I_{IL} = 10$ pA (for the CMOS), $I_{IH} = 10$ pA (for the CMOS), and $I_{CEXI} =$ 100 μA (for the TTL). From Eqs. (3.99) and (3.100), we get

$$R_{P(min)} = \frac{5 - 0.4}{2\text{ mA} - 1 \times 10\text{ pA}} \approx 2.3\text{ k}\Omega$$

$$R_{P(max)} = \frac{5 - 3.5}{1 \times 100\ \mu\text{A} - 1 \times 10\text{ pA}} \approx 15\text{ k}\Omega$$

3.14.2 CMOS Driving TTL

If the CMOS device drives the TTL device, Q_1 in Fig. 3.58(a) will be replaced by a MOSFET. The current-sinking capability of the CMOS must be considered when it drives a medium-power TTL. The TTL device typically requires no more than $I_{IL} = 0.18$ mA in the 0 input state and a maximum of $I_{IH} = 10\ \mu$A in the 1 input state. The CMOS must be capable of sourcing and sinking these currents while maintaining the voltage output levels required by the TTL gates.

Current Sourcing

In high-state operation, V_{DD} is normally connected to the output through one or more p-channel devices, which must be able to source the total load leakage current of TTL load stages. The I_{OH} of the CMOS stage must match the leakage currents NI_{IH} (for the logic 1 state) with a fan-out of N.

Current Sinking

When the output of the CMOS is in the low state (at V_{OL}), an *n*-channel device is on and the output is approximately at ground potential. The CMOS device sinks the current flowing from the TTL input load stage. The I_{OL} of the CMOS stage must match the input currents NI_{IL} (for the logic 0 state) with a fan-out of *N*.

KEY POINT OF SECTION 3.14

- The load of a logic gate is often another logic gate. It is essential that the logic gates be compatible in terms of supply voltage, logic levels, and noise immunity.

3.15 Comparison of Logic Gates

Broadly speaking there are three logic families: TTL, ECL, and CMOS. Each family has its advantages and disadvantages. The delay-power product (DP), which is the product of power dissipation P_D and propagation delay t_{pd}, is the key parameter for high-speed switching. Figure 3.59 compares delay time with power dissipation for various logic families. Note that the first generation of the CMOS family is labeled

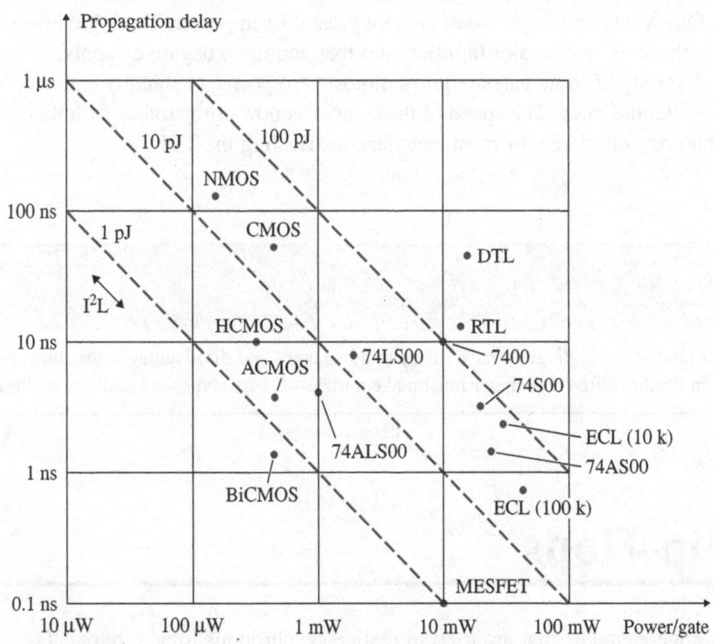

FIGURE 3.59 Comparison of delay time and power dissipation for logic families

TABLE 3.8 Comparison of logic families

Parameter	TTL 74LSxx	ECL 10K	CMOS 400B
V_{OH}, in V	3.4	−0.9	5
V_{OL}, in V	0.25	−1.7	0
NM_H, in V	1.4	0.36	2.25
NM_L, in V	0.6	0.33	2.25
P_D/gate	2 mW	24 mW	1.5 μW × f_{clk}
t_{pd}, in ns	9.5 at 15 pF	2	30 + 1.7/pF

CMOS; HCMOS refers to high-speed CMOS and ACMOS to advanced-type CMOS. The diode-transistor logic (DTL) and resistor-transistor logic (RTL) gates were developed earlier and have the highest delay-power product. Integrated-injection logic (I^2L) gates have low power but are relatively slower. The MESGET (gallium arsenide) gates are extremely fast because of the high mobility of electrons in GaAs—typically five to six times higher than that of electrons in *n*-type silicon. Gate delays as low as 20 ps to 100 ps have been reported, but this speed must be weighed against the power dissipation, which is typically 10 mW.

The typical values of various parameters for the TTL, ECL, and CMOS families are shown in Table 3.8. We can draw the following conclusions from these data:

- The TTL (LS) family of logic gates has high speed and low power dissipation and is compatible with many applications. In addition, the technology is mature and hence low in cost.
- The ECL family has very high speed but at a great cost in power. The logic levels of these gates differ from those of other major families, and they require a negative supply.
- The CMOS family of logic gates requires almost zero power in standby and very low power at moderate switching rates. The speed of these gates is now comparable with that of the TTL. This is the technology of choice for most new designs, rivaling the TTL.

KEY POINT OF SECTION 3.15

- Different families of logic gates have different advantages and disadvantages that must be taken into account in design. Often a designer must make a trade-off between speed and power dissipation.

3.16 Flip-Flops

Flip-flops are storage elements that are used to realize synchronous logic circuits. They save the state of the machine from one clock cycle to the next. They are also used to break up any feedback loops around a cyclic logic circuit to prevent the logic from oscillations often known as *race* conditions. Flip-flops have signal inputs and a clock signal. Most flip-flops also have a differential output, which is generally denoted as Q and $\bar{Q}$. The output values of a flip-flop change only just after a clock edge. This change is dependent on the input signals just prior to the clock changes. Some flip-flops also have asynchronous set (S) and reset (R) inputs. If either of these inputs becomes high, it controls the internal storage and the output of the flip-flop. There are commercially available flip-flops [9].

The common types of flip-flops are SR flip-flops, D flip-flops, JK flip-flops, and toggle or T flip-flops. D flip-flops are preferable for integrated circuit applications. The flip-flops often realized in one of two approaches: a master–slave approach with two cascaded clocked latches, or an edge-sensitive approach. However, the edge-sensitive type is not generally used in integrated circuit design because it requires clock waveforms having fast rise and fall times.

3.16.1 D Flip-Flop

This type of flip-flop is often used in integrated circuits. The characteristic of the D flip-flop is shown in Table 3.9. This flip-flop is clocked and its output becomes equal to whatever value of D had just before the clock changed. Figure 3.60 shows a logic diagram of a possible realization of a D flip-flop with two cross-coupled NOR latches. The first latch is clocked by *Clk*, whereas the second latch is clocked by the opposite phase by $\overline{Clk}$. Thus, it is not possible for a signal to propagate from the input through both latches at one time. The first latch acts as the master latch and the second one as the slave latch. While one of the latches is activated, the other latch is deactivated and put in the hold mode. This technique of using a cascade of two latches that are clocked on the opposite phases is called a master–slave approach. This approach is commonly used in ICs to realize flip-flops.

Let us consider initially that D and Clk are both "l"s. The master latch is being accessed, but the slave is isolated. The value of A will be $\bar{D}$ and B will be D. If the input D changes, both A and B will reflect that change. But the output Q (or $\bar{Q}$) of the flip-flop will not change since the second or slave latch is isolated by the $\overline{Clk}$ = "0" inputs to the AND gates e and f.

When Clk goes low, the master latch becomes isolated from the D input by AND gates a and b both have "0" outputs. The contents of the master latch reflect the D input just before the clock went low. After

TABLE 3.9 Characteristic table for a D flip-flop

D	Q_{n+1}
0	0
1	1

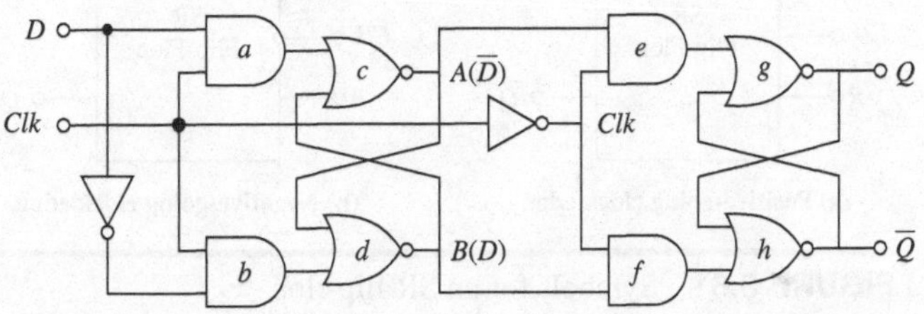

FIGURE 3.60 Logic diagram of a master–slave D flip-flop.

one inverter delay, when *Clk* goes to a "0", $\overline{Clk}$ will go to a "1", which now connects the slave latch to the master. The *Q* output of the slave latch will change to $\bar{A}$, and $\bar{Q}$ will be equal to $\bar{B}$. After *Clk* has become low, Q_{n+} becomes equal to the value of *D* just before *Clk* has become low. If *D* changes after *Clk* has become low, its change is ignored as the master is now isolated.

3.16.2 SR Flip-Flop

Figure 3.61(a) shows the symbol for an SR flip-flop. The output changes just after a positive-going clock edge. The small circle on the clock input as shown in Fig. 3.61(b) indicates that its outputs change just after a negative-going clock edge. SR flip-flops are not very popular in ICs because they have an indeterminate state when both inputs are high.

Figure 3.61 shows a master–slave realization of an RS flip-flop, which is very similar to the flip-flop in Fig. 3.60. The characteristic is described in Table 3.10. This type of flip-flop can be set, reset, or left in its previous state. If *S* is a "1" while *Clk* is high, then the next state Q_{n+1} becomes a "1" after *Clk* goes low. If *R* is a "1" while *Clk* is high, then Q_{n+1} becomes a "0". If neither *S* nor *R* goes high while *Clk* is a "1", then Q_{n+1} keeps the same value Q_n of the current state that it had just before *Clk* goes low. The state in which both *S* and *R* are high at the time *Clk* goes low is considered to be an illegal stage and the state of the output is considered to be indeterminate. It could be in either state "0" or "1".

The SR flip-flop of Fig. 3.62 has a major limitation such that the output it can be affected by the characteristics of the inputs at the instance just before *Clk* goes low and also at any other time when *Clk* is high. This limitation can be overcome by edge-sensitive flips-flops so that the output depends on the values of *S* and *R* just when the *Clk* goes low. The master–slave SR flip-flops can have problems with noise, whereas edge-sensitive SR flip-flops have problems with clock and timing requirements. As a result, SR flip-flops are not recommended for general use inside 1Cs.

TABLE 3.10 Characteristic Table of an SR flip-flop

S	*R*	Q_{n+1}
0	0	Q_n
0	1	0
1	0	1
1	1	Indeterminate

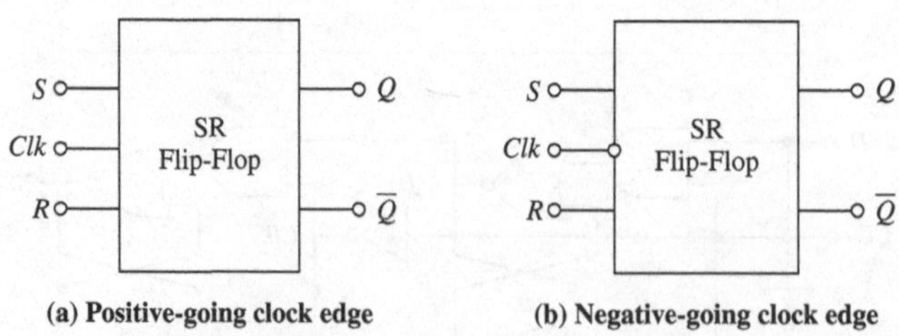

FIGURE 3.61 Symbols for an SR flip-flop

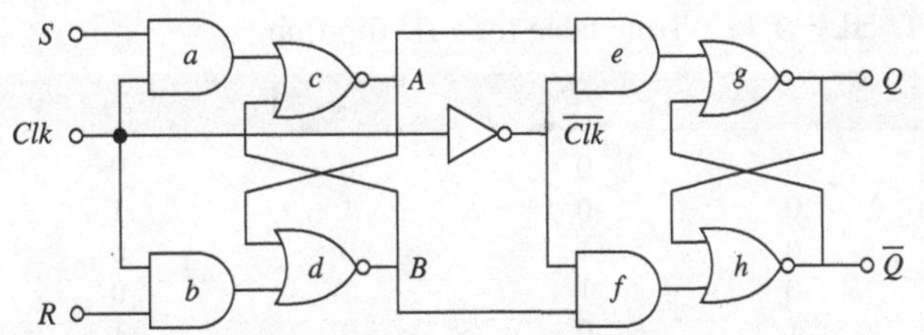

FIGURE 3.62 Logic diagram of a master–slave SR flip-flop

3.16.3 JK Flip-Flop

Although the D flip-flops are more commonly used in ICs, the JK flip-flops are also used in ICs. The JK flip-flop is often used for synchronous machines or counters. The characteristics of the JK flip-flop are similar to that of an SR flip-flop, except the "1 1" input state is now defined. This additional state makes this flip-flop useful for counters. The characteristic of a JK flip-flop is shown in Table 3.11, and it exhibits the following characteristics:

- If *J* is a "1" and *K* is a "0", the flip-flop will be set.
- If *J* is a "0" and *K* is a "1", the flip-flop will be reset.
- If both *J* and *K* are "0"s, the state will not change.
- If both *J* and *K* are "1"s, the flip-flop will change state or toggle.

TABLE 3.11 Characteristic table of an SR flip-flop

J	K	Q_{n+1}
0	0	Q_n
0	1	0
1	0	1
1	1	$\overline{Q}_n$

The state of the slave latch just before the master latch goes into latch mode can be described with the truth table as shown in Table 7.12 for Q_{n+1} as a function of *J*, *K*, and Q_n. This table includes all possible values of Q_n. The logic function for the implementation of a *JK* flip-flop can be found by applying OR functions for all the input values that give a "1" for Q_{n+1}, i.e.,

$$Q_{n+1} = Q_n\overline{J}\overline{K} + J\overline{K}\overline{Q_n} + J\overline{K}Q_n + JK\overline{Q_n} \tag{3.101}$$

Each of these terms in Eq. (3.101) is called a prime implicant for the logic function for Q_{n+1}. Equation (3.101) can be simplified by using Karnaugh map reduction [10, 11] as shown in Table 3.13 and is given by

$$Q_{n+1} = J\overline{Q_n} + \overline{K}Q_n \tag{3.102}$$

Equation (3.102) is the basis for integrated-circuit realizations of JK flip-flops. A D flip-flop can be converted into a JK flip-flop as shown in Table 3.63 by adding a 2 to 1 multiplexer to the input of a D flip-flop. This multiplexor inputs $\overline{K}$ if Q_n is a "1", otherwise it inputs *J*.

TABLE 3.12 Truth table for a JK flip-flop

J	*K*	Q_n	Q_{n+1}
0	0	0	0
0	0	1	1
0	1	0	0
0	1	1	0
1	0	0	1
1	0	1	1
1	1	0	1

TABLE 3.13 Karnaugh map reduction of the truth table of the JK flip-flop

J	KQ_n			
	00	01	11	10
0	0	1	0	0
1	1	1	0	1

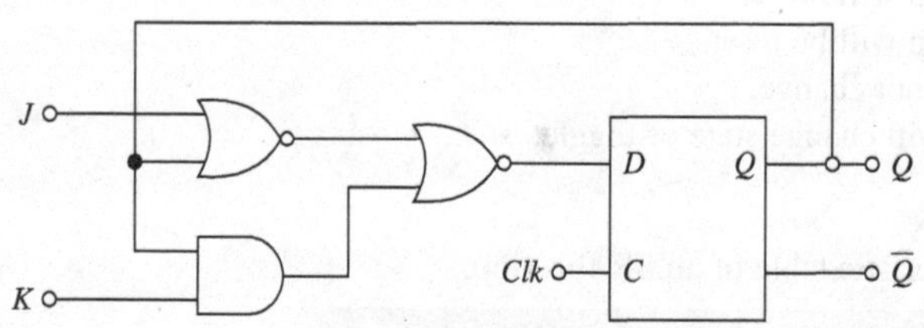

FIGURE 3.63 Logic implementation of a JK flip-flop by using a D flip-flop

3.16.4 CMOS Flip-Flops

A CMOS D flip-flop can be realized by cascading two latches that are clocked on opposite phases—one master latch and one slave latch. The CMOS implementation [10] of the master–slave D flip-flop in Fig. 3.60 is shown in Fig. 3.64. Synchronous set and reset inputs are added. If either of these asynchronous inputs is not required, the unnecessary transistors used for the other asynchronous input can be removed. When the state of the master is changing its state, it is necessary to pull one side low before the other side can go high. Therefore, it is necessary to size the transistors correctly so that they behave as clocked latches. Two additional inverters as shown in Fig. 3.64 are often connected to buffer the output of the flip-flops so that they can drive large capacitive loads.

The CMOS implementation [10] of the JK flip-flop in Fig. 3.63 is shown in Fig. 3.65. This is a slightly modified CMOS flip-flop in Fig. 3.64. A few more of transistors are added to realize the JK function. The asynchronous set or reset are omitted to simplify the circuit. However, these can be added, if required. It should be noted that the load transistors of the master latch are made smaller to allow the *n*-channel drive circuit consisting of three series transistors. This ensures that driver circuit takes over the control when changing the state of the latch.

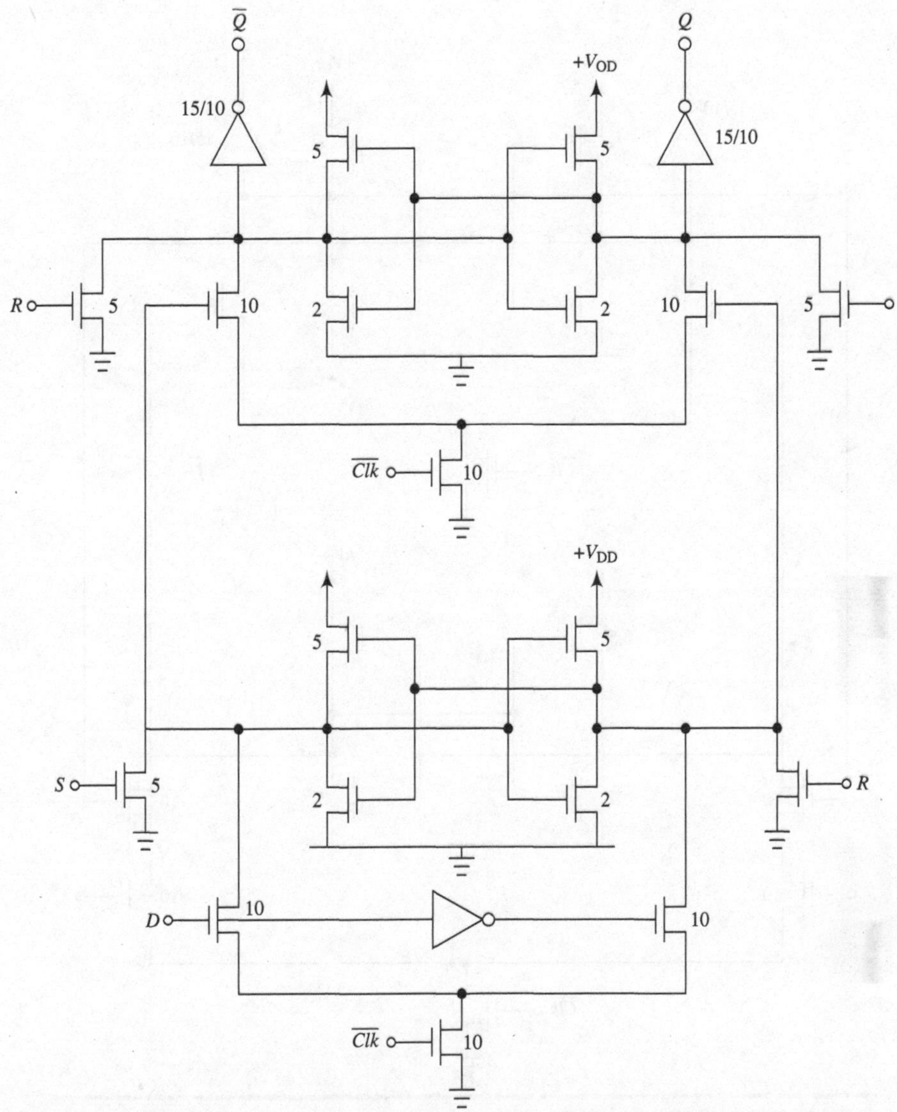

FIGURE 3.64 A CMOS D flip-flop with asynchronous set/reset and output buffers

KEY POINT OF SECTION 3.16

- Flip-flops are storage elements that are used to realize synchronous logic circuits. They save the state of the machine from one clock cycle to the next.
- The common types of flip-flops are SR flip-flops, D flip-flops, JK flip-flops, and toggle or T flip-flops. However, D-flip-flops are preferable for integrated circuits.

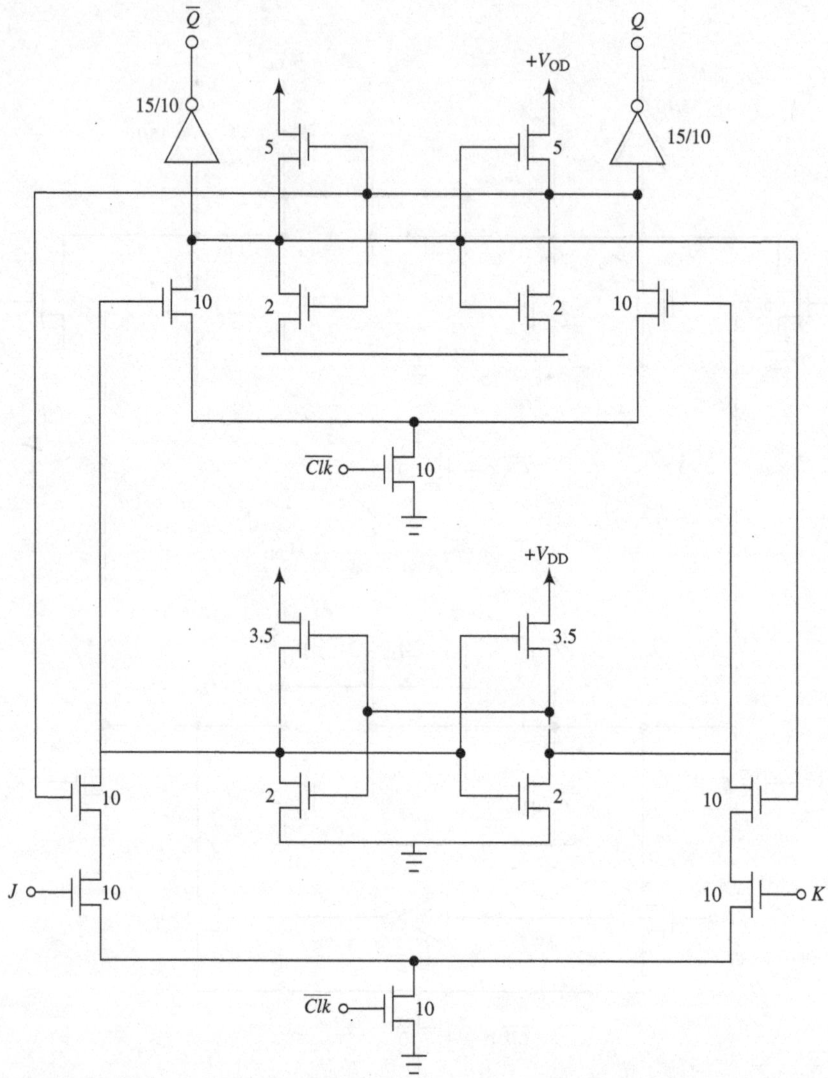

FIGURE 3.65 A CMOS JK flip-flop

3.17 Design of Logic Circuits

In the design examples in the preceding sections, we calculated the component values of MOS and bipolar inverters and found their performance parameters, such as V_{OL}, V_{OH}, t_{pd}, NM_L, and NM_H. Since bipolar and MOS gates are available in IC circuits, designs are seldom done using discrete devices. However, these examples provided an idea of the inside operation of the gates. Circuit design using gates usually involves applications of logic gates, and the following points should be kept in mind:

- Logic gates are highly nonlinear, and their output is considered as either high or low. The width of the undefined input voltage range should be kept to a minimum so that the noise margin becomes as large as possible.
- The output voltage levels of a gate must be compatible with the input voltage levels of the same or similar gates; that is, a design must address interfacing issues.
- The output of one gate must be capable of driving the inputs of more than one gate; that is, fan-out of a gate should be as large as possible for a circuit consisting of multiple gates.
- The logic gate should consume only as much power as is needed to operate at the required speed. Also, the DC supply must meet the design requirement.

EXAMPLE 3.11

D

Designing a clock circuit

(a) Design a clock circuit to produce a pulse output at a frequency of $f_{clk} = 100$ kHz. The DC supply voltage is 5 V.

(b) Use PSpice/SPICE to check your design by plotting the output voltage.

SOLUTION

(a) Given that $T = 1/f_{clk} = 1/100$ kHz $= 10$ μs, the design steps are carried out as follows:

Step 1. Choose the circuit topology. Figure 3.60(a) shows a circuit that uses inverters; NOR gates could be used instead.

Step 2. Choose the gate types. Although use of CMOS inverters would simplify the analysis, we will use bipolar inverters of type 7404. We will derive a general expression for the clock frequency that can be applied to CMOS inverters also. From the data sheet, the inverter parameters are $V_{CC} = 5$ V, $V_{OH} = 3.4$ V, $V_{OL} = 0.5$ V, $V_{IH} = 2$ V, $V_{IL} = 0.8$ V, $I_{IL} = 0.1$ mA, and $I_{IH} = 20$ μA.

Step 3. Analyze the circuit. Suppose that, at $t = 0$, v_{O1} (that is, the output of inverter U_1) is high at V_{OH}, v_O is low at V_{OL}, and v_I (the input of inverter U_1) is at $V_{IH} - V_{OH}$ (from the previous cycle). Capacitor C will charge exponentially, with a time constant of $\tau = RC$. The equivalent circuit during charging is shown in Fig. 3.60(b). The input voltage v_I will have a DC component and an exponentially decaying component. Thus,

$$v_I(t) - V_{OH} = RI_{IL} + [v_I(0^+) - v_O(0^+)]e^{-t/\tau} = RI_{IL} + (V_{IH} - 2V_{OH})e^{-t/\tau} \qquad (\text{for } 0 \le t \le T_2)$$

At $t = T_2$, v_I rises to V_{IL}, and we can find T_2 from the condition $v_I(t = T_2) = V_{IL}$:

$$T_2 = RC\ \ln\left(\frac{V_{IH} - 2V_{OH}}{V_{IL} - V_{OH} - RI_{IL}}\right) \tag{3.101}$$

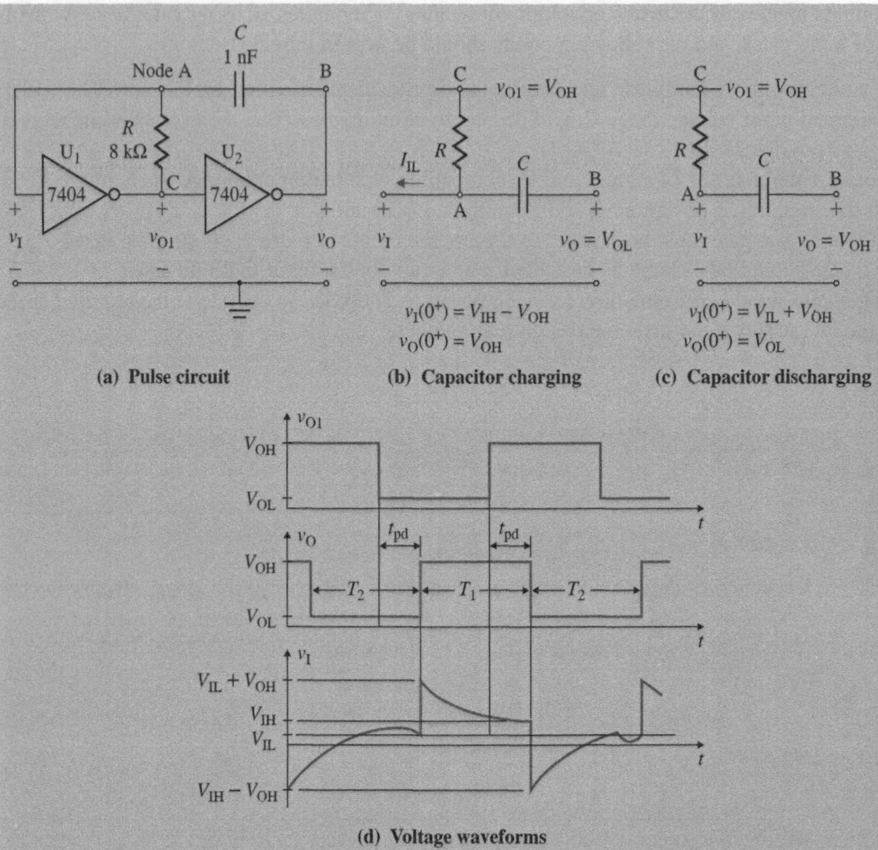

FIGURE 3.60 Pulse circuit using inverters

where I_{IL} is the input current in the logic low state. When v_I reaches V_{IL} at $t = T_2$, v_{O1} goes low to V_{OL} and v_O rises to V_{OH} after a delay time of t_{pd}. The capacitor feeds this change in v_O back to the input side. As a result, v_I jumps from V_{IL} to $V_{IL} + V_{OH}$.

Now, with $v_I = V_{IL} + V_{OH}$, $v_{O1} = V_{OL}$, and $v_O = V_{OH}$, the capacitor C will discharge exponentially, with a time constant of $\tau = RC$. The equivalent circuit during discharging is shown in Fig. 3.60(c). The input voltage v_I will have a DC component of V_{OL} and a decaying component. Redefining the time origin $t = 0$ (at $t = T_2$) gives the input voltage v_I:

$$v_I(t) = V_{OL} + [v_I(0^+) - v_O(0^+)]e^{-t/\tau} = V_{OL} + (V_{OH} + V_{IL} - V_{OL})e^{-t/\tau} \qquad (\text{for } 0 \le t \le T_1)$$

At $t = T_1$, v_I falls to V_{IH}, and we can find T_1 from the condition $v_I(t = T_1) = V_{IH}$:

$$T_1 = RC \ln\left(\frac{V_{OH} + V_{IL} - V_{OL}}{V_{IH} - V_{OL}}\right) \qquad (3.102)$$

When v_I falls to V_{IH} at $t = T_1$, v_{O1} becomes high at V_{OH} and v_O falls to V_{OL} after a delay time of t_{pd}. Because the capacitor feeds this change in v_O back to the input side, v_I falls from V_{IH} to $V_{IH} - V_{OH}$, and the cycle is repeated. The voltage waveforms are shown in Fig. 3.60(d). The period T of the output voltage is given by

$$T = T_1 + T_2 = RC\left(\ln\frac{V_{OH} + V_{IL} - V_{OL}}{V_{IH} - V_{OL}} + \ln\frac{V_{IH} - 2V_{OH}}{V_{IL} - V_{OH} - RI_{IL}}\right) \tag{3.103}$$

NOTE: For a CMOS gate, $V_{OL} \approx 0$, $V_{OH} = V_{DD}$, $I_{IL} \equiv 0$, and $V_{IL} = V_{IH} = V_t$ (threshold voltages of the NMOS). Thus, Eq. (3.103) becomes

$$T = T_1 + T_2 = RC\left(\ln\frac{V_{DD} + V_{IL}}{V_{IH}} + \ln\frac{V_{IH} - 2V_{OH}}{V_{IL} - V_{OH}}\right) \tag{3.104}$$

Step 4. Find the values of R and C. Using Eq. (3.103), we get

$$10\ \mu\text{A} = RC\left(\ln\frac{3.4 + 0.8 - 0.5}{2 - 0.5} + \ln\frac{2 - 2 \times 3.4}{0.8 - 3.4 - R \times 0.1\ \text{mA}}\right)$$

Choose a suitable value of C: Let $C = 1$ nF. Solving for R by iteration, we get $R \approx 8$ kΩ. Then $\tau = RC = 8$ μs, $T_1 = 7.22$ μs, and $T_2 = 2.76$ μs.

(b) The PSpice simulation is shown in Fig. 3.61. The PSpice plots shown in Fig. 3.62 give $T = 7.53$ μs, $T_1 = 6.41$ μs, and $T_2 = 1.12$ μs. Thus, a higher value of R (say, $R = 10$ kΩ) will be needed.

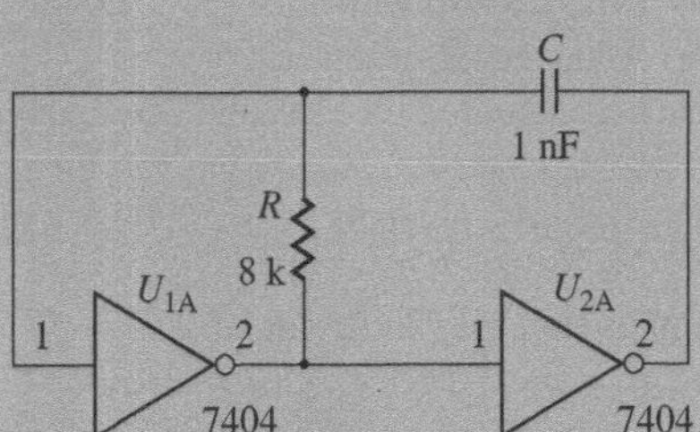

FIGURE 3.61 PSpice schematic for Example 3.11

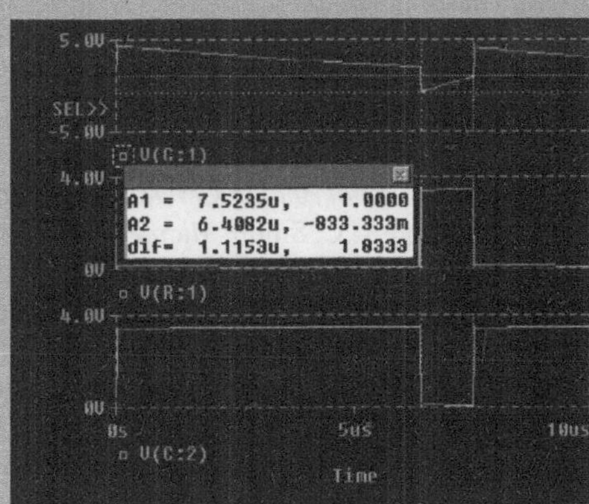

FIGURE 3.62 PSpice plots for clock circuit: $v_{O1} \equiv$ V(R:1) and $v_O \equiv$ V(C:2) for Example 3.11

Summary

Electronic circuits are used to perform various logic functions, such as NOT, AND, OR, NOR, and NAND. The transistors are operated as on and off switches. This chapter analyzed the characteristics of NMOS and CMOS inverters, which are the basis of digital circuits and have applications in NAND and NOR gates. The use of a depletion-load transistor allows the logic level to be the same as the supply voltage and also gives the shortest NMOS switching times. NMOS and CMOS transmission gates are commonly used in steering logic function. The power consumption of CMOS gates is very small, essentially zero. In general, CMOS gates have many advantages over NMOS gates.

DTL gates are simple, but their switching speed is low because they include saturated transistors. TTL gates are better than DTL gates in terms of switching speed, component density, and fan-out. I^2L gates have higher component density, higher fan-out, and lower power dissipation than TTL gates. Schottky TTL gates minimize transistor saturation and have higher switching speeds than standard TTL gates. In ECL gates, the transistors are operated in the active region, and saturation is completely avoided, thereby reducing propagation delay time. However, ECL gates have more power dissipation than TTL and I^2L gates. Each gate has its advantages and limitations, and the applications engineer has to decide which types to use in a specific application.

Flip-flops are storage elements that are used to realize synchronous logic circuits. They save the state of the machine from one clock cycle to the next. Flip-flops have signal inputs and a clock signal. Most flip-flops also have a differential output that is almost always denoted Q and $\bar{Q}$. The output values of a flip-flop change only just after a clock edge. The common types of flip-flops are SR flip-flops, D flip-flops, JK flip-flops, and toggle or T flip-flops. However, D-flip-flops are preferable for integrated circuits.

References

1. D. A. Hodges and H. G. Jackson, *Analysis and Design of Digital Integrated Circuits*. New York: McGraw-Hill, 1988.
2. G. M. Glasford, *Digital Electronic Circuits*. Englewood Cliffs, NJ: Prentice Hall, 1988.
3. J. E. Ayers, *Digital Integrated Circuits: Analysis and Design*. Boca Raton, FL: CRC Press, 2004.
4. F. Maloberti, *Analog Design for CMOS VLSI Systems*. Boston: Kluwer Academic, 2001.
5. J. M. Rabaey, A. Chandrakasan, and B. Nikolic, *Digital Integrated Circuits: A Design Perspective*. Upper Saddle River, NJ: Pearson Education, 2003.
6. B. Razavi, *Design of Analog CMOS Integrated Circuits*. New York: McGraw-Hill, 2001.
7. A. S. Sedra and K. C. Smith, *Microelectronic Circuits*. New York: Oxford Univerity Press, 2004.
8. K. Gopalan, *Introduction to Digital Microelectronic Circuits*. Chicago: R. Irwin, 1996.
9. Datasheet Catalog, http://www.DatasheetCatalog.com. A datasheet source for electronic components and semiconductors. Accessed November 2012.
10. K. Martin, *Digital Integrated Circuit Design*. New York, NY: Oxford University Press, 2000, Chapter 7: Latches, Flip-Flops, and Synchronous System Design.
11. M. M. Mano, *Digital Design*. Upper Saddle River, NJ: Prentice-Hall Inc, 2007, 4th ed.

Review Questions

1. What are the advantages and disadvantages of an NMOS inverter with a resistive load?
2. What are the intervals of operation of an NMOS inverter with a resistive load?
3. What are the advantages and disadvantages of an NMOS inverter with a saturated load?
4. What are the intervals of operation of an NMOS inverter with a saturated load?
5. What is the effect of the ratio K_D/K_L on the transfer characteristic of an NMOS inverter with a saturated load?
6. What are the advantages and disadvantages of an NMOS inverter with a depletion load?
7. What are the intervals of operation of an NMOS inverter with a depletion load?
8. What is the effect of the ratio K_D/K_L on the transfer characteristic of an NMOS inverter with a depletion load?
9. What is the function of an NMOS transmission gate?
10. What are the advantages and disadvantages of a CMOS inverter?
11. What are the intervals of operation of a CMOS inverter?
12. What is the function of a CMOS transmission gate?
13. What are the differences between CMOS and NMOS transmission gates?
14. What are the advantages and disadvantages of CMOS and NMOS gates?
15. What is the saturation of a transistor?
16. What is the overdrive factor of a transistor?
17. What is the forced β_F of a transistor?
18. What is delay time?
19. What is the saturating charge of a transistor?
20. What is the storage time of a transistor?
21. What is the fan-out of a gate?
22. What is the noise margin of a gate?
23. What is the pull-down resistance of a gate?
24. What is a multiemitter transistor?
25. What are the advantages and limitations of a TTL gate?
26. What is a totem-pole output stage?
27. What is the active-biased recovery circuit?
28. What are the advantages and limitations of high-speed TTL gates?
29. What is a Schottky clamped transistor?
30. What are the advantages and limitations of Schottky TTL gates?
31. What are the advantages and limitations of ECL gates?
32. What is the purpose of connecting the collectors of an ECL gate to the ground?
33. What problems result from saturation in transistor gates?
34. What is a SR flip-flop?
35. What is the major limitation of a SR flip-flop?
36. What is a D flip-flop?
37. What is a JK flip-flop?
38. Why do flip-flops use master–slave approach?
39. Why are the MOSFETs in CMOS-type flip-flops sized differently?

Problems

The symbol **D** indicates that a problem is a design problem. The symbol **P** indicates that you can check the solution to a problem using PSpice/SPICE or Electronics Workbench.

3.4 *Performance Parameters of Logic Gates*

3.1 An inverter, as shown in Fig. 3.8, drives identical inverters and has $V_{OL} = 0.4$ V, $I_{OL} = 1$ mA, $V_{OH} = 2.4$ V, $I_{OH} = 100$ μA, and $V_{CC} = 5$ V. The input currents drawn by each load inverter are $I_{IL} = 0.1$ mA (at logic low) and $I_{IH} = 10$ μA (at logic high). **D**

a. If there are three load inverters, determine the value of pull-up resistance R_P that will ensure a logic 1 output of $V_{OH} = 2.4$ V.

b. If $R_P = 2.5$ kΩ, find the fan-out N.

3.2 The inverter shown in Fig. 3.10(a) has $V_{OL} = 0.4$ V, $V_{OH} = 2.4$ V, $V_{CC} = 5$ V, $R_{on} = 200$ Ω, $R_{off} \approx \infty$, $R_P = 2.5$ kΩ, $C_L = 2$ pF, and $f_{clk} = 100$ kHz.

a. Find the delay times for the output voltage to rise from 0.4 V to 2.4 V and to fall from 5 V to 0.4 V.

b. Find the power dissipation P_D.

3.5 *NMOS Inverters*

3.3 The parameters of the NMOS inverter in Fig. P3.3 are $V_{tL} = 1.5$ V, $V_{DD} = 5$ V, $R_D = 1.5$ kΩ, and $K_L = 0.5$ mA/V^2.

a. Determine the input voltage at the transition point $V_{I(tran1)}$.

b. Calculate the output voltage v_O, the drain current i_D, and the power dissipation P_D for $v_I = 2.5$ V and for $v_I = 5$ V.

FIGURE P3.3

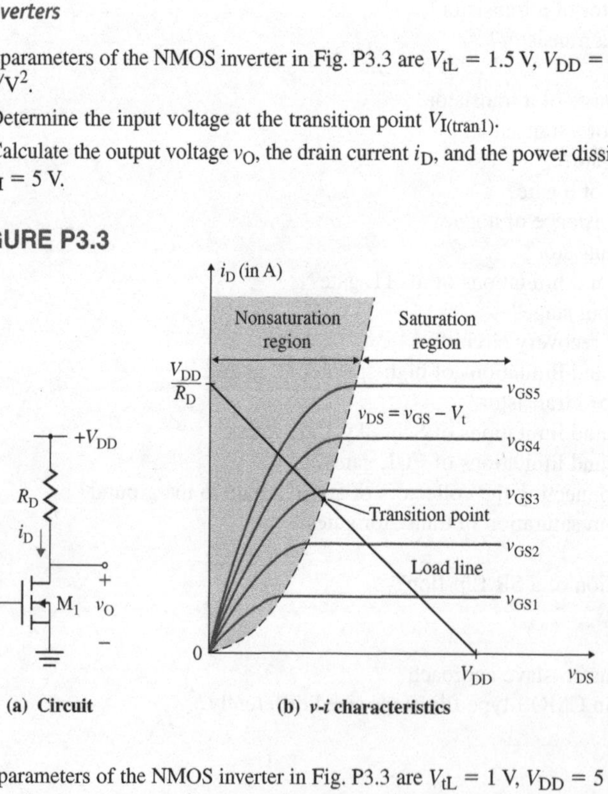

(a) Circuit (b) *v-i* characteristics

3.4 The parameters of the NMOS inverter in Fig. P3.3 are $V_{tL} = 1$ V, $V_{DD} = 5$ V, and $R_D = 1.5$ kΩ. Determine the conduction parameter K_L so that the output voltage v_O is 0.5 V at $v_I = 5$ V.

3.5 The parameters of the NMOS inverter in Fig. 3.12(a) with a saturated load are $V_{tL} = V_{tD} = 1.5$ V, $V_{DD} = 5$ V, $K_L = 0.1$ mA/V^2, and $K_D = 50$ μA/V^2.

a. Determine the input voltage at the transition point $V_{I(tran)}$.

b. Calculate the output voltage, the drain current, and the power dissipation P_D for $v_I = 2.5$ V and for $v_I = 5$ V.

3.6 The parameters of the NMOS inverter in Fig. 3.12(a) with a saturated load are $V_{tL} = V_{tD} = 1$ V and $V_{DD} = 5$ V. Determine the ratio K_D/K_L so that **(a)** $v_O = 0.25$ V at $v_I = 5$ V and **(b)** $v_O = 0.25$ V at $v_I = 4.5$ V.

3.7 The parameters of the NMOS inverter in Fig. 3.12(a) with an enhancement load are $V_{tL} = V_{tD} = 1$ V, $V_{DD} = 5$ V, $K_L = 0.25$ mA/V^2, and $K_D = 1$ mA/V^2.

a. Determine the input voltage at the transition point $V_{I(tran)}$.

b. Calculate the output voltage and the drain current for $v_I = 2$ V and for $v_I = 5$ V.

3.8 D P Design an enhancement-load NMOS inverter, as shown in Fig. 3.12(a), to obtain a noise margin of $NM_L \geq 0.9$ V. The threshold voltages are $V_{tL} = V_{tD} = 1$ V, and $K_L = 40$ μA/V^2. The supply voltage is $V_{DD} = 5$ V. Assume $V_{OH} = V_{IH} = 5$ V, load capacitance $C_L = 2$ pF, and clock frequency $f_{clk} = 1$ MHz.

a. Find the design parameter $K_R = K_D/K_L$, neglecting the body effect.

b. Calculate NM_L if the body effect is included. Assume $\gamma = 0.5\sqrt{V}$ and $2\phi_f = 0.6$ V.

c. Calculate the low-to-high propagation time t_{pLH}.

d. Calculate the high-to-low propagation time t_{pHL}.

e. Calculate the delay-power product (DP).

3.9 The parameters of the NMOS inverter in Fig. 3.16(a) with a depletion load are $V_{tL} = -1.5$ V, $V_{tD} = 1$ V, $V_{DD} = 5$ V, $K_D = 1.5$ mA/V^2, and $K_L = 0.1$ mA/V^2.

a. Determine the input and output voltages at the first and second transition points.

b. Calculate the output voltage, the drain current, and the power dissipation P_D for $v_I = 2.5$ V and for $v_I = 5$ V.

3.10 The parameters of the NMOS inverter in Fig. 3.16(a) with a depletion load are $V_{tL} = -1.5$ V, $V_{tD} = 1$ V, and $V_{DD} = 5$ V. Determine the ratio K_D/K_L so that **(a)** $v_O = 0.25$ V at $v_I = 5$ V and **(b)** $v_O = 0.25$ V at $v_I = 4.5$ V.

3.11 The parameters of the NMOS inverter in Fig. 3.16(a) with a depletion load are $V_{tD} = 1$ V, $V_{DD} = 5$ V, $K_D = 1.5$ mA/V^2, and $K_L = 0.1$ mA/V^2. Determine the value of V_{tL} so that $v_O = 0.15$ V at $v_I = 5$ V.

3.12 The parameters of the NMOS inverter in Fig. 3.16(a) with a depletion load are $V_{tL} = -2$ V, $V_{tD} = 1$ V, $V_{DD} = 5$ V, $K_D = 1$ mA/V^2, and $K_L = 0.25$ mA/V^2.

a. Determine the input and output voltages at the first and second transition points.

b. Calculate the output voltage and the drain currents for $v_I = 1.5$ V and for $v_I = 5$ V.

3.13 D P Design a depletion-load NMOS inverter, as shown in Fig. 3.16(a), to obtain a noise margin of $NM_H \geq 2.5$ V. The threshold voltages are $V_{tL} = -1.5$ V, $V_{tD} = 1$ V, and $K_L = 40$ μA/V^2. The supply voltage is $V_{DD} = 5$ V. Assume $V_{OH} = V_{DD} = 5$ V, load capacitance $C_L = 2$ pF, and frequency $f_{clk} = 1$ MHz.

a. Find the design parameter $K_R = K_D/K_L$, neglecting the body effect.

b. Calculate NM_L if the body effect is neglected.

c. Calculate NM_H if the body effect is included. Assume $\gamma = 0.5\sqrt{V}$ and $2\phi_f = 0.6$ V.

d. Calculate the low-to-high propagation time t_{pLH}.

e. Calculate the high-to-low propagation time t_{pHL}.

f. Calculate the delay-power product (DP).

3.6 NMOS Logic Circuits

3.14 The parameters of the NMOS transmission gate in Fig. 3.23 are $V_t = 1.5$ V, $K = 0.1$ mA/V^2, $C_L = 5$ pF, and $V_G = 5$ V.

a. What is the steady-state output voltage if the input voltage is changed in the following sequence: $v_I = 0$, $v_I = 5$ V, and $v_I = 3$ V?

b. What would the steady-state output voltage be if v_G were switched to 0?

3.15 An NMOS *RS* flip-flop circuit is shown in Fig. P3.15. The input conditions, which are sequential from time 1 to time 4, are also listed in Fig. P3.15. Determine the state of each transistor (on or off) at each time and the logic outputs at Q_A and Q_B for each input condition.

FIGURE P3.15

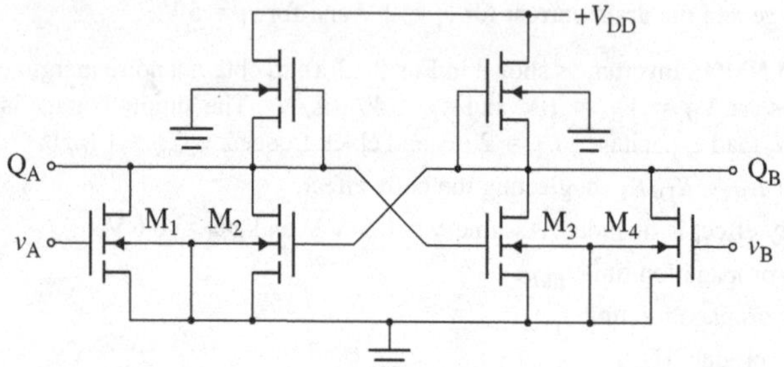

Time	v_A	v_B
1	0	0
2	0	1
3	1	0
4	1	1

3.7 *CMOS Inverters*

3.16 The parameters of the CMOS inverter in Fig. 3.26(a) are $V_{tN} = 1.5$ V, $V_{tP} = -1.5$ V, $V_{DD} = 5$ V, $K_n = 1$ mA/V^2, and $K_p = 100$ mA/V^2.

a. Determine the input and output voltages at the first and second transition points.

b. Calculate the output voltage and the drain currents for $v_I = 1.5$ V and for $v_I = 4.5$ V.

3.17 The parameters of the CMOS inverter in Fig. 3.26(a) are $V_{tN} = 1.5$ V, $V_{tP} = -1.5$ V, $V_{DD} = 5$ V, $K_n = 0.2$ mA/V^2, and $K_p = 100$ mA/V^2.

a. Determine the input and output voltages at the first and second transition points.

b. Calculate the output voltage and the drain currents for $v_I = 1.5$ V and for $v_I = 4.5$ V.

3.18 The parameters of the CMOS inverter in Fig. 3.26(a) are $V_{tN} = 1$ V, $V_{tP} = -1$ V, $V_{DD} = 5$ V, $K_n = 1$ mA/V^2, and $K_p = 1$ mA/V^2.

a. Determine the input and output voltages at the first and second transition points.

b. Calculate the output voltage and the drain currents for $v_I = 1.5$ V and for $v_I = 3.5$ V.

3.19 The parameters of the CMOS inverter in Fig. 3.26(a) are $V_{tN} = 1.5$ V, $V_{tP} = -1.5$ V, and $V_{DD} = 5$ V. Determine the ratio K_n/K_p so that the transition occurs at an input voltage of $v_I = V_{I(tran1)} = V_{DD}/2$.

3.20 The parameters of the CMOS inverter in Fig. 3.26(a) are $V_{tN} = 1.5$ V, $V_{tP} = -1$ V, and $V_{DD} = 5$ V. Calculate the output voltages if the input voltage is varied from 0 to 5 V with a step of 0.25 V. Assume $K_n/K_p = 1$.

3.21 The parameters of the CMOS inverter in Fig. 3.26(a) are $V_{tN} = 1.5$ V, $V_{tP} = -1.5$ V, and $V_{DD} = 5$ V. Calculate the output voltages if the input voltage is varied from 0 to 5 V with a step of 0.5 V. The ratio $K_n/K_p = 1$.

3.22 The parameters of the CMOS inverter in Fig. 3.26(a) are $V_{tN} = 1$ V, $V_{tP} = -1$ V, $V_{DD} = 5$ V, $K_n = 1$ mA/V^2, and $K_p = 1$ mA/V^2.

a. Determine the input and output voltages at the first and second transition points.

b. Calculate the output voltages and the drain currents for $v_I = 1.5$ V and for $v_I = 3.5$ V.

3.23 Design a CMOS inverter, as shown in Fig. 3.26(a), to operate at a transition voltage $V_M = 2.5$ V. The threshold voltages are $V_{tP} = -1.5$ V, $V_{tN} = 1$ V, and $K_p = 40$ μA/V^2. The supply voltage is $V_{DD} = 5$ V. Assume $V_{IH} = V_{OH} = 5$ V, load capacitance $C_L = 2$ pF, and frequency $f_{clk} = 1$ MHz.

a. Find the design parameter $K_R = K_n/K_p$, neglecting the body effect.

b. Calculate NM_L and NM_H.

c. Calculate the propagation delay t_{pd}.

d. Calculate the delay-power product (DP).

3.8 *CMOS Logic Circuits*

3.24 The parameters of the CMOS transmission gate in Fig. 3.29 are $V_{tN} = 1.5$ V, $V_{tP} = -1.5$ V, $V_{DD} = 5$ V, $K_n = 0.1$ mA/V^2, $K_p = 0.1$ mA/V^2, $C_L = 2$ pF, $\bar{v}_G = 5$ V, and $v_G = 0$.

a. What is the steady-state output voltage if the input voltage is changed in the following sequence: $v_I = 0$, $v_I = 5$ V, and $v_I = 3$ V?

b. What would the steady-state output voltage be if the gate voltages were switched to $\bar{v}_G = 0$ and $v_G = 5$ V?

3.25 The parameters of the CMOS transmission gate in Fig. 3.29 are $V_{tN} = 1.5$ V, $V_{tP} = -1.5$ V, $V_{DD} = 5$ V, $K_n = 0.1$ mA/V^2, $K_p = 0.1$ mA/V^2, $C_L = 2$ pF, $\bar{v}_G = 5$ V, and $v_G = 0$. Use PSpice/SPICE to plot the transfer characteristic v_O versus v_I if v_I is varied from 0 to 5 V with a step increment of 0.5 V. Indicate the regions over which the *n*-channel and the *p*-channel transistors are either turned on or cut off.

3.26 A CMOS *RS* flip-flop circuit is shown in Fig. P3.26. The input conditions, which are sequential from time 1 to time 4, are also listed. Determine the state of each transistor (on or off) at each time and the logic outputs at Q_A and Q_B for each input condition.

FIGURE P3.26

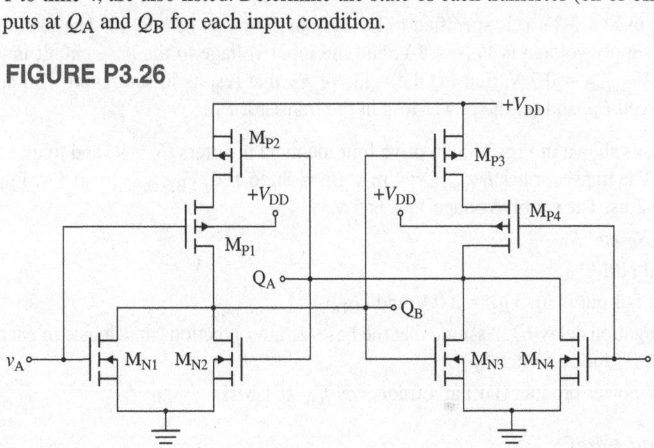

Time	v_A	v_B
1	0	0
2	0	1
3	1	0
4	1	1

3.27 A CMOS logic circuit is shown in Fig. P3.27. The input conditions, which are sequential from time 1 to time 4, are also listed. Determine the state of each transistor (on or off) at each time and the logic output v_O.

FIGURE P3.27

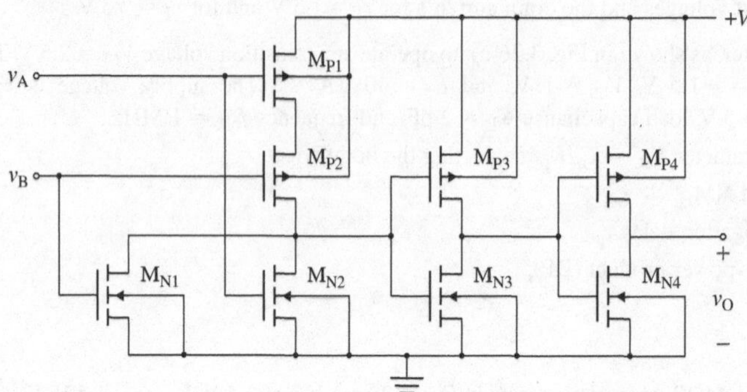

Time	v_A	v_B
1	0	0
2	0	1
3	1	0
4	1	1

3.10 *BJT Inverters*

For Probs. 3.28–3.30, assume that the PSpice/SPICE model parameters are

```
CCS=2PF TF=0.1NS TR=10NS VJC=0.85 VAF=50
```

(β_F is as specified).

3.28 The bipolar transistor in Fig. 3.33(a) is specified to have β_F in the range of 10 to 50. The load resistance is R_C = 10 kΩ. The DC supply voltage is V_{CC} = 5 V, and the input voltage to the base circuit is v_I = 5 V. If $V_{CE(sat)}$ = 0.1 V and $V_{BE(sat)}$ = 0.7 V, find **(a)** the value of R_B that results in saturation with an overdrive factor of 5, **(b)** the forced β_F, and **(c)** the power loss in the transistor P_D.

3.29 The bipolar transistor in Fig. 3.33(a) is specified to have β_F in the range of 8 to 40. The load resistance is R_C = 1 kΩ. The DC supply voltage is V_{CC} = 5 V, and the input voltage to the base circuit is v_I = 5 V. If $V_{CE(sat)}$ = 0.2 V and $V_{BE(sat)}$ = 0.7 V, find **(a)** the value of R_B that results in saturation with an overdrive factor of 5, **(b)** the forced β_F, and **(c)** the power loss in the transistor P_D.

3.30 D P Design a BJT inverter, as shown in Fig. 3.35, to drive four identical inverters (N = 4) and to give V_{OH} = 2.5 V and NM_L = 0.2 V. The transistor has $I_{C(max)}$ = 4 mA, β_F = 80 to 120, $V_{BE(cut\text{-}in)}$ = 0.5 V, $V_{BE(sat)}$ = 0.7 V, t_d = 2 ns, and t_s = 2 ns. The supply voltage V_{CC} is 5 V.

a. Find the values of R_C and R_B.

b. Calculate k_{ODF} and NM_H.

c. Find the maximum fan-out N for V_{IH} = 2.0 V and k_{ODF} = 1.

d. Calculate the propagation delay t_{pd}. Assume that the base–emitter junction capacitance of each load transistor is C_B = 1.5 pF (during switching).

e. Calculate the delay-power product (DP) at a frequency f_{clk} = 1 MHz.

3.11 *Transistor-Transistor Logic Gates*

For Probs. 3.31–3.42, assume that the PSpice/SPICE model parameters are

```
CCS=2PF TF=0.1NS TR=10NS VJC=0.85 VAF=50
```

(β_F is as specified).

3.31 **a.** Design the gate of the circuit in Fig. P3.31. It has two inputs and feeds one similar gate with one input. Assume $V_{CC} = 5$ V, $V_D = 0.7$ V, $V_{CE(sat)} = 0.1$ V, $V_{BE} = 0.7$ V, $\beta_{F(forced)} = 10$, and $I_{C(sat)} = 1$ mA.

b. If the output voltage must be maintained at more than 30% of V_{CC}, what is the maximum value of fan-out N?

FIGURE P3.31

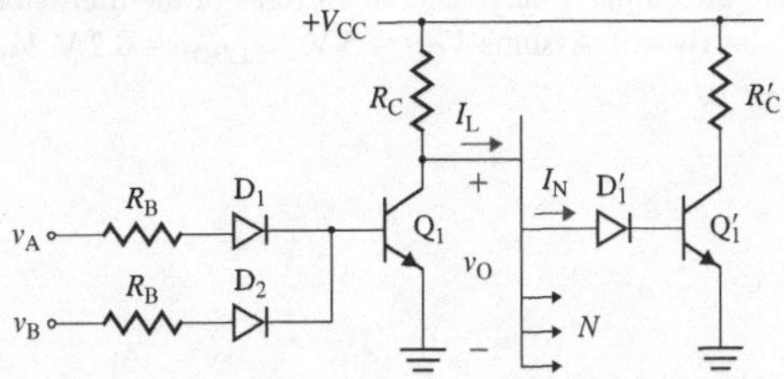

3.32 **a.** Design the gate of the circuit in Fig. P3.31. It has two inputs and feeds one similar gate with one input. Assume $V_{CC} = 5$ V, $V_D = 0.7$ V, $V_{CE(sat)} = 0.2$ V, $V_{BE} = 0.7$ V, $\beta_{F(forced)} = 4$, and $I_{C(sat)} = 1$ mA.

b. If the output voltage must be maintained at more than 30% of V_{CC}, what is the maximum value of fan-out N? From the transfer function, calculate NM_L and NM_H.

3.33 Design the TTL NAND gate of the circuit in Fig. P3.33. It has two inputs and feeds three similar NAND gates. Assume $V_{CC} = 5$ V, $V_D = 0.7$ V, $V_{CE(sat)} = 0.2$ V, $V_{BE} = 0.7$ V, $\beta_{F(forced)} = 10$, and $I_{C(sat)} = 1$ mA. From the transfer function, calculate NM_L and NM_H.

FIGURE P3.33

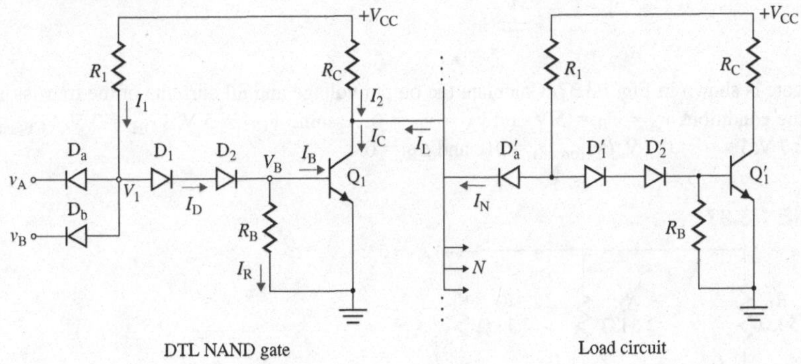

3.34 A TTL gate is shown in Fig. P3.34.

a. Calculate the currents I_1, I_2, I_C, and I_B and the voltages V_I and V_O for input conditions $v_A = v_B = 0$ and $v_A = v_B = 5$ V.

b. Calculate the fan-out for the output low condition. Assume $I_{NO} = 50$ μA and $I_{LO} = 200$ μA.

FIGURE P3.34

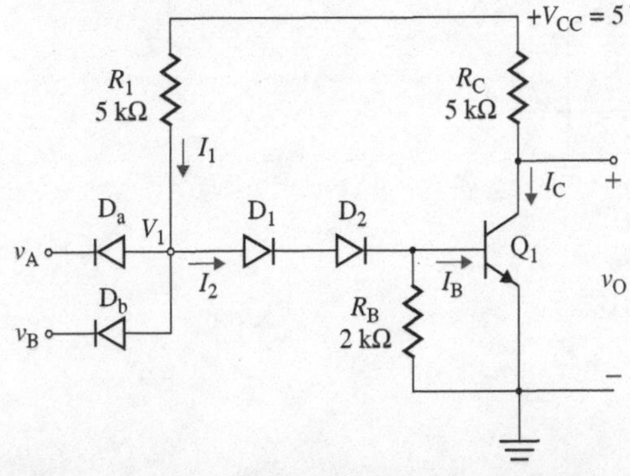

3.35 Design the TTL NAND gate of the circuit in Fig. 3.38. It has two inputs and feeds two similar NAND gates. The output voltage at output high is $V_{OH} = 3.5$ V, and $I_{N1} = 650$ μA. Assume $V_{CC} = 5$ V, $V_D = 0.7$ V, $V_{CE(sat)} = 0.2$ V, $V_{BE} = 0.7$ V, $\beta_{F(forced)} = 10$, $\beta_R = 0.2$, $I_{C3(sat)} = 1$ mA, and $I_{C4(max)} = 2$ mA. From the transfer function, calculate NM_L and NM_H.

3.36 A TTL gate is shown in Fig. P3.36. Calculate the output voltage and all currents of the transistors and fan-out for the conditions $v_A = v_B = 5$ V and $v_A = v_B = 0$. Assume $V_{CC} = 5$ V, $V_{CE(sat)} = 0.2$ V, $V_{BE} = 0.7$ V, $V_{BC1} = 0.7$ V, $\beta_{F(forced)} = 10$, and $\beta_R = 0.4$.

FIGURE P3.36

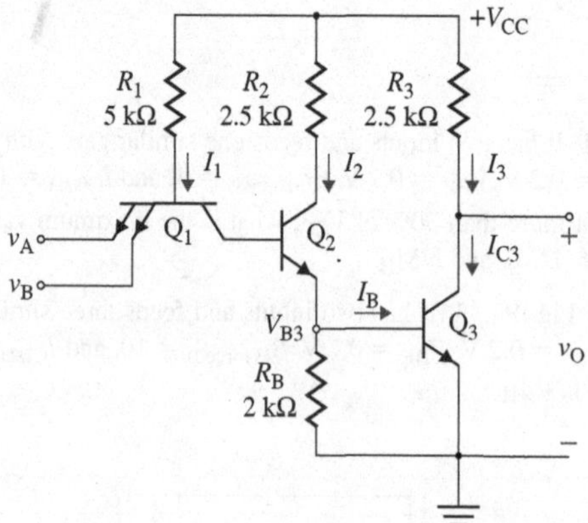

3.37 A TTL gate is shown in Fig. P3.37. Calculate the output voltage and all currents of the transistors and fan-out for the conditions $v_A = v_B = 5$ V and $v_A = v_B = 0$. Assume $V_{CC} = 5$ V, $V_{OH} = 3$ V, $V_{CE(sat)} = 0.2$ V, $V_{BE} = 0.7$ V, $V_{BC1} = 0.7$ V, $\beta_{F(forced)} = 10$, and $\beta_R = 0.4$.

FIGURE P3.37

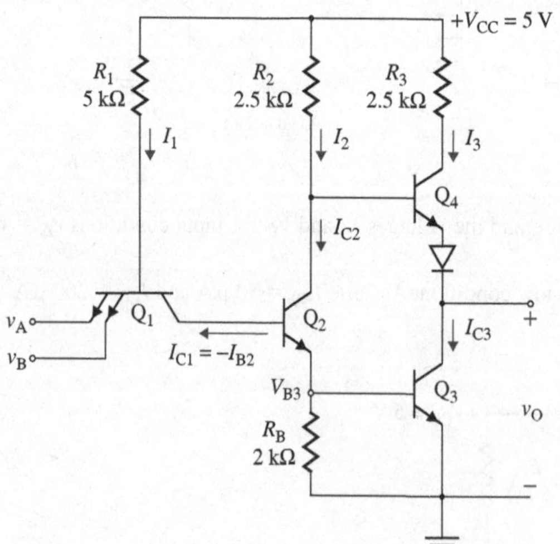

3.38 Design the TTL NAND gate of the circuit in Fig. 3.36. It has two inputs and feeds four identical NAND gates. The output voltage at output high is $V_{OH} = 2.8$ V, and $I_{N1} = 72.5$ μA. Assume $V_{CC} = 5$ V, $V_D = 0.7$ V, $V_{CE(sat)} = 0.3$ V, $V_{BE} = 0.8$ V, $\beta_{F(forced)} = 20$, $\beta_R = 0.1$, $I_{C3(sat)} = 2$ mA, and $I_{C4(max)} = 2$ mA. From the VTC, find NM_L and NM_H.

3.39 Design the high-speed TTL NAND gate of the circuit in Fig. 3.43. It has two inputs and feeds four similar NAND gates. The output voltage at output high is $V_{OH} = 3.5$ V, and $I_{N1} = 650$ μA. Assume $V_{CC} = 5$ V, $V_{CE(sat)} = 0.2$ V, $V_{BE} = 0.7$ V, $\beta_{F(forced)} = 10$, $\beta_R = 0.2$, $I_{C3(max)} = 1$ mA, $I_{C4(max)} \approx 2$ mA, and $I_{BR} = 0.7$ mA. From the VTC, calculate NM_L and NM_H.

3.40 Design the TTL NAND gate of the circuit in Fig. 3.43. It has two inputs and feeds four identical NAND gates. The output voltage at output high is $V_{OH} = 2.8$ V, and $I_{N1} = 65$ μA. Assume $V_{CC} = 5$ V, $V_D = 0.7$ V, $V_{CE(sat)} = 0.3$ V, $V_{BE} = 0.7$ V, $\beta_{F(forced)} = 40$, $\beta_R = 0.1$, $I_{C3(sat)} = 2$ mA, $I_{C4(max)} = 2$ mA, and $I_{BR} = 700$ μA. From the VTC, calculate NM_L and NM_H.

3.41 A Schottky clamped transistor is shown in Fig. P3.41. If the input current is $I_I = 2$ mA, calculate I_D, I_B, and I_C for the conditions $I_L = 5$ mA and $I_L = 20$ mA. Assume $\beta_F = 25$, $V_{BE} = 0.7$ V, and $V_D = 0.3$ V.

FIGURE P3.41

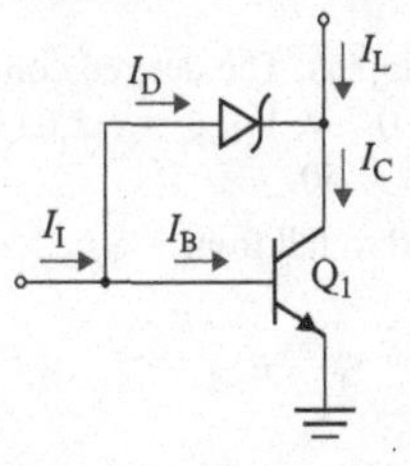

3.42 A Schottky clamped transistor is shown in Fig. P3.42.

a. If the input current is $I_I = 2$ mA, calculate I_D, I_B, and I_C for the condition $I_L = 0$.

b. Determine the maximum value of load current I_L that the transistor can sink and still remain at the edge of saturation. Assume $\beta_F = 25$, $V_{BE} = 0.7$ V, $V_{CE(sat)} = 0.2$ V, and $V_D = 0.3$ V.

FIGURE P3.42

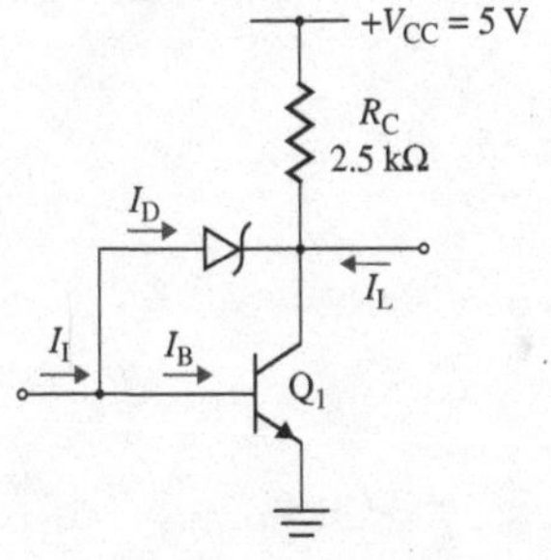

3.12 *Emitter-Coupled Logic OR/NOR Gates*

3.43 **a.** Design the ECL OR/NOR gate of the circuit in Fig. 3.50. It has two inputs. The desired collector currents are $I_{C1} = 2$ mA and $I_{C4} = I_{C5} = 1$ mA. At logic 1, $V_{OR} = -0.7$ V, $V_{NOR} = -1.5$ V, and v_A (or v_B) $= V_{OR} = -0.7$ V. Assume $V_{EE} = -5.2$ V, $V_{BE} = 0.7$ V, and $\beta_F = 50$.

b. Calculate the maximum fan-out with similar ECL gates if V_{OR} is allowed to fall from $V_{OR(max)} = -0.7$ V to $V_{OR(min)} = -0.75$ V.

3.44 Repeat Prob. 3.43 if $V_{CC} = 3$ V, $V_{EE} = -3.2$ V, $V_{OR} = 0.5$ V, $V_{NOR} = 1.5$ V, and v_A (or v_B) $= V_{OR} = 0.5$ V.

D

3.45 An ECL logic gate with one input is shown in Fig. P3.45. Calculate the resistances R_{C1}, R_{C2}, R_E, and R_4. The desired emitter current of all transistors is 2 mA. At logic 1, the outputs are $V_{OR} = 2.5$ V and $V_{NOR} = 3.5$ V. Assume $V_{CC} = 5$ V, $V_{EE} = 0$, $V_{BE} = 0.7$ V, and $\beta_F = 200$. From the transfer function, calculate NM_L and NM_H.

D P

FIGURE P3.45

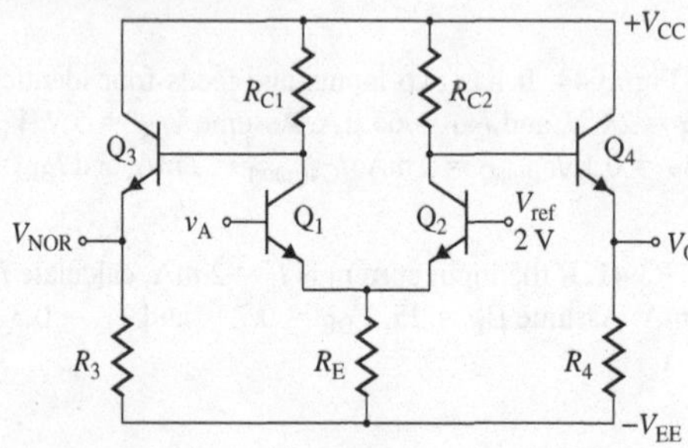

3.46 a. Design the ECL OR/NOR gate of the circuit in Fig. 3.50. It has two inputs. The desired collector currents are $I_{C1} = 2$ mA and $I_{C4} = I_{C5} = 1$ mA. At logic high, $V_{OR} = -0.7$ V, $V_{NOR} = -1.63$ V, and v_A (or v_B) $= V_{OR} = -0.7$ V. Assume $V_{EE} = -5.2$ V, $V_{BE} = 0.7$ V, and $\beta_F = 50$.

D P

b. Calculate the maximum fan-out with similar ECL gates if V_{OR} is allowed to fall from $V_{OR(max)} = -0.7$ V to $V_{OR(min)} = -0.75$ V.

c. From the VTC, calculate NM_L and NM_H.

3.13 BiCMOS Inverters

3.47 The BiCMOS inverter shown in Fig. 3.54 has $V_{DD} = 5$ V, $V_{BE} = 0.7$ V, $R_1 = 2$ kΩ, $R_2 = 6$ kΩ, $C_1 = 2$ pF, and $\beta_F = 150$. The threshold voltages are $V_{tP} = -1.5$ V, $V_{tN} = 1.5$ V, and $K_p = K_n = 40\ \mu A/V^2$. Assume $V_{OL} = 0$ and $V_{OH} = 5$ V.

a. Calculate t_{pLH}, t_{pHL}, and t_{pd}.

b. Use PSpice/SPICE to plot the VTC and transient response.

CHAPTER 4

INTEGRATED ANALOG CIRCUITS AND APPLICATIONS

Learning Outcomes

After completing this chapter, students should be able to do the following:

- Design and analyze precision rectifiers.
- Design and analyze nonlinear voltage limiters.
- List the differences between comparators and op-amps.
- Apply comparators and op-amps in generating waveforms such as the zero-crossing detector, Schmitt trigger, and square-wave, triangular-wave, and sawtooth-wave generators.
- Describe the internal structure and the principle of operation of commonly used analog-integrated circuits such as the NE/SE-566 VCO, the 555 timer, NE/SE-565 PLL, Teledyne 9400 V/F and F/V converters, LF198 sample-and-hold amplifier, MC1408 and NE/SE-5018 D/A converters, and NE/SE-5034 A/D converter.
- Apply some ICs and design circuits for waveform generation, conversion, detection, or signal processing.

Symbols and Their Meanings

Symbol	**Meaning**
A, A_f	Open-loop and closed-loop voltage gains
β	Feedback factor

Symbol	**Meaning**
f_o, T, k	Output frequency, period, and duty cycle
f_L	Lock frequency
i_S, i_f	Input signal and feedback currents
v_S, v_O	Input and output signal voltages
v_+, v_-	Noninverting and inverting terminal voltages
v_{st}, v_{CN}	Switching and control signal voltages
$v_{S(max)}$, $v_{O(max)}$	Maximum input and output voltages
V_L, V_H	Reference low and high voltages
V_{ref}, V_{sat}	Reference and saturation voltages
V_{Lt}, V_{Ht}	Low and high threshold voltages
$+V_{th}$, $-V_{th}$	Positive and negative threshold voltages
V_{CC}, V_{EE}	Positive and negative DC supply voltages

4.1 Introduction

Electronic circuits are commonly employed in generating waveforms of various shapes for control and interfacing purposes. This chapter examines the basic principles of operation of circuits used to generate waveforms and the applications of some commonly used ICs [1–3].

4.2 Circuits with Op-Amps and Diodes

Many applications—such as peak signal detectors, precision rectifiers, comparator circuits, and limiters—require nonlinear functions. A diode, however, typically has a finite voltage drop of 0.7 V, which distorts the output voltage of a circuit with one or more diodes, especially for low-voltage signals. Op-amp circuits with diodes can reduce the effect of diode drop and are used for precision signal processing.

Let us take the op-amp circuit shown in Fig. 4.1(a) in which the diode provides a unidirectional current flow. If the input voltage v_S is negative, the output voltage v_{O1} of the op-amp becomes negative, causing the diode to be reverse biased; no current flows through the load R_L because the current i_i flowing into the op-amp is zero. Thus, the output voltage is zero: $v_O = 0$ for $v_S \leq 0$. But the voltage v_{O1} will reach the negative saturation limit of the op-amp. If the input voltage v_S is positive, the output voltage v_{O1} of the op-amp becomes positive, causing the diode to be forward biased and supply the load current i_L. Since $v_d \approx 0$ and $i_i = 0$, the output voltage is $v_O = v_S$ for $v_S \geq 0$. The transfer characteristic of v_O versus v_S is shown in Fig. 4.1(b).

For the diode to start conduction, only a very small input voltage is required: $v_{S(min)} = V_D/A_o$, where V_D is the diode drop (typically 0.7 V) and A_o is the open-circuit gain of the op-amp (typically 2×10^5). This circuit exhibits the characteristic of a diode. However, the effect of the diode drop is negligible. Thus, this circuit is called a *superdiode*. The drawback of this circuit is that, for negative values of v_S, the voltage v_{O1} will swing to the negative saturation limit, thereby slowing the speed of op-amp operation.

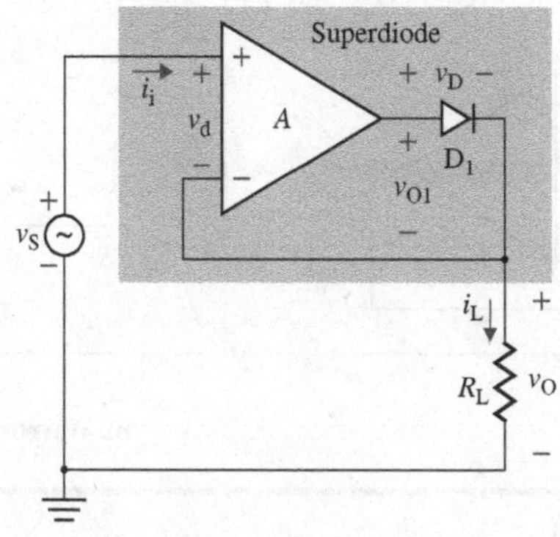

(a) Superdiode

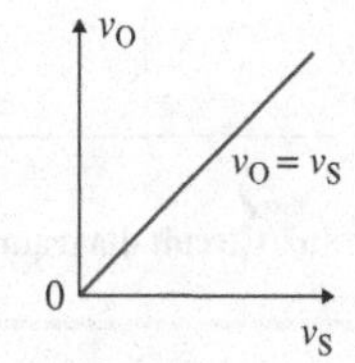

(b) Transfer characteristic

FIGURE 4.1 Superdiode

4.2.1 Most Positive Signal Detectors

The value of the most positive of a number of input signals can be detected by the circuit shown in Fig. 4.2. The inputs to the unity follower are the input signals to be detected. The diode that has the highest positive signal will conduct, and that signal will appear on the output of the circuit. The current source I_{DC} keeps the diode current constant irrespective of the value of the input signal, and it maintains a constant diode drop. As a result, the voltage drop across the conducting diode does not vary with variations in the input signal. Without the current source, the current and voltage of the conducting diode will change with the level of the most positive voltage, and the output voltage will vary. The unity follower acts as a buffer stage, providing a very high resistance at the diodes and a very low resistance at the output. If each diode is replaced by the superdiode shown in Fig. 4.1, there will be no voltage drop at the diodes; however, this arrangement will increase the complexity of the circuit.

4.2.2 Precision Peak Voltage Detectors

A precision peak voltage detector is used in many applications, such as monitoring the peak temperature of a day (or month) or initiating the action of a device if the signal exceeds a reference value. The circuit of Fig. 4.3 will detect the peak input voltage. The superdiode circuit allows the capacitor to charge to the

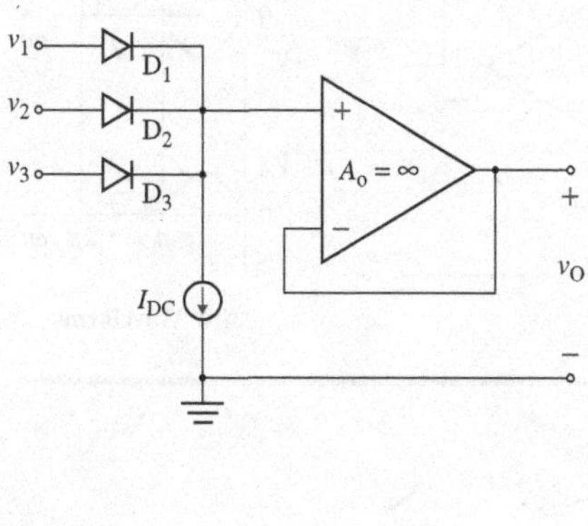

FIGURE 4.2 Positive signal detector

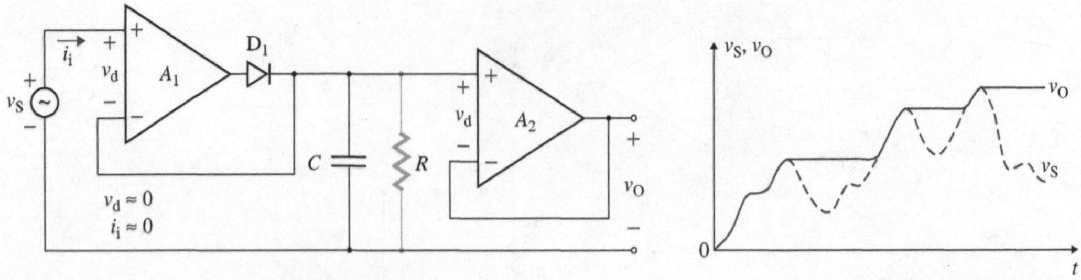

(a) Circuit diagram

(b) Waveforms for voltages

FIGURE 4.3 Precision peak voltage detector

peak input signal. The unity follower offers a high resistance to the capacitor and supplies the capacitor voltage to the output without discharging any charge of the capacitor. Resistance R, when connected, will allow the capacitor to discharge slowly so that the circuit can adjust to a lower input voltage. Otherwise, the capacitor will maintain its previous higher voltage and will not indicate the correct peak voltage.

4.2.3 Precision Half-Wave Rectifiers

A diode requires a minimum voltage, typically 0.7 V, to conduct. In a single-phase half-wave rectifier, one diode conducts. If the input voltage is less than 0.7 V, the output of the rectifier will be zero. Therefore, diode rectifiers are not suitable for rectification of low voltage. An op-amp circuit with two diodes, as shown in Fig. 4.4(a), can rectify a very small voltage in the range of microvolts. The circuit operation can be divided into two intervals: interval 1 and interval 2. We will consider the circuit operation with a sinusoidal input voltage $v_S = V_m \sin \omega t$.

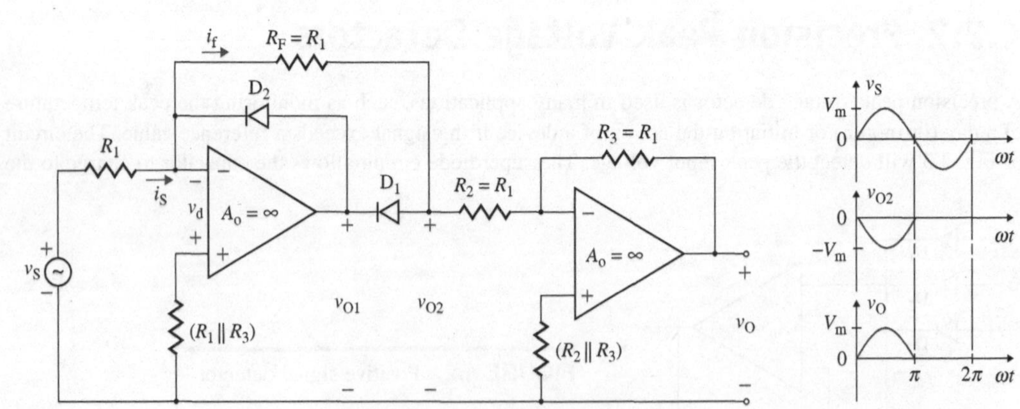

(a) Circuit

(b) Waveforms

FIGURE 4.4 Precision half-wave rectifier

During interval 1, $0 \le \omega t \le \pi$. The input voltage is positive. The voltage v_{O1} at the output of the first op-amp is negative, and diode D_2 is off. Diode D_1 conducts, and current i_f through R_F equals input current i_S. Since the current flowing into the op-amp is zero, $i_S = i_f$ and

$$i_S = i_f = \frac{v_S}{R_I}$$

$$v_d = R_F i_f + v_{O2}$$

For an ideal op-amp, $v_d = 0$, and voltage v_{O2} becomes

$$v_{O2} = -R_F i_f = -R_I i_F = -R_I i_S \quad (\text{for } R_F = R_I)$$

$$= -v_S$$

Thus, the output voltage at the output of the second inverting op-amp is

$$v_O = -v_{O2} = v_S \quad (\text{for } v_S \ge 0, R_3 = R_2 = R_1, \text{ and } v_S \ge 0)$$

During interval 2, $\pi \le \omega t \le 2\pi$. The input voltage is negative. The voltage v_{O1} at the output of the first op-amp is positive, and diode D_2 conducts. As a result, the voltage v_{O1} is clamped to approximately the voltage of one diode. Diode D_1 remains off, and no current flows through R_F. The voltage v_{O2} becomes zero. Thus, the output voltage at the output of the second op-amp is $v_O = -v_{O2} = 0$.

The output voltage v_O is almost independent of the diode characteristics. This is because diode D_1 is included in series with the op-amp. Since the op-amp gain is very high, tending to infinity, the effect of the diode and its voltage drop becomes insignificant. The output waveforms are shown in Fig. 4.4(b).

If the directions of the diodes are reversed, the output voltage will correspond to the negative part of the input voltage; there is no need for the second inverting op-amp. This arrangement is shown in Fig. 4.5(a). When the input voltage v_S is positive, the voltage v_{O1} becomes negative, making diode D_2 conduct and diode D_1 turn off. As a result, the output voltage becomes zero; that is, the output voltage $v_O = 0$ for $v_S \ge 0$. On the other hand, when the input voltage v_S is negative, the voltage v_{O1} becomes positive, making diode D_2 turn off and diode D_1 conduct. As a result, the output voltage will be equal and opposite to the input voltage; that is, the output voltage $v_O = -v_S$ for $v_S < 0$. The transfer characteristic is shown in Fig. 4.5(b), and the voltage waveforms are shown in Fig. 4.5(c).

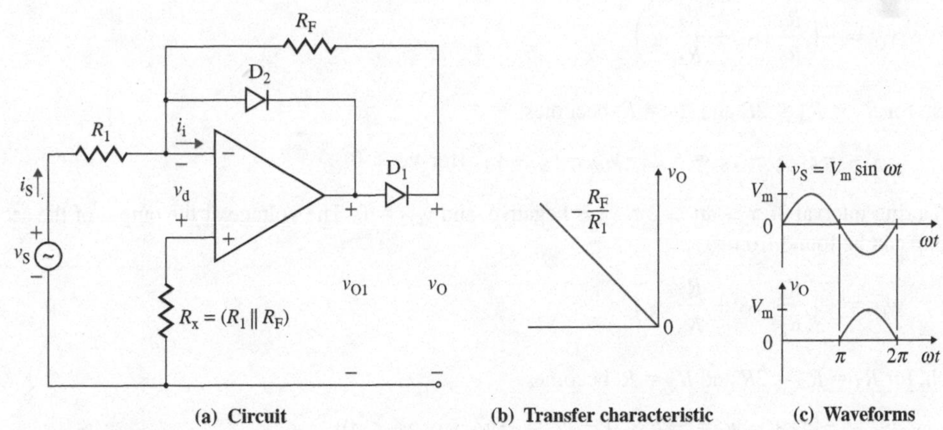

FIGURE 4.5 Alternate precision half-wave rectifier

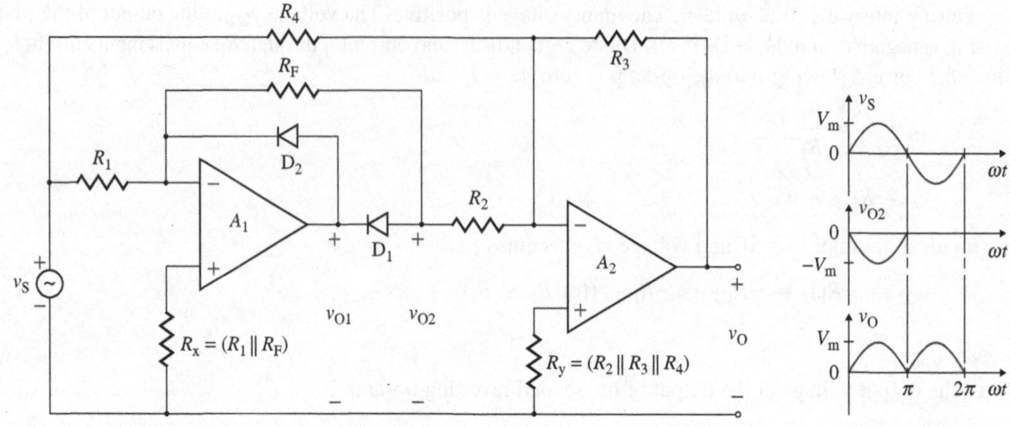

(a) Circuit (b) Waveforms

FIGURE 4.6 Precision full-wave rectifier

4.2.4 Precision Full-Wave Rectifiers

The half-wave rectifier of Fig. 4.5(a) can be modified to operate as a precision full-wave rectifier if we use the following algebraic relationship:

$$v_O = 2v_S - v_S = v_S \quad \text{(for the positive interval of the input voltage)}$$

This situation is shown in Fig. 4.6(a). Let us consider the case with $R_1 = R_2 = R_F = R$ and $R_3 = R_4 = 2R$. We will divide the circuit operation into two intervals, interval 1 and interval 2, and use a sinusoidal input voltage $v_S = V_m \sin \omega t$.

During interval 1, $0 \leq \omega t \leq \pi$. v_S is positive, and $v_{O2} = -v_S$. The voltage at the output of the second op-amp can be found from

$$v_O = -\left(\frac{R_3}{R_2} v_{O2} + \frac{R_3}{R_4} v_S\right) \tag{4.1}$$

which, for $R_3 = R_4 = 2R$ and $R_2 = R$, becomes

$$v_O = -2v_{O2} - v_S = -2(-v_S) - v_S = v_S \quad (\text{for } v_S \geq 0)$$

During interval 2, $\pi \leq \omega t \leq 2\pi$. v_S is negative, and $v_{O2} = 0$. The voltage at the output of the second op-amp can be found from

$$v_O = -\left(\frac{R_3}{R_2} v_{O2} + \frac{R_3}{R_4} v_S\right) \tag{4.2}$$

which, for $R_3 = R_4 = 2R$ and $R_2 = R$, becomes

$$v_O = -2v_{O2} - v_S = -2 \times 0 - v_S = -v_S \quad (\text{for } v_S < 0)$$

Thus, the output voltage is the inverted version of the input voltage, as shown in Fig. 4.6(b).

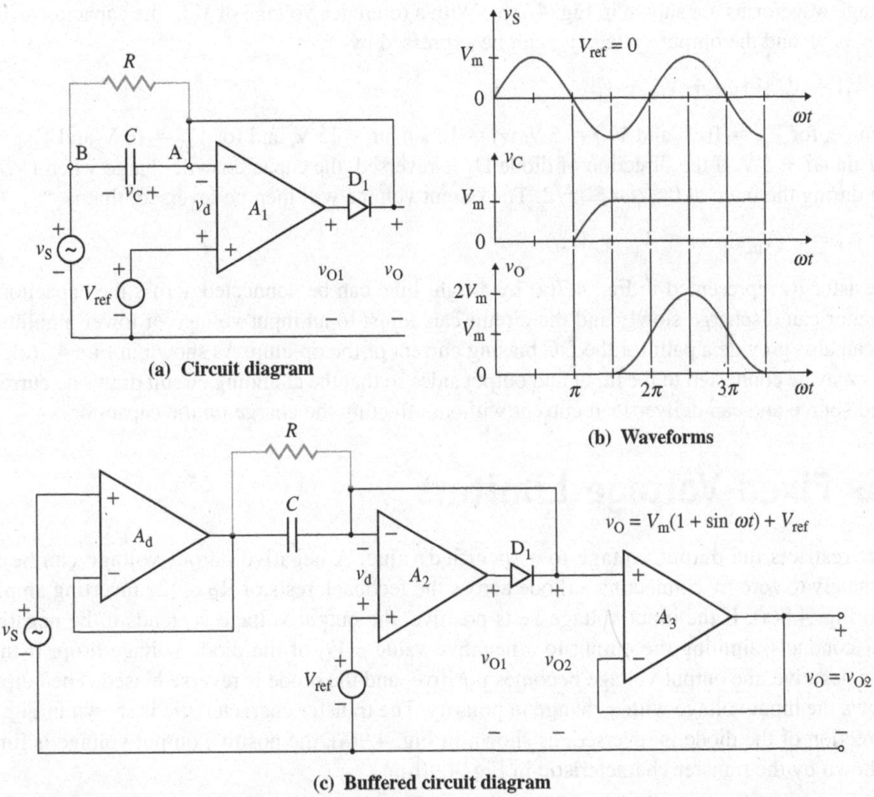

FIGURE 4.7 Precision clamping circuit

4.2.5 Precision Clamping Circuits

A precision clamping circuit is sometimes used in signal processing to add just enough DC voltage to the input voltage so that the sum never crosses the zero level. Figure 4.7(a) shows a superdiode circuit with an input voltage v_S and a DC reference voltage V_{ref}. As long as the op-amp voltage v_d is positive, v_{O1} will be positive, and diode D_1 will conduct, thereby making the (−) terminal behave as a virtual ground. To consider the circuit operation, let us take a sinusoidal input voltage $v_S = V_m \sin \omega t$ and $V_{ref} = 0$.

During the interval $0 \leq \omega t \leq \pi$, v_S is positive, v_d is negative, v_{O1} is negative, and diode D_1 is reverse biased. No current flows through the capacitor, and it cannot charge. During the interval $\pi < \omega t \leq 3\pi/2$, v_S is negative, v_d is positive, v_{O1} is positive, and diode D_1 conducts. The (−) terminal behaves as a virtual ground. A charging current flows through the capacitor, with point A being at a higher potential than point B. The capacitor is charged to the peak negative voltage V_m. For the interval $\omega t > 3\pi/2$,

$$v_d = -(V_m + v_S) = -V_m(1 + \sin \omega t)$$

and diode D_1 remains off; that is, the output voltage v_O is

$$v_O = -v_d = V_m(1 + \sin \omega t)$$

The voltage waveforms are shown in Fig. 4.7(b). With a reference voltage of V_{ref}, the capacitor will charge to $(V_m + V_{ref})$, and the output voltage v_O can be expressed by

$$v_O = V_m \sin \omega t + V_m + V_{ref} \tag{4.3}$$

For example, for $V_m = 10$ V and $V_{ref} = 5$ V, $v_O = 10 \sin \omega t + 15$ V, and for $V_m = 10$ V and $V_{ref} = -5$ V, $v_O = 10 \sin \omega t + 5$ V. If the direction of diode D_1 is reversed, the capacitor will charge when v_d becomes negative during the interval $0 \leq \omega t \leq \pi/2$. The output voltage will then be reversed; that is,

$$v_O = -(V_m \sin \omega t + V_m + V_{ref})$$

A resistor R, represented in Fig. 4.7(a) by a light line, can be connected across the capacitor so that the capacitor can discharge slowly and the circuit can adjust to an input voltage of lower amplitude. The resistor can also provide a path for the DC biasing current of the op-amp. As shown in Fig. 4.7(c), voltage followers may be connected to the input and output sides so that the clamping circuit draws no current from the signal source and can deliver load current without affecting the charge on the capacitor.

4.2.6 Fixed-Voltage Limiters

A limiter restricts the output voltage to a specified value. A negative output voltage can be limited approximately to zero by connecting a diode across the feedback resistor R_F of the inverting amplifier as shown in Fig. 4.8(a). If the input voltage v_S is positive, the output voltage v_O tends to be negative, and diode D_1 conducts, limiting the output to a negative value $-V_D$ of the diode voltage drop. If the input voltage is negative, the output voltage becomes positive, and the diode is reverse biased. The output voltage follows the input voltage with a change in polarity. The transfer characteristic is shown in Fig. 4.8(b). If the direction of the diode is reversed, as shown in Fig. 4.9(a), the positive output voltage is limited to V_D, as shown by the transfer characteristic in Fig. 4.9(b).

4.2.7 Adjustable Voltage Limiters

Output voltage can be limited to an adjustable level by choosing resistors with appropriate values. A circuit that limits the negative output voltage is shown in Fig. 4.10(a). If the output is positive, diode D_1 is reverse biased, and the circuit operates as an inverting amplifier. If the output voltage is negative, its transfer

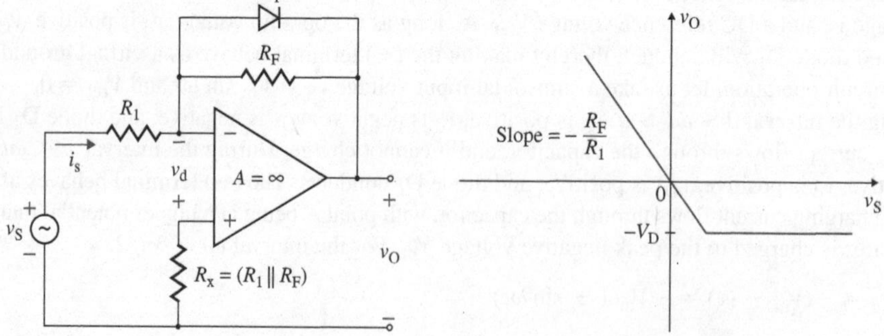

(a) Circuit (b) Transfer characteristic

FIGURE 4.8 Negative voltage limiter

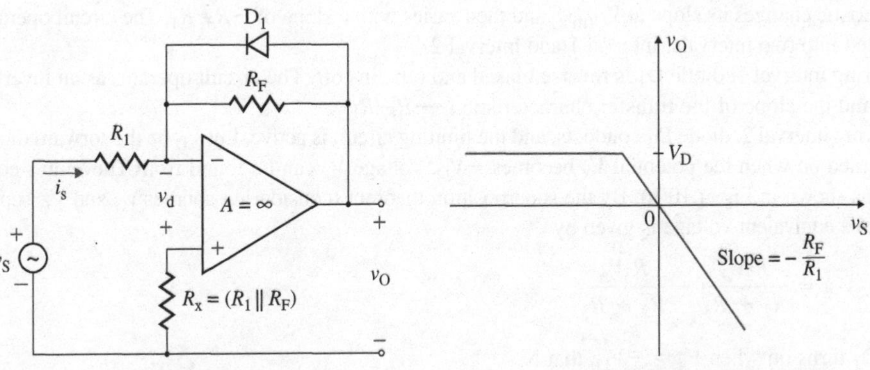

(a) Circuit (b) Transfer characteristic

FIGURE 4.9 Positive voltage limiter

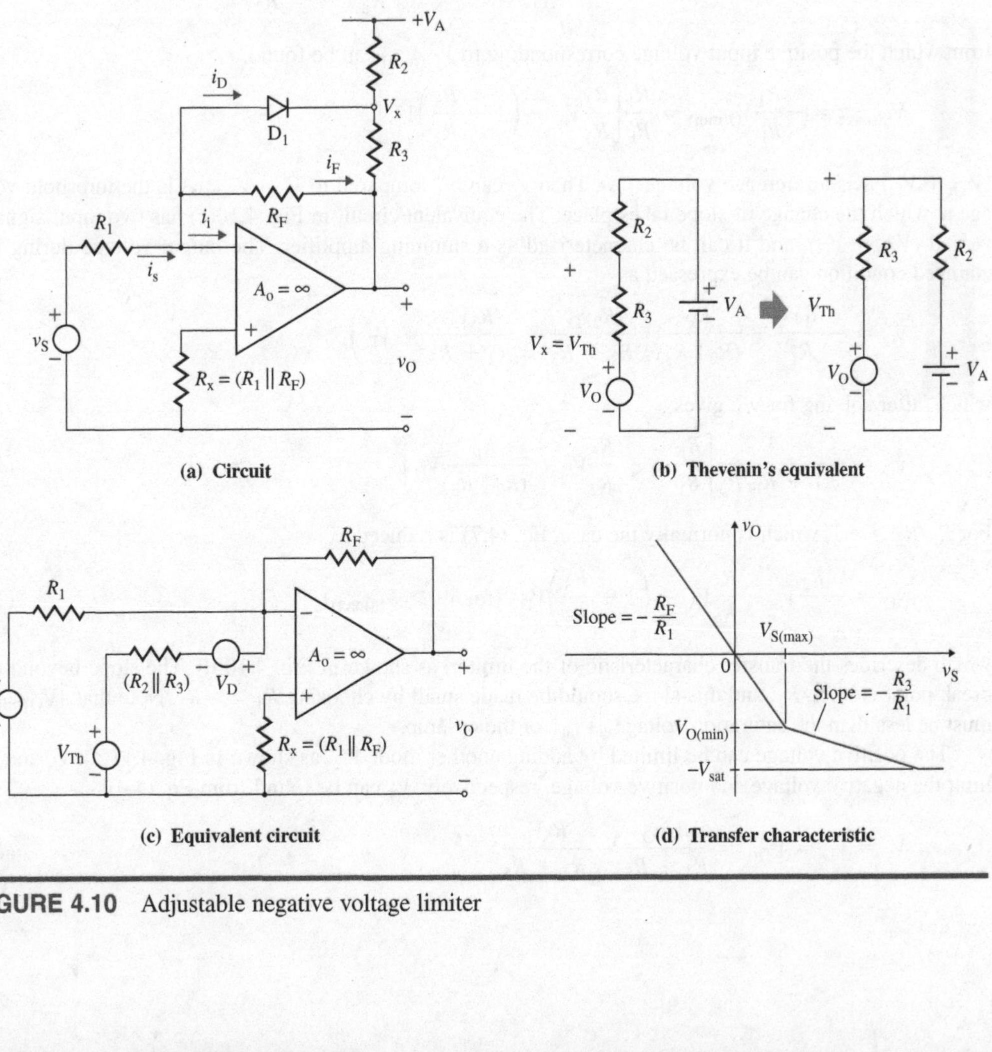

(c) Equivalent circuit (d) Transfer characteristic

FIGURE 4.10 Adjustable negative voltage limiter

characteristic changes its slope at $V_{O(min)}$ and then varies with a slope of $-R_3/R_1$. The circuit operation can be divided into two intervals: interval 1 and interval 2.

During interval 1, diode D_1 is reverse biased and remains off. The circuit operates as an inverting amplifier, and the slope of the transfer characteristic is $-R_F/R_1$.

During interval 2, diode D_1 conducts, and the limiting circuit is active. Let V_D be the forward diode drop. D_1 is turned on when the potential V_x becomes $-V_D$. Voltage V_x can be found from Thevenin's equivalent circuit, as shown in Fig. 4.10(b). By the superposition theorem (considering sources v_O and V_A separately), Thevenin's equivalent voltage is given by

$$V_{Th} = \frac{R_2 v_O}{R_2 + R_3} + \frac{R_3 V_A}{R_2 + R_3}$$

Diode D_1 turns on when $V_x = -V_D$; that is,

$$V_x = -V_D = V_{Th} = \frac{R_2 v_O}{R_2 + R_3} + \frac{R_3 V_A}{R_2 + R_3} \tag{4.4}$$

which gives the negative clamping output voltage $V_{O(min)}$ as

$$V_{O(min)} = v_O = -V_D - (V_A + V_D)\frac{R_3}{R_2} = -V_D\left(1 + \frac{R_3}{R_2}\right) - V_A\frac{R_3}{R_2} \tag{4.5}$$

from which the positive input voltage corresponding to $V_{O(min)}$ can be found:

$$V_{S(max)} = -\frac{R_1}{R_F}V_{O(min)} = \frac{R_1}{R_F}\left[\frac{R_3}{R_2}V_A + \left(1 + \frac{R_3}{R_2}\right)V_D\right] \tag{4.6}$$

$(V_{Th} + V_D)$ acts a reference voltage V_{ref}. Then v_S can be compared to V_{ref}. $V_{S(max)}$ is the threshold voltage at which the change of slope takes place. The equivalent circuit in Fig. 4.10(c) has two input signals, v_S and $(V_{Th} + V_D)$, and it can be characterized as a summing amplifier. The output voltage during the clamped condition can be expressed as

$$v_O = -\frac{R_F v_S}{R_1} - \frac{R_F}{(R_2 \parallel R_3)}\left(\frac{R_2 v_O}{R_2 + R_3} + \frac{R_3 V_A}{R_2 + R_3} + V_D\right)$$

which, after solving for v_O, gives

$$v_O = -\frac{1}{1 + R_F/R_3}\left[\frac{R_F}{R_1}v_S + \frac{R_F}{R_2}V_A + \frac{R_F}{(R_2 \parallel R_3)}V_D\right] \tag{4.7}$$

For $R_F/R_3 >> 1$, which is normally the case, Eq. (4.7) is reduced to

$$v_O = -\frac{R_3}{R_1}v_S - \frac{R_3}{R_2}V_A - \left(1 + \frac{R_3}{R_2}\right)V_D \quad (\text{for } v_S > V_{S(max)}) \tag{4.8}$$

which describes the transfer characteristic of the limiter, as shown in Fig. 4.10(d). The slope beyond the break point is $-R_3/R_1$, and this slope should be made small by choosing $R_1 >> R_3$. Note that $|V_{O(min)}|$ must be less than the saturation voltage $|V_{sat}|$ of the op-amp.

The positive voltage can be limited by adding another diode D_2, as shown in Fig. 4.11(a). V_x and V_y limit the negative voltage and positive voltage, respectively. V_y can be found from Eq. (4.4):

$$V_y = V_D = V_{Th1} = \frac{R_4 v_O}{R_4 + R_5} - \frac{R_5 V_B}{R_4 + R_5} \tag{4.9}$$

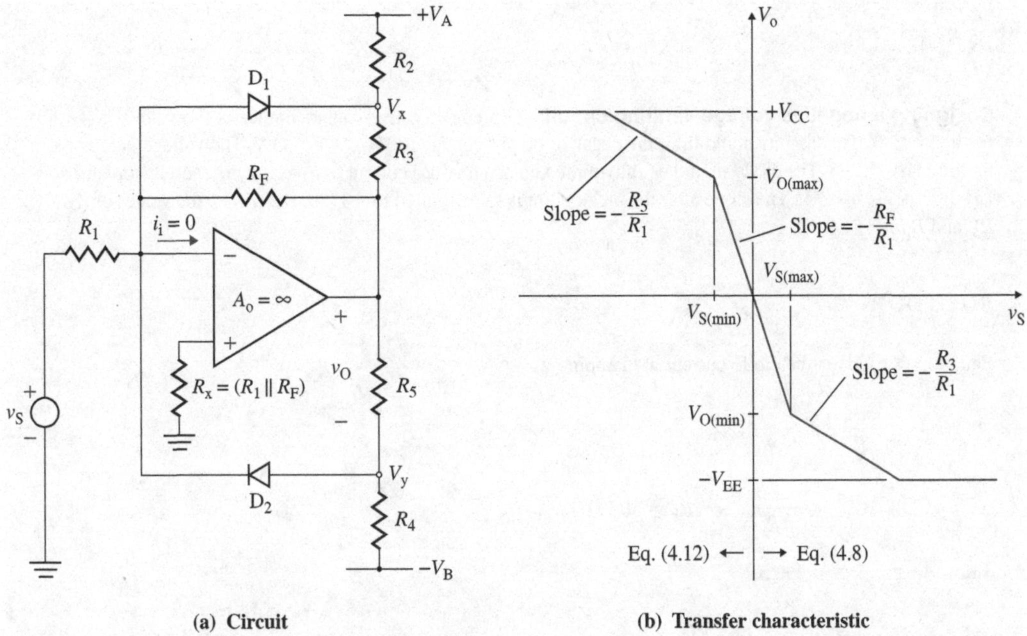

FIGURE 4.11 Adjustable positive and negative voltage limiter

Similarly, the positive clamping voltage can be found from Eq. (4.5):

$$V_{O(max)} = v_O = V_D + (V_B + V_D)\frac{R_5}{R_4} = V_D\left(1 + \frac{R_5}{R_4}\right) + V_B\frac{R_5}{R_4} \tag{4.10}$$

The negative input voltage corresponding to $V_{O(max)}$ can be found from

$$V_{S(min)} = -\frac{R_1}{R_F}V_{O(max)} = -\frac{R_1}{R_F}\left[\frac{R_5}{R_4}V_B + \left(1 + \frac{R_5}{R_4}\right)V_D\right] \tag{4.11}$$

If V_{Th1} is Thevenin's equivalent voltage due to V_B, $(V_{Th1} - V_D)$ acts as reference voltage V_{ref1}. Then v_S can be compared to V_{ref1}. $V_{S(min)}$ is the threshold voltage at which the change of slope takes place. Using Eq. (4.7), we can find the output voltage during the positive voltage clamping to be

$$v_O = -\frac{R_5}{R_1}v_S + \frac{R_5}{R_4}V_B + \left(1 + \frac{R_5}{R_4}\right)V_D \quad (\text{for } v_S < V_{S(min)}) \tag{4.12}$$

which describes the transfer characteristic of an adjustable positive and negative voltage limiter, as shown in Fig. 4.11(b). This is a practical limiter and is commonly used. It is also called a *soft limiter* since the output voltage will increase slightly if the input voltage is increased beyond the break points. All DC supply voltages of the limiter are generally made the same magnitude; that is, $V_A = V_B = V_{CC} = V_{EE}$.

EXAMPLE 4.1

D

Designing a negative voltage–limiting circuit Design a negative voltage limiter as shown in Fig. 4.10(a) for $V_A = 12$ V. The circuit should limit the negative output voltage to $V_{O(min)} = -5$ V. The voltage gain without limiting is $A_f = -4$. The diode should be fully turned on at a forward current of $i_D = 0.1$ mA, and its forward voltage drop is $V_D = 0.7$ V. The slope after the break point is to be limited to $-1/20$. Determine the values of R_1, R_2, R_3, and R_F.

SOLUTION

Equation (4.6) gives the diode current at clamping as

$$i_D = \frac{V_{S(max)}}{R_1} = -\frac{1}{R_F}V_{O(min)}$$

so $$0.1 \times 10^{-3} = -\frac{-5}{R_F} \quad \text{or} \quad R_F = 50 \text{ k}\Omega$$

Since $A_f = -4 = -R_F/R_1$,

$$R_1 = \frac{R_F}{4} = \frac{50 \text{ k}}{4} = 12.5 \text{ k}\Omega$$

Since slope $S = -R_3/R_1 = -1/20$,

$$R_3 = \frac{R_1}{20} = \frac{12.5 \text{ k}}{20} = 625 \ \Omega$$

From Eq. (4.5),

$$-5 = -0.7 - (12 + 0.7)\frac{R_3}{R_2}$$

$$\frac{R_3}{R_2} = 0.3386$$

$$R_2 = \frac{R_3}{0.3386} = \frac{625}{0.3386} = 1846 \ \Omega$$

EXAMPLE 4.2

Finding the limiting voltages of an op-amp limiting circuit The adjustable limiter in Fig. 4.11(a) has $R_1 = 15$ kΩ, $R_F = 60$ kΩ, $R_2 = 4$ kΩ, $R_3 = 1$ kΩ, $R_4 = 5$ kΩ, $R_5 = 1$ kΩ, $V_A = 15$ V, $-V_B = -15$ V, and $V_D = 0.7$ V. Determine **(a)** the positive clamping voltage $V_{O(max)}$ and the corresponding input voltage $V_{S(min)}$, **(b)** the negative clamping voltage $V_{O(min)}$ and the corresponding input voltage $V_{S(max)}$, and **(c)** the output voltage when the input voltage is $v_S = 5$ V.

SOLUTION

$R_1 = 15\ \text{k}\Omega$, $R_F = 60\ \text{k}\Omega$, $R_2 = 4\ \text{k}\Omega$, $R_3 = 1\ \text{k}\Omega$, $R_4 = 5\ \text{k}\Omega$, $R_5 = 1\ \text{k}\Omega$, $V_A = 15$ V, $-V_B = -15$ V, $V_D = 0.7$ V, and $v_S = 5$ V.

(a) From Eq. (4.10),

$$V_{O(max)} = 0.7 + (15 + 0.7) \times \frac{1\ \text{k}}{5\ \text{k}} = 3.84\ \text{V}$$

From Eq. (4.11),

$$V_{S(min)} = -V_{O(max)} \frac{R_1}{R_F} = -3.84 \times \frac{15\text{k}}{60\ \text{k}} = -0.96\ \text{V}$$

(b) From Eq. (4.5),

$$V_{O(min)} = -0.7 - (15 + 0.7) \times \frac{1\ \text{k}}{4\ \text{k}} = -4.625\ \text{V}$$

From Eq. (4.6),

$$V_{S(max)} = -V_{O(min)} \frac{R_1}{R_F} = 4.625 \times \frac{15\ \text{k}}{60\ \text{k}} = 1.15625\ \text{V}$$

(c) From Eq. (4.8),

$$v_O = -5 \times \frac{1\ \text{k}}{15\ \text{k}} - 15 \times \frac{1\ \text{k}}{4\ \text{k}} - \left(1 + \frac{1\ \text{k}}{4\ \text{k}}\right) \times 0.7 = -4.958\ \text{V}$$

EXAMPLE 4.3

D

Designing an adjustable voltage-limiting circuit

(a) Design an adjustable limiter as shown in Fig. 4.11(a) to satisfy the following specifications: $V_A = 12$ V, $-V_B = -12$ V, $V_{O(min)} = -5$ V, $V_{O(max)} = 6$ V, voltage gain $A_f = -4$, slope after break with a positive input (for $v_S > 0$) $= S_1 = -1/20$, and slope after break with a negative input (for $v_S < 0$) $= S_2 = -1/25$. Assume that diodes are fully on at a diode current of $i_D = 0.1$ mA and the corresponding on-state diode voltage is $V_D = 0.7$ V.

(b) Use PSpice/SPICE [4] to plot the transfer characteristic v_O versus $v_S = V(6)$. Assume $V_{CC} = 12$ V and $-V_{EE} = -12$ V.

SOLUTION

(a) The steps in completing the design are as follows:

Step 1. The diode current can be related to $V_{O(min)}$ and R_F by

$$i_D = -\frac{V_{O(min)}}{R_F} \quad \text{or} \quad 0.1\ \text{mA} = -\frac{-5}{R_F}$$

which gives $R_F = 50\ \text{k}\Omega$.

Step 2. The voltage gain A_f is given by

$$A_f = -\frac{R_F}{R_1} \quad \text{or} \quad -4 = -\frac{50\ \text{k}\Omega}{R_1}$$

which gives $R_1 = 12.5\ \text{k}\Omega$.

Step 3. The slope after break with a positive input (for $v_S > 0$) is

$$S_1 = -\frac{R_3}{R_1} \quad \text{or} \quad -\frac{1}{20} = -\frac{R_3}{12.5\ \text{k}\Omega}$$

which gives $R_3 = 625\ \Omega$.

Step 4. The slope after break with a negative input (for $v_S < 0$) is

$$S_2 = -\frac{R_5}{R_1} \quad \text{or} \quad -\frac{1}{25} = -\frac{R_5}{12.5\ \text{k}\Omega}$$

which gives $R_5 = 500\ \Omega$.

Step 5. The value of R_2 can be found from Eq. (4.5):

$$V_{O(\min)} = -V_D - (V_A + V_D)\frac{R_3}{R_2} \quad \text{or} \quad -5 = -0.7 - (12 + 0.7)\frac{625}{R_2}$$

which gives $R_2 = 1846\ \Omega$.

Step 6. The value of R_4 can be found from Eq. (4.10):

$$V_{O(\max)} = V_D + (V_B + V_D)\frac{R_5}{R_4} \quad \text{or} \quad 6 = 0.7 + (12 + 0.7)\frac{500}{R_4}$$

which gives $R_4 = 1198\ \Omega$.

(b) The limiter circuit for PSpice simulation is shown in Fig. 4.12. The transfer characteristic (for voltage-controlled voltage source model) is shown in Fig. 4.13, which gives $V_{S(\max)} = 1.41$ V, $V_{S(\min)} = -1.72$ V, $V_{O(\max)} = 6.7$ V (expected value is 6 V), and $V_{O(\min)} = -5.5$ V (expected value is -5 V). The discrepancies between the design values and the PSpice results may be attributed to the use of an ideal op-amp versus a macromodel and the use of a specific diode versus a diode of type D1N4148. If we run the simulation using voltage-controlled voltage source model, the PSpice and calculated values will agree more closely.

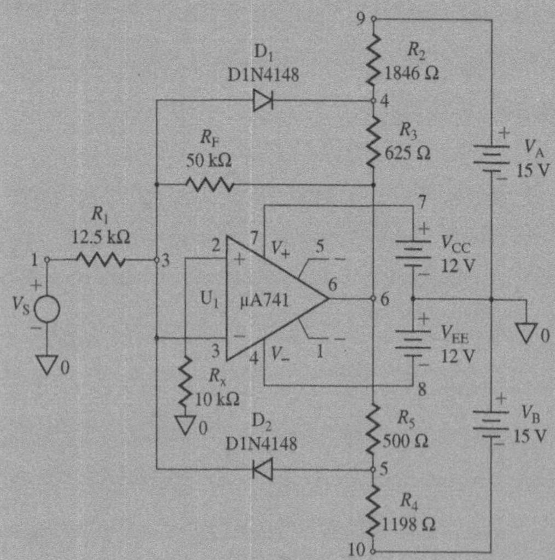

FIGURE 4.12 Limiter circuit for PSpice simulation

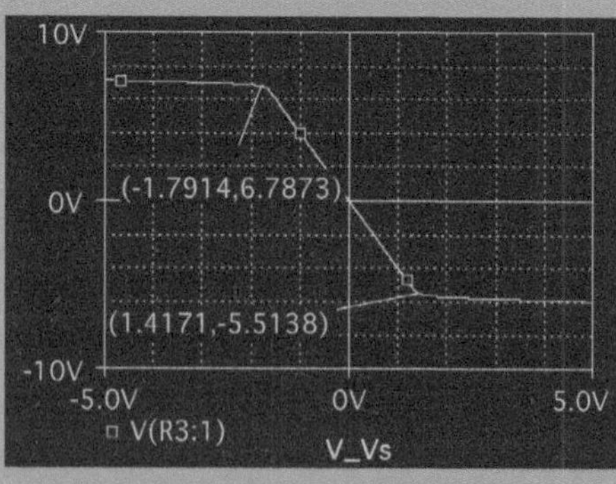

FIGURE 4.13 PSpice transfer characteristic for Example 4.3

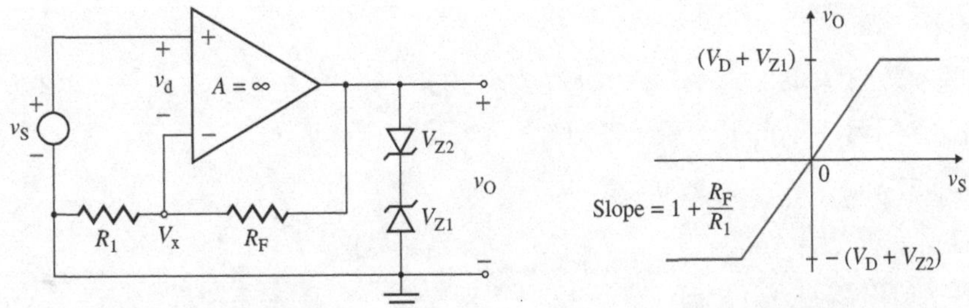

(a) Circuit (b) Transfer characteristic

FIGURE 4.14 Output voltage–clamping circuit with zener diodes

4.2.8 Zener Voltage Limiters

The output can be limited by adding two zener diodes across the output terminals of a noninverting or an inverting amplifier. An arrangement for a noninverting amplifier circuit is shown in Fig. 4.14(a). Zener diodes limit the output voltage between $-(V_{Z2} + V_D)$ and $(V_{Z1} + V_D)$, where V_D is the voltage drop of a zener diode in the forward direction.

$$v_O = \left(1 + \frac{R_F}{R_1}\right) v_S$$

If the output voltage is to be limited to a maximum voltage $V_{O(max)} = v_O = (V_{Z1} + V_D)$ and a minimum voltage $V_{O(min)} = v_O = -(V_{Z2} + V_D)$, it can be expressed as

$$v_O = \begin{cases} V_{Z1} + V_D = V_{O(max)} & (\text{for } v_S > V_{S(max)}) \\ -(V_{Z2} + V_D) = V_{O(min)} & (\text{for } v_S < V_{S(min)}) \\ \left(1 + \dfrac{R_F}{R_1}\right) v_S & (\text{for } V_{S(min)} \le v_S \le V_{S(min)}) \end{cases}$$

The transfer characteristic is shown in Fig. 4.14(b).

EXAMPLE 4.4

Designing a zener voltage–clamping circuit Design an output voltage–clamping circuit as shown in Fig. 4.14(a) so that the normal slope of the transfer characteristic is $S = v_O/v_S = 10$, $V_{O(max)} = 5.7$ V, and $V_{O(min)} = -7.7$ V. Determine the zener voltages V_{Z1} and V_{Z2}. Assume $V_D = 0.7$ V.

SOLUTION

$V_{O(max)} = 5.7$ V, and $V_{O(min)} = -7.7$ V. Since $S = 1 + R_F/R_1 = 10$,

$$\frac{R_F}{R_1} = 10 - 1 = 9$$

If $R_1 = 5\ \text{k}\Omega$,

$$R_F = 9 \times 5\ \text{k} = 45\ \text{k}\Omega$$

Since $V_{O(max)} = (V_{Z1} + V_D)$,

$$V_{Z1} = V_{O(max)} - V_D = 5.7 - 0.7 = 5\ \text{V}$$

Since $V_{O(min)} = -(V_{Z2} + V_D)$,

$$V_{Z2} = -(V_{O(min)} + V_D) = -(-7.7 + 0.7) = 7\ \text{V}$$

4.2.9 Hard Limiters

The limiter in Fig. 4.11(a) can be made to switch between $V_{O(min)}$ and $V_{O(max)}$ if the input voltage v_S becomes greater than zero or less than zero, respectively. This can be accomplished by making the feedback resistance R_F in Fig. 4.11(a) very large, tending to infinity. That is, for $R_F = \infty$, the gain of the circuit becomes very large, tending to infinity, and the output voltage becomes $V_{O(min)}$ if $v_S > 0$ and $V_{O(max)}$ if $v_S < 0$. The circuit shown in Fig. 4.15(a) compares the input signal v_S with the reference signal $V_{ref} = 0$, and if the input signal is greater than or less than zero, the output changes its value. This circuit is known as a *hard limiter*, although it exhibits finite slopes beyond the breaks.

The equations given earlier for adjustable voltage limiters can be applied to the limiter in Fig. 4.15(a), except that $R_F = \infty$ and the gain $A_f = -R_F/R_1$ is very large, tending to infinity. The transfer characteristic is shown in Fig. 4.15(b). The voltage limiting can also be accomplished by connecting two zener diodes, as shown in Fig. 4.16(a), for which $V_{O(max)} = (V_{Z1} + V_D)$ and $V_{O(min)} = -(V_{Z2} + V_D)$. Practical circuits exhibit finite slopes beyond the breaks because of the finite resistances of the zener diodes. These finite slopes are represented in Fig. 4.16(b) by light lines.

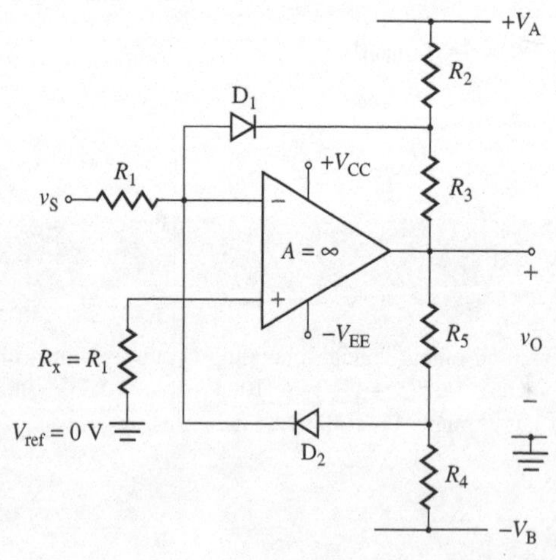

(a) Circuit

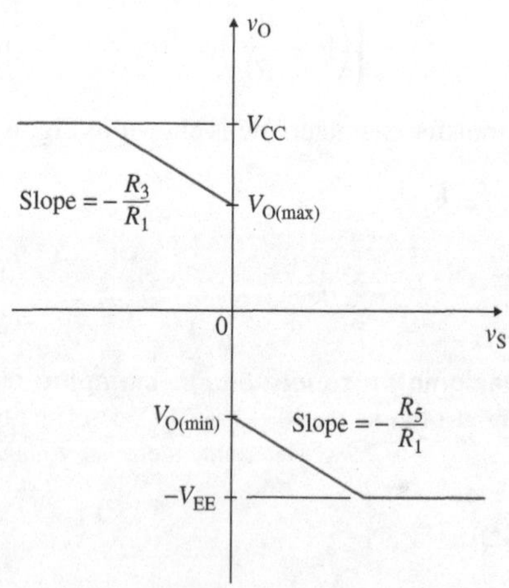

(b) Transfer characteristic

FIGURE 4.15 Hard limiter

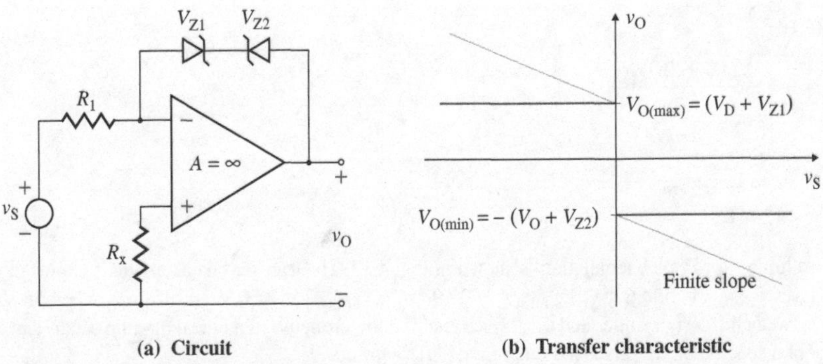

FIGURE 4.16 Zener hard limiter

EXAMPLE 4.5

D

Designing a hard voltage limiter

(a) Design a hard limiter like the one in Fig. 4.15(a). The circuit should limit the negative output voltage to $V_{O(min)} = -5$ V and the positive output voltage to $V_{O(max)} = 5$ V. The magnitude of the slopes after the break points should be less than or equal to 1/20. The diode drop is $V_D = 0.7$ V. The DC supplies are given by $V_A = V_B = 12$ V. Determine the values of R_1, R_2, R_3, R_4, and R_5.

(b) Use PSpice/SPICE to plot the transfer characteristic. Assume $V_{CC} = 12$ V, $-V_{EE} = -12$ V, and $v_S = -6$ V to 6 V.

SOLUTION

(a) Choose $R_1 = 10$ kΩ. Since the slope $|S_1| = R_3/R_1 = 1/20$,

$$R_3 = \frac{R_1}{20} = \frac{10\text{ k}\Omega}{20} = 500\ \Omega$$

From Eq. (4.5),

$$-5 = -0.7 - \frac{(12 + 0.7)R_3}{R_2}$$

$$\frac{R_3}{R_2} = 0.3386$$

$$R_2 = 1477\ \Omega$$

Since the slope $|S_2| = R_5/R_1 = 1/20$,

$$R_5 = \frac{R_1}{20} = \frac{10\text{ k}\Omega}{20} = 500\ \Omega$$

From Eq. (4.10),

$$5 = 0.7 + \frac{(12 + 0.7)R_5}{R_4}$$

$$\frac{R_5}{R_4} = 0.3386$$

$$R_4 = 1477\ \Omega$$

(b) The hard limiter for PSpice simulation is shown in Fig. 4.17. The transfer characteristic is shown in Fig. 4.18, which gives $V_{S(max)} = 55.86$ mV, $V_{S(min)} = -27.93$ mV, $V_{O(max)} = 4.50$ V, and $V_{O(min)} = -4.55$ V. Discrepancies between the design values and the PSpice results come from use of an ideal op-amp versus a macromodel and use of a specific diode versus a diode of type D1N4148.

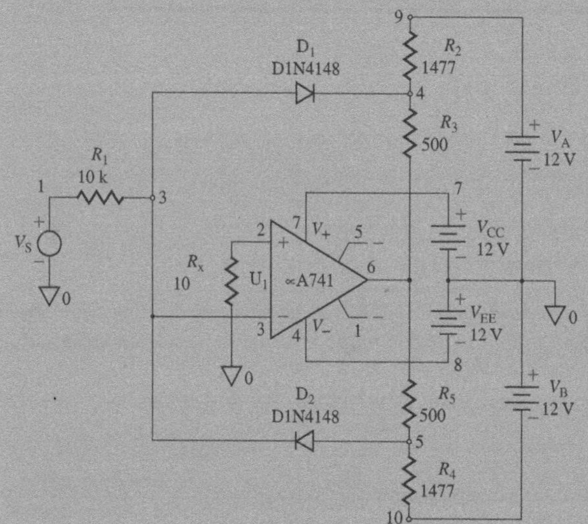

FIGURE 4.17 Comparator circuit for PSpice simulation

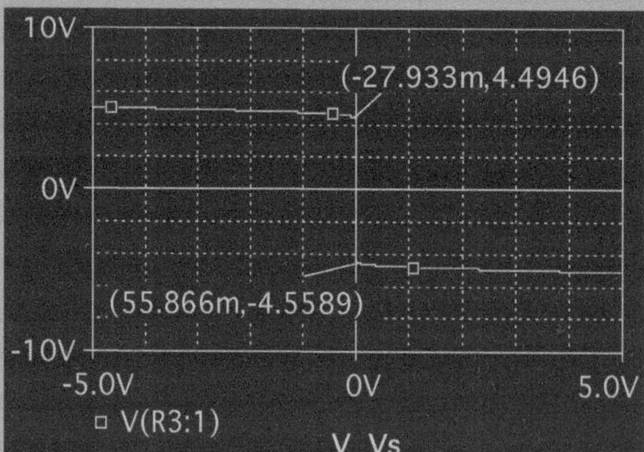

FIGURE 4.18 PSpice transfer characteristic for Example 4.5

KEY POINTS OF SECTION 4.2

- A diode has a voltage drop, typically 0.7 V, which affects the waveforms of voltage and currents in a circuit. However, when a diode is placed inside the feedback path of an op-amp circuit, the effective voltage drop becomes negligible, on the order of microvolts. The diode is then called a superdiode. Superdiodes are generally used for precision signal processing.
- The output voltage of an op-amp circuit can be clamped to a certain level by connecting a diode in parallel with the feedback resistance R_F. If we make $R_F = \infty$, the voltage gain $A_f = \infty$ and the output voltage swings from V_{sat} to $-V_{sat}$ as the input voltage v_S crosses zero.

4.3 Comparators

A comparator compares a signal voltage v_S on one input terminal with a known voltage, called the *reference voltage* V_{ref}, on the other input terminal. The symbol of a comparator, which is similar to that of an op-amp, is shown in Fig. 4.19(a). A comparator gives a digital output voltage v_O. Thus, it can be considered a simple one-bit analog-to-digital (A/D) converter, which produces a digital 1 output ($v_O = V_H$) if the input voltage v_S is above the reference level V_{ref} and a digital 0 output ($v_O = V_L$) if the input voltage v_S falls below the reference level V_{ref}. The output levels V_L and V_H may be of opposite polarity (i.e., V_H positive and V_L negative or vice versa), or both V_L and V_H may be either positive or negative. The transfer characteristic of an ideal comparator is shown in Fig. 4.19(b). The output may be symmetric or asymmetric.

A practical comparator has a finite voltage gain in the range from 3000 to 200,000 and takes a finite amount of time (in the range from 10 ns to 1 μs) to make a transition from one level to another (e.g., V_L to V_H). The transfer characteristic of a practical comparator is shown in Fig. 4.19(c). The input voltage swing required to produce the output voltage transition is in the range of about 0.1 mV to 4 mV. The output of a comparator must switch rapidly between the levels. The bandwidth must be wide because the wider the bandwidth, the faster the switching speed will be. Some typical parameters (listed here for the LM111 comparator) are as follows:

- Operates from a single 5-V power supply
- Input current: 150 nA (maximum)
- Offset current: 20 nA (maximum)
- Differential input voltage: ±30 V
- Voltage gain: 200 V/mV (typical)

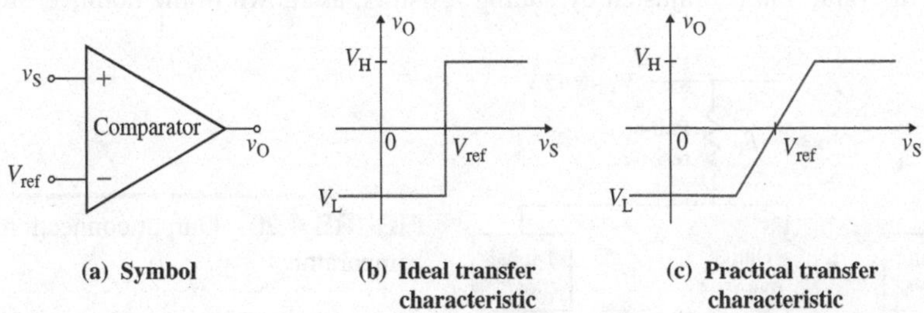

FIGURE 4.19 Symbol and transfer characteristics of a comparator

TABLE 4.1 Comparators versus op-amps

Op-Amps	Comparators
Operation is in closed-loop mode. To avoid an unstable oscillatory response, a sacrifice is usually made in bandwidth, rise time, and slew rate.	Operation is in open-loop mode. No sacrifice is needed in frequency characteristics, and a very fast response time can be obtained.
The output voltage is designed to be zero when the differential input voltage is zero.	The output voltage operates between two fixed output levels: V_L (low) and V_H (high).
The output voltage saturates about 1 V or 2 V away from the positive and negative supply voltages (V_{CC} and V_{EE}).	The low- and high-output levels can be changed for ease in interfacing with digital logic circuits.

4.3.1 Comparators versus Op-Amps

A comparator is designed to operate under open-loop conditions, usually as a switching device, whereas an op-amp is generally operated under closed-loop conditions as a linear amplifier. Otherwise, comparators are very similar to op-amps. Like an op-amp, a comparator has an offset voltage (typically 4 mV), a biasing current (typically 150 nA), and an offset current (typically 20 nA). The characteristics of comparators and op-amps are listed in Table 4.1.

4.3.2 Output-Side Connection

Comparators are often used as an interface between digital and analog signals. The power supply at the analog side (V_{CC} and V_{EE}, typically ± 15 V) is different from that at the digital side (V'_{CC}, typically 0 to 5 V). Comparators generally have an open-collector output stage, which allows separate power supplies for the analog and digital parts. A block diagram of the LM111 comparator is shown in Fig. 4.20. The output terminals are the collector and the emitter of an *npn* transistor. If the input voltage to the transistor is low, the transistor is off, and the output logic level is 1; that is, the output is high (5 V), and the current flows through the pull-up resistor R_P to the digital load. On the other hand, if the input voltage to the transistor is high, the transistor is driven into saturation, and the output logic level is 0; that is, the output is low at the saturation voltage of the transistor (typically 0.2 V), and no current flows through the pull-up resistor R_P to the digital load.

4.3.3 Threshold Comparators

The voltage at which a comparator changes from one level to another is called the *crossover* (or *threshold*) voltage. Its value can be adjusted by adding resistors, as shown in the noninverting comparator

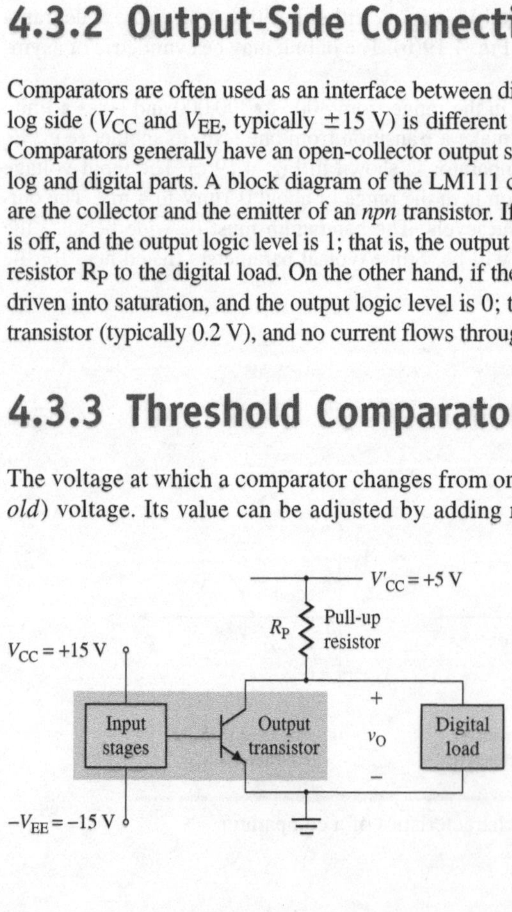

FIGURE 4.20 Output connection of LM111 comparator

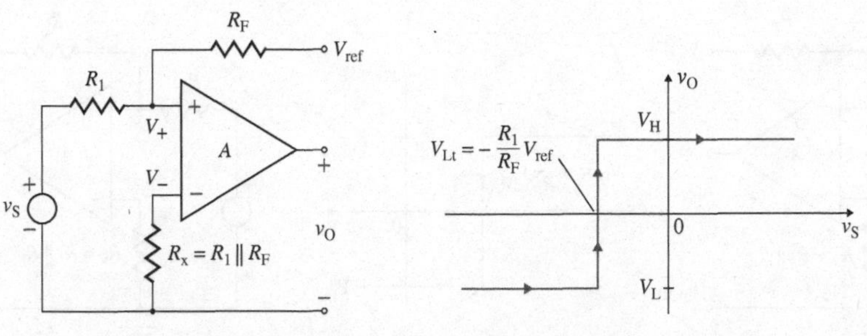

(a) Circuit (b) Transfer characteristic

FIGURE 4.21 Noninverting threshold comparator

in Fig. 4.21(a). From the superposition theorem, the voltage V_+ at the noninverting terminal is given by

$$V_+ = \frac{R_1}{R_1 + R_F}V_{ref} + \frac{R_F}{R_1 + R_F}v_S \tag{4.13}$$

Ideally, the crossover will occur when $V_+ = 0$; that is,

$$R_1 V_{ref} + R_F v_S = 0$$

which gives the low threshold voltage of the comparator $V_{Lt} = v_S$ (for changing from low to high) as

$$V_{Lt} = -\frac{R_1}{R_F}V_{ref} \tag{4.14}$$

Thus, the output voltage becomes high (V_H) at the positive saturation voltage ($+V_{sat}$) when $V_+ > 0$ (i.e., $v_S > V_{Lt}$). The transfer characteristic is shown in Fig. 4.21(b).

If the input signal v_S is connected to the inverting terminal, as shown in Fig. 4.22(a), the output will change from high (V_H) to low (V_L). This situation is shown in Fig. 4.22(b). The high threshold voltage of the comparator $V_{Ht} = v_S$ (for changing from high to low) is given by

$$V_{Ht} = \frac{R_1}{R_1 + R_F}V_{ref} \tag{4.15}$$

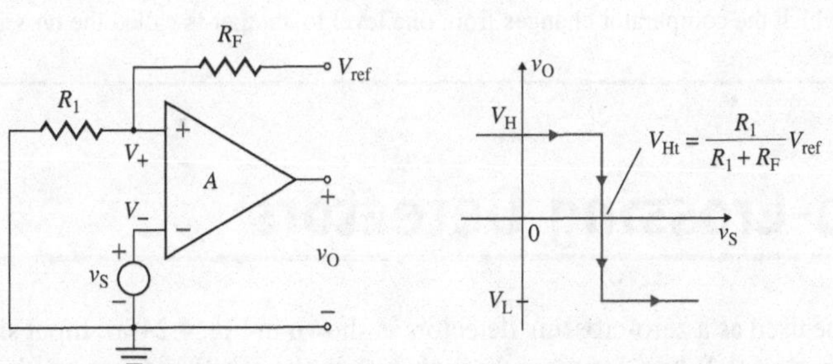

(a) Circuit (b) Transfer characteristic

FIGURE 4.22 Inverting configuration for noninverting threshold comparator

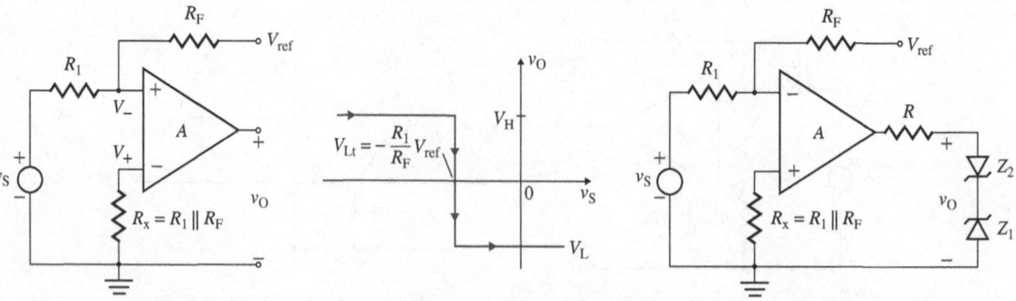

(a) Circuit (b) Transfer characteristic (c) Output voltage limiting

FIGURE 4.23 Inverting threshold comparator

Thus, the output voltage becomes low (V_L) at the negative saturation voltage ($-V_{sat}$) when $v_S > V_+$ (i.e., $v_S > V_{Ht}$). The transfer characteristic is shown in Fig. 4.22(b).

If both the input signal v_S and the reference signal V_{ref} are connected to the inverting terminal, as shown in Fig. 4.23(a), the output will be the inversion of the output in Fig. 4.21(b). That is, the output will change from high (V_H) to low (V_L) when the input is $v_S = V_{Lt}$. This situation is shown in Fig. 4.23(b).

In both inverting and noninverting configurations, the output voltage is limited to the saturation voltage of the comparator. The output voltage can, however, be set to specified limits by external limiters such as zener diodes connected across the output terminals of Figs. 4.21(a), 4.22(a), and 4.23(a). This approach is illustrated in Fig. 4.23(c), in which resistance R is connected to limit the current through the zener diodes.

KEY POINTS OF SECTION 4.3

- A comparator compares a signal voltage on one input terminal with a known voltage, called the *reference voltage*, on the other input terminal. It is designed to operate under open-loop conditions, usually as a switching device. Comparators are often used as an interface between digital and analog signals.
- A comparator can be used as a *threshold comparator* in either inverting or noninverting mode. The voltage at which the comparator changes from one level to another is called the *crossover* (or *threshold*) voltage.

4.4 Zero-Crossing Detectors

A comparator can be used as a zero-crossing detector, as shown in Fig. 4.24(a). Input signal v_S is compared with a reference signal of 0. When v_S passes through zero in the positive direction, the output v_O is driven into negative saturation ($-V_{sat}$) as a result of the very high gain of the comparator. Conversely, when v_S passes through zero in the negative direction, the output is driven into positive saturation ($+V_{sat}$). The input and output waveforms are shown in Fig. 4.24(b).

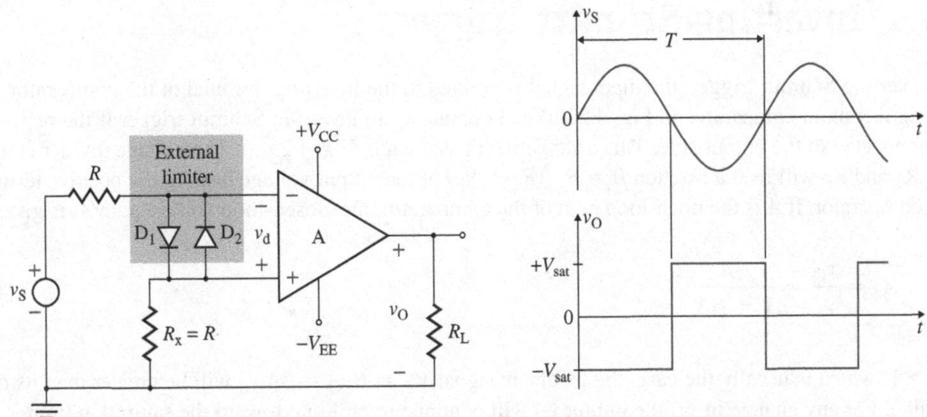

(a) Circuit (b) Waveforms

FIGURE 4.24 Zero-crossing detector

Diodes D_1 and D_2 in Fig. 4.24(a) protect the comparator from damage due to excessive input voltage v_S. Because of these diodes, the differential input voltage v_d of the comparator is clamped to approximately 0.7 V or −0.7 V. These diodes, called *clamp diodes*, are external to the comparator. It is up to the designer to determine if the diodes are needed to protect the circuit. Resistance R is connected in series with input signal v_S to limit the current through D_1 and D_2. Resistance R_x is used to reduce the effect of comparator offset problems.

If v_S is such that it crosses zero very slowly, then v_O may not switch quickly from one saturation voltage to the other. Instead, v_O may fluctuate between two saturation voltages $+V_{sat}$ and $-V_{sat}$ as a result of input offset voltage or noise signals at the comparator input terminals. Therefore, a zero-crossing detector will not be suitable for a low-frequency signal or a signal with noise superimposed on it.

KEY POINT OF SECTION 4.4

- A zero-crossing detector is a special application of a comparator in which the input signal is compared with a reference signal of 0.

4.5 Schmitt Triggers

A *Schmitt trigger* compares a regular or irregular waveform with a reference signal and converts the waveform to a square or pulse wave. A Schmitt trigger is often known as a *squaring circuit*. It is also known as a *bistable multivibrator* because it has two stable states, low and high. It can remain in one state indefinitely; it moves to the other stable state only when a triggering signal is applied. Schmitt triggers can be classified into two types depending on the type of op-amp configuration used: inverting or noninverting.

4.5.1 Inverting Schmitt Trigger

In an inverting Schmitt trigger, the input signal is applied to the inverting terminal of the comparator. The inverting threshold comparator in Fig. 4.22(a) can operate as an inverting Schmitt trigger if the resistance R_F is connected to the output side. This arrangement is shown in Fig. 4.25(a). The voltage divider consisting of R_1 and R_F will feed a fraction $\beta = R_1/(R_1 + R_F)$ of the output voltage back to the positive terminal of the comparator. If A is the open-loop gain of the comparator, the closed-loop voltage gain A_f is given by

$$A_f = \frac{v_O}{v_S} = \frac{-A}{1 - \beta A} \tag{4.16}$$

If $\beta A > 1$, which is usually the case, the feedback signal $V_+ = \beta v_O = \beta A_f v_S$ will be greater than its original value. For any change in v_S, the ouptut v_O will continue to build up toward the saturation limit.

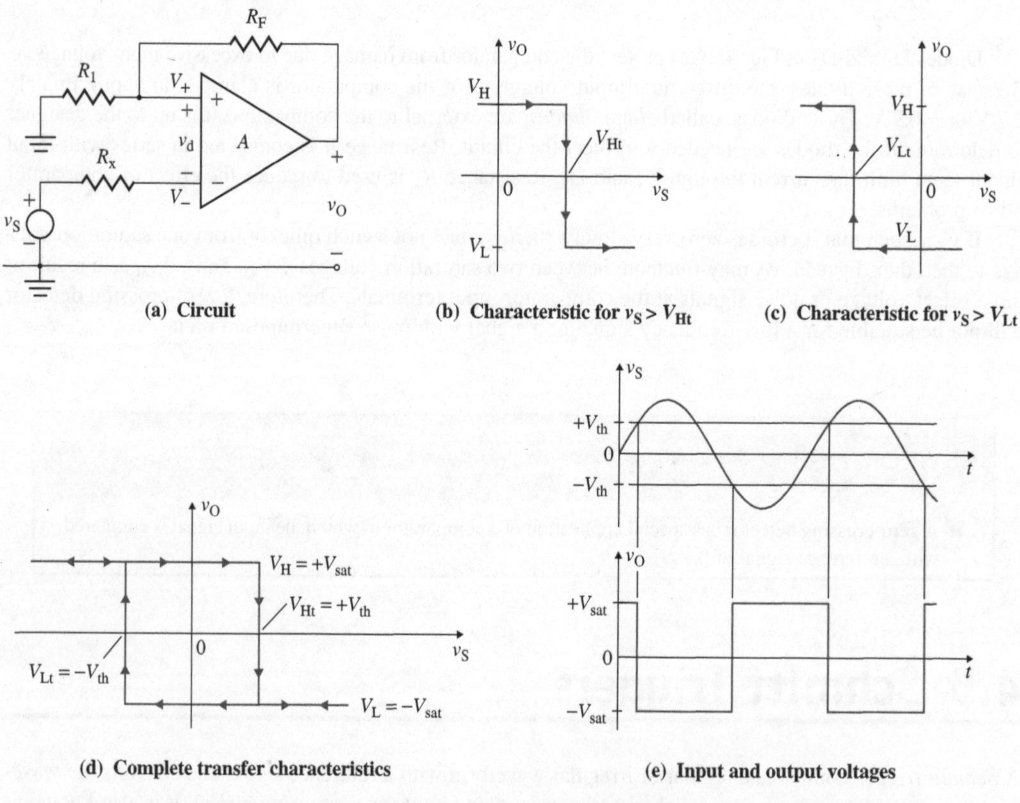

FIGURE 4.25 Schmitt trigger

Transfer Characteristics

To start with, let us assume that v_S is negative and the output is at the positive saturation voltage, $V_H = +V_{sat}$. If v_S is increased from 0, there will be no change in the output until v_S reaches a value of $v_S = V_+ = \beta V_{sat}$. If v_S begins to exceed $V_{Ht} = \beta V_{sat}$, the differential voltage v_d, which will be negative, will be amplified by the voltage gain A of the comparator; that is, v_O will be negative, thereby making V_+ negative also. The result will be an increase in the magnitude of the differential voltage v_d, and v_O will become more negative. This regenerative process will continue until eventually the comparator saturates, with its output voltage equal to the negative saturation voltage, $v_O = V_L = -V_{sat}$, and $V_+ = -\beta V_{sat}$. Increasing v_S beyond $v_S = \beta V_{sat}$ will have no effect on the state of the output voltage. The transfer characteristic for increasing v_S is shown in Fig. 4.25(b).

If v_S is decreased further while the output is low, there will be no change in the output until v_S goes negative, with a value of $v_S = V_+ = -\beta V_{sat}$. If v_S begins to exceed $V_{Lt} = -\beta V_{sat}$, the differential voltage v_d, which will be positive, will be amplified by the gain of the comparator; that is, V_+ will be positive. The result will be an increase in the differential voltage v_d, and v_O will become more positive. This regenerative process will continue until eventually the comparator saturates, with its output voltage equal to the positive saturation voltage, $v_O = V_H = +V_{sat}$, and $V_+ = \beta V_{sat}$. Decreasing v_S further ($v_S \leq -\beta V_{sat}$) will have no effect on the state of the output voltage. The transfer characteristic for decreasing v_S is shown in Fig. 4.25(c).

The complete transfer characteristics are shown in Fig. 4.25(d). A Schmitt trigger exhibits a *hysteresis*, or *deadband*, condition. That is, when the input of the Schmitt trigger exceeds $V_{Ht} = +V_{th}$, its output switches from $+V_{sat}$ to $-V_{sat}$, and when the input goes below $V_{Lt} = -V_{th}$, the output reverts to its original state, $+V_{sat}$.

Every time input voltage v_S exceeds certain levels, called the *positive (or upper) threshold voltage* $+V_{th}$ and the *negative (or lower) threshold voltage* $-V_{th}$, it changes the state of output voltage v_O. If the input signal is a sine wave, the output will be a square wave, as shown in Fig. 4.25(e). $+V_{th}$ and $-V_{th}$ are given by

$$+V_{th} = V_{Ht} = \frac{R_1}{R_1 + R_F}(+V_{sat}) \tag{4.17}$$

$$-V_{th} = V_{Lt} = \frac{R_1}{R_1 + R_F}(-V_{sat}) \tag{4.18}$$

where $V_{sat} = |+V_{sat}| = |-V_{sat}|$ and $V_{th} = |+V_{th}| = |-V_{th}|$.

Effect of Positive Feedback

R_F provides positive feedback. As soon as the output voltage begins to change, positive feedback increases the differential voltage v_d, which, in turn, further changes the output voltage. Once a transition is initiated by a change in the input signal v_S, the positive feedback forces the comparator to complete the transition from one state to another rapidly and to operate in saturation, either positive or negative. Positive feedback leads to rapid transition of the output. Oscillations, which normally occur in the active region and hence prevail for a short time, are avoided.

EXAMPLE 4.6

D

Designing a Schmitt trigger with a hysteresis band

(a) Design a Schmitt trigger as in Fig. 4.25(a) so that $V_{th} = |+V_{th}| = |-V_{th}| = 5$ V. Assume $V_{sat} = |-V_{sat}| = 14$ V.

(b) Use PSpice/SPICE [4] to plot the hysteresis characteristic for $v_S = 10 \sin(800\pi t)$.

SOLUTION

(a) The steps required to design the Schmitt trigger are as follows:

Step 1. Find the values of R_1 and R_F. From Eq. (4.17),

$$1 + \frac{R_F}{R_1} = \frac{V_{sat}}{V_{th}} \tag{4.19}$$

$$= \frac{14}{5} = 2.8$$

so $R_F/R_1 = 2.8 - 1 = 1.8$. Let $R_1 = 10$ kΩ; then

$$R_F = 1.8 \times R_1 = 18 \text{ k}\Omega \quad \text{(use a 20-k}\Omega\text{ potentiometer)}$$

Step 2. Choose the value of offset minimizing resistance R_x:

$$R_x = R_1 \| R_F = 10 \text{ k}\Omega \| 18 \text{ k}\Omega = 6.43 \text{ k}\Omega$$

(b) The circuit for PSpice simulation is shown in Fig. 4.26. The comparator is simulated by the PSpice macromodel of the LM111. To obtain the negative output voltage swing, terminal 1 is connected to the negative power supply instead of to the ground.

The transfer characteristic and the output voltage v_O = V(U1:OUT) (using EX16-6.SCH) are shown in Fig. 4.27[(a) and (b)], respectively. The simulated values are $V_{Ht} = 5.02$ V (expected value is 5 V), $V_{Lt} = -5.28$ V (expected value is −5 V), $V_H = 14.05$ V (expected value is 14 V), and $V_L = -14.83$ V (expected value is −14 V).

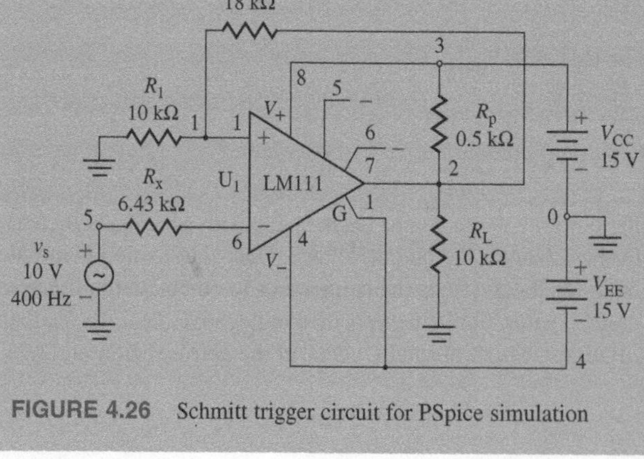

FIGURE 4.26 Schmitt trigger circuit for PSpice simulation

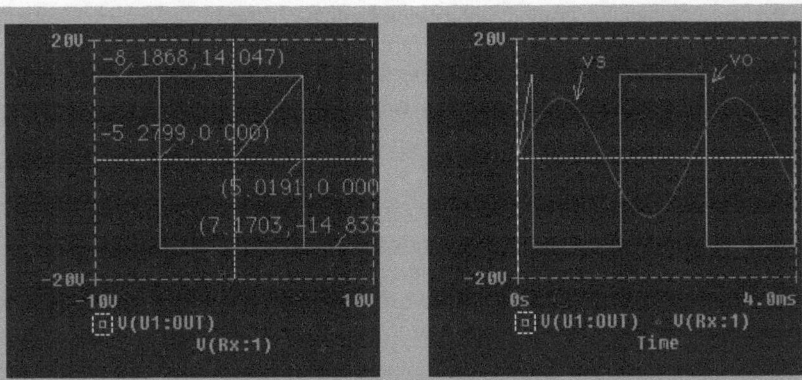

(a) Transfer characteristic **(b) Input and output voltages**

FIGURE 4.27 Transfer characteristic and output voltage for Example 4.6

At the beginning, the output voltage varies linearly with the input voltage until the input reaches the threshold level, after which the transfer characteristic follows the normal hysteresis band. As a result, the transfer characteristic starts from the origin ($v_O = 0$ and $v_S = 0$). When the output transistor is off, the pull-up resistor R_P forms a potential divider with the load resistor R_L. As a result, the positive output voltage will depend on R_P, whose value should be made small compared to that of R_L. Notice that the transition from low to high and vice versa is very sharp because of the high slew rate of the comparator.

4.5.2 Noninverting Schmitt Trigger

In a noninverting Schmitt trigger, the input signal is applied to the noninverting terminal of the comparator, and the transfer characteristics are inverted. The noninverting threshold comparator in Fig. 4.21(a) can be operated as a noninverting Schmitt trigger if the resistance R_F is connected to the output side. This arrangement is shown in Fig. 4.28(a). Resistance R_F will feed a current signal, a fraction $\beta = 1/R_F$ of the output voltage, back to the positive terminal of the comparator, providing positive shunt-shunt feedback. As soon

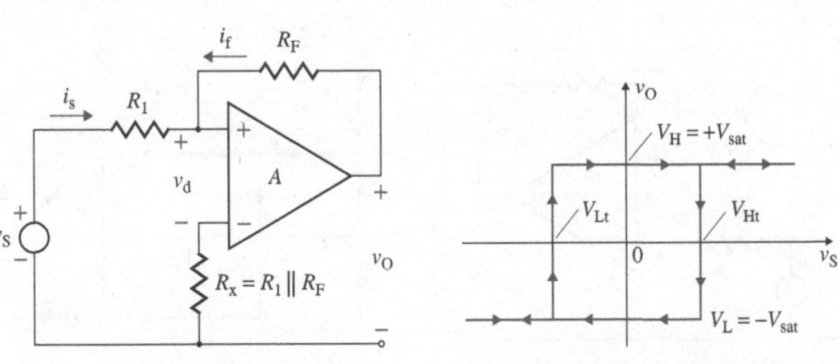

(a) Circuit **(b) Transfer characteristics**

FIGURE 4.28 Noninverting Schmitt trigger

as the output voltage begins to change, the positive shunt-shunt feedback increases the feedback current i_f, which in turn increases the differential voltage v_d and hence further changes the output voltage. Once a transition is initiated by a change in the input signal v_S, positive feedback forces the comparator to complete the transition from one state to another rapidly and to operate in saturation, either positive or negative.

Transfer Characteristics

To start with, let us assume that v_S is negative and the output is at the negative saturation voltage, $V_L = -V_{sat}$. If v_S is increased from a relatively large negative value, there will be no change in the output until v_S reaches a value of $v_S = V_{Lt} = -V_{sat}R_1/R_F$. If v_S begins to exceed V_{Lt}, the differential voltage v_d, which will be positive, will be amplified by the voltage gain A of the comparator. That is, v_O will be positive, thereby making v_+ positive also. The result will be an increase in the magnitude of the differential voltage v_d, and v_O will become more positive. This regenerative process will continue until eventually the comparator saturates, with its output voltage equal to the positive saturation voltage, $V_H = +V_{sat}$. Increasing v_S further ($v_S \geq V_{Lt}$) will have no effect on the state of the output voltage.

If v_S is decreased while the output is high, there will be no change in the output until v_S decreases to a value of $v_S = V_{Ht} = +V_{sat}R_1/R_F$. If v_S begins to decrease beyond V_{Ht}, the differential voltage v_d will be negative and will be amplified by the gain of the comparator. As a result, v_O will become more negative. This regenerative process will continue until eventually the comparator saturates, with its output voltage equal to the negative saturation voltage $-V_{sat}$. Decreasing v_S further ($v_S \leq V_{Ht}$) will have no effect on the state of the output voltage. The complete transfer characteristics are shown in Fig. 4.28(b).

4.5.3 Schmitt Trigger with Reference Voltage

The *switching voltage* of a Schmitt trigger circuit is defined as the average of V_{Lt} and V_{Ht}. For the circuits in Figs. 4.25(a) and 4.28(a), $V_{Lt} = -V_{Ht}$, and hence the switching voltage V_{st}, which is the width of the hysteresis band, is zero; that is, $V_{st} = (V_{Lt} + V_{Ht})/2 = 0$. However, some applications require shifting the crossover voltage in either the positive or the negative direction along the v_S-axis. This can be accomplished by adding a reference voltage V_{ref} to the circuit in Fig. 4.25(a), as shown in Fig. 4.29(a) for a non-

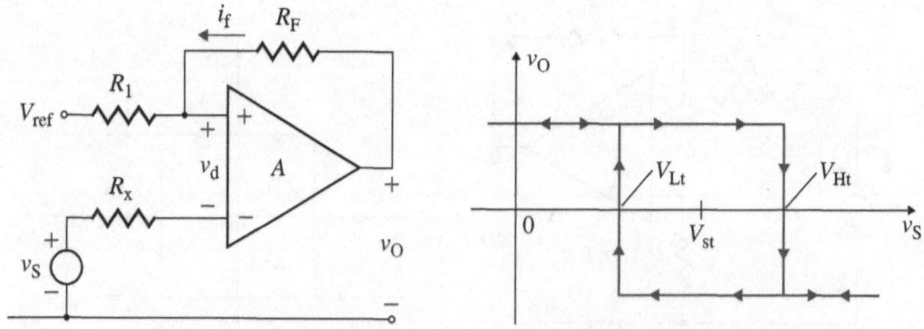

(a) Circuit (b) Transfer characteristics

FIGURE 4.29 Noninverting Schmitt trigger with reference voltage

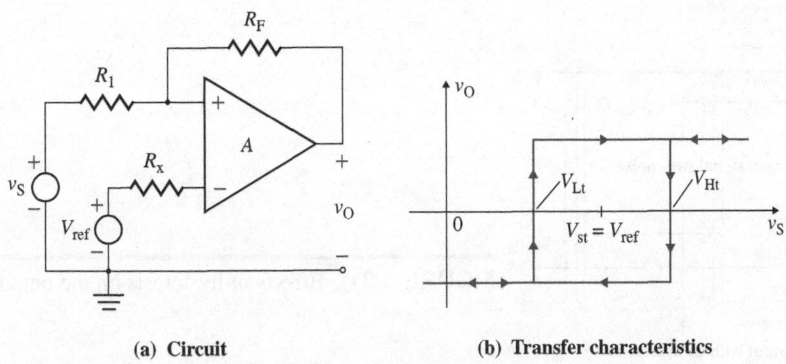

FIGURE 4.30 Noninverting Schmitt trigger with reference voltage

inverting Schmitt trigger. The complete transfer characteristics are shown in Fig. 4.29(b). Assuming that V_{Lt} and V_{Ht} are symmetric about the zero-axis, the switching voltage is given by

$$V_{st} = \frac{R_F}{R_1 + R_F} V_{ref} \tag{4.20}$$

Thus, the upper and lower crossover voltages become

$$V_{Ht} = V_{st} + \frac{R_1}{R_1 + R_F}(+V_{sat}) \tag{4.21}$$

$$V_{Lt} = V_{st} + \frac{R_1}{R_1 + R_F}(-V_{sat}) \tag{4.22}$$

The direction of the hysteresis loop in Fig. 4.29(b) can be reversed by applying a reference voltage V_{ref} to the circuit in Fig. 4.28(a). This arrangement is shown in Fig. 4.30(a), and the corresponding transfer characteristics are shown in Fig. 4.30(b).

4.5.4 Effects of Hysteresis on the Output Voltage

To understand the effect of the *hysteresis*, or *deadband*, condition, consider a sinusoidal signal with a noise signal superimposed on it, as shown in Fig. 4.31(a). If there is no hysteresis, the output voltage will change to its saturation limit when the input signal v_S crosses zero, as shown in Fig. 4.31(b). However, if the output is made to change when the input signal exceeds specified voltage limits V_{Ht} and V_{Lt}, there will be less switching of the output voltage, as shown in Fig. 4.31(c). As a result, any unwanted signal (e.g., noise) will not make the output change. Also, a deadband can be used to reduce the number of contact-bounces in a system such as a temperature-control system, in which the heating element turns on or off when the temperature falls below or rises above the set value.

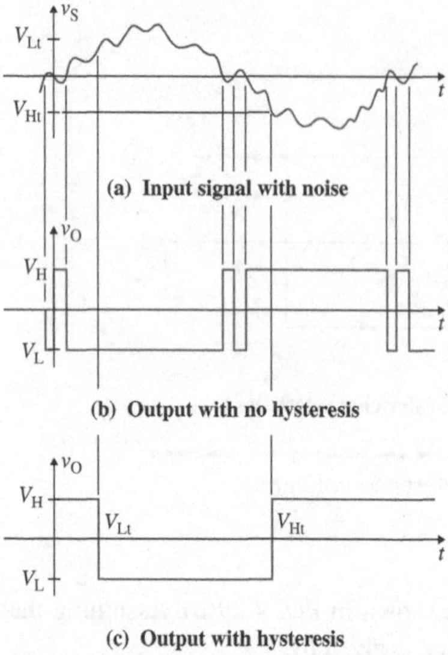

FIGURE 4.31 Effects of hysteresis on the output voltage

EXAMPLE 4.7

D **Designing a Schmitt trigger with a shifted hysteresis band**

(a) Design a Schmitt trigger as in Fig. 4.29(a) so that $V_{Ht} = 7$ V and $V_{Lt} = 3$ V. Assume $V_{sat} = |-V_{sat}| = 14$ V and an input frequency of $f = 400$ Hz. Determine the values of R_1, R_F, and V_{ref}.

(b) Use PSpice/SPICE to plot the hysteresis characteristic for $v_S = 10 \sin (800\pi t)$. Use the macromodel of the μA741 op-amp.

SOLUTION

(a) The Schmitt trigger can be designed using the following steps:

Step 1. Find the values of R_1 and R_F. From Eqs. (4.21) and (4.22), we can find the input width of the hysteresis band (HB) as follows:

$$\text{HB} = V_{Ht} - V_{Lt} = \frac{2R_1}{R_1 + R_F} V_{sat}$$

which gives

$$1 + \frac{R_F}{R_1} = \frac{2V_{sat}}{V_{Ht} - V_{Lt}} = \frac{2 \times 14}{7 - 3} = 7$$

Let $R_1 = 10$ kΩ; then $R_F = (7 - 1) \times R_1 = 60$ kΩ.

Step 2. Determine the value of reference voltage V_{ref}.

$$V_{st} = \frac{R_F}{R_1 + R_F} V_{ref} = \frac{V_{Ht} + V_{Lt}}{2} = \frac{7 + 3}{2} = 5 \text{ V}$$

which gives $V_{ref} = 5.83$ V.

(b) We will use op-amp μA741 rather than comparator LM111 to illustrate the advantage of a comparator in a Schmitt trigger. The circuit for PSpice simulation is shown in Fig. 4.32.

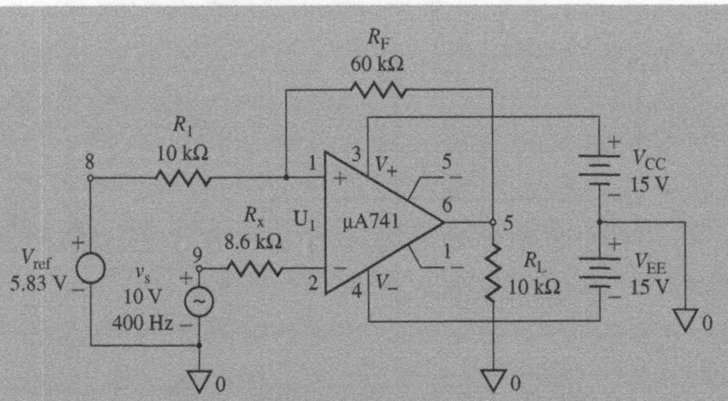

FIGURE 4.32 Schmitt trigger circuit for PSpice simulation

The transfer characteristic and the output voltage (using EX16-7.SCH) are shown in Fig. 4.33[(a) and (b)], which gives $V_{Ht} = 7.03$ V (expected value is 7 V), $V_{Lt} = 3.03$ V (expected value is 3 V), and $V_{sat} = 14.6$ V (expected value is 14 V). The transition from low to high and vice versa is not very sharp, as a result of the low slew rate of the op-amp (compared to that of a comparator, shown in Fig. 4.27).

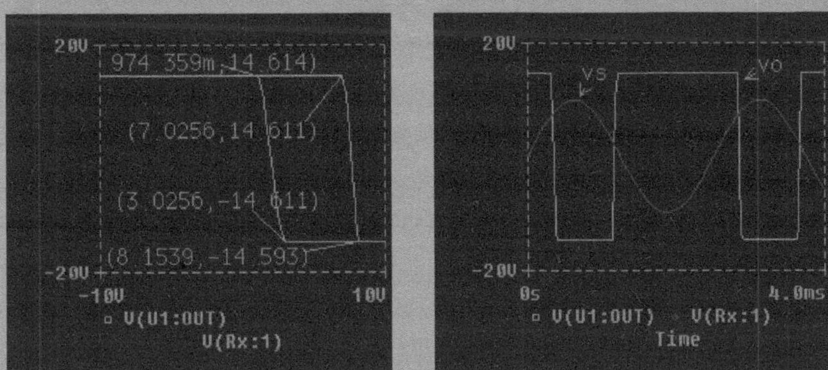

(a) Transfer characteristic

(b) Input and output voltages

FIGURE 4.33 Transfer characteristic and input and output voltages for Example 4.7

KEY POINTS OF SECTION 4.5

- A *Schmitt trigger* compares a regular or irregular waveform with a reference signal and converts the waveform to a square or pulse wave. A Schmitt trigger is often known as a *squaring circuit*. It is also known as a *bistable multivibrator* because it has two stable states, low and high. Schmitt triggers can be classified into two types depending on the type of op-amp configuration used: inverting or noninverting.
- A Schmitt trigger exhibits a *hysteresis*, or *deadband*, condition; that is, when the input of the Schmitt trigger exceeds $+V_{th}$, its output switches from $+V_{sat}$ to $-V_{sat}$, and when the input goes below $-V_{th}$, the output reverts to its original state, $+V_{sat}$.
- When a general-purpose op-amp (e.g., μA741) is used to generate the characteristic of a hysteresis loop, the transition from low to high and vice versa is not as sharp as it is with a comparator.

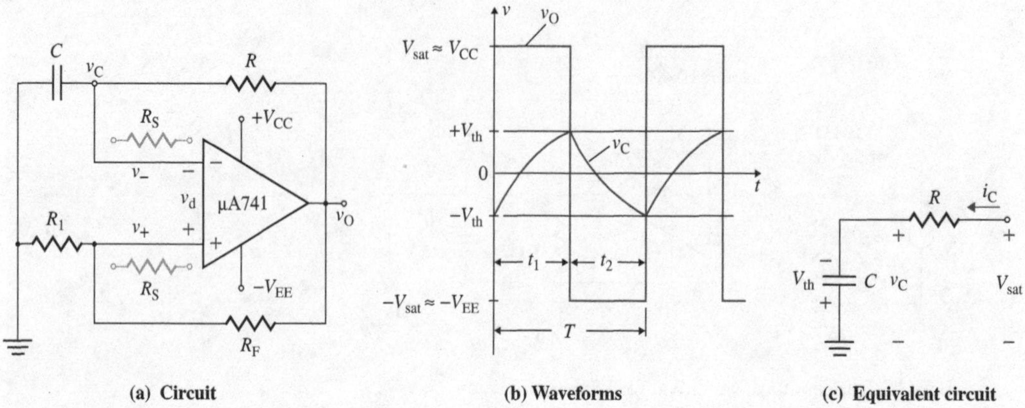

FIGURE 4.34 Square-wave generator

4.6 Square-Wave Generators

A square wave can be generated if the output of an op-amp is forced to swing repetitively between positive saturation $+V_{sat}$ and negative saturation $-V_{sat}$. This can be accomplished by connecting a bistable multivibrator (or Schmitt trigger) with an *RC* circuit in the feedback loop. The circuit implementation is shown in Fig. 4.34(a). This square-wave generator is also called a *free-running* or *astable multivibrator* because the output does not have any stable state. The output of the op-amp will be in either positive saturation or negative saturation, depending on whether the differential input voltage v_d is positive or negative.

Assuming that the voltage across capacitor C is zero, the voltage at the inverting terminal is zero initially; that is, $v_- = 0$ at the instant the DC supply voltages V_{CC} and V_{EE} are turned on. However, at the same instant, the voltage v_+ at the noninverting terminal will have a very small value that will depend on the output offset voltage V_{OO}; that is,

$$v_d = v_+ - v_- = v_+ = V_{OO}$$

which has a positive value. But v_d will be amplified as a result of the very large gain of the op-amp (typically 2×10^5) and will drive the output of the op-amp to positive saturation $+V_{sat}$. Capacitor C will start charging toward $+V_{sat}$ through R. However, as soon as the voltage across C, which is equal to v_-, is slightly more than v_+, then $v_d = v_+ - v_-$ will become negative, and the output of the op-amp will switch to negative saturation $-V_{sat}$. The operation of the circuit can be divided into two modes: mode 1 for $v_d > 0$ and mode 2 for $v_d < 0$.

During mode 1, $v_d > 0$ and the output voltage of the op-amp is at positive saturation $+V_{sat}$. The voltage v_+ becomes

$$v_+ = \frac{R_1}{R_1 + R_F}(+V_{sat}) \tag{4.23}$$

Capacitor C will again start charging toward $+V_{sat}$ through R. As soon as the voltage across C is slightly more than v_+, then $v_d = v_+ - v_-$ will become negative, and the output of the op-amp will be forced to switch to negative saturation $-V_{sat}$.

During mode 2, $v_d < 0$ and the op-amp's output voltage is at negative saturation $-V_{sat}$. The voltage v_+ follows the voltage divider rule and can be found from

$$v_+ = \frac{R_1}{R_1 + R_F}(-V_{sat}) \tag{4.24}$$

As long as v_d is negative, the output will remain in negative saturation. The capacitor will discharge and then recharge toward $-V_{sat}$ through R. When the voltage across C is slightly more negative than v_+, then $v_d = v_+ - v_-$ will become positive, and the output of the op-amp will be forced to switch to positive saturation $+V_{sat}$. Then mode 1 begins again, and the cycle is repeated.

Voltage v_+ acts as the reference. The capacitor voltage v_- tries to follow v_+. However, as soon as the magnitude of v_- becomes slightly greater than that of v_+, v_+ switches its polarity. As a result, the output swings from positive to negative and vice versa. The waveforms of the output voltage and capacitor voltage are shown in Fig. 4.34(b).

Assuming that $+V_{sat}$ is the output voltage and the capacitor has an initial voltage of $-V_{th}$ during mode 1, the equivalent circuit during the charging period is as shown in Fig. 4.34(c). Using KVL, we can write

$$V_{sat} = Ri_C + \frac{1}{C}\int i_C dt - V_{th}$$

which gives the charging current $i_C(t)$ as

$$i_C(t) = \frac{V_{sat} + V_{th}}{R} e^{-t/RC} \tag{4.25}$$

where $$V_{th} = \frac{R_1}{R_1 + R_F} V_{sat} \tag{4.26}$$

The capacitor voltage $v_C(t)$ can be found from

$$v_C(t) = V_{sat} - (V_{sat} + V_{th})e^{-t/RC} \tag{4.27}$$

At $t = t_1$, the capacitor is recharged to V_{th}. That is, $v_C(t = t_1) = V_{th}$. From Eq. (4.27), we get

$$V_{th} = V_{sat} - (V_{sat} + V_{th})e^{-t_1/RC} \tag{4.28}$$

which gives

$$t_1 = -RC \ln\frac{V_{sat} - V_{th}}{V_{sat} + V_{th}} = -RC \ln\frac{V_{sat} - R_1V_{sat}/(R_1 + R_F)}{V_{sat} + R_1V_{sat}/(R_1 + R_F)}$$

$$= -RC \ln\frac{R_F}{2R_1 + R_F} = RC \ln\frac{2R_1 + R_F}{R_F} \tag{4.29}$$

The period T of the output voltage is given by

$$T = t_1 + t_2 = 2t_1 = 2RC \ln\frac{2R_1 + R_F}{R_F} = 2RC \ln\left(1 + \frac{2R_1}{R_F}\right) \tag{4.30}$$

Thus, the frequency of the output voltage f_o is given by

$$f_o = \frac{1}{T} = \frac{1}{2RC \ln(1 + 2R_1/R_F)} \tag{4.31}$$

Equation (4.31) shows that the output frequency depends not only on the time constant $\tau = RC$ but also on the relationship between R_1 and R_F. If $R_F \approx 1.164R_1$, Eq. (4.31) is reduced to

$$f_o = \frac{1}{2RC} \tag{4.32}$$

▶ **NOTES**

1. The inputs of the op-amp are subjected to large differential voltages. To prevent excessive differential current flow to the op-amp, a series resistance R_S, typically on the order of 100 kΩ, can be connected to each of the inverting and noninverting terminals of the op-amp.
2. The peak-to-peak output voltage can be reduced by connecting a pair of zener diodes back to back at the output terminal.

EXAMPLE 4.8

D

Designing a square-wave generator

(a) Design the square-wave generator shown in Fig. 4.34(a) so that $f_o = 5$ kHz. Assume $+V_{sat} = |-V_{sat}| = 14$ V.
(b) Use PSpice/SPICE to check your design. Use the macromodel of the LM111 for sharper transition.

SOLUTION

(a) The steps used to design the square-wave generator are as follows:

Step 1. Choose the value of R_1: Let $R_1 = 10$ kΩ.

Step 2. To simplify the design, choose $R_F = 1.164R_1$ and find R_F:

$$R_F = 1.164R_1 = 1.164 \times 10\ \text{k}\Omega = 11.64\ \text{k}\Omega \quad \text{(use a 20-k}\Omega\text{ potentiometer)}$$

Step 3. Choose a value of C: Let $C = 0.01$ μF.

Step 4. Find the value of R from Eq. (4.32):

$$R = \frac{1}{2Cf_o} = \frac{1}{2 \times 0.01\ \mu\text{F} \times 5\ \text{kHz}} = 10\ \text{k}\Omega$$

(b) The circuit for PSpice simulation is shown in Fig. 4.35.

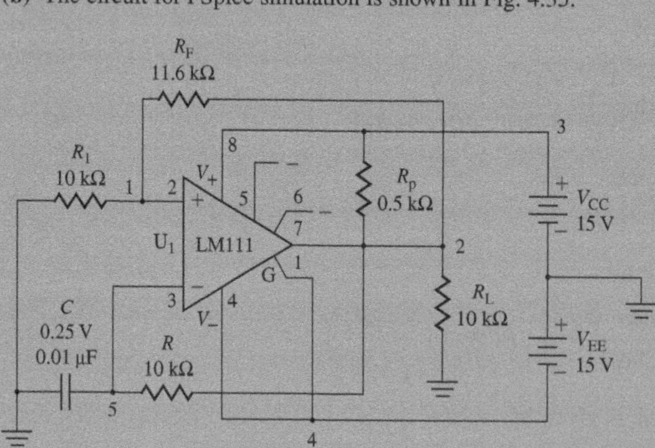

FIGURE 4.35 Square-wave generator for PSpice simulation

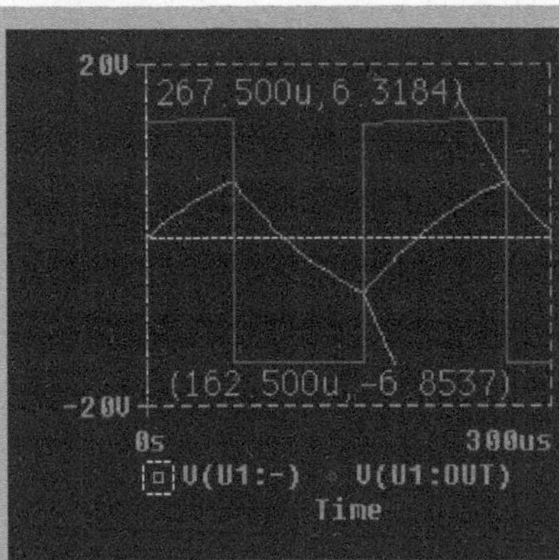

FIGURE 4.36 Output voltage waveforms for Example 4.8

The output waveforms (using EX16-8.SCH) are shown in Fig. 4.36. The results are $V_{sat} = 13.53$ V (expected value is 14 V), $-V_{sat} = -14.83$ V (expected value is -14 V), $V_{th} = 6.3$ V, $-V_{th} = -6.8$ V, and $T = 2 \times (267.5 - 162.5) = 210\ \mu s$ (expected value is 200 μs). The output voltage does not change sharply from one state to another. It is not a full square wave because of the slew rate and bandwidth limits of the op-amp.

KEY POINT OF SECTION 4.6

- A square wave can be generated by forcing the output of an op-amp to swing repetitively between positive saturation $+V_{sat}$ and negative saturation $-V_{sat}$. This is accomplished by connecting a bistable multivibrator (or Schmitt trigger) in the feedback loop. This square-wave generator is also called a *free-running* or *astable multivibrator.*

4.7 Triangular-Wave Generators

A triangular-wave generator can be produced by integrating the square-wave output of a Schmitt trigger circuit. This can be accomplished by cascading an integrator with a bistable multivibrator (or Schmitt trigger), as shown in Fig. 4.37(a), which consists of a comparator A_1 and an integrator A_2. Comparator A_1 continuously compares the noninverting input v_+ at point P with the inverting input v_- (i.e., 0 V). Thus, the differential voltage is $v_d = v_+ - v_- = v_+$. Because of the very large gain of the op-amp, the output of A_1 will be at negative saturation $-V_{sat}$ or positive saturation $+V_{sat}$ when v_+ goes slightly below or above 0, respectively.

To examine the principle of operation of the triangular-wave generator, let us assume that when the supply voltages V_{CC} and $-V_{EE}$ are switched on, the voltage at the noninverting terminal begins slightly above 0 as a result of the input offset voltage of the op-amp. Because of the high gain of the op-amp, the output of A_1 will be switched to positive saturation $+V_{sat}$. Therefore, the output of the op-amp is forced to swing repetitively between positive saturation $+V_{sat}$ and negative saturation $-V_{sat}$. The output voltages of A_1 and A_2 are shown in Fig. 4.37(b). The operation of the circuit can be divided into two modes.

During mode 1, $v_+ > 0$ and the output of A_1 is at positive saturation $+V_{sat}$, which is the input to the inverting integrator A_2. The output of A_2 will be a negative-going ramp. Thus, one side of R_F will be at

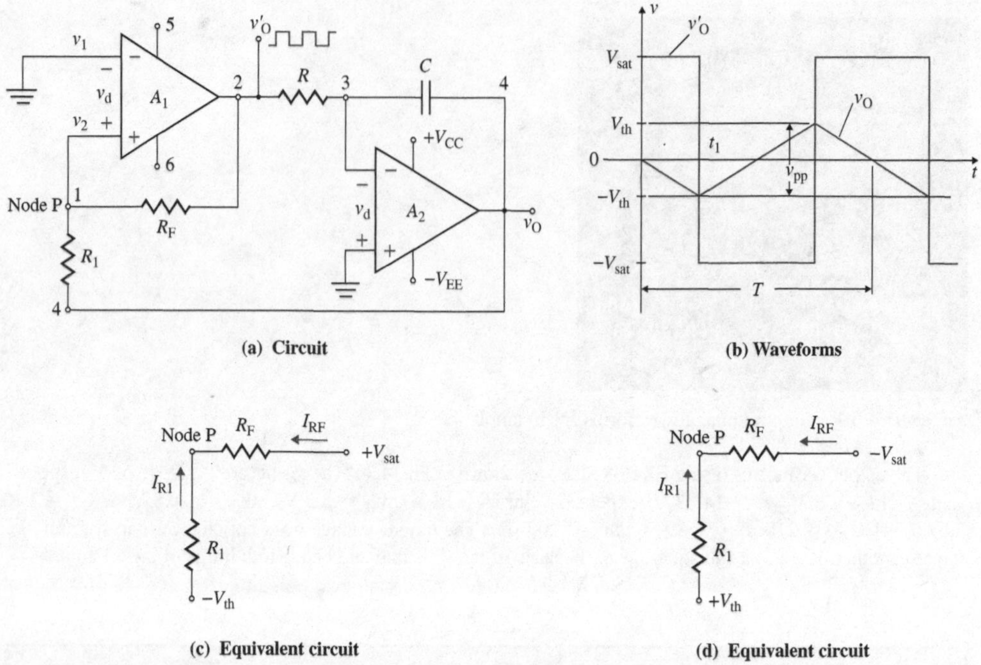

FIGURE 4.37 Triangular-wave generator

$+V_{sat}$, and one side of R_1 will be at the negative-going ramp of A_2. When the negative-going ramp exceeds a certain value $-V_{th}$, the voltage at point P will be slightly below 0, and the output of A_1 will switch from positive saturation $+V_{sat}$ to negative saturation $-V_{sat}$.

The equivalent circuit for determining the condition under which the circuit switches over to negative saturation is shown in Fig. 4.37(c). Since the current flowing into the op-amp is negligible, $I_{R1} = -I_{RF}$; that is,

$$\frac{-V_{th}}{R_1} = \frac{-V_{sat}}{R_F}$$

which gives the condition for $v_+ < 0$ as

$$-V_{th} = \frac{R_1}{R_F}(-V_{sat}) \tag{4.33}$$

During mode 2, $v_+ < 0$ and the output of A_1 is at negative saturation $-V_{sat}$, which is the input to the integrator A_2. The output of A_2 will be a positive-going ramp. When the positive-going ramp exceeds a certain value $+V_{th}$, the voltage at point P will be slightly above 0, and the output of A_1 will switch from negative saturation $-V_{sat}$ to positive saturation $+V_{sat}$. When this occurs, mode 1 starts again, and the cycle is repeated.

The equivalent circuit for determining the condition under which the circuit switches over to positive saturation is shown in Fig. 4.37(d). Neglecting the current flowing into the op-amp, we have $I_{R1} = -I_{RF}$; that is,

$$\frac{V_{th}}{R_1} = \frac{V_{sat}}{R_F}$$

which gives the condition for $v_+ > 0$ as

$$V_{th} = \frac{R_1}{R_F}(+V_{sat}) \tag{4.34}$$

where $V_{sat} = |+V_{sat}| = |-V_{sat}|$. The peak-to-peak output amplitude of the triangular wave is given by

$$v_{pp} = V_{th} - (-V_{th}) = \frac{2R_1}{R_F}V_{sat} \tag{4.35}$$

The period and the frequency of the output voltage can be determined from the time that is required to charge the capacitor from $-V_{th}$ to V_{th} or from V_{th} to $-V_{th}$. Let us consider the capacitor voltage during mode 2, when the input voltage to the integrator is $-V_{sat}$; that is,

$$v_C(t) = -\frac{1}{RC}\int v_O' \, dt - V_{th} = -\frac{1}{RC}\int (-V_{sat}) \, dt - V_{th}$$

$$= \frac{V_{sat}}{RC}t - V_{th} \tag{4.36}$$

At half period, $t = t_1 = T/2$ and $v_C(t = T/2) = V_{th}$, and Eq. (4.36) yields

$$V_{th} = \frac{V_{sat}}{RC}\left(\frac{T}{2}\right) - V_{th}$$

which gives the period as

$$T = \frac{4RCV_{th}}{V_{sat}} \tag{4.37}$$

Substituting the value of V_{th} from Eq. (4.34), we have for the period of the triangular wave

$$T = \frac{4RC}{V_{sat}} \times \frac{R_1 V_{sat}}{R_F} = \frac{4RCR_1}{R_F} \tag{4.38}$$

which gives the frequency of oscillation as

$$f_o = \frac{R_F}{4RCR_1} \tag{4.39}$$

EXAMPLE 4.9

D

Designing a triangular-wave generator

(a) Design the triangular-wave generator shown in Fig. 4.37(a) so that $f_o = 4$ kHz and $V_{th} = |-V_{th}| = 5$ V. Assume $+V_{sat} = |-V_{sat}| = 14$ V.

(b) Use PSpice/SPICE to check your design.

SOLUTION

(a) The steps used to design the triangular-wave generator are as follows:

Step 1. Find the values of R_1 and R_F. Equation (4.34) gives

$$\frac{R_1}{R_F} = \frac{V_{th}}{V_{sat}} = \frac{5}{14} = 0.36$$

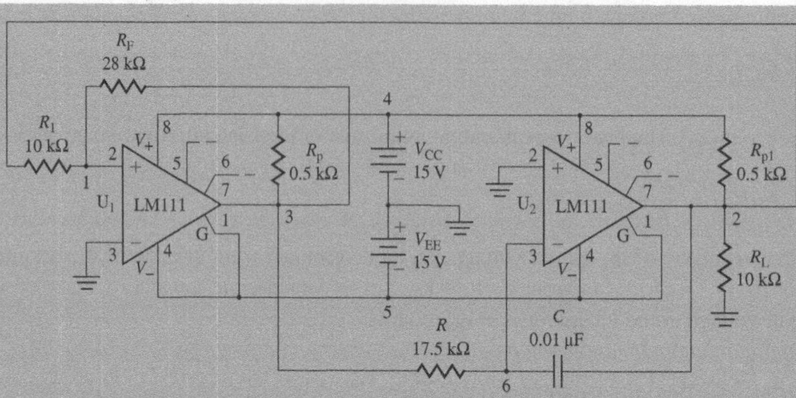

FIGURE 4.38 Triangular-wave generator for PSpice simulation

Let $R_1 = 10\text{ k}\Omega$; then

$$R_F = \frac{R_1}{0.36} = \frac{10\text{ k}\Omega}{0.36} = 28\text{ k}\Omega \quad \text{(use a 30-k}\Omega\text{ potentiometer)}$$

Step 2. Choose a suitable value of C: Let $C = 0.01\ \mu\text{F}$.

Step 3. Find the value of R. Equation (4.39) gives

$$R = \frac{R_F}{4f_oCR_1} \tag{4.40}$$

$$= \frac{28\text{ k}\Omega}{4 \times 4\text{ kHz} \times 0.01\ \mu\text{F} \times 10\text{ k}\Omega} = 17.5\text{ k}\Omega \quad \text{(use a 20-k}\Omega\text{ potentiometer)}$$

(b) The circuit for PSpice simulation is shown in Fig. 4.38. The op-amp is simulated by PSpice macromodel LM111.

The PSpice plots of the voltages at the output of amplifiers A_1 and A_2 (using EX16-9.SCH) are shown in Fig. 4.39. The results are $V_{sat} = 14.4$ V (expected value is 14 V), $-V_{sat} = -14.8$ V (expected value is -14 V), $V_{th} = 5.3$ V (expected value is 5 V), $-V_{th} = -5.2$ V (expected value is -5 V), and $T = 312.1 - 62.6 = 249.5\ \mu\text{s}$ (expected value is 250 μs). If the integrator time constant $\tau = RC$ is made much smaller than the period of the output waveform, the triangular wave can be made very close to a sine wave.

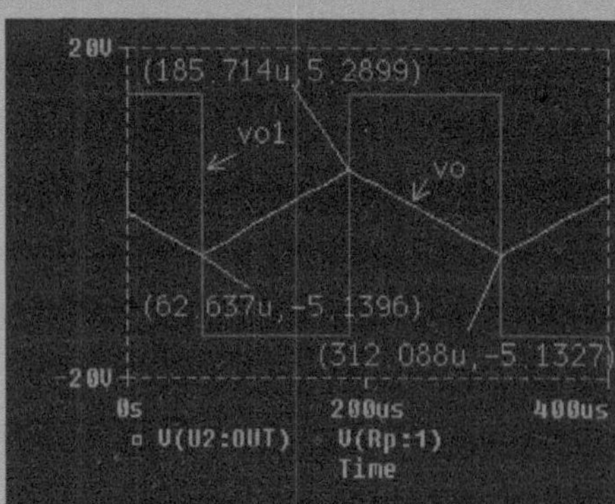

FIGURE 4.39 Output waveforms for Example 4.9

KEY POINTS OF SECTION 4.7

- A triangular-wave generator can be produced by integrating the square-wave output of a Schmitt trigger. This can be accomplished by cascading an integrator with a bistable multivibrator (or Schmitt trigger).
- If the integrator time constant $\tau = RC$ is made much smaller than the period of the output waveform, the triangular wave can be made very close to a sine wave.

4.8 Sawtooth-Wave Generators

In a triangular waveform, the rise time is always equal to the fall time. That is, the same amount of time is required for a triangular wave to swing from $-V_{th}$ to $+V_{th}$ as from $+V_{th}$ to $-V_{th}$. On the other hand, a sawtooth waveform has unequal rise and fall times. The rise time may be faster than the fall time or vice versa. The triangular-wave generator in Fig. 4.37(a) may be converted to a sawtooth generator by adding a variable DC voltage V_{ref} to the noninverting terminal of the op-amp. The addition of V_{ref}, which acts as a reference signal for the integrator A_2, can be accomplished by using a potentiometer, as shown in Fig. 4.40(a). The voltage waveforms are shown in Fig. 4.40(b). As with the triangular-wave generator, the operation of the circuit can be divided into two modes.

During mode 1, $v_+ > 0$ and the output of A_1 is at positive saturation $+V_{sat}$, which is the input signal to the integrator A_2. The equivalent circuit for the operation of the integrator is shown in Fig. 4.40(c). At the beginning of this mode, the output voltage is V_{th}, and the voltage at the inverting terminal is $v_1 \approx V_{ref}$. Thus, the initial voltage on the capacitor is

$$v_C(t = 0) = v_1 - v_O = V_{ref} - V_{th}$$

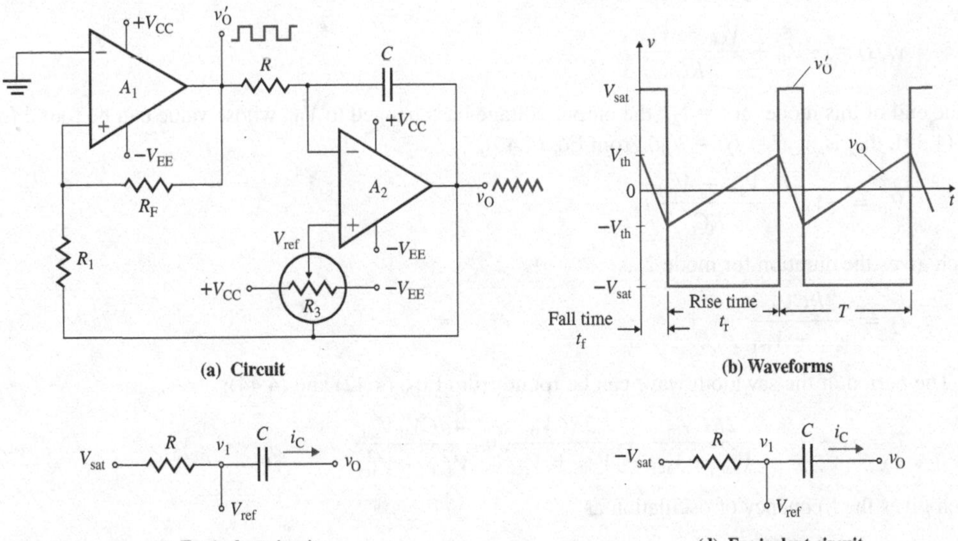

FIGURE 4.40 Sawtooth-wave generator

The instantaneous voltage across the capacitor $v_C(t)$ is given by

$$V_{ref} - v_O(t) = \frac{1}{C}\int i_C\,dt + v_C(t=0) = \frac{1}{C}\int \frac{V_{sat} - V_{ref}}{R}\,dt + V_{ref} - V_{th}$$

which gives the instantaneous output voltage as

$$v_O(t) = V_{th} - \frac{V_{sat} - V_{ref}}{RC}\,t \tag{4.41}$$

At the end of this mode (at $t = t_1$), the output voltage has changed to $-V_{th}$, whose value can be found from Eq. (4.33). That is, $v_O(t = t_1) = -V_{th}$. From Eq. (4.41),

$$-V_{th} = V_{th} - \frac{V_{sat} - V_{ref}}{RC}\,t_1$$

which gives the duration of mode 1 as

$$t_1 = \frac{2RCV_{th}}{V_{sat} - V_{ref}} \tag{4.42}$$

During mode 2, $v_+ < 0$ and the output of A_1 is at negative saturation $-V_{sat}$, which is the input to the integrator A_2. The equivalent circuit for the operation of the integrator is shown in Fig. 4.40(d). At the beginning of this mode, the output voltage is $-V_{th}$, and the initial voltage on the capacitor is

$$v_C(t=0) = v_1 - v_O = V_{ref} + V_{th}$$

If we redefine the time origin $t = 0$ as the beginning of this mode, the instantaneous voltage across the capacitor is given by

$$V_{ref} - v_O(t) = \frac{1}{C}\int i_C\,dt + v_C(t=0) = \frac{1}{C}\int \frac{-V_{sat} - V_{ref}}{R}\,dt + V_{ref} + V_{th}$$

which gives the instantaneous output voltage as

$$v_O(t) = -V_{th} + \frac{V_{sat} + V_{ref}}{RC}\,t \tag{4.43}$$

At the end of this mode (at $t = t_2$), the output voltage has changed to V_{th}, whose value can be found from Eq. (4.34); that is, $v_O(t = t_2) = V_{th}$. From Eq. (4.43),

$$V_{th} = -V_{th} + \frac{V_{sat} + V_{ref}}{RC}\,t_2$$

which gives the duration for mode 2 as

$$t_2 = \frac{2RCV_{th}}{V_{sat} + V_{ref}} \tag{4.44}$$

The period of the sawtooth wave can be found from Eqs. (4.42) and (4.44):

$$T = t_1 + t_2 = \frac{2RCV_{th}}{V_{sat} - V_{ref}} + \frac{2RCV_{th}}{V_{sat} + V_{ref}} = \frac{4RCV_{th}V_{sat}}{V_{sat}^2 - V_{ref}^2} \tag{4.45}$$

which gives the frequency of oscillation as

$$f_o = \frac{V_{sat}^2 - V_{ref}^2}{4RCV_{th}V_{sat}} \tag{4.46}$$

Duty cycle k, which is defined as the ratio of t_1 to T, can be determined from Eqs. (4.42) and (4.45) as follows:

$$k = \frac{t_1}{T} = \frac{2RCV_{th}}{V_{sat} - V_{ref}} \times \frac{V_{sat}^2 - V_{ref}^2}{4RCV_{th}V_{sat}} = \frac{V_{sat} + V_{ref}}{2V_{sat}}$$

$$= \frac{1}{2}\left[1 + \frac{V_{ref}}{V_{sat}}\right] \tag{4.47}$$

▶ **NOTE** This circuit allows a designer more control over the shape of the output waveform because k, T, and V_{th} can be varied.

EXAMPLE 4.10

D **Designing a sawtooth-wave generator** Design the sawtooth-wave generator shown in Fig. 4.40(a) so that $f_o = 4$ kHz, $V_{th} = 5$ V, and the circuit has a duty cycle of $k = 0.25$. Assume $V_{sat} = |-V_{sat}| = 14$ V.

SOLUTION

The steps used to design the sawtooth-wave generator are as follows:

Step 1. Find the value of V_{ref} required to obtain the desired duty cycle k. From Eq. (4.47),

$$V_{ref} = (2k - 1)V_{sat} \tag{4.48}$$

$$= (2 \times 0.25 - 1) \times 14 = -7 \text{ V}$$

Step 2. Find the values of R_1 and R_F. Equation (4.34) gives

$$\frac{R_1}{R_F} = \frac{V_{th}}{V_{sat}} = \frac{5}{14} = 0.36$$

Let $R_1 = 10$ kΩ; then

$$R_F = \frac{R_1}{0.36} = \frac{10 \text{ k}\Omega}{0.36} = 28 \text{ k}\Omega \quad \text{(use a 30-k}\Omega\text{ potentiometer)}$$

Step 3. Choose a suitable value of C: Let $C = 0.01$ μF.

Step 4. Find the value of R. Equation (4.46) gives

$$R = \frac{V_{sat}^2 - V_{ref}^2}{4f_o CV_{th}V_{sat}} \tag{4.49}$$

$$= \frac{14^2 \text{ V}^2 - (-7)^2 \text{ V}^2}{4 \times 4 \text{ kHz} \times 0.01 \text{ μF} \times 5 \text{ V} \times 14 \text{ V}} = 13.1 \text{ k}\Omega \quad \text{(use a 15-k}\Omega\text{ potentiometer)}$$

KEY POINT OF SECTION 4.8

- A sawtooth waveform has unequal rise and fall times. The rise time may be faster than the fall time or vice versa. A triangular-wave generator can be converted to a sawtooth-wave generator by adding a variable DC voltage V_{ref} to the noninverting terminal of the op-amp.

4.9 Voltage-Controlled Oscillators

A *voltage-controlled oscillator* (VCO) is an oscillator circuit in which the oscillation frequency is controlled by an externally applied voltage [5, 6]. A linear relationship between the oscillation frequency f_o and the control voltage v_C is often required in a VCO. A VCO is also called a *voltage-to-frequency* (V/F) *converter*. VCOs are employed in many applications such as frequency modulation (FM), tone generation, and frequency-shift keying (FSK).

To convert a voltage to a frequency, a capacitor is normally charged and then discharged at a constant current whose value depends on an externally applied voltage. The charging begins when the capacitor voltage falls to a lower threshold voltage V_L. Similarly, the discharging begins when the capacitor voltage rises to an upper threshold voltage V_H.

Figure 4.41(a) shows a simplified block diagram of the operation of a VCO. The current sources are used to charge and discharge capacitor C_1. The input to the Schmitt trigger is the capacitor voltage; the output of the Schmitt trigger has two threshold voltage switching levels V_L and V_H, which control the closing and opening of the current switch. Thus, depending on the capacitor voltage $v_C(t)$, the current switch connects the capacitor either to the top current source for charging or to the bottom current source for discharging. The waveform of the capacitor voltage is shown in Fig. 4.41(b); it has two modes, charging and discharging.

4.9.1 Charging Mode

During charging mode, the capacitor is charged by the top current source I_Q from the lower threshold voltage V_L to the upper trigger level V_H. The time required to charge the capacitor from V_L to V_H is given by

$$\Delta t_1 = \frac{C_1}{I_Q}\Delta v_C = \frac{C_1}{I_Q}(V_H - V_L) \tag{4.50}$$

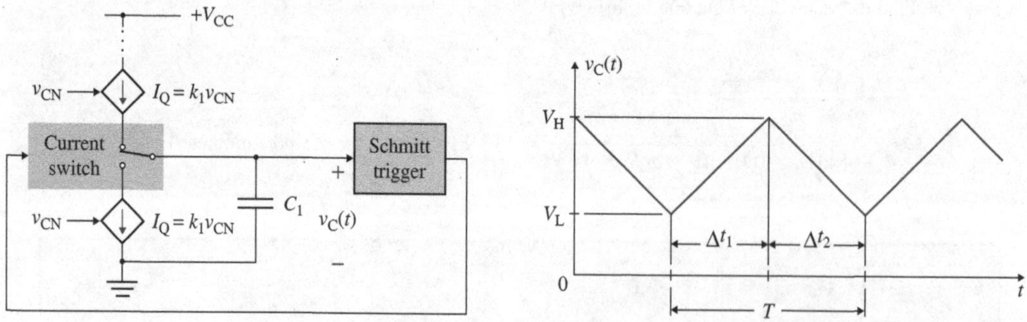

(a) Block diagram (b) Waveform of capacitor voltage

FIGURE 4.41 Principle of a voltage-controlled oscillator

4.9.2 Discharging Mode

The discharging mode begins when the capacitor is charged to the upper trigger level V_H. At this point the current switch disconnects the top current source from C_1 and connects the bottom current source to C_1. The capacitor is then discharged by the bottom current source down to the lower trigger level V_L. The time required to discharge the capacitor from V_H to V_L is given by

$$\Delta t_2 = -\frac{C_1}{I_Q}\Delta v_C = -\frac{C_1}{I_Q}(V_L - V_H) = \frac{C_1}{I_Q}(V_H - V_L) \tag{4.51}$$

As long as the charging and discharging currents are maintained at the same magnitude of I_Q, $\Delta t_1 = \Delta t_2$. The period of oscillation T is given by

$$T = \Delta t_1 + \Delta t_2 = \frac{2C_1(V_H - V_L)}{I_Q} \tag{4.52}$$

which gives the frequency of oscillation f_o as

$$f_o = \frac{1}{T} = \frac{I_Q}{2C_1(V_H - V_L)} \tag{4.53}$$

Let us assume that the voltage-controlled current sources have a linear voltage-to-current transfer relationship; that is,

$$I_Q = G_m(v_{CN} + V_{CO}) \tag{4.54}$$

where G_m = transconductance of the current source, in A/V
v_{CN} = applied control voltage, in V
V_{CO} = constant voltage, in V

Therefore, the oscillation frequency will be a linear function of the control voltage v_{CN}; that is,

$$f_o = \frac{I_Q}{2C_1(V_H - V_L)} = \frac{G_m(v_{CN} + V_{CO})}{2C_1(V_H - V_L)} \tag{4.55}$$

which gives the voltage-to-frequency transfer coefficient K_{vf} as

$$K_{vf} = \frac{df_o}{dv_{CN}} = \frac{G_m}{2C_1(V_H - V_L)} \tag{4.56}$$

4.9.3 Circuit Implementation

A circuit implementation for the charging and discharging of the capacitor is shown in Fig. 4.42. It can be divided into two parts: a voltage-controlled current source and a current switch.

Voltage-Controlled Current Source

The voltage-controlled current source consists of an *npn* transistor Q_1 and a *pnp* transistor Q_2. The voltage at the emitter of Q_2 is given by

$$V_{E2} = v_{CN} - V_{BE1} + V_{BE2}$$

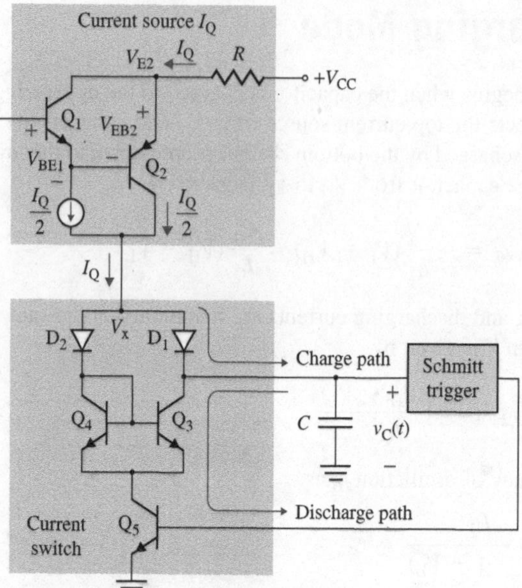

FIGURE 4.42 Circuit implementation

which gives the current source I_Q flowing through R as

$$I_Q = \frac{V_{CC} - V_{E2}}{R} = \frac{V_{CC} - v_{CN} + V_{BE1} - V_{EB2}}{R} \tag{4.57}$$

Since V_{BE1} (for an *npn* transistor) $\approx V_{EB2}$ (for a *pnp* transistor) within the range from 10 mV to 50 mV, I_Q in Eq. (4.57) can be approximated by

$$I_Q \approx \frac{V_{CC} - v_{CN}}{R} \tag{4.58}$$

which gives a linear relationship between the current source I_Q and the control voltage v_{CN}.

Current Switch

The current switch consists of diodes D_1 and D_2 and transistors Q_3, Q_4, and Q_5. Transistor Q_5 is controlled by the Schmitt trigger and is operated as an on or off switch. When Q_5 is off, capacitor C is charged by the current source I_Q via diode D_1. Diode D_2 and transistors Q_3 and Q_4 are off. When Q_5 is turned on (in saturation) by the Schmitt trigger, D_2, Q_3, and Q_4 are also turned on. As a result, the current source I_Q flows through D_2 and Q_4 instead of via diode D_1. The voltage at the anode of diode D_2 becomes

$$V_x = V_{D2(\text{anode})} = V_{D2} + V_{BE4} + V_{CES(\text{sat})} = 0.6 + 0.6 + 0.2 = 1.4 \text{ V}$$

However, the voltage at the cathode of diode D_1 is the capacitor voltage $v_C(t)$, which will be larger than 1.4 V. Thus, diode D_1 will be reverse biased (i.e., off). Capacitor C will discharge through Q_3 at a rate of I_Q. The current through Q_3 is equal to the current through Q_4. Thus, transistor Q_5 will carry a current of $2I_Q$.

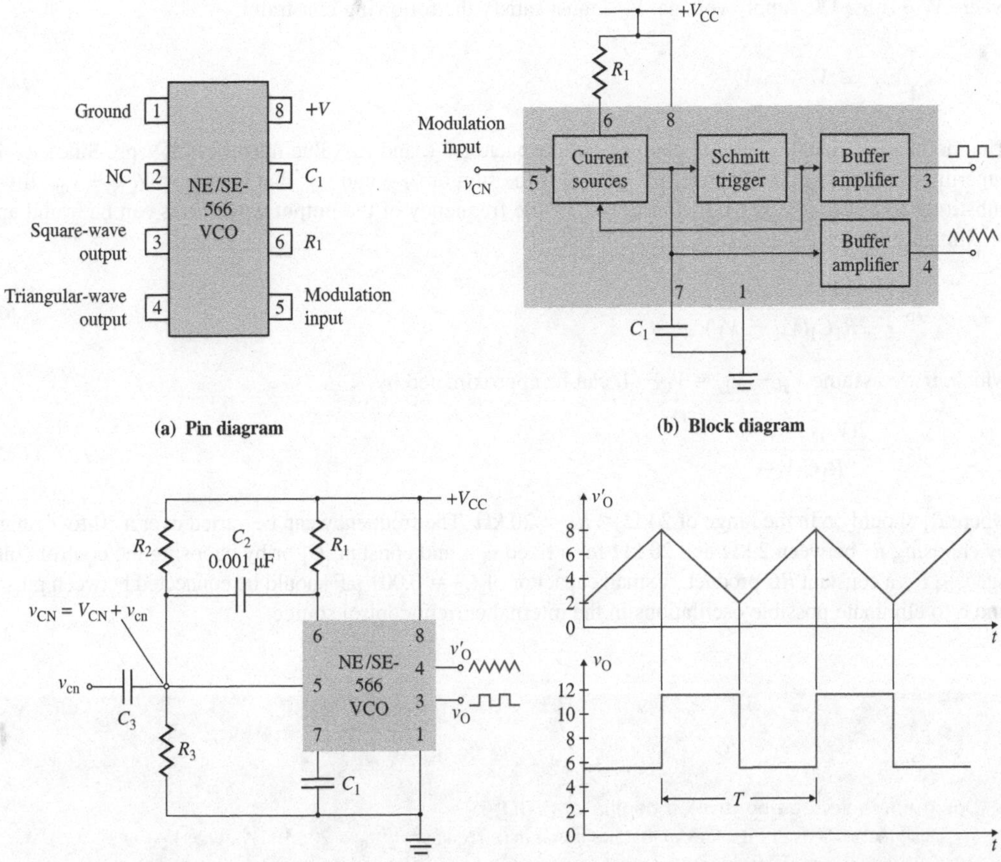

(a) Pin diagram (b) Block diagram (c) Circuit (d) Output waveforms

FIGURE 4.43 Voltage-controlled oscillator NE/SE-566 (Courtesy of Philips Semiconductors)

4.9.4 The NE/SE-566 VCO

A typical example of a VCO-integrated circuit is the NE/SE-566 VCO, whose pin diagram is shown in Fig. 4.43(a) and whose internal block diagram is shown in Fig. 4.43(b). This VCO produces simultaneous square-wave and triangular-wave outputs at frequencies of up to 1 MHz. Both outputs are buffered so that the output impedance of each is 50 Ω. The typical amplitude of the square wave is 5.4 V pp (peak to peak), and that of the triangular wave is 2.4 V pp. A typical connection diagram is shown in Fig. 4.43(c), and the typical outputs are shown in Fig. 4.43(d).

The output frequency is determined by an external resistor R_1, capacitor C_1, and the voltage v_{CN} applied to control terminal 5. The nominal DC value V_{CN} of v_{CN} is set by the voltage divider formed by R_2 and R_3 and must be in the range of

$$V_{CN} = \frac{R_3}{R_2 + R_3} V_{CC} \tag{4.59}$$

where V_{CC} is the DC supply voltage. V_{CN} must satisfy the following constraint:

$$\frac{3}{4}V_{CC} \leq V_{CN} \leq V_{CC} \tag{4.60}$$

The modulating signal v_{cn} is AC-coupled with capacitor C_3, and its value must be <3 V pp. Since v_{cn} is superimposed on V_{CN}, the control voltage v_{CN} is the sum of V_{CN} and v_{cn}; that is, $v_{CN} = V_{CN} + v_{cn}$. If we substitute $I_Q = (V_{CC} - v_{CN})/R_1$ in Eq. (4.53), the frequency of the output waveforms can be found approximately from

$$f_o = \frac{V_{CC} - v_{CN}}{2R_1C_1(V_H - V_L)} \tag{4.61}$$

which, if we assume $V_H - V_L = V_{CC}/4$, can be approximated by

$$f_o \approx \frac{2(V_{CC} - v_{CN})}{R_1C_1V_{CC}} \tag{4.62}$$

where R_1 should be in the range of $2\ \text{k}\Omega < R_1 < 20\ \text{k}\Omega$. The frequency can be varied over a 10-to-1 range by choosing R_1 between 2 kΩ and 20 kΩ for a fixed v_{CN} and constant C_1 or by choosing the control voltage v_{CN} for a constant RC product. A small capacitor of $C_2 = 0.001\ \mu\text{F}$ should be connected between pins 5 and 6 to eliminate possible oscillations in the internal current control source.

EXAMPLE 4.11

D

Designing a voltage-controlled oscillator (VCO)

(a) Design a VCO as in Fig. 4.43(c) that has a nominal frequency of $f_o = 20$ kHz. Assume $V_{CC} = 12$ V.

(b) Calculate the modulation in the output frequencies if v_{CN} is varied by ±10% because of the modulating signal v_{cn}.

SOLUTION

(a) The steps used to design the VCO are as follows:

Step 1. Find the limiting values of V_{CN}. From Eq. (4.60), $9\ \text{V} \leq V_{CN} \leq 12$ V.

Step 2. Choose a suitable value of C_1: Let $C_1 = 0.001\ \mu\text{F}$.

Step 3. Choose a value of R_1 between 2 kΩ and 20 kΩ: Let $R_1 = 10$ kΩ.

Step 4. Find the value of v_{CN}. From Eq. (4.62), v_{CN} is given by

$$v_{CN} = V_{CC}\left(1 - \frac{R_1C_1f_o}{2}\right) \tag{4.63}$$

$$= 12\left(1 - \frac{10\ \text{k}\Omega \times 0.001\ \mu\text{F} \times 20\ \text{kHz}}{2}\right) = 12 \times 0.9 = 10.8\ \text{V}$$

which falls within the range specified in step 1. If the calculated value falls outside of the specified range, choose a different value for R_1 and/or C_1 and recalculate v_{CN}.

Step 5. Find the values of R_2 and R_3. Letting $V_{CN} = v_{CN} = 10.8$ V from step 4, Eq. (4.59) gives

$$1 + \frac{R_2}{R_3} = \frac{V_{CC}}{V_{CN}} \tag{4.64}$$

$$= \frac{12}{10.8} = 1.11$$

so $R_2/R_3 = 1.11 - 1 = 0.11$. Let $R_3 = 100$ kΩ; then

$$R_2 = 0.11 \times R_3 = 11 \text{ k}\Omega$$

(b) For a 10% increase in v_{CN},

$$v_{CN}^{+} = V_{CN} + v_{cn} = 1.1 \times v_{CN} = 1.1 \times 10.8 = 11.88 \text{ V}$$

The corresponding value of output frequency can be calculated from Eq. (4.62):

$$f_{o1} \approx \frac{2 \times (12 - 11.88) \text{ V}}{10 \text{ k}\Omega \times 0.001 \text{ }\mu\text{F} \times 12 \text{ V}} = 2 \text{ kHz}$$

For a 10% decrease in v_{CN},

$$v_{CN}^{-} = V_{CN} - v_{cn} = 0.9 \times V_C = 0.9 \times 10.8 = 9.72 \text{ V}$$

The output frequency is

$$f_{o2} \approx \frac{2 \times (12 - 9.72) \text{ V}}{10 \text{ k}\Omega \times 0.001 \text{ }\mu\text{F} \times 12 \text{ V}} = 38 \text{ kHz}$$

Thus, the change in the output frequency is

$$\Delta f_o = f_{o2} - f_{o1} = 38 \text{ kHz} - 2 \text{ kHz} = 36 \text{ kHz}$$

Using Eq. (4.62), we can find the V/F transfer coefficient K_{vf}:

$$K_{vf} = \frac{df_o}{dv_{CN}} = -\frac{2}{R_1 C_1 V_{CC}} \tag{4.65}$$

$$= -\frac{2}{10 \text{ k}\Omega \times 0.001 \text{ }\mu\text{F} \times 12 \text{ V}} = -16.67 \text{ kHz/V}$$

Thus, for $\Delta v_{CN} = 2v_{CN} = -2 \times 0.1 \times 10.8 = -2.16$ V,

$$\Delta f_o = f_{o2} - f_{o1} = K_{vf}\Delta v_{CN} = -16.67 \text{ kHz/V} \times (-2.16 \text{ V}) = 36.01 \text{ kHz}$$

which is the same as the value obtained by calculating individual frequencies.

NOTE: If the modulating signal is a sine wave so that $v_{cn} = V_m \sin \omega t$, then the control voltage becomes $v_{CN} = V_{CN} + v_m \sin \omega t$. During the positive half-cycle of the modulating signal, the control voltage will increase and the frequency f_o of the output voltage will decrease. However, during the negative half-cycle of the modulating signal, the control voltage will decrease and the frequency f_o of the output voltage will increase.

KEY POINTS OF SECTION 4.9

- A *voltage-controlled oscillator* (VCO) is an oscillator circuit in which the oscillation frequency is controlled by an externally applied voltage.
- To convert a voltage to frequency, a capacitor is normally charged and then discharged at a constant current whose value depends on an externally applied voltage. The output of a Schmitt trigger controls the charging and discharging time of the capacitor.
- The NE/SE-566 VCO can produce simultaneous square-wave and triangular-wave outputs at frequencies of up to 1 MHz. The output frequency is determined by an external resistor R_1, capacitor C_1, and the voltage v_{CN} applied to control terminal 5.

4.10 The 555 Timer

The 555 timer, introduced by Signetics Corporation in early 1970, is one of the most versatile integrated circuits. The 555 is a monolithic timing circuit that can produce highly accurate and highly stable delays or oscillation. It is used in many applications, such as monostable and astable multivibrators, digital logic probes, analog frequency meters, tachometers, infrared transmitters, and burglar and toxic gas alarms. Several versions of the 555 timer are available from various manufacturers. In addition to looking briefly at its internal structure, we consider two common applications of the 555 timer: monostable multivibrator and astable multivibrator.

4.10.1 Functional Block Diagram

The pin diagram of a 555 timer is shown in Fig. 4.44(a); its functional block diagram is shown in Fig. 4.44(b). The timer consists of two comparators CM_1 and CM_2, an *RS* flip-flop, a discharge transistor Q_1, and a resistive voltage divider string. The voltage divider sets the voltage at the inverting terminal of CM_1 to $2V_{CC}/3$ and the voltage at the noninverting terminal of CM_2 to $V_{CC}/3$.

The reset, threshold, and trigger inputs control the state of the flip-flop. If the reset input is low, the Q output of the flip-flop is low, and $\overline{Q}$ is high. With $\overline{Q}$ high, current flows through the base of transistor Q_1, and the transistor is switched on (in saturation). This generally provides a path for an external capacitor to discharge its voltage.

The reset input has the highest priority in setting the state of the flip-flop. Thus, Q is low if the reset input is low, regardless of the inputs to the comparators. If the reset is not in use, then it is connected to the positive DC supply V_{CC} so that it does not affect the state of the flip-flop.

If the trigger input becomes lower than the voltage at the noninverting input of CM_2 (i.e., $<V_{CC}/3$), the output of CM_2 (i.e., the S input to the flip-flop) will be high. As a result, the Q output of the flip-flop will be set to high. Thus, $\overline{Q}$ will be low, and the discharge transistor Q_1 will be off.

If the threshold input becomes higher than the voltage at the inverting input of CM_1 (i.e., $>2V_{CC}/3$), the output of CM_1 will be high. As a result, the Q output of the flip-flop will be reset to low. Thus, $\overline{Q}$ will be high, and the discharge transistor Q_1 will be on (in saturation), providing a discharge path.

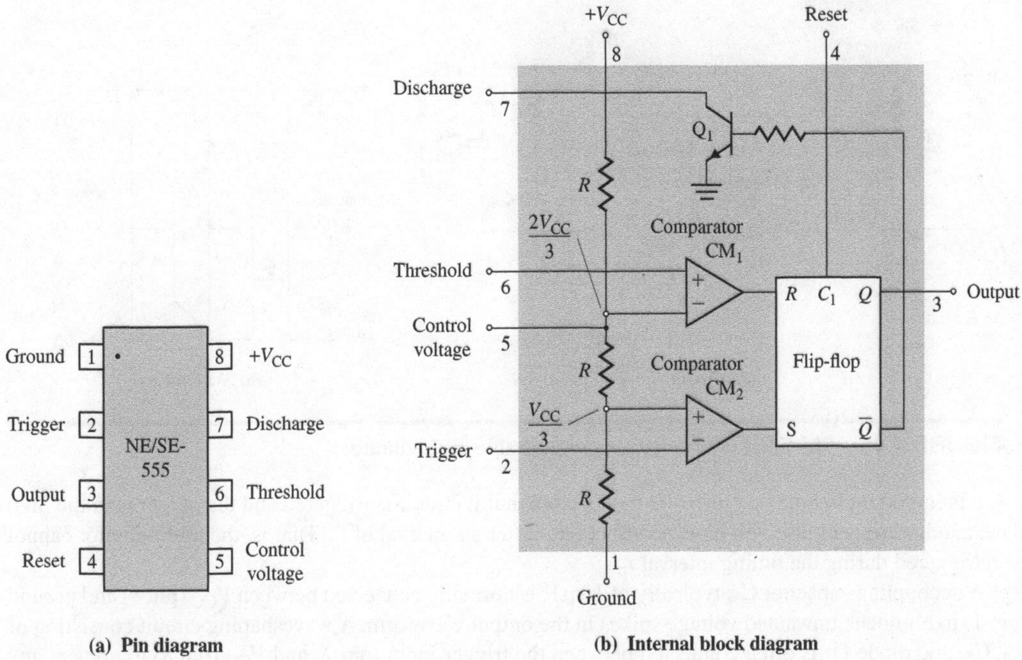

FIGURE 4.44 Functional block diagram of 555 timer (Courtesy of Philips Semiconductors)

4.10.2 Monostable Multivibrators

A monostable multivibrator is a *one-shot* pulse-generating circuit. Normally, its output is zero—that is, at logic low level in the stable state. This circuit has only one stable state at output low—hence the name *monostable*. The circuit configuration of the 555 timer for monostable operation is shown in Fig. 4.45(a). The discharge and threshold terminals are connected together. The external pulse v_I is applied to the trigger terminal through coupling capacitor C_2.

If the external pulse v_I is high, the Q output of the flip-flop is low; that is, the output of the timer is low. At the negative edge of the trigger signal, the flip-flop will be set to high. Thus, the output will be switched to high ($\approx V_{CC}$). It will remain high until the capacitor is charged to the threshold voltage of $2V_{CC}/3$, at which time the flip-flop will be reset and the output will return to zero.

The duration of the output pulse (t_p) is determined by the RC network connected externally to the 555 timer. At the end of the timing interval t_p, the output automatically reverts to its stable state of logic low. The output remains low until another negative-going trigger pulse is applied. Then the cycle repeats. The waveforms for the trigger input voltage $v_I(t)$, output voltage $v_O(t)$, and capacitor voltage $v_C(t)$ are shown in Fig. 4.45(b).

The width of the triggering pulse must be smaller than the expected pulse width of the output waveform. Also, the trigger pulse must be a negative-going signal and have an amplitude larger than $V_{CC}/3$. The time during which the output remains high is given by

$$t_p = 1.1RC \tag{4.66}$$

where R and C are the external resistance and capacitance, respectively.

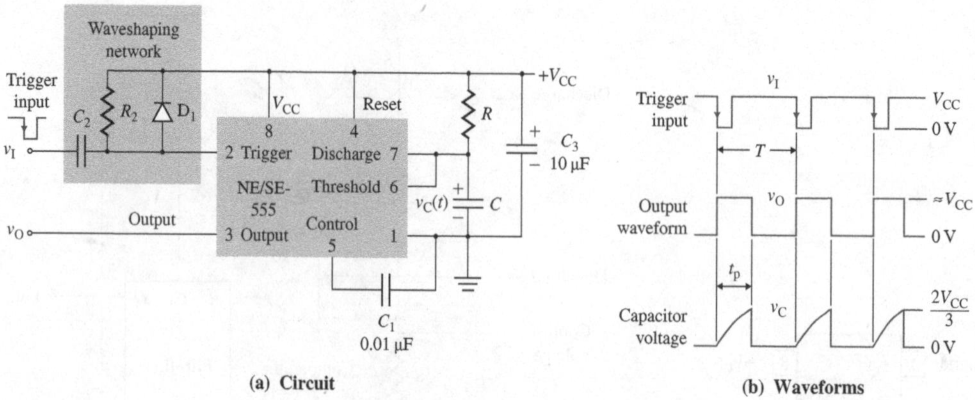

FIGURE 4.45 555 timer connected as a monostable multivibrator

It is important to note that once the monostable multivibrator is triggered and the output is in the high state, another trigger pulse will have no effect until after an interval of t_p. That is, the multivibrator cannot be retriggered during the timing interval t_p.

A decoupling capacitor C_3, typically of 10 μF, is normally connected between V_{CC} (pin 8) and ground (pin 1) to eliminate unwanted voltage spikes in the output waveform. A waveshaping circuit consisting of R_2, C_2, and diode D_1 is often connected between the trigger input (pin 2) and V_{CC} (pin 8) to prevent any possible mistriggering by positive pulse edges. This circuit is shown in Fig. 4.45(a) within the shaded area. The values of R_2 and C_2 should be selected so that the output pulse width t_p is much larger than the time constant R_2C_2. This condition is generally met by maintaining the following relationship:

$$t_p = 10R_2C_2 \tag{4.67}$$

EXAMPLE 4.12

D **Designing a monostable multivibrator** Design a monostable multivibrator as in Fig. 4.45(a) so that t_p = 5 ms. Assume V_{CC} = 12 V.

SOLUTION

The steps used to design a monostable multivibrator are as follows:

Step 1. Choose a suitable value of C: Let C = 0.1 μF.

Step 2. Find the value of R. From Eq. (4.66),

$$R = \frac{t_p}{1.1C} = \frac{5\text{ ms}}{1.1 \times 0.1\ \mu\text{F}} = 45.5\text{ k}\Omega \quad \text{(use a 50-k}\Omega\text{ potentiometer)}$$

Step 3. Choose a suitable value of C_2: Let C_2 = 0.01 μF.

Step 4. Find the value of R_2. From Eq. (4.67),

$$R_2 = \frac{t_p}{10C_2} = \frac{5\text{ ms}}{10 \times 0.01\ \mu\text{F}} = 50\text{ k}\Omega$$

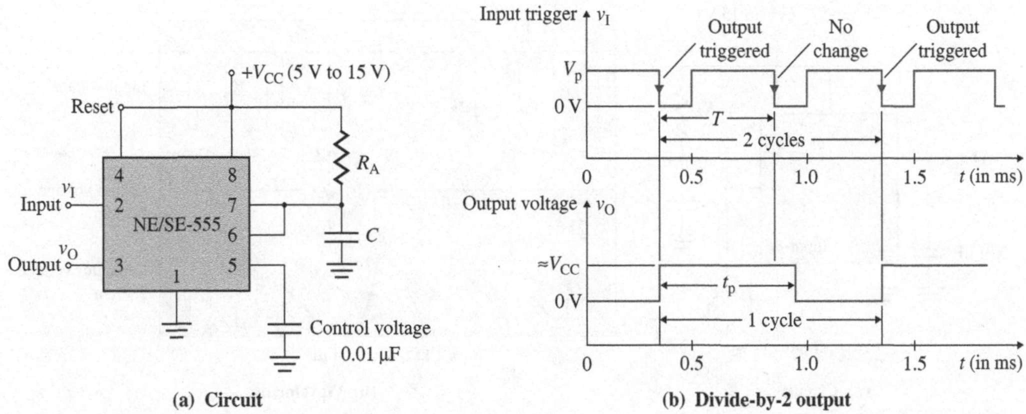

FIGURE 4.46 Monostable multivibrator as a frequency divider

4.10.3 Applications of Monostable Multivibrators

Monostable multivibrators can be used in many applications such as frequency dividers, missing-pulse detectors, and pulse stretchers.

Frequency Divider

If the frequency of the input signal is known, adjusting the length of the timing cycle t_p will permit a monostable multivibrator to be used as a frequency divider. The circuit configuration for a frequency divider is shown in Fig. 4.46(a). This application makes use of the fact that the monostable multivibrator cannot be retriggered during the timing interval.

For a monostable multivibrator to be used as a divide-by-2 circuit, the timing interval must be slightly larger (say, by 20%) than the period of the trigger signal T, as shown in Fig. 4.46(b); that is, $t_p = 1.2T$. At the first falling edge, the output is set to high. However, at the second falling edge, the capacitor is still charging and has not yet reached the threshold value of $2V_{CC}/3$. As a result, the second triggering signal has no effect on the output, and the output is set by the alternate trigger pulses.

For a monostable multivibrator to be used as a divide-by-3 circuit, t_p must be slightly larger than twice the period of the trigger signal. Thus, for a divide-by-n circuit, t_p must be slightly larger than $(n - 1)T$; that is, $t_p = [0.2 + (n - 1)]T$. For example, for a divide-by-2 circuit, if $f_o = 5$ kHz and $T = 1/f_o = 200$ µs, then $t_p = 1.2T = 240$ µs.

Missing-Pulse Detector

In some applications, a train of regular pulses is needed for the normal operation of a circuit or system. Any missing pulse may cause malfunction. The configuration of a monostable multivibrator that can detect any missing pulse is shown in Fig. 4.47(a).

The timing cycle is continuously reset by the input pulse train, but the time duration is not enough to complete the timing cycle. When the trigger pulse becomes low, transistor Q_1 is turned on, and it provides a discharge path for capacitor C. As a result, the capacitor voltage cannot reach the threshold voltage $2V_{CC}/3$. A change in frequency, or a missing pulse, allows completion of the timing cycle so that v_C

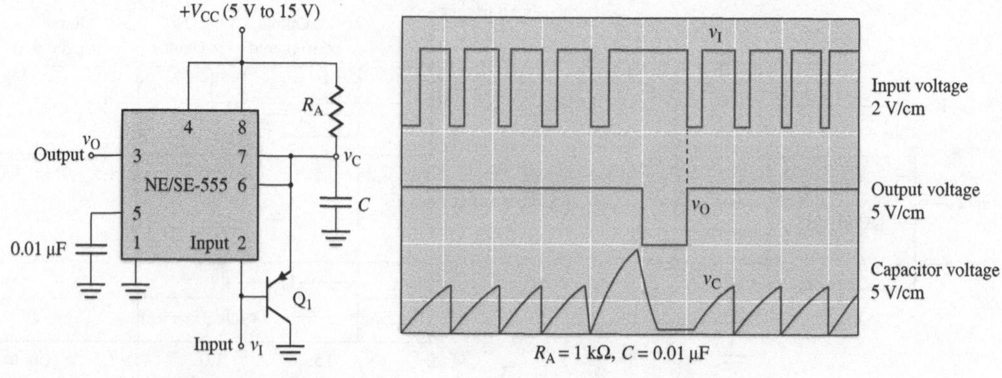

(a) Circuit (b) Waveforms

FIGURE 4.47 Monostable multivibrator as a missing-pulse detector

reaches $2V_{CC}/3$, causing a change in the output level due to the discharge of C through the internal transistor Q_1 in Fig. 4.44(b). The time delay t_p should be slightly longer than the normal time between input pulses. The waveforms are shown in Fig. 4.47(b).

Pulse Widener

A narrow pulse is not desirable for driving an LED display, as the flashing of the LED will not be visible to the eyes if the on time is infinitesimally small compared to the off time. A narrow pulse can be widened by a monostable multivibrator. This application is possible because of the fact that the timing interval t_p is longer than the negative pulse width of the trigger input.

The circuit configuration for a pulse widener is shown in Fig. 4.48(a). At the falling trigger edge, the output will be high, and it will remain high until the capacitor voltage reaches $2V_{CC}/3$ after time t_p. The waveforms of the input and output voltages are shown in Fig. 4.48(b). Since the output pulse can be viewed as the stretched version of the narrow input signal, this configuration is also known as a *pulse stretcher*.

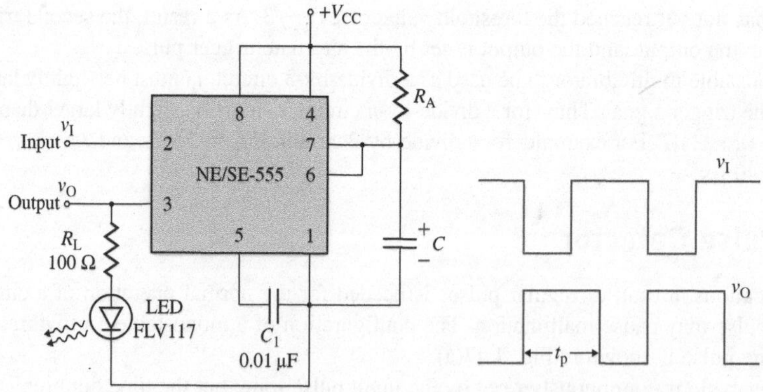

(a) Circuit (b) Waveforms

FIGURE 4.48 Monostable multivibrator as a pulse widener

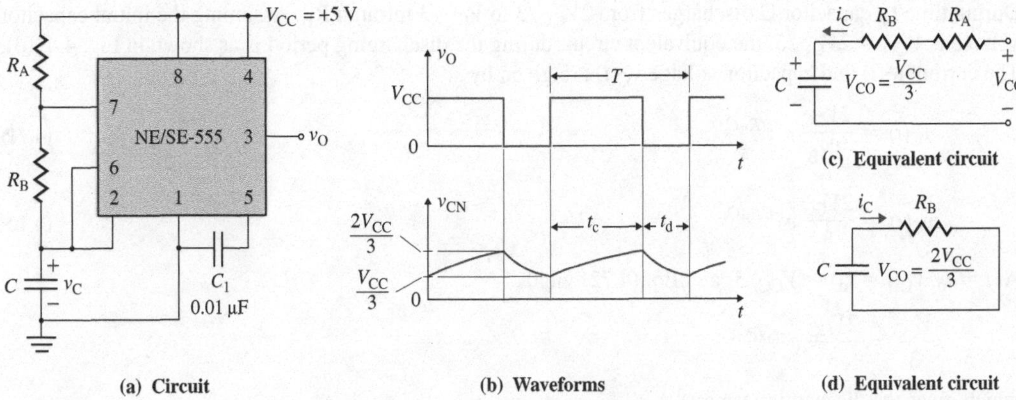

FIGURE 4.49 555 timer connected as an astable multivibrator

4.10.4 Astable Multivibrators

An astable multivibrator is a rectangular-wave-generating circuit. Because this circuit does not require an external trigger to change the state of the output, it is often called a *free-running multivibrator*. A 555 timer connected as an astable multivibrator is shown in Fig. 4.49(a). The duration of the high or low output is determined by resistors R_A and R_B and capacitor C. When the output is high, capacitor C starts charging toward V_{CC} through R_A and R_B. As soon as the capacitor voltage equals $2V_{CC}/3$, the output switches to low, and capacitor C discharges through R_B and the internal circuit of the timer. When the capacitor voltage equals $V_{CC}/3$, the output goes high, and the capacitor charges through R_A and R_B. Then the cycle repeats. The waveforms for the output voltage and the voltage across the capacitor are shown in Fig. 4.49(b).

The capacitor is periodically charged and discharged between $2V_{CC}/3$ and $V_{CC}/3$. Assuming the initial capacitor voltage is $V_{CO} = V_{CC}/3$, the equivalent circuit during the charging period is as shown in Fig. 4.49(c). The charging current $i_C(t)$ and the capacitor voltage $v_C(t)$ are given by

$$i_C(t) = \frac{2V_{CC}}{3(R_A + R_B)} e^{-t/(R_A+R_B)C} \tag{4.68}$$

$$v_C(t) = V_{CC} - \frac{2V_{CC}}{3} e^{-t/(R_A+R_B)C} \tag{4.69}$$

At $t = t_c$, $v_C(t = t_c) = 2V_{CC}/3$, and Eq. (4.69) yields

$$\frac{2V_{CC}}{3} = V_{CC} - \frac{2V_{CC}}{3} e^{-t_c/(R_A+R_B)C}$$

which gives the charging time t_c as

$$t_c = C(R_A + R_B)\ln(2) = 0.69C(R_A + R_B) \tag{4.70}$$

During time t_d, capacitor C discharges from $2V_{CC}/3$ to $V_{CC}/3$ through R_B. Assuming the initial capacitor voltage is $V_{CO} = 2V_{CC}/3$, the equivalent circuit during the discharging period is as shown in Fig. 4.49(d). The current $i_C(t)$ and capacitor voltage $v_C(t)$ are given by

$$i_C(t) = \frac{2V_{CC}}{3R_B} e^{-t/R_BC} \tag{4.71}$$

$$v_C(t) = \frac{2V_{CC}}{3} e^{-t/R_BC} \tag{4.72}$$

At $t = t_d$, $v_C(t = t_d) = V_{CC}/3$, and Eq. (4.72) yields

$$\frac{V_{CC}}{3} = \frac{2V_{CC}}{3} e^{-t_d/R_BC}$$

which gives the discharging time t_d as

$$t_d = CR_B \ln(2) = 0.69CR_B \tag{4.73}$$

Thus, the period of the output waveform is given by

$$T = t_c + t_d = 0.69C(R_A + R_B) + 0.69CR_B = 0.69C(R_A + 2R_B) \tag{4.74}$$

and the frequency of the output voltage is therefore given by

$$f_o = \frac{1}{T} = \frac{1}{0.69C(R_A + 2R_B)} = \frac{1.45}{C(R_A + 2R_B)} \tag{4.75}$$

Duty cycle k, which is the ratio of charging time t_c to period T, can be found from Eqs. (4.70) and (4.74):

$$k = \frac{t_c}{T} = \frac{R_A + R_B}{R_A + 2R_B} \tag{4.76}$$

Thus, the duty cycle k can be set by selecting R_A or R_B.

EXAMPLE 4.13

D

Designing an astable multivibrator Design an astable multivibrator as in Fig. 4.49(a) so that $k = 75\%$ and $f_o = 2.5$ kHz. Assume $V_{CC} = 12$ V. Use PSpice to verify by simulating, EX4-13.SCH.

SOLUTION

$k = 75\% = 0.75$, and $T = 1/f_o = 1/2.5\text{ kHz} = 400\ \mu\text{s}$. The steps used to design an astable multivibrator are as follows:

Step 1. Find the charging time t_c and the discharging time t_d:

$$t_c = kT \tag{4.77}$$

$$= 0.75 \times 400\ \mu\text{s} = 300\ \mu\text{s}$$

$$t_d = (1 - k)T \tag{4.78}$$

$$= (1 - 0.75) \times 400\ \mu\text{s} = 100\ \mu\text{s}$$

Step 2. Choose a suitable value of C: Let $C = 0.1\ \mu F$.

Step 3. Find the value of R_B. From Eq. (4.73),

$$R_B = \frac{t_d}{0.69C} \tag{4.79}$$

$$= \frac{100\ \mu s}{0.69 \times 0.1\ \mu F} = 1449\ \Omega$$

Step 4. Find the value of R_A. From Eq. (4.70),

$$R_A = \frac{t_c}{0.69C} - R_B \tag{4.80}$$

$$= \frac{300\ \mu s}{0.69 \times 0.1\ \mu F} - 1449 = 2899\ \Omega$$

EXAMPLE 4.14

D **PSpice/SPICE simulation of an astable multivibrator** An astable multivibrator can be used as a voltage-controlled oscillator (VCO) if an external control voltage is applied to terminal 5. This arrangement is shown in Fig. 4.50 for $v_C = 6 + 4 \sin(2000\pi t)$. Use PSpice/SPICE to plot the output voltage $v_O(t)$ from 0 to 2 ms with an increment of 10 ns.

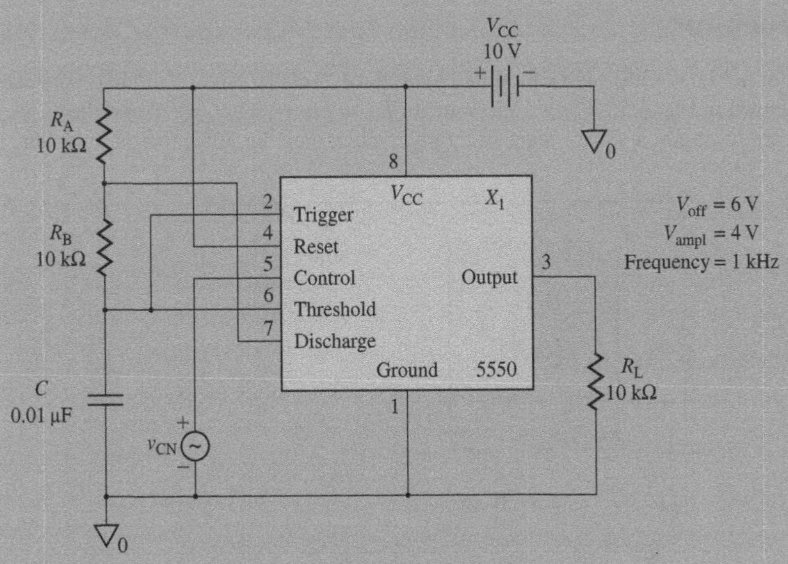

FIGURE 4.50 Astable multivibrator as a VCO for PSpice simulation

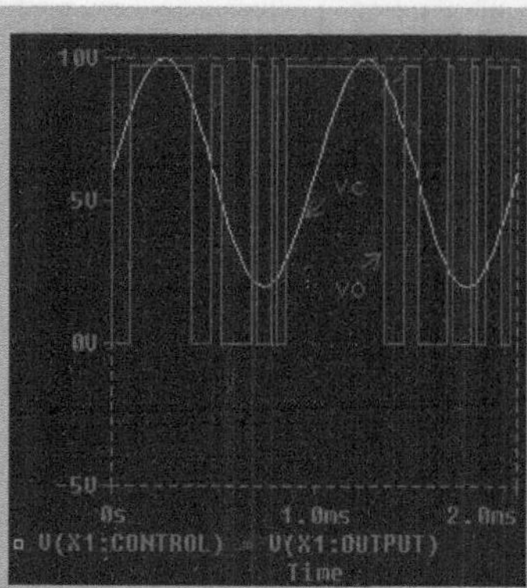

FIGURE 4.51 Control and output voltages for Example 4.14

SOLUTION

The PSpice plots of the control and output voltages are shown in Fig. 4.51. As expected, the frequency becomes lower—that is, the period becomes larger—as the control voltage increases in magnitude.

4.10.5 Applications of Astable Multivibrators

As examples of applications of an astable multivibrator, we consider a square-wave generator, a ramp generator, and an FSK modulator.

Square-Wave Generator

The astable multivibrator in Fig. 4.49(a) can be modified to produce a square wave. A diode is connected across resistor R_B, as shown in Fig. 4.52. For a finite value of R_A in Eq. (4.70), $t_c > t_d$, and the duty cycle is $k > 50\%$. However, to obtain a square wave, the duty cycle k must be 50%; that is, the value of

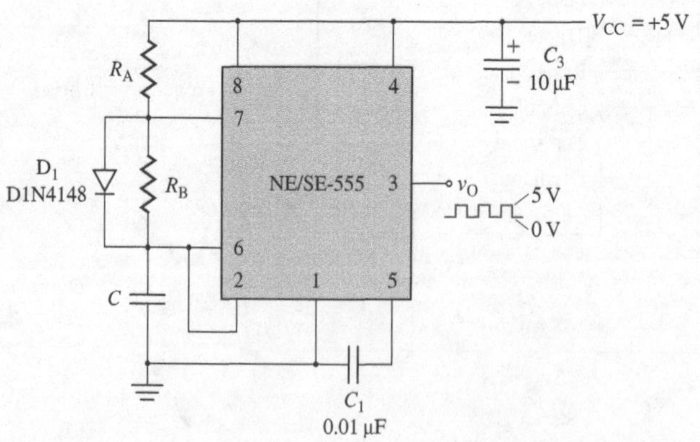

FIGURE 4.52 555 timer connected as a square-wave generator

R_A should be set to zero. With $R_A = 0$, terminal 7 is connected directly to V_{CC}. During the discharging time, capacitor C discharges through the internal circuit of the timer, and an external current is applied by V_{CC} to the same internal path. The current flowing through the internal path may be large enough to damage the timer. This situation can be avoided by connecting a diode across R_B so that R_B is bypassed during the charging time. In that case, Eqs. (4.74) and (4.75) are reduced, respectively, to

$$T = 0.69C(R_A + R_B) \quad \text{(for } k = 0.5) \tag{4.81}$$

$$f_o = \frac{1}{T} = \frac{1.45}{C(R_A + R_B)} \quad \text{(for } k = 0.5) \tag{4.82}$$

▶ **NOTE** We need both R_A and D_1. Making $R_A = 0$, the circuit will not work.

EXAMPLE 4.15

D

Designing a square-wave generator Design a square-wave generator as in Fig. 4.52 so that $k = 50\%$ and $f_o = 2.5$ kHz. Assume $V_{CC} = 12$ V. Use PSpice to verify by simulating, EX4-15.SCH.

SOLUTION

$k = 50\% = 0.5$, and $T = 1/f_o = 1/2.5\text{ kHz} = 400\ \mu\text{s}$. The steps used to design the square-wave generator are as follows:

Step 1. Find the charging time t_c and the discharging time t_d:

$$t_c = t_d = kT \tag{4.83}$$

$$= 0.5 \times 400\ \mu\text{s} = 200\ \mu\text{s}$$

Step 2. Choose a suitable value of C: Let $C = 0.1\ \mu$F.

Step 3. Find the value of R_B. From Eq. (4.73),

$$R_B = \frac{t_d}{0.69C} \tag{4.84}$$

$$= \frac{200\ \mu\text{s}}{0.69 \times 0.1\ \mu\text{F}} = 2899\ \Omega$$

Step 4. Find the value of R_A. From Eq. (4.82),

$$R_A = \frac{1.45}{Cf_o} - R_B \tag{4.85}$$

$$= \frac{1.45}{0.1\ \mu\text{F} \times 2.5\text{ kHz}} - 2899\ \Omega = 2901\ \Omega$$

Ramp Generator

The astable multivibrator in Fig. 4.49(a) can be used as a free-running ramp generator. This is accomplished by charging the capacitor through a constant current source and discharging through the internal circuit of the timer. That is, resistors R_A and R_B are replaced by a current source, as shown in Fig. 4.53(a). The waveforms for the output voltage and the capacitor voltage are shown in Fig. 4.53(b).

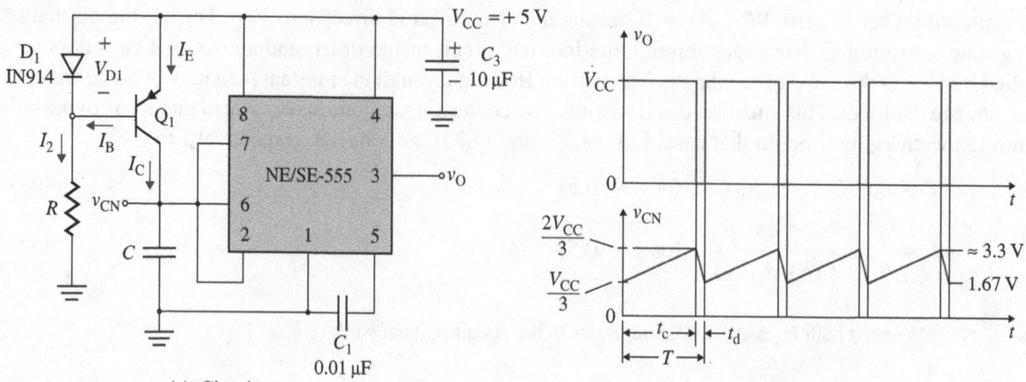

(a) Circuit (b) Waveforms

FIGURE 4.53 555 timer connected as a ramp generator

The collector current, which is the charging current, is given by

$$I_C = I_E - I_B$$

Assuming that the voltage drop of diode D_1 is approximately equal to the base–emitter voltage V_{BE} of the transistor, the diode current $I_D \approx I_E$. Thus,

$$I_C = I_D - I_B = I_2$$

$$= \frac{V_B}{R} = \frac{V_{CC} - V_{BE}}{R} \tag{4.86}$$

The capacitor charges from $V_{CC}/3$ to $2V_{CC}/3$ at a constant current of I_C. For a charging time of t_c, the change in the capacitor voltage Δv_C is given by

$$\Delta v_C = \frac{2V_{CC}}{3} - \frac{V_{CC}}{3} = \frac{1}{C}\int_0^{t_c} I_C\,dt = \frac{1}{C} I_C t_c$$

which gives the charging time t_c as

$$t_c = \frac{CV_{CC}}{3I_C} \tag{4.87}$$

The charging time is related to the duty cycle k and period T by

$$kT = t_c$$

$$= \frac{CV_{CC}}{3I_C} \tag{4.88}$$

Therefore, the free-running frequency of the ramp generator is given by

$$f_o = \frac{1}{T} = \frac{3kI_C}{CV_{CC}} \tag{4.89}$$

$$= \frac{3k(V_{CC} - V_{BE})}{CRV_{CC}} \tag{4.90}$$

If the discharging time t_d of the capacitor is negligible compared to its charging time t_c, then $k \approx 1$ and the free-running frequency becomes

$$f_o = \frac{3(V_{CC} - V_{BE})}{CRV_{CC}} \tag{4.91}$$

EXAMPLE 4.16

D **Designing a ramp generator** Design a ramp generator using the circuit in Fig. 4.53(a) so that $k = 50\%$ and $f_o = 2.5$ kHz. Assume $V_{CC} = 12$ V, $V_{BE} = 0.7$ V, and a transistor of $\beta_F = 150$.

SOLUTION

$k = 50\% = 0.5$, and $T = 1/f_o = 1/2.5$ kHz $= 400$ μs. The steps used to design the ramp generator are as follows:

Step 1. Find the charging time t_c and the discharging time t_d:

$$t_c = t_d = kT \tag{4.92}$$

$$= 0.5 \times 400 \text{ μs} = 200 \text{ μs}$$

Step 2. Choose a suitable value of C: Let $C = 0.1$ μF.

Step 3. Find the value of R. From Eq. (4.90),

$$R = \frac{3k(V_{CC} - V_{BE})}{V_{CC}Cf_o} \tag{4.93}$$

$$= \frac{3 \times 0.5 \times (12 - 0.7)\text{ V}}{12 \times 0.1 \text{ μF} \times 2.5 \text{ kHz}} = 5.65 \text{ kΩ}$$

Step 4. Find the collector current I_C of the transistor. From Eq. (4.86),

$$I_C = \frac{V_{CC} - V_{BE}}{R} = \frac{(12 - 0.7)\text{ V}}{5.65 \text{ kΩ}} = 2 \text{ mA}$$

Step 5. Find the current I_D through the diode:

$$I_D = I_E = I_C + I_B = I_C\frac{1 + \beta_F}{\beta_F} = 2 \text{ mA} \times \frac{1 + 150}{150} \approx 2.01 \text{ mA}$$

FSK Modulator

In computer peripheral and radio (wireless) communication, the binary data or code is transmitted by means of a carrier frequency that shifts between two preset frequencies. This technique for data transmission is called *frequency-shift keying* (FSK). The frequency shift is usually accomplished by driving a VCO with the binary data signal so that the 0 to 1 states (commonly called *space* and *mark*) of the binary data signal produce two frequencies, known as the *space* and *mark frequencies*. For example, when teletypewriter information is transmitted using a modulator/demodulator (*modem*, for short), a 1070-Hz (for mark) and 1270-Hz (for space) pair will represent the original signal, whereas a 2025-Hz (for mark) and 2250-Hz (for space) pair will represent the answer signal.

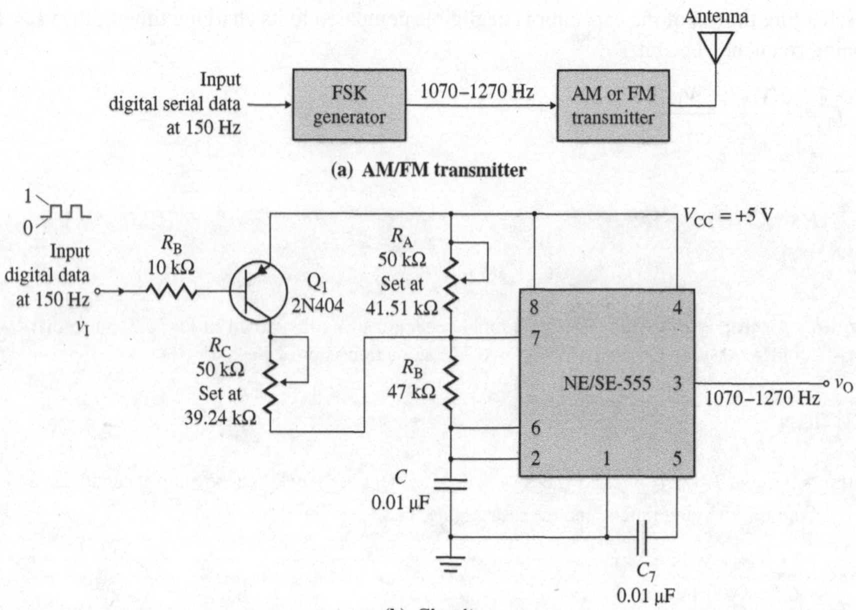

(b) Circuit

FIGURE 4.54 Astable multivibrator as an FSK modulator (Courtesy of Philips Semiconductors)

FSK modulators are often used in AM/FM transmitters, as shown in Fig. 4.54(a). The 555 astable multivibrator can be used as an FSK generator; the connection is shown in Fig. 4.54(b). The on or off condition of transistor Q_1 will depend on the input signal. Thus, the output frequency depends on the logic state of the digital input signal. A signal frequency of 150 Hz is commonly used for data transmission.

When the input signal is 1, transistor Q_1 is off, and the 555 operates in its normal mode as an astable multivibrator. Thus, the output frequency corresponding to logic 1 can be found from Eq. (4.75) as

$$f_{o(\text{mark})} = \frac{1.45}{C(R_A + 2R_B)}$$

The values for C, R_A, and R_B can be selected so as to give 1070 Hz.

When the input signal is 0, transistor Q_1 is turned on (in saturation). As a result, R_C is effectively connected in parallel with R_B. This reduces the charging time of capacitor C and increases the output frequency. Thus, the output frequency corresponding to logic 0 can be found from

$$f_{o(\text{space})} = \frac{1.45}{C(R_A \parallel R_C + 2R_B)}$$

The value for R_C can be selected so as to give 1270 Hz.

Thus, with properly selected values of C, R_A, R_B, and R_C, the 555 astable multivibrator can produce frequencies of 1070 Hz and 1270 Hz corresponding to 1 and 0, respectively. The difference between the FSK signals of 1270 Hz and 1070 Hz (i.e., 200 Hz) is called the *frequency shift*.

KEY POINTS OF SECTION 4.10

- The 555 timer is one of the most versatile ICs. It is used as a monostable or astable multivibrator in many applications.
- A monostable multivibrator is a *one-shot* pulse-generating circuit. This circuit has only one stable state at output low—hence the name *monostable*. The 555 can be configured as a monostable multivibrator, which can be used as a frequency divider, a missing-pulse detector, or a pulse widener.
- An astable multivibrator is a rectangular-wave-generating circuit. Because it does not require an external trigger to change the state of the output, it is often called a *free-running* multivibrator. The 555 can be configured as an astable multivibrator, which can be used as a square-wave generator, a ramp generator, or an FSK modulator.

4.11 Phase-Lock Loops

The phase-lock loop (PLL) is one of the fundamental building blocks of electronic circuits used in such applications as motor-speed controllers, FM stereo decoders, tracking filters, frequency-synthesized transmitters and receivers, and FSK decoders. A block diagram of a phase-lock loop is shown in Fig. 4.55(a). The loop consists of a phase detector, a low-pass filter, and a VCO.

The phase detector (or comparator) compares the phase of the input voltage with that of the VCO output voltage and produces a DC or low-frequency voltage proportional to their phase difference. The output of the phase detector, which is called the *error voltage*, is applied to a low-pass filter. The filter removes any high-frequency components and produces a smooth DC voltage. This DC voltage is then applied to the control input of the VCO, whose output frequency is proportional to the DC value. If the frequency of the input voltage shifts slightly, the phase difference between the input signal and the VCO output voltage will begin to increase with time. This will change the control voltage to the VCO in such a way as to bring the VCO frequency back to the same value as the input voltage. The VCO frequency is continuously adjusted until it is equal to the input frequency.

The operation of a PLL involves three modes: a *free-running mode*, a *capture mode*, and a *phase-lock mode*. During the free-running mode, there is no input frequency (or voltage), and the VCO runs at a fixed frequency corresponding to zero-applied input voltage. This frequency is called the *center*, or *free-running*, *frequency* f_O. Once an input frequency is applied, the VCO frequency starts to change, and the PLL is said

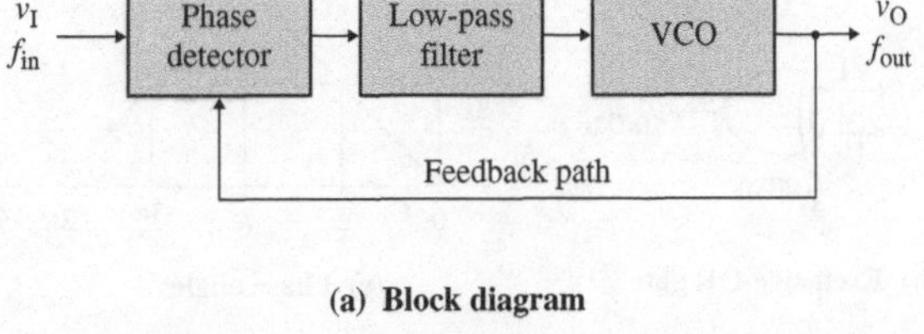

(a) Block diagram

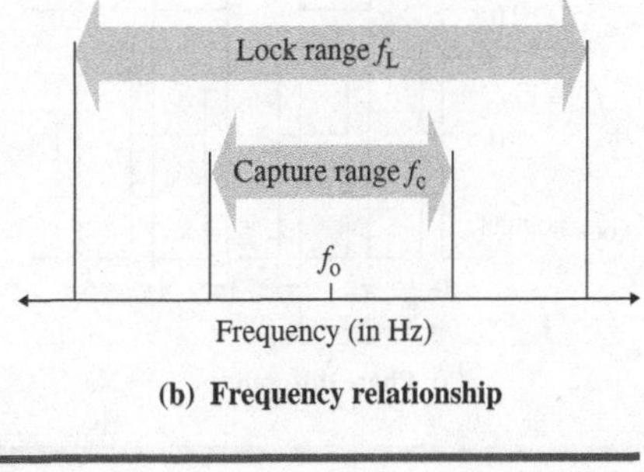

(b) Frequency relationship

FIGURE 4.55 Block diagram of a phase-lock loop

to be in the capture mode. The VCO frequency changes continuously to match the input frequency. When the input frequency becomes equal to the output frequency, the PLL is said to be in the phase-lock mode. The feedback loop maintains the lock when the input signal frequency changes.

The center frequency f_o is the free-running frequency of the VCO. The *lock range* f_L is defined as the range of input frequencies around the center frequency for which the loop can maintain lock. The *capture range* f_C is defined as the range of input frequencies around the center frequency for which the loop will become locked from an unlocked condition. The relations among f_o, f_L, and f_C are shown in Fig. 4.55(b).

4.11.1 Phase Detector

A phase detector takes two input voltages and produces a DC voltage proportional to their phase difference. To understand the principle of operation, consider two voltages v_{I1} and v_{I2}, as shown in Fig. 4.56(a), with a phase difference of ϕ. An output voltage is obtained when they differ in phase—that is, when only one input is high. A phase detector can be implemented using either an exclusive-OR gate, as shown in Fig. 4.56(b), or an analog multiplier [7]. Integrating the output voltage will give an average output voltage, which will be a linear function of the phase difference ϕ, as shown in Fig. 4.56(c). The average output voltage $V_{O(DC)}$ can be expressed as

$$V_{O(DC)} = \begin{cases} \dfrac{V_{CC}}{\pi}\phi & (\text{for } 0 \le \phi \le \pi) \\ \dfrac{V_{CC}}{\pi}(2\pi - \phi) & (\text{for } \pi \le \phi \le 2\pi) \end{cases} \tag{4.94}$$

The phase difference can also be detected by using an edge-triggered *RS* flip-flop. Two input signals are shown in Fig. 4.57(a). If these signals are passed through an edge-triggered *RS* flip-flop as shown in Fig. 4.57(b), the output voltage will be as shown in Fig. 4.57(a). Integrating the output voltage will give an average output voltage, as shown in Fig. 4.57(c). The average output voltage $V_{O(DC)}$ is given by

$$V_{O(DC)} = \frac{V_{CC}}{2\pi}\phi \quad (\text{for } 0 \le \phi \le 2\pi) \tag{4.95}$$

Phase detectors can be broadly divided into two types: digital detectors and analog detectors. The digital detectors are simple to implement with digital devices. However, they are sensitive to the harmonic

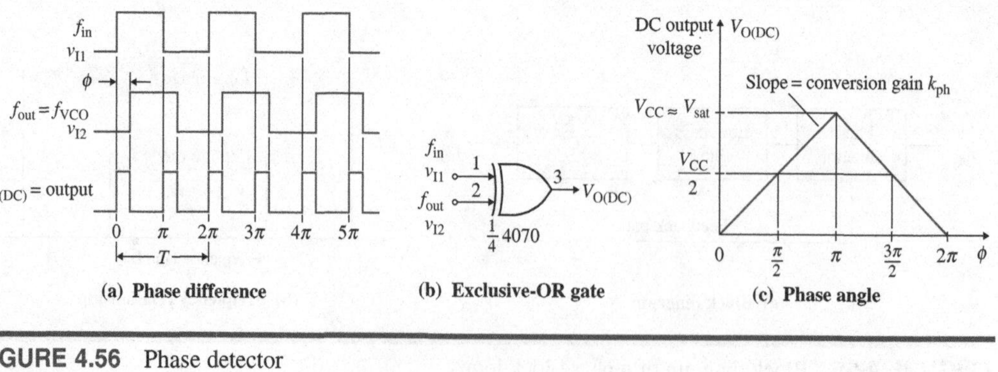

FIGURE 4.56 Phase detector

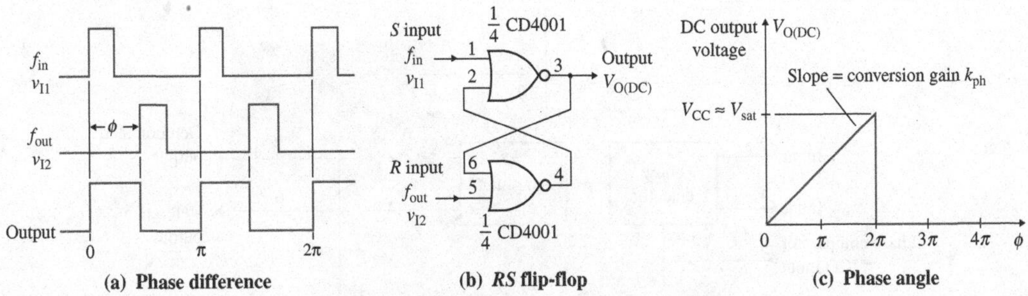

FIGURE 4.57 Edge-triggered phase detector

content of the input signal and changes in the duty cycles of the input signal and the VCO output voltage. Analog detectors are monolithic types such as CMOS MC4344/4044. They respond only to transitions in the input signals. Thus, sensitivity to harmonic content and duty cycle is not a problem. The output voltage is independent of variations in the amplitude and duty cycle of the input waveform. Analog detectors are generally preferred over digital detectors, especially in applications in which accuracy is a critical factor.

4.11.2 Integrated Circuit PLL

The NE/SE-565 PLL is one of the most commonly used IC devices. The elements of the PLL in Fig. 4.55(a) are built into the 565 IC. The internal block diagram of the 565 is shown in Fig. 4.58(a), and the pin configuration is shown in Fig. 4.58(b). A typical connection diagram for the NE/SE-565 PLL is shown in Fig. 4.58(c). A small capacitor C_3, typically of 0.001 μF, is connected between pins 7 and 8 to eliminate possible oscillations. The center frequency of the PLL is given approximately by

$$f_o \approx \frac{1.2}{4R_1C_1} \tag{4.96}$$

where R_1 and C_1 are an external resistance and a capacitance connected to pins 8 and 9, respectively. C_1 can have any value, but R_1 must have a value between 2 kΩ and 20 kΩ. A capacitor C_2 is connected between pins 7 and 10 to form a first-order low-pass filter with an internal resistance of 3.6 kΩ. The filter capacitor C_2 should be large enough to eliminate variations in the demodulated output voltage at pin 7 in order to stabilize the VCO frequency.

The 565 PLL can typically lock to and track an input signal over a bandwidth of ±60% of the center frequency f_o. The lock range f_L is given by

$$f_L = \frac{8f_o}{V_{CC} - V_{EE}} \tag{4.97}$$

where V_{CC} and $-V_{EE}$ are the positive and negative power supplies, in volts, respectively. The capture range f_c is given by

$$f_c = \left[\frac{f_L}{2\pi \times 3.6 \times 10^3 C_2}\right]^{1/2} \quad (C_2 \text{ in farads}) \tag{4.98}$$

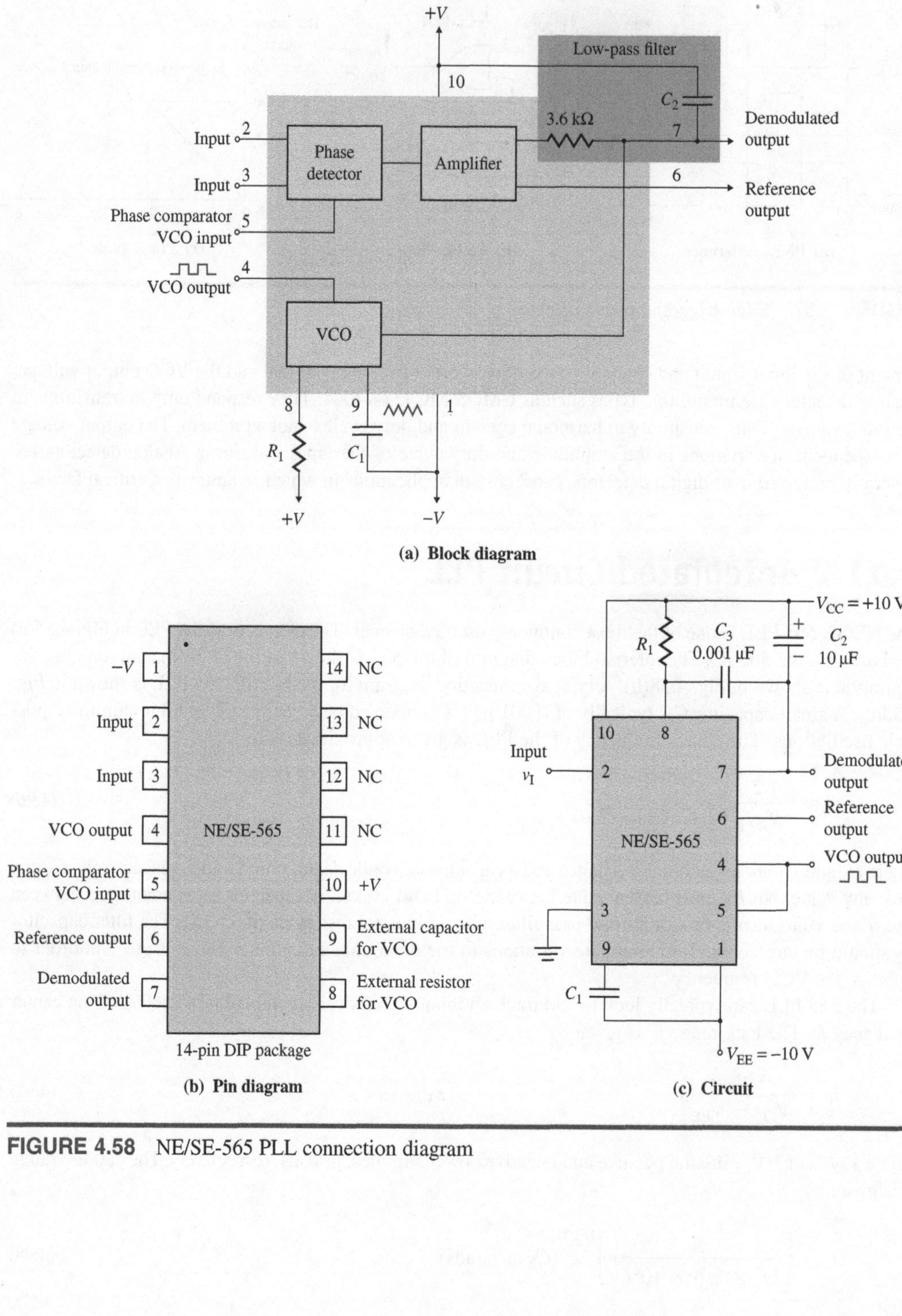

FIGURE 4.58 NE/SE-565 PLL connection diagram

EXAMPLE 4.17

D **Designing a PLL** Design a PLL as shown in Fig. 4.58(c) so that $f_o = 2.5$ kHz and $f_c = 50$ Hz. Assume $V_{CC} = -V_{EE} = 12$ V.

SOLUTION

The steps used to design the 565 PLL are as follows:

Step 1. Choose a suitable value of C_1: Let $C_1 = 0.01$ μF.

Step 2. Find the value of R_1. From Eq. (4.96),

$$R_1 = \frac{1.2}{4C_1 f_o} = \frac{1.2}{4 \times 0.01\ \mu\text{F} \times 2.5\ \text{kHz}} = 12\ \text{k}\Omega$$

Step 3. Find the lock range f_L. From Eq. (4.97),

$$f_L = \frac{8 \times 2.5\ \text{kHz}}{12 - (-12)} = 833\ \text{Hz}$$

Step 4. Find the value of C_2. From Eq. (4.98),

$$C_2 = \frac{f_L}{2\pi \times 3.6 \times 10^3 f_c^2} = \frac{833}{2\pi \times 3.6 \times 10^3 \times 50^2} = 14.17\ \mu\text{F}$$

Choose $C_2 = 14$ μF.

4.11.3 Applications of the 565 PLL

As examples of applications of the 565 PLL, we consider a frequency multiplier, an FSK demodulator, and an SCA (subsidiary carrier authorization) decoder [8].

Frequency Multiplier

A block diagram of a frequency multiplier using the 565 PLL is shown in Fig. 4.59(a). A frequency divider is inserted between the VCO and the phase detector. Since the output frequency of the divider is locked to the input frequency, the VCO will actually be running at a multiple of the input frequency; that is, $f_o = Nf_{in}$, where N is an integer. The amount of multiplication desired can be obtained by selecting the appropriate divide-by network. A typical connection of the 565 PLL to give an output frequency of $f_o = 5f_{in}$ is shown in Fig. 4.59(b).

To set up the circuit, you must know the frequency limits of the input signal. The free-running frequency of the VCO then can be adjusted by means of R_1 and C_1 so that the output frequency of the divider is midway between the input frequency limits. That is, for $f_{in} = 400$ Hz to 4 kHz, the output frequency would be $f_o = 2$ kHz to 20 kHz, with a midway frequency of $f_{o(mid)} = 11$ kHz. The filter capacitance C_2 should be large enough (typically 10 μF) to eliminate variations in the demodulated output voltage (at pin 7) in order to stabilize the VCO frequency. The output of the VCO will be a square wave whose frequency will be the multiple of the input frequency as long as the loop is in lock. The input and output waveforms are shown in Fig. 4.59(c).

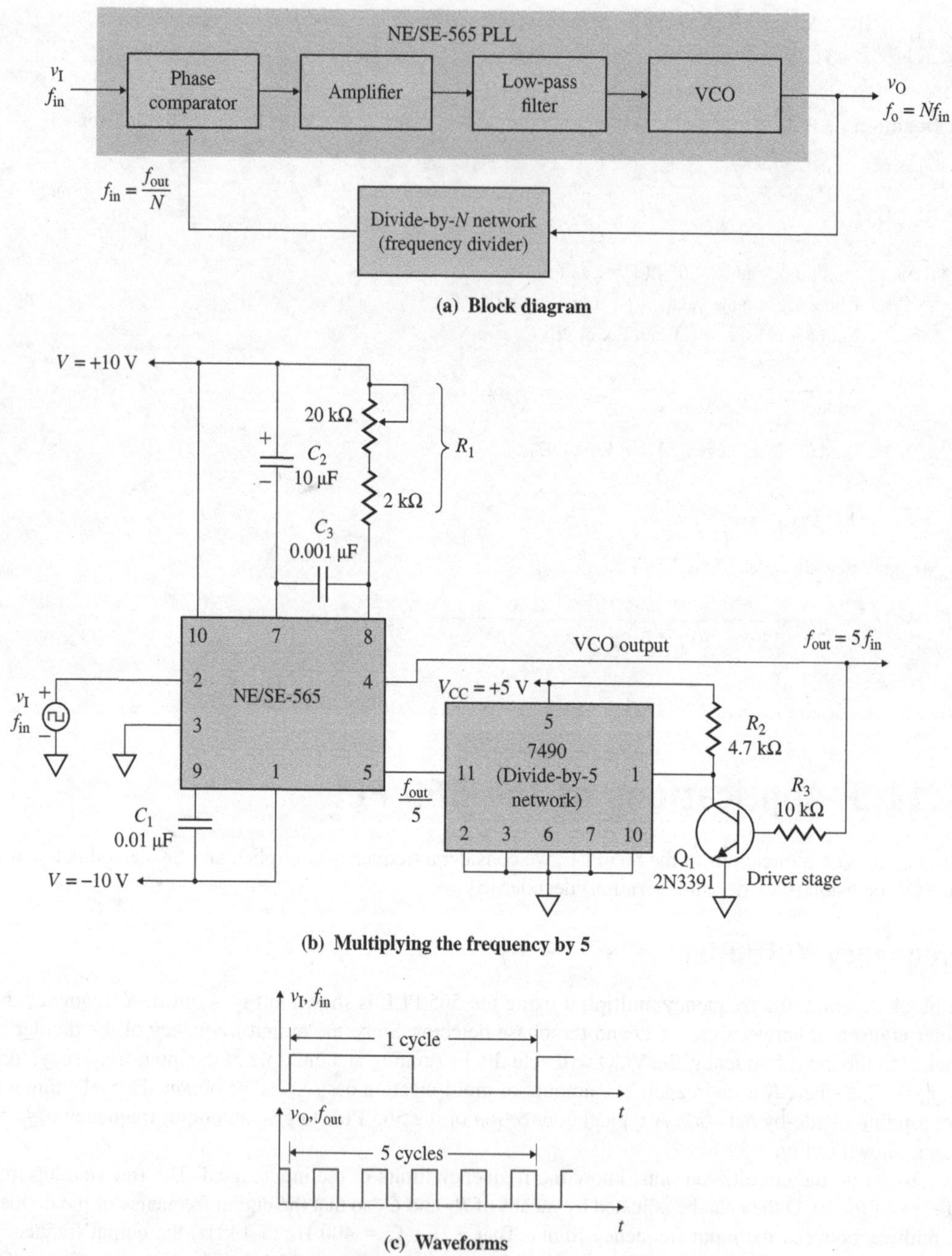

FIGURE 4.59 565 PLL as a frequency multiplier

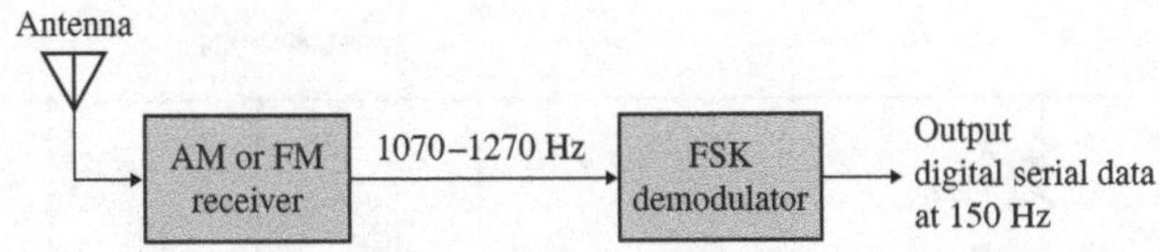

(a) AM/FM receiver

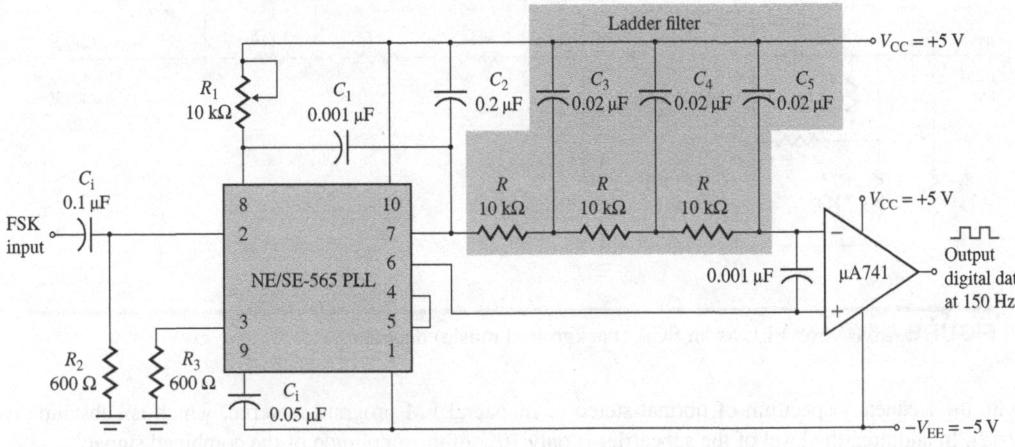

(b) Circuit

FIGURE 4.60 565 PLL as an FSK demodulator

FSK Demodulator

An FSK demodulator is often used in AM/FM receivers, as shown in Fig. 4.60(a). One very useful application of the 565 PLL is as an FSK demodulator to receive FSK signals of 1070 Hz and 1270 Hz; the configuration is shown in Fig. 4.60(b). As the signal appears at the input, the PLL locks to the input frequency and tracks it between the two frequencies, with a corresponding DC shift at the output. The input signal is connected through a coupling capacitor C_i to block the DC level from the FSK receiver. Both input terminals are connected to the ground through identical resistors R_2 and R_3.

The loop filter capacitor C_2 determines the dynamic characteristics of the demodulator, and its value should be smaller than usual to eliminate overshoot on the output pulse. A three-stage *RC*-ladder low-pass filter is used to remove the carrier component from the output. The high cutoff frequency of the ladder filter—that is, $f_H = 1/2\pi RC$—should be approximately halfway between the maximum keying rate (150 Hz) and twice the input frequency (2×1070 Hz is approximately 2200 Hz).

The output signal of 150 Hz can be made logic compatible by connecting a voltage comparator between the output and pin 6 of the 565 PLL. R_2 and C_1 determine the free-running frequency of the VCO. The free-running frequency is adjusted with R_1 so as to produce a slightly positive voltage with $f_o = 1070$ Hz.

SCA (Background Music) Decoder

Some FM stations are authorized by the FCC to broadcast uninterrupted background music for commercial use, using a frequency-modulated subcarrier of 67 kHz. This frequency was chosen so as not to interfere

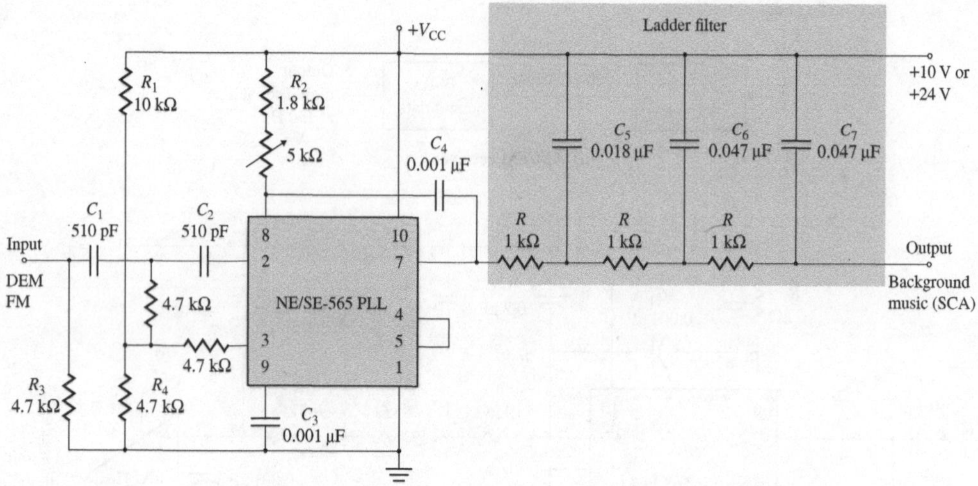

FIGURE 4.61 565 PLL as an SCA (background music) decoder

with the frequency spectrum of normal stereo or monaural FM program material, which is substantially lower. In addition, the level of the subcarrier is only 10% of the amplitude of the combined signal.

A PLL may be used to recover the SCA (subsidiary carrier authorization, or storecast music) signal from the combined signal of many commercial FM broadcast stations. This application involves demodulation of a frequency-modulated subcarrier of the main channel. The SCA signal can be filtered out and demodulated by the 565 PLL without the use of any resonant circuits. A connection diagram is shown in Fig. 4.61. The PLL is tuned to 67 kHz with a 5-kΩ potentiometer; only approximate tuning is required, since the loop will seek the signal.

The demodulated output (pin 7) is passed through a three-stage low-pass filter to provide deemphasis and attenuate the high-frequency noise that often accompanies SCA transmissions. Since no capacitor is provided directly at pin 7, the circuit operates as a first-order loop. The demodulated output signal is on the order of 50 mV, and the frequency response extends to 7 kHz. By connecting the circuit of Fig. 4.61 to a point between the FM discriminator and the deemphasis filter of a commercial-band (home) FM receiver and tuning the receiver to a station that broadcasts an SCA signal, one can obtain hours of commercial-free background music.

KEY POINTS OF SECTION 4.11

- A phase-lock loop (PLL) consists of a phase detector, a low-pass filter, and a voltage-controlled oscillator. PLLs find applications as frequency multipliers, FSK demodulators, and SCA (background music) decoders.
- The operation of a PLL involves three modes: a free-running mode, a capture mode, and a phase-lock mode. The VCO frequency is continuously adjusted until it is equal to the input frequency. When the input frequency becomes equal to the output frequency, the PLL is said to be in the phase-lock mode.
- The 565 PLL can typically lock to and track an input signal over a bandwidth of ±60% of the center frequency.

4.12 Voltage-to-Frequency and Frequency-to-Voltage Converters

The NE/SE-566 VCO discussed in Sec. 4.9 can be used as a voltage-to-frequency converter. In many applications, it is also necessary to convert frequency to voltage. The TelCom 9400 series converters can be used as either *voltage-to-frequency* (V/F) or *frequency-to-voltage* (F/V) *converters*, and they can produce pulse and square-wave outputs with a frequency range of 1 Hz to 100 kHz. For V/F conversion, the device accepts an analog input signal and generates an output pulse train whose frequency is linearly proportional to the input voltage. For F/V conversion, the device accepts any input frequency waveform and provides a linearly proportional voltage output. The complete V/F or F/V conversion requires only the addition of two capacitors, three resistors, and a reference voltage. The 9400 series consists of CMOS devices and bipolar devices that can operate on single or dual supply voltages.

4.12.1 V/F Converter

The 9400 V/F converter operates on the principle of charge balancing. The functional block diagram is shown in Fig. 4.62(a). The input voltage v_I is converted to a current I_{in} by the input resistor R_{in}. This current $I_{in} = v_I/R_{in}$ is then converted to a charge by the internal integrating capacitor C_{int} and gives a linearly decreasing voltage v_{O3} at the output of the op-amp integrator; that is,

$$v_{O3} = -\frac{I_{in}}{C_{int}}t = -\frac{v_I}{R_{in}C_{int}}t \qquad (4.99)$$

As soon as voltage v_{O3} falls below the threshold level of the threshold detector, the switch closes and causes reference voltage V_{ref} to be applied to reference capacitor C_{ref} for a time long enough to charge the capacitor to reference voltage V_{ref}. This action also reduces the charge on the integrating capacitor by a fixed amount ($q = C_{ref}V_{ref}$), causing the integrator output to step up by a certain amount. At the end of the charging period, C_{ref} is shorted out, dissipating the charge stored on the reference capacitor so that the system is ready to repeat the cycle when the output passes again through the zero axis.

The continued charging of the integrating capacitor C_{int} by the input voltage is balanced out by fixed charges from the reference voltage. As the input voltage is increased, the number of reference pulses required to maintain balance increases, causing the output frequency to increase also. Since each charge increment is fixed (i.e., $q = C_{ref}V_{ref}$), the increase in frequency with voltage is linear. The output frequency f_o is related to the input voltage by

$$f_o = \frac{v_I}{|V_{ref}|\, R_{in}C_{ref}} \qquad (4.100)$$

where v_I = input voltage
$|V_{ref}|$ = reference voltage
C_{ref} = reference capacitance

The pin diagram of the 9400 V/F converter is shown in Fig. 4.62(b), and its internal block diagram is shown in Fig. 4.62(c). The threshold detector senses the output of the integrator. The output of the detector triggers a 3-μs network when its input voltage passes through the threshold. The nominal threshold of the

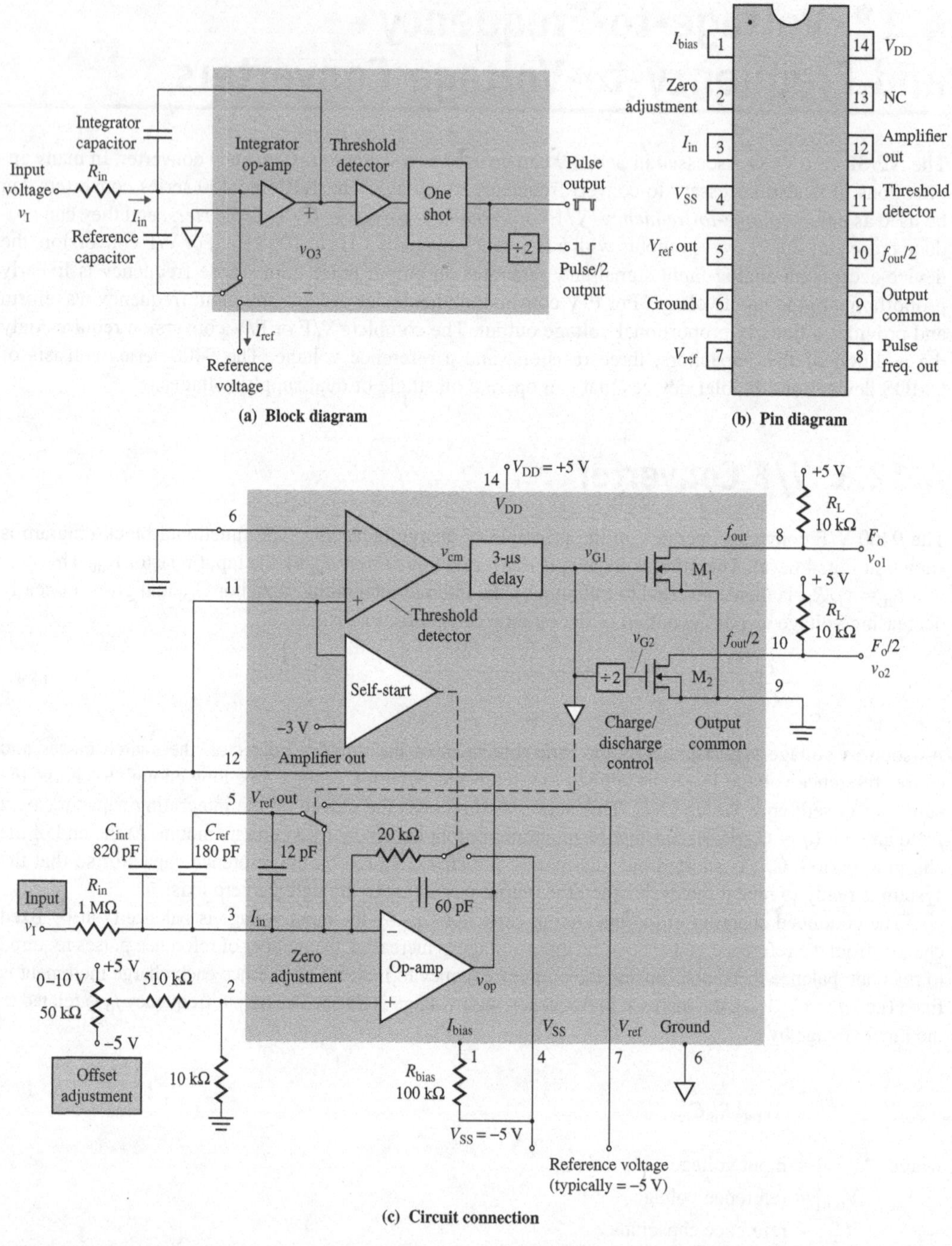

(a) Block diagram

(b) Pin diagram

(c) Circuit connection

FIGURE 4.62 9400 V/F converter (Courtesy of TelCom Semiconductor, Inc.)

detector is halfway between the power supplies, or $(V_{DD} + V_{SS})/2 \pm 400$ mV. The output of the 3-μs network is applied to the output transistor M_1, the divide-by-2 network, and the C_{ref} charge/discharge control circuit.

The self-start circuit ensures that the V/F converter operates properly when the power is first applied. If the integrator output is below the threshold voltage (i.e., 0 V) of the threshold detector and C_{ref} is already charged, then a positive voltage step will not occur when the power is turned on. The integrator output will continue to decrease until it crosses the −3.0-V threshold of the self-start comparator. When this happens, an internal resistor of 20 kΩ is connected to the op-amp integrator input, thereby forcing the output to become positive. As soon as the op-amp output becomes positive, the self-start circuit is disabled, and the 9400 operates in its normal mode.

Pulse f_{out} $(=f_o)$ output is an open-drain *n*-channel FET that provides a pulse waveform whose frequency is proportional to the input voltage v_I. Pulse $f_{out}/2$ $(=f_o/2)$ output is an open-drain *n*-channel FET that provides a square wave whose frequency is one-half that of the pulse waveform. This output will change state on the rising edge of f_o. Both f_o and $f_o/2$ outputs require a pull-up resistor and interface directly with MOS, CMOS, and TTL logic circuits.

The waveforms of the V/F converter are shown in Fig. 4.63. Three microseconds after the output V_d of the detector switches to low, the output V_{G1} of the delay circuit switches from low to high. When V_{G1}

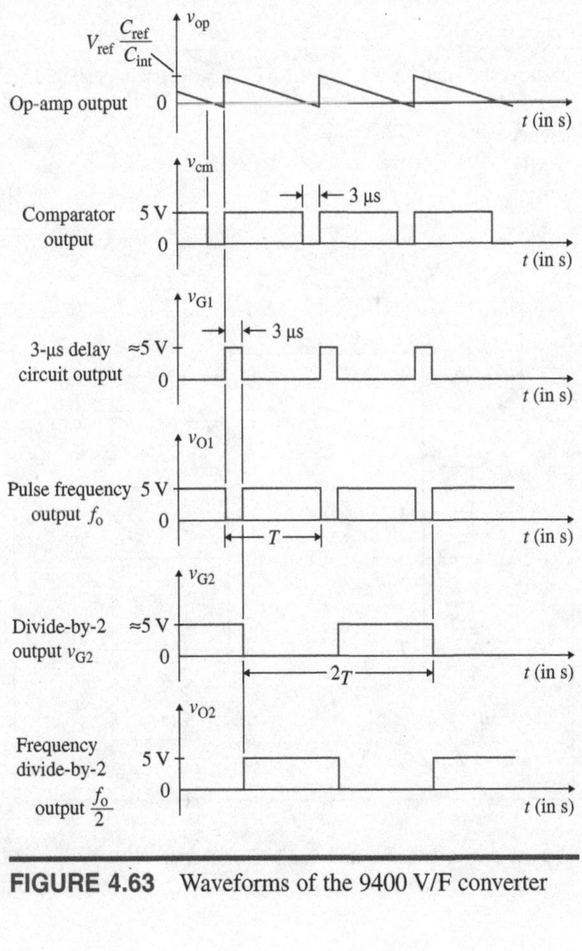

FIGURE 4.63 Waveforms of the 9400 V/F converter

is low, transistor M_1 is off and f_{out} is high (i.e., 5 V). The divide-by-2 network is a negative-edge-triggered flip-flop whose output V_{G2} is a complement (inversion) of V_{G1}. Thus, transistor M_2 will be on, and $f_{out}/2$ will be low. With V_{G1} low, the charge/discharge control is disabled and capacitor C_{ref} remains discharged (i.e., shorted out).

EXAMPLE 4.18

D

Designing a V/F converter Using the 9400 as shown in Fig. 4.64, design a V/F converter so that $f_o = 5$ kHz at $v_I = 5$ V. The input voltage v_I can vary between 10 mV and 10 V. Assume $V_{DD} = -V_{SS} = 5$ V.

SOLUTION

The steps used to design the V/F converter are as follows:

Step 1. Choose V_{DD} and V_{SS} such that

$$4\text{ V} \le V_{DD} \le 7.5\text{ V} \quad \text{and} \quad -7.5\text{ V} \le V_{SS} \le -4\text{ V}$$

Choose $V_{DD} = 5$ V and $V_{SS} = -5$ V.

Step 2. Choose the capacitors such that $C_3 = C_4 = 0.1$ µF. These capacitors should be close to pins 4 and 14, respectively.

Step 3. Choose the reference voltage $V_{ref} = V_{SS} = -5$ V.

Step 4. Choose $R_{in} = 1$ MΩ.

Step 5. Choose $R_{bias} = 100$ kΩ.

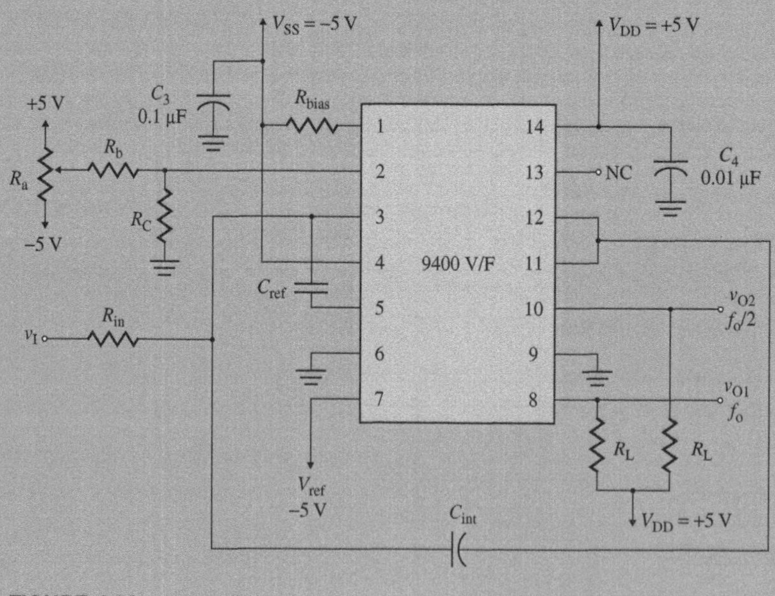

FIGURE 4.64 TelCom 9400 converter connected as a V/F converter

Step 6. Choose the pull-up resistance $R_L = 10\ \text{k}\Omega$.

Step 7. Choose C_{ref} such that $C_{ref} < 500$ pF. From Eq. (4.100),

$$C_{ref} = \frac{v_I}{|V_{ref}|R_{in}f_o} = \frac{5\ \text{V}}{5\ \text{V} \times 1\ \text{M}\Omega \times 5\ \text{kHz}} = 200\ \text{pF}$$

C_{ref} should be located as close as possible to pins 3 and 5. Glass-film capacitors are recommended for high accuracy.

Step 8. Choose C_{int} such that $4C_{ref} \le C_{int} \le 10C_{ref}$. Assume

$$C_{int} = 5 \times C_{ref} = 5 \times 200\ \text{pF} = 1\,\text{nF}$$

C_{int} should be located as close as possible to pins 3 and 12.

Step 9. Determine the values of the offset resistors. Since $R_c \le R_a \le R_b$, assume R_a is a 50-kΩ potentiometer, $R_b = 450\ \text{k}\Omega$, and $R_c = 10\ \text{k}\Omega$.

Step 10. Determine the minimum frequency corresponding to the minimum input voltage $v_I = 10$ mV. From Eq. (4.100),

$$f_{o(min)} = \frac{10\ \text{mV}}{5\ \text{V} \times 1\ \text{M}\Omega \times 200\ \text{pF}} = 10\ \text{Hz}$$

Step 11. Set the input voltage to the minimum value $v_I = 10$ mV, and adjust potentiometer R_a to obtain the corresponding minimum output frequency $f_{o(min)} = 10$ Hz.

Step 12. Determine the maximum frequency corresponding to the maximum input voltage $v_I = 10$ V. From Eq. (4.100),

$$f_{o(max)} = \frac{10\ \text{V}}{5\ \text{V} \times 1\ \text{M}\Omega \times 200\ \text{pF}} = 10\ \text{kHz}$$

Step 13. Set the input voltage to the maximum value $v_I = 10$ V, and adjust R_{in}, V_{ref}, or C_{ref} to obtain the corresponding maximum output frequency $f_{o(max)} = 10$ kHz.

4.12.2 F/V Converter

When used as an F/V converter, the 9400 generates an output voltage that is linearly proportional to the input frequency f_{in}. The internal block diagram of the 9400 F/V converter is shown in Fig. 4.65(a). The input signal is differentiated by an *RC* network whose output is then applied to the (+) input of the threshold detector (i.e., at pin 11). The threshold detector has about ±200 mV of hysteresis. Each time the input to the detector at pin 11 crosses zero in the negative direction, its output goes to low. Three microseconds later, the charge/discharge circuit is enabled, instantaneously connecting the reference capacitor C_{ref}, which remains discharged, to the reference voltage V_{ref}. This causes a precise amount of charge ($q = C_{ref}V_{ref}$) to be dispensed into the op-amp's summing junction. This charge in turn flows through feedback resistor R_{int} and generates a voltage pulse at the output of the op-amp. Capacitor C_{int} across R_{int} averages these pulses into a DC signal that is linearly proportional to the input frequency. The waveforms of the F/V converter are shown in Fig. 4.65(b). For charge q dispensed to capacitor C_{int} in time period T, we get

$$q = iT = \frac{V_O}{R_{int}}T$$

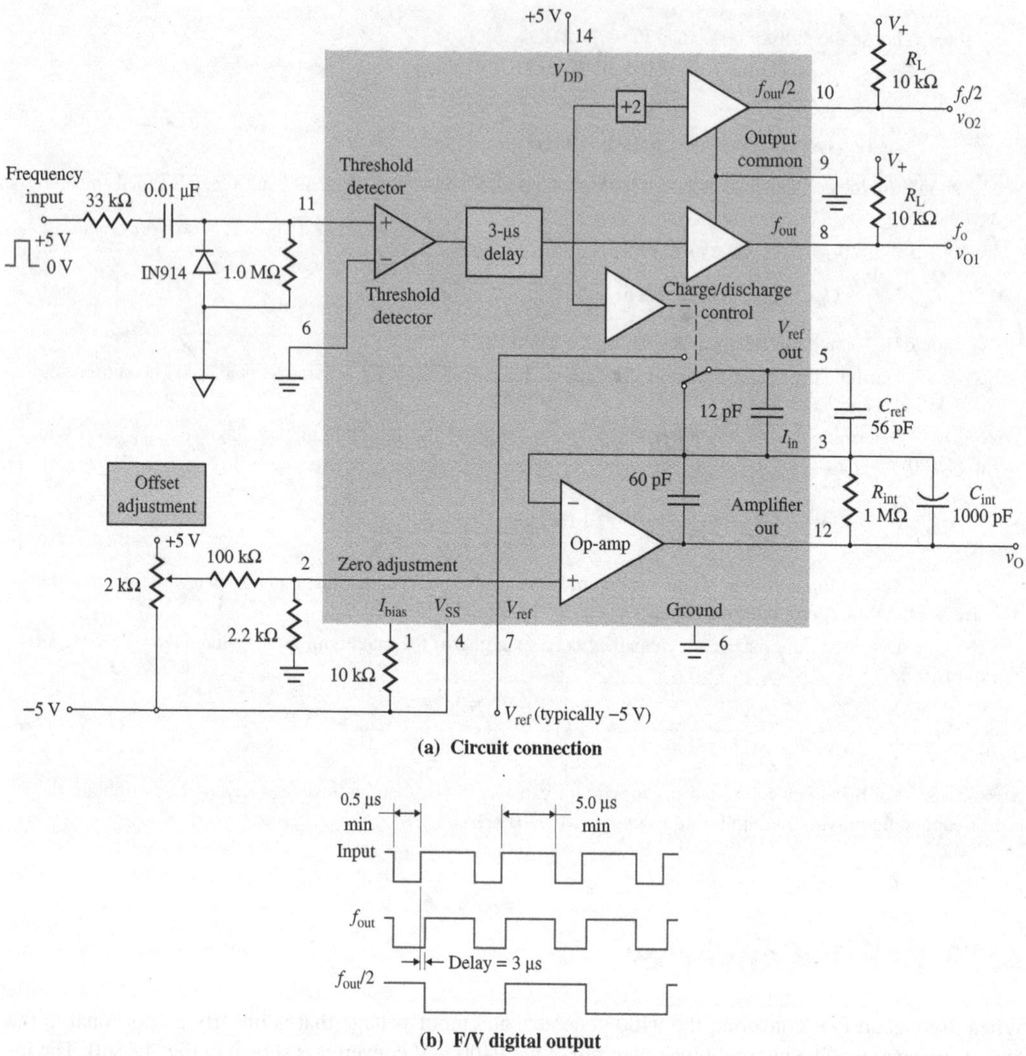

FIGURE 4.65 Internal block diagram of 9400 F/V converter (Courtesy of TelCom Semiconductor, Inc.)

which, for $q = C_{ref}V_{ref}$, relates average output voltage V_O to input frequency f_{in} as follows:

$$V_O = |V_{ref}|\,R_{int}C_{ref}f_{in} \tag{4.101}$$

where f_{in} = input frequency, in hertz

$|V_{ref}|$ = reference voltage, in volts

R_{int} = internal integrating resistance, in ohms

C_{ref} = reference capacitance, in farads

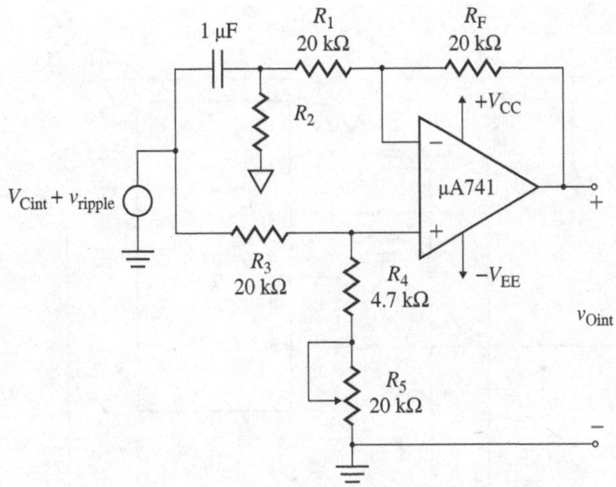

FIGURE 4.66 Ripple elimination in an F/V converter

The F/V converter will accept any input wave shape. However, the positive pulse width of the detector input (pin 11) must be at least 5 μs, and the negative pulse width must be greater than 0.5 μs. When the input frequency is less than 1 kHz, the duty cycle should be greater than 20% to ensure that C_{ref} is fully charged and discharged.

The output voltage V_O will have a certain amount of ripple, which is inversely proportional to C_{int} and the input frequency f_{in}. Therefore, for low frequencies, C_{int} can be increased in the range from 1 μF to 100 μF to reduce the ripple. To eliminate the ripple on V_O, an op-amp circuit in the common-mode configuration may be connected at the output of the F/V converter. This arrangement is shown in Fig. 4.66. Since the AC ripple content appears at both the (+) and the (−) terminals of the op-amp, the AC ripple will be canceled, and the output will have only DC voltage.

EXAMPLE 4.19

D **Designing an F/V converter** Using the 9400 as shown in Fig. 4.67, design an F/V converter so that $V_O = 2.5$ V at $f_{in} = 5$ kHz. The input frequency f_{in} can vary between 0 and 10 kHz. Assume $V_{DD} = 5$ V, $-V_{SS} = 0$.

SOLUTION

The steps used to design the F/V converter are as follows.

Step 1. Choose V_{DD} and V_{SS} such that

$$4\text{ V} \le V_{DD} \le 7.5\text{ V}$$

Choose $V_{DD} = 5$ V.

Step 2. Choose the capacitors such that $C_4 = 0.1$ μF. These capacitors should be close to pins 4 and 14, respectively.

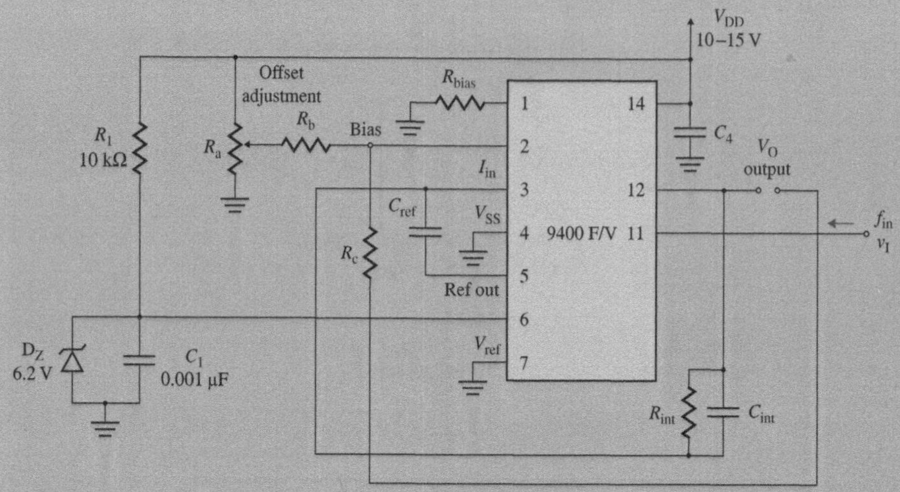

FIGURE 4.67 TelCom 9400 converter connected as an F/V converter

Step 3. Choose the reference voltage $V_{ref} = 0$ V.
Step 4. Choose $R_{int} = 1$ MΩ.
Step 5. Choose $R_{bias} = 100$ kΩ.
Step 6. Choose the pull-up resistance $R_L = 10$ kΩ.
Step 7. Choose C_{ref}. From Eq. (4.101),

$$C_{ref} = \frac{V_O}{|V_{ref}|R_{int}f_{in}} = \frac{2.5\text{ V}}{5 \times 1\text{ M}\Omega \times 5\text{ kHz}} = 100\text{ pF}$$

C_{ref} should be located as close as possible to pins 3 and 5. Glass-film capacitors are recommended for high accuracy.

Step 8. Choose C_{int}. Let

$$C_{int} = 10 \times C_{ref} = 10 \times 100\text{ pF} = 1\text{ nF}$$

C_{int} should be located as close as possible to pins 3 and 12. Since the amount of ripple on the output voltage is inversely proportional to C_{int} and the input frequency, C_{int} can be increased to lower the ripple. Acceptable values of C_{int} for low frequencies are 1 μF to 100 μF.

Step 9. Determine the values of the offset resistors. Since $R_c \le R_a \le R_b$, assume R_a is a 50-kΩ potentiometer, $R_b = 450$ kΩ, and $R_c = 10$ kΩ.

Step 10. With no input signal applied ($f_{in} = 0$), adjust the potentiometer R_a to obtain the minimum output voltage $V_{O(min)} = 0$.

Step 11. Determine the maximum output voltage corresponding to the maximum input frequency $f_{in} = 10$ kHz. From Eq. (4.101),

$$V_{O(max)} = 5 \times 1\text{ M}\Omega \times 100\text{ pF} \times 10\text{ kHz} = 5\text{ V}$$

Step 12. Set the input frequency to the maximum value $f_{in} = 10$ kHz, and adjust C_{ref} so that V_O is approximately 2.5 V.

KEY POINTS OF SECTION 4.12

- The TelCom 9400 converter can be used as either a voltage-to-frequency (V/F) converter or a frequency-to-voltage (F/V) converter, and it can produce pulse and square-wave outputs with a frequency range of 1 Hz to 100 kHz.
- In a V/F converter, the input voltage is converted to charge by an op-amp integrator and gives a linearly decreasing output voltage. As soon as this voltage falls below a threshold level, a threshold detector causes the output to step up by a certain amount so that the system is ready to repeat the cycle when the output passes again through the zero axis.
- In an F/V converter, a precise amount of charge dispensed into the op-amp's summing junction generates a voltage pulse at the output of the op-amp. A capacitor is then used to average these pulses into a DC signal that is linearly proportional to the input frequency.

4.13 Sample-and-Hold Circuits

Sample-and-hold (SAH) circuits are used as an interface between an analog signal and a digital circuit in a wide variety of applications such as data acquisition, analog-to-digital (A/D) conversion, and synchronous data demodulation. An SAH circuit is used to *sample* an analog signal at a particular instant of time and *hold* the value of the sample as long as required. The sampling instants and hold duration are determined by a logic control signal. The hold duration depends on the type of application. For example, in A/D conversion, samples must be held long enough for the conversion to be completed.

The principle of operation of a sample-and-hold circuit can be explained with Fig. 4.68(a), which shows a circuit consisting of capacitor C, switch S_1, and internal resistor R_s. The capacitor is used to hold the sample. The switch provides a means of rapidly charging the capacitor to the sample voltage and then removing the input so that the capacitor can retain the desired voltage. When the control signal v_{CN} is high, the switch is closed. If the time constant R_sC is very small, the output voltage v_O will be very close to the

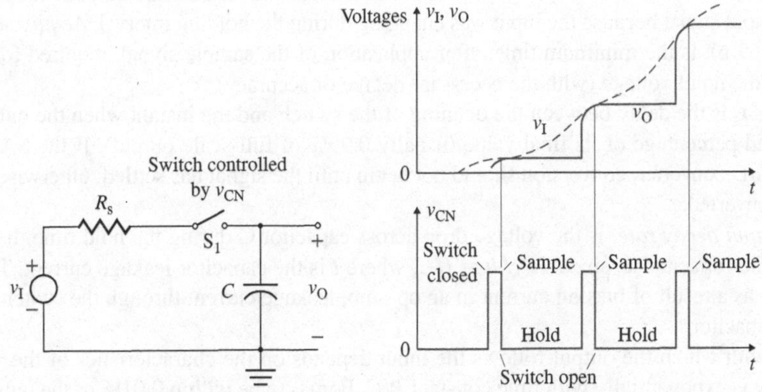

(a) Simple switch (b) Waveforms

FIGURE 4.68 Principle of a sample-and-hold circuit

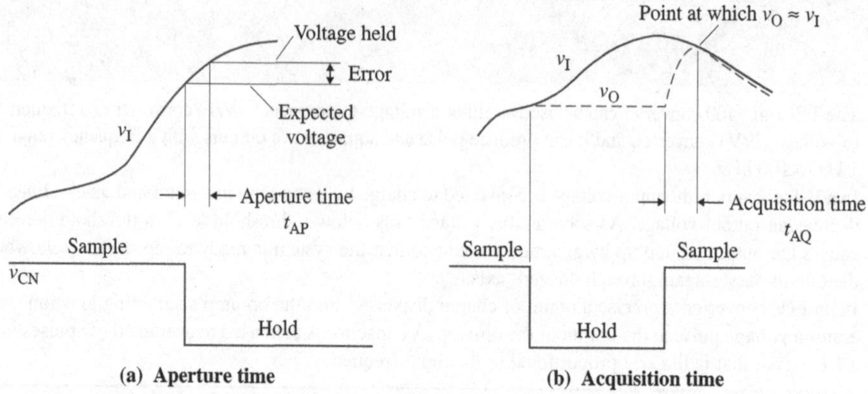

FIGURE 4.69 Aperture time and acquisition time

input voltage v_I and will be equal to it at the instant the control signal becomes low and the switch is opened. The idealized waveforms for output voltage v_O, input voltage v_I, and control voltage v_{CN} are shown in Fig. 4.68(b).

In practice, the capacitor can neither charge instantly nor hold a constant voltage. Also, the switch cannot open and close instantaneously. As a result, the practical output waveform will differ from the ideal one. Among the important specifications given by the manufacturers of SAH circuits are aperture time, acquisition time, settling time, and drop.

Aperture time t_{AP}, shown in Fig. 4.69(a), is the maximum time required for the SAH circuit to open. It is the delay between application of the control signal to open the switch and the instant when the switch actually does open. This time depends on the type of switch, but typically ranges from 4 μs to 20 μs. The t_{AP} of FET switches is in the range of 50 ns to 100 ns. The aperture time should be much less than the sampling period (i.e., the reciprocal of the sampling rate). Since the input signal is changing continuously, the hold voltage will change slightly during the aperture time, causing an error in the hold voltage.

Once the switch is closed for sampling, it takes a finite amount of time for the output voltage to become identical to the input signal because the input was changing during the holding interval. *Acquisition time* t_{AQ}, shown in Fig. 4.69(b), is the minimum time, after application of the sample signal, required for the output voltage to reach the input voltage (with the necessary degree of accuracy).

Settling time t_s is the delay between the opening of the switch and the instant when the output reaches within a specified percentage of its final value (usually 0.99% of full-scale output). If the SAH circuit is followed by an A/D converter, conversion should not begin until the signal has settled; otherwise, the wrong signal will be converted.

Drop, or *output decay rate*, is the voltage drop across capacitor C during the hold time. It is inversely proportional to the capacitance, since $dv_O/dt = I/C$, where I is the capacitor leakage current. This leakage current can arise as a result of biasing current in an op-amp, leakage current through the switch, or internal leakage in the capacitor.

The speed with which the output follows the input depends on the characteristics of the input signal v_I. v_O will follow v_I exponentially with time constant R_sC. For v_O to be within 0.01% of the output, its time period should be approximately $9R_sC$. In addition, the signal source must be able to supply the charging current required by capacitor C. Usually, the analog signal is buffered from the switch by a unity-gain op-amp follower in order to ensure a low value of R_s.

4.13.1 SAH Op-Amp Circuits

An SAH circuit can be implemented using an op-amp and a switch, as shown in Fig. 4.70(a). When switch S_1 is closed, the circuit operates as an *RC* filter. For a step input voltage of V_I, the output voltage $v_O(t)$ can be found from

$$v_O(t) = -\frac{R_F}{R_1} V_I(1 - e^{-t/R_F C}) \tag{4.102}$$

For v_O to reach V_I in the shortest time, the time constant $R_F C$ must be shorter than the sample interval so that the output can track the input. When switch S_1 is opened, the capacitor will hold its voltage of $-V_I$. To minimize the output voltage drop, the op-amp should have a low input biasing current (as does an op-amp with a FET input stage, for example). Also, a high-quality capacitor with a low leakage current should be used.

Switch S_1 can be replaced by a transistor M_1 (i.e., a *p*-type depletion MOSFET), as shown in Fig. 4.70(b). If the control voltage v_{CN} is low (say, 0 V), the FET M_1 will be on (i.e., the switch will be closed), and the capacitor will be in sample mode, charging to V_I. If the control voltage v_{CN} is high (say, +5 V), the FET will be off (i.e., the switch will be open), and the capacitor will be in hold mode.

Diode D_1 clamps the voltage at node A to 0.7 V. When M_1 is on, the diode effectively becomes connected across the FET (i.e., between its drain and source). Since the voltage drop across the FET is low, the voltage across the diode will also be small—much less than 0.7 V. Thus, the diode will have no effect during the sampling time.

4.13.2 SAH Integrated Circuits

SAH integrated circuits such as the LF198 use BiFET technology to obtain ultrahigh DC accuracy (within 0.01%) with fast signal acquisition (4 μs) and a low drop (3 mV/s). The functional block diagram of the LF198 is shown in Fig. 4.71(a), and its connection diagram is shown in Fig. 4.71(b). The manufacturers give curves showing the variation in acquisition time t_{AQ} with hold capacitance C_h. For example, t_{AQ} = 4 μs for hold capacitance C_h = 1000 pF, and t_{AQ} = 20 μs for C_h = 0.01 μF.

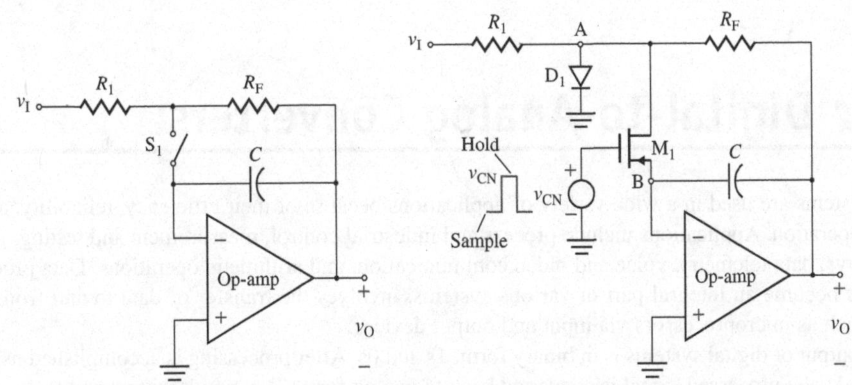

(a) Integrator (b) Integrator with FET switch

FIGURE 4.70 Inverting SAH op-amp circuit

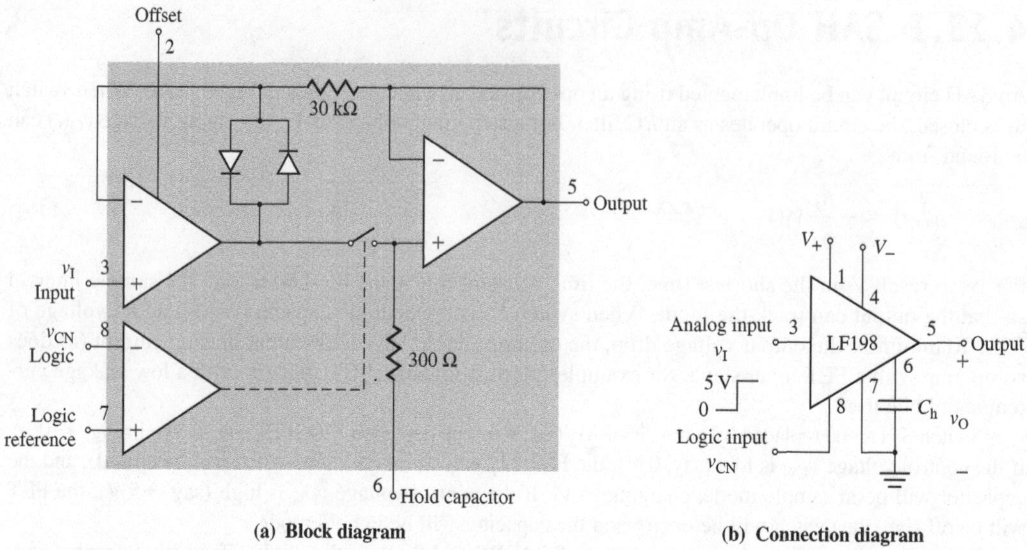

FIGURE 4.71 Sample-and-hold LF198 (Courtesy of National Semiconductor, Inc.)

KEY POINTS OF SECTION 4.13

- An SAH circuit uses a capacitor to sample an analog signal at a particular instant of time and hold the value of the sample as long as required. A switch is closed to charge the capacitor rapidly to the sample voltage and then is opened to remove the input so that the capacitor can retain the desired voltage.
- The specifications of an SAH circuit include acquisition time, aperture time, settling time, and drop, or output decay rate.

4.14 Digital-to-Analog Converters

Digital systems are used in a wide variety of applications because of their efficiency, reliability, and economical operation. Applications include process and industrial control, measurement and testing, graphics and displays, data telemetry, voice and video communication, and arithmetic operations. Data processing, which has become an integral part of various systems, involves the transfer of data to and from digital devices such as microprocessors via input and output devices.

The output of digital systems is in binary form: 1s and 0s. After processing is accomplished using digital methods, the processed signal is converted back to analog form. The circuit that performs this conversion is called a *digital-to-analog* (D/A) *converter*. A D/A system normally contains four separate parts: a reference quantity; a set of binary switches to simulate binary coefficients $B_0, \ldots, B_N$; a resistive network; and an output summing means.

4.14.1 Weighted-Resistor D/A Converter

A simple D/A converter is shown in Fig. 4.72(a). This converter can convert a 4-bit parallel digital word ($B_0B_1B_2B_3$) to an analog voltage that is proportional to the binary number corresponding to the digital word. Four switches are used to simulate the binary inputs. (In practice, a 4-bit binary counter may be used instead.) The logic voltages, which represent the individual bits B_0, B_1, B_2, and B_3, are used to operate switches S_0, S_1, S_2, and S_3, respectively. When a B is a 1, the corresponding switch is connected to reference voltage V_{ref}; when a B is a 0, the corresponding switch is grounded.

The inverting terminal of the op-amp is at virtual ground (i.e., $V_d \approx 0$), so the total current I_S is given by

$$I_S = V_{ref}\left(\frac{B_3}{R_3} + \frac{B_2}{R_2} + \frac{B_1}{R_1} + \frac{B_0}{R_0}\right)$$

Since the current flowing into the op-amp is negligible, $I_S \approx I_F$. Thus, the analog output voltage is given by

$$V_O = -R_F I_F = -R_F V_{ref}\left(\frac{B_3}{R_3} + \frac{B_2}{R_2} + \frac{B_1}{R_1} + \frac{B_0}{R_0}\right) \tag{4.103}$$

Resistors are weighted so that successive resistor values are related by a factor of 2 and the value of each individual resistor is inversely proportional to the numerical significance of the appropriate binary digit. That is,

LSB (least significant bit) → $R_0 = \dfrac{R}{2^0} = R$

$R_1 = \dfrac{R}{2^1} = \dfrac{R}{2}$

$R_2 = \dfrac{R}{2^2} = \dfrac{R}{4}$

MSB (most significant bit) → $R_3 = \dfrac{R}{2^3} = \dfrac{R}{8}$

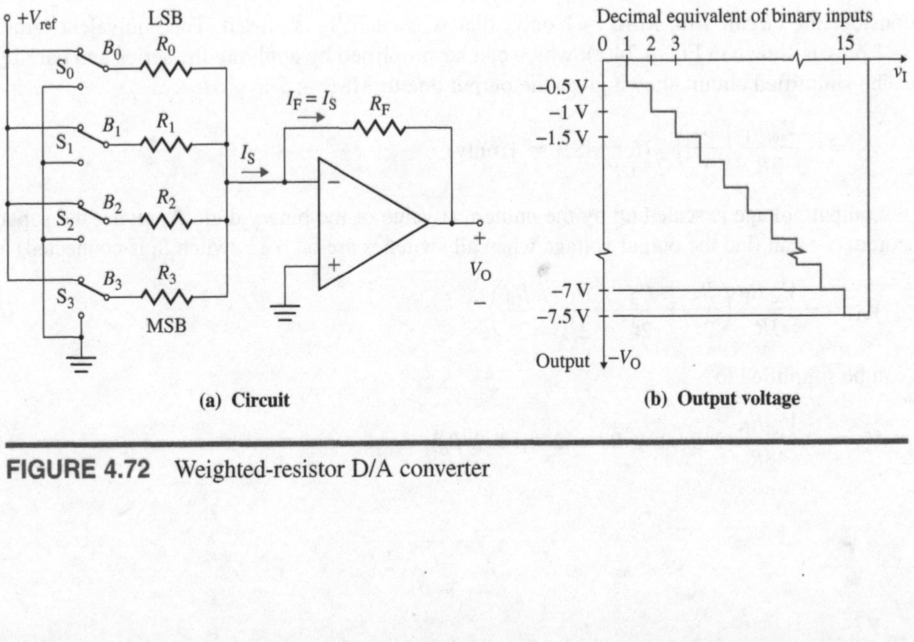

FIGURE 4.72 Weighted-resistor D/A converter

Substituting these weighted resistor values into Eq. (4.103) gives the analog output voltage V_O as

$$V_O = -\frac{V_{ref}R_F}{R}(2^3B_3 + 2^2B_2 + 2^1B_1 + 2^0B_0) \tag{4.104}$$

where $B_i = 1$ if switch S_i is connected to V_{ref} and $B_i = 0$ if switch S_i is grounded. For an input of $B_3B_2B_1B_0 = 1111$, $V_O = -15V_{ref}R_F/R$; for $B_3B_2B_1B_0 = 0110$, $V_O = -6V_{ref}R_F/R$; and for $B_3B_2B_1B_0 = 0001$, $V_O = -V_{ref}R_F/R$. Thus, the output V_O is directly proportional to the numerical value of the binary number $B_3B_2B_1B_0$. Since there are 16 (i.e., 2^4) combinations of the binary inputs B_3, B_2, B_1, and B_0, the analog output will have 16 possible corresponding values. For $V_{ref} = 5$ V and $R = 10R_F$, Eq. (4.104) gives V_O as

$$V_O = -0.5 \times (2^3B_3 + 2^2B_2 + 2^1B_1 + 2^0B_0)$$

The plot is shown in Fig. 4.72(b). The major disadvantage of this D/A converter is the wide variety of resistor values required to weight the network. If the resistor values change in response to temperature changes, it will be difficult to obtain identical tracking characteristics. As a result, the accuracy and the stability of the D/A will be degraded.

4.14.2 *R*-2*R* Ladder Network D/A Converter

The *R*-2*R* D/A ladder converter, shown in Fig. 4.73(a), has only two resistor values *R* and 2*R*, rather than a wide range of resistor values. The plot of the output (which is known as a "resistance ladder") is shown in Fig. 4.73(b). Figure 4.73(c) exhibits the property that the equivalent resistance, looking into any of the terminals *X*, *Y*, S_3, S_2, S_1, or S_0 with the remainder of the terminals grounded, is 3*R*.

Consider the circuit with LSB = 1 only; that is, switch S_0 is closed. The equivalent circuit for LSB = 1 only is shown in Fig. 4.74(a). Successive Thevenin's conversions lead, through the circuits shown in Fig. 4.74[(b) and (c)], to the final circuit shown in Fig. 4.74(d), which gives the output due to LSB = 1 as

$$V_O = -\frac{V_{ref}R_F}{3R}\left(\frac{B_0}{2^4}\right) \quad \text{(for LSB = 1 only)}$$

Now consider the circuit with MSB = 1 only; that is, switch S_3 is closed. The equivalent circuit for MSB = 1 only is shown in Fig. 4.74(e), which can be simplified by applying the series and parallel rule for *R*s. The simplified circuit shown gives the output due to MSB = 1 as

$$V_O = -\frac{V_{ref}R_F}{3R}\left(\frac{B_3}{2^1}\right) \quad \text{(for MSB = 1 only)}$$

Thus, the output voltage is scaled up by the numerical value of the binary digit. Applying the superposition theorem, we can find the output voltage when all switches are on (i.e., switch S_i is connected) as

$$V_O = -\frac{V_{ref}R_F}{3R}\left(\frac{B_3}{2^1} + \frac{B_2}{2^2} + \frac{B_1}{2^3} + \frac{B_0}{2^4}\right) \tag{4.105}$$

which can be simplified to

$$V_O = -\frac{V_{ref}R_F}{48R}(2^3B_3 + 2^2B_2 + 2^1B_1 + 2^0B_0) \tag{4.106}$$

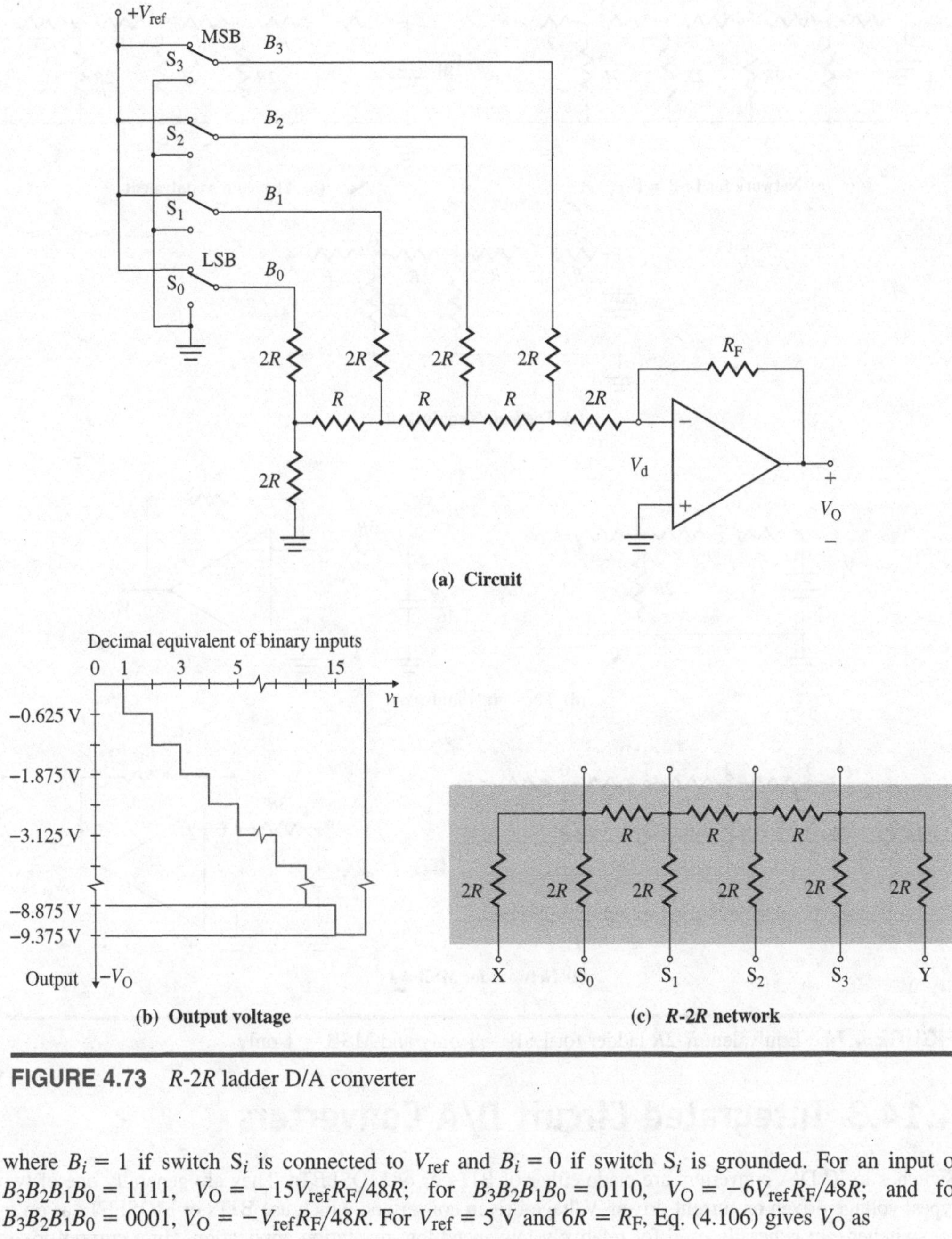

FIGURE 4.73 *R*-2*R* ladder D/A converter

where $B_i = 1$ if switch S_i is connected to V_{ref} and $B_i = 0$ if switch S_i is grounded. For an input of $B_3B_2B_1B_0 = 1111$, $V_O = -15V_{ref}R_F/48R$; for $B_3B_2B_1B_0 = 0110$, $V_O = -6V_{ref}R_F/48R$; and for $B_3B_2B_1B_0 = 0001$, $V_O = -V_{ref}R_F/48R$. For $V_{ref} = 5$ V and $6R = R_F$, Eq. (4.106) gives V_O as

$$V_O = -\left(\frac{5}{48}\right) \times (2^3B_3 + 2^2B_2 + 2^1B_1 + 2^0B_0)$$

whose plot is shown in Fig. 4.73(b) for $V_{O(max)} = 10$ V.

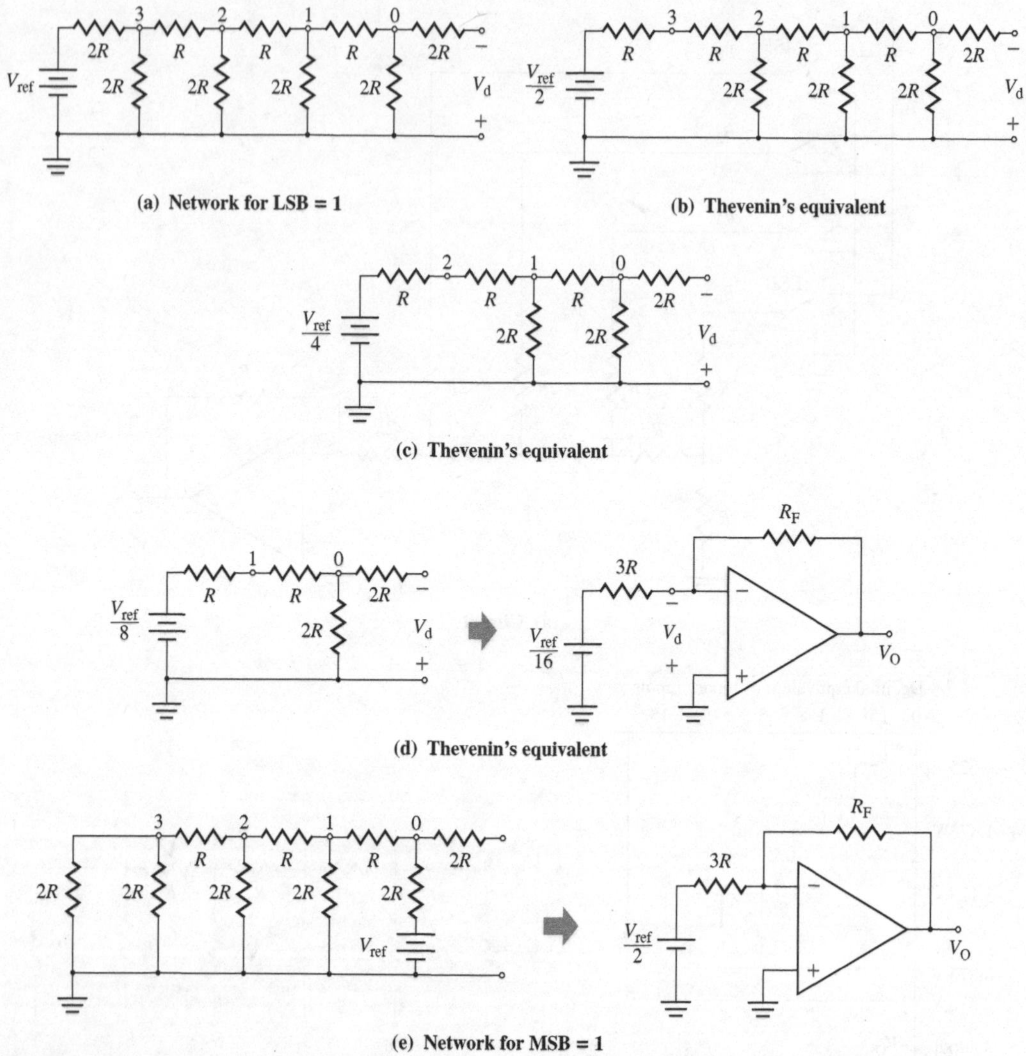

FIGURE 4.74 Equivalent *R*-2*R* ladder for LSB = 1 only and MSB = 1 only

4.14.3 Integrated Circuit D/A Converters

Switches in IC D/A converters are made either of BJTs or of MOSFETs. They are generally one of two types: voltage driven or current driven. Voltage-driven converters, which use BJTs or MOSFETs as on or off switches, are generally used for relatively low-speed low-resolution applications. In a current-driven converter, switching is accomplished using emitter-coupled logic (ECL) current switches, which do not saturate but are driven from the active region to cutoff. This type of converter is capable of much faster operation than the voltage-driven type. IC D/A converters of 8, 10, 12, 14, and 16 bits are commercially available with either a current output, a voltage output, or both a current and a voltage output.

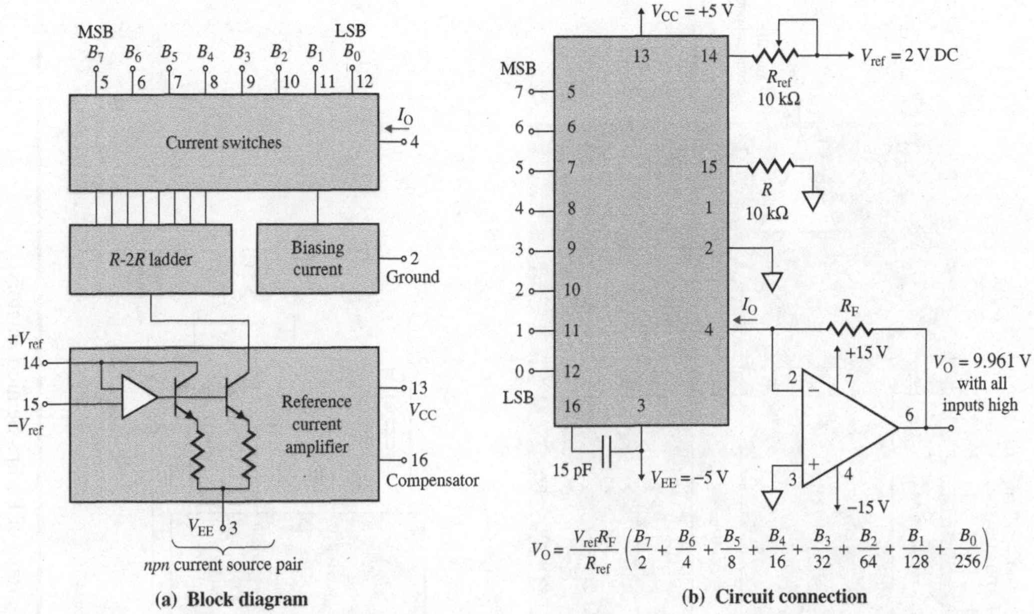

FIGURE 4.75 MC1408 D/A converter with current output

The MC1408 is an example of a D/A converter with current output. It is a low-cost, high-speed converter designed for use in applications where the output current is a linear product of an 8-bit digital word and an analog reference voltage. Its internal block diagram, shown in Fig. 4.75(a), consists of four parts: current switches, an *R*-2*R* ladder, a biasing current network, and a reference current amplifier. The connection diagram is shown in Fig. 4.75(b). The output current is converted to a voltage by an instantaneous current-to-voltage (I/V) op-amp converter.

The NE/SE-5018 is an example of a D/A converter with voltage output. It gives an output voltage that is a linear product of an 8-bit digital word and an analog reference voltage. Its internal block diagram is shown in Fig. 4.76(a). A typical configuration of the 5018 is shown in Fig. 4.76(b).

The manufacturer's specifications for a D/A converter normally include the following parameters. *Resolution* is determined by the number of input bits of the D/A converter. An 8-bit converter has 2^8 possible output levels, so its resolution is $1/2^8 = 1/256 = 0.39\%$. For a 4-bit converter, the resolution is $1/2^4 = 1/16 = 6.25\%$. Thus, resolution is the value of the LSB. *Accuracy* is defined in terms of the maximum deviation of the D/A output from an ideal straight line drawn from zero to full-scale output. *Nonlinearity*, or *linearity error*, is the difference between the actual output of the D/A converter and its ideal straight-line output. The error is normally expressed as a percentage of the full-scale range. *Gain error* is any error in gain, usually caused by deviations in the feedback resistor on the I/V converter. *Offset error* is any error caused by the fact that the output of the D/A converter is not zero when the binary inputs are all zero. This error stems from input offsets (in voltages and currents) of the op-amp as well as the D/A converter. *Settling time* is the time required for the output of the D/A converter to reach within $\pm 1/2$ LSB of the final value for a given digital input—that is, go from zero to full-scale output. *Stability* is a measure of the independence of the converter parameters from variations in external conditions such as temperature and supply voltage.

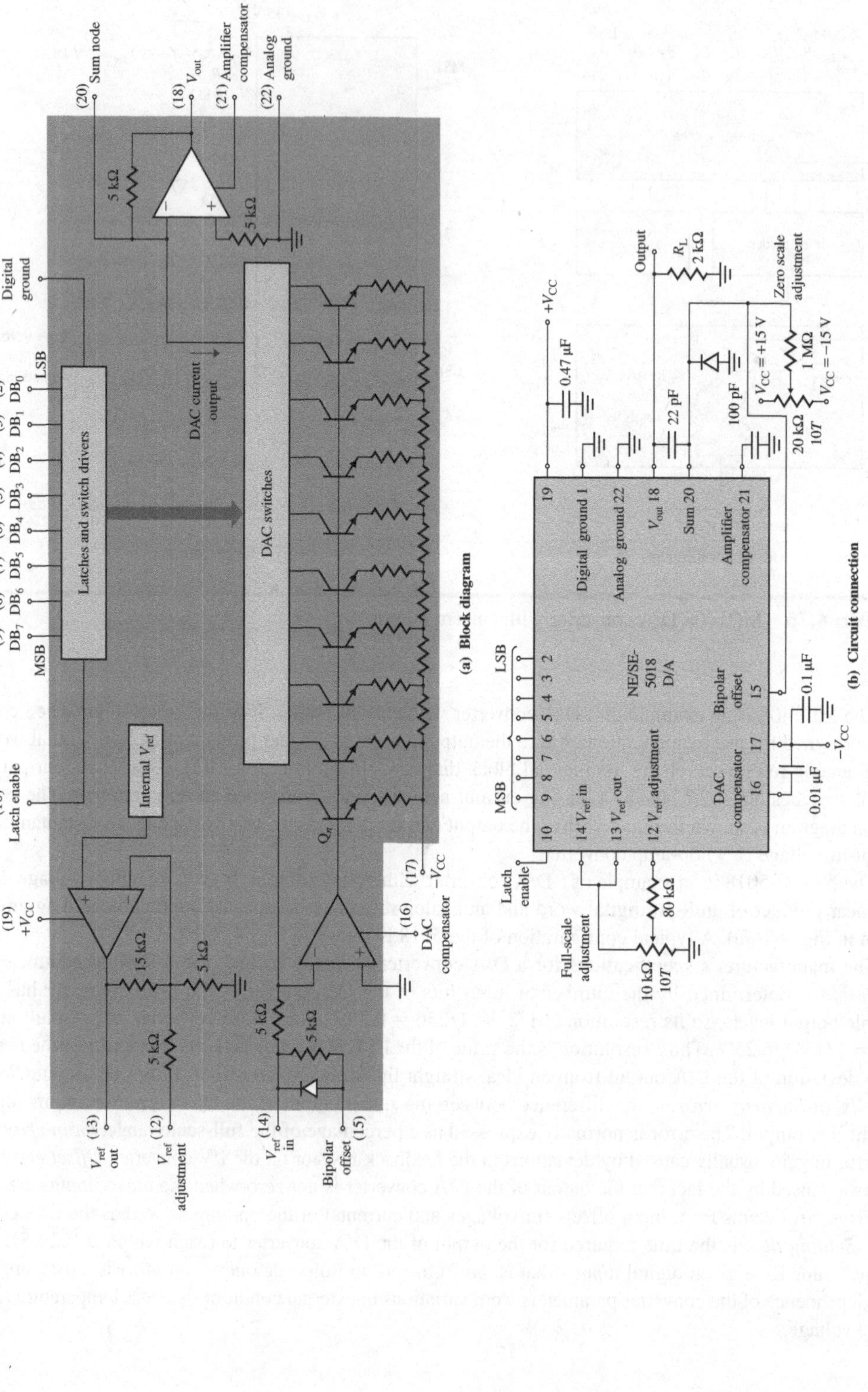

FIGURE 4.76 NE/SE-5018 D/A converter with voltage output (Courtesy of Philips Semiconductors)

KEY POINTS OF SECTION 4.14

- A D/A converter can convert a digital word to an analog voltage that is proportional to the binary number corresponding to the digital word.
- The specifications for a D/A converter include resolution, accuracy, nonlinearity (or linearity error), gain error, offset error, settling time, and stability. The resolution of an N-bit converter is $1/2^N$.

4.15 Analog-to-Digital Converters

A large number of physical devices generate output signals that are analog or continuous variables; examples include temperature and pressure gauges and flow transducers. For digital processing, the input signal must be converted into a binary form of 1s and 0s. The circuit that performs this conversion is called an analog-to-digital (A/D) converter. There are many types of A/D converters, depending on the type of conversion technique used, such as counting, tracking (up-down), successive approximation, single-ramp integrating, or dual-ramp integrating. The successive-approximation technique is the one most commonly used, mainly because it offers excellent trade-offs in resolution, speed, accuracy, and cost.

4.15.1 Successive-Approximation A/D Converter

A successive-approximation A/D converter operates by successively dividing in half the voltage range of the converter. The simplified block diagram of a 4-bit A/D converter is shown in Fig. 4.77(a). The converter consists of five parts: an analog comparator, a 4-bit register that has independent set and reset capability for each stage, a 4-bit D/A converter, a ring counter, and a logic control. The ring counter provides a timing (or clock) signal to control the operation of the converter. The logic control synchronizes the operation of the converter with the clock. The combination of the logic control, 4-bit register, and ring counter is often known as the *successive-approximation register* (SAR).

The comparator converts analog voltages to digital signals. It has two inputs, V_a and V_b, and gives a binary voltage. If $V_a > V_b$, the output is high (logic 1); if $V_a < V_b$, the output is low (logic 0). Thus, the comparator output V_{com} is

$$V_{com} = \text{sgn}\,(V_a - V_b) = \begin{cases} 1 & \text{for } V_a > V_b \\ 0 & \text{for } V_a < V_b \end{cases}$$

An SAH circuit is commonly used to hold the input voltage constant during the conversion process. There is no need for an SAH circuit if the input signal varies slowly enough and has a low enough noise level that the input will not change during the conversion.

The algorithm for the operation of a successive-approximation A/D converter can be best described by an example. The steps in converting an analog voltage of, say, 10 V are as follows:

Step 1. The first pulse from the ring counter sets the D/A converter, 4-bit register, and ring counter so that MSB = 1 and all others are 0; that is, $B_3 = 1$, and $B_2 = B_1 = B_0 = 0$. Thus, for $B_3B_2B_1B_0 = 1000$, the output V_b of the D/A is 8 V, which is compared by the comparator. If $V_a \geq 8$ V, the MSB in the register (B_3) is maintained at 1; otherwise, it is set to 0. At the end of step 1, $B_3 = 1$ for $V_a = 10$ V.

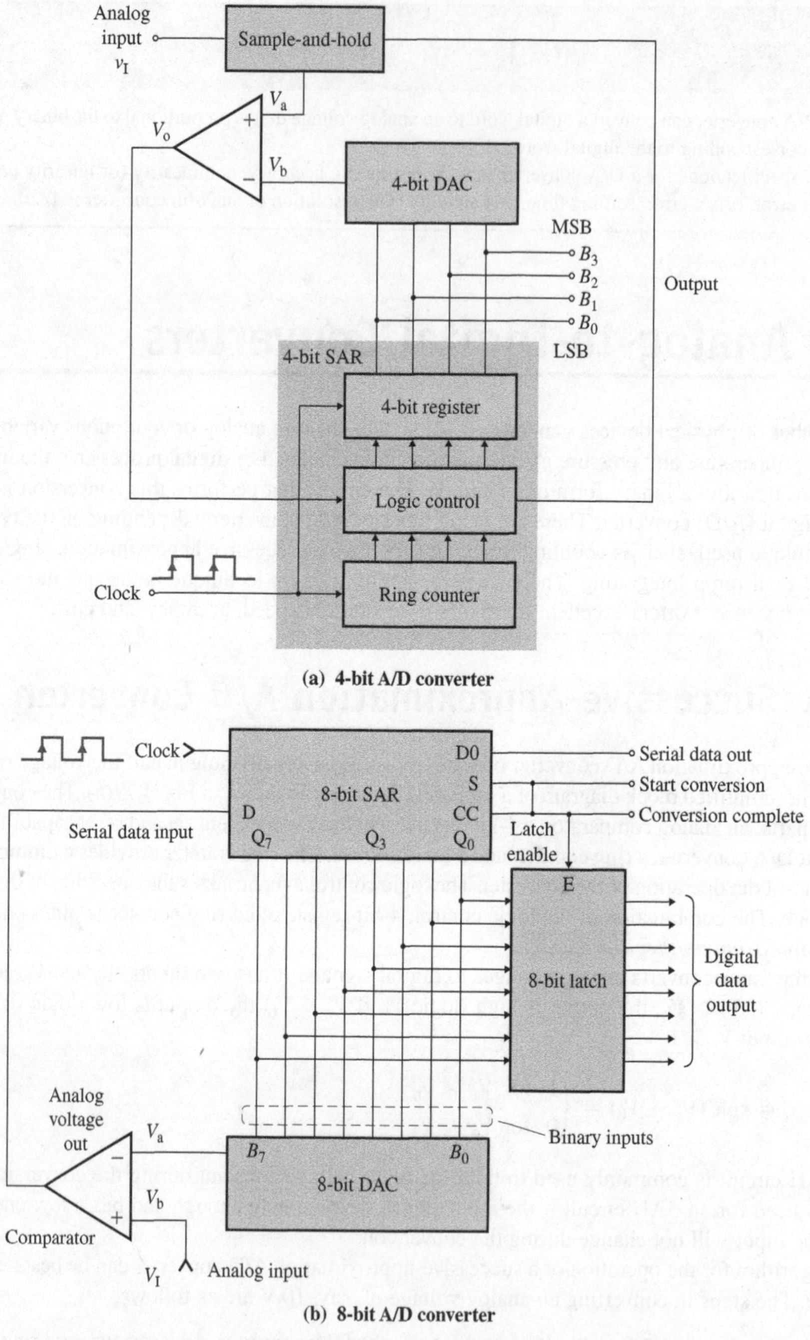

(a) 4-bit A/D converter

(b) 8-bit A/D converter

FIGURE 4.77 Successive-approximation A/D converter

Step 2. The second pulse from the ring counter sets $B_2 = 1$. B_1 and B_0 remain at 0, and B_3 remains at either 1 or 0, depending on the condition in step 1; that is, $B_3 = B_2 = 1$, and $B_1 = B_0 = 0$. Thus, for $B_3B_2B_1B_0 = 1100$, the output V_b of the D/A is 12 V, which is compared by the comparator. If $V_a \geq 12$ V, the B_2 in the register is maintained at 1; otherwise, it is set to 0. At the end of step 2, $B_2 = 0$ for $V_a = 10$ V.

Step 3. The third pulse from the ring counter sets $B_1 = 1$. B_0 remains at 0. B_2 and B_3 remain as they were at the end of step 2. That is, $B_3 = 1$, $B_2 = 0$, $B_1 = 1$, and $B_0 = 0$. Thus, for $B_3B_2B_1B_0 = 1010$, the output V_b of the D/A is 10 V, which is compared by the comparator. If $V_a \geq 10$ V, the B_1 in the register is maintained at 1; otherwise, it is set to 0. At the end of step 3, $B_1 = 1$ for $V_a = 10$ V.

Step 4. The fourth pulse from the ring counter sets $B_0 = 1$. B_3, B_2, and B_1 remain as they were at the end of step 3. That is, $B_3 = 1$, $B_2 = 0$, $B_1 = 1$, and $B_0 = 1$. Thus, for $B_3B_2B_1B_0 = 1011$, the output V_b of the D/A is 11 V, which is compared by the comparator. If $V_a \geq 11$ V, the B_0 in the register is maintained at 1; otherwise, it is set to 0. That is, $B_0 = 0$ for $V_a = 10$ V.

At the end of the fourth step, the desired number, which is in the counter, will give Read output. The results of the conversion steps are shown in Table 4.2. For an N-bit A/D converter, the conversion process will take N clock periods. That is, for an 8-bit A/D converter and a 10-MHz clock, the conversion will take $8/(10 \times 10^6) = 8 \times 10^{-7} = 800$ ns.

The successive-approximation technique can be extended to the higher-bit converter shown in Fig. 4.77(b), in which the SAR also performs the functions of logic control and ring counter. The conversion complete (CC) signal enables the latch. Digital data appear at the output of the latch and are also available serially as the SAR determines each bit. The cycle of the conversion process is normally repeated continuously, and the CC signal is connected to the Start-Conversion input.

4.15.2 Integrated Circuit A/D Converters

There are many types of IC A/D converters, such as the integrating A/D converter, the integrating A/D converter with three-stage outputs, and the tracking A/D converter with latched output. Also, the output can be in straight binary, binary-coded decimal (BCD), complementary binary (1s or 2s), or sign-magnitude binary form.

The NE5034 is an example of an IC A/D converter. Its internal block diagram is shown in Fig. 4.78(a). It is a high-speed microprocessor-compatible 8-bit A/D converter that uses the successive-approximation technique. It includes a comparator, a reference D/A converter, an SAR, an internal clock, and three-stage buffers all on the same chip. The connection diagram for the NE5034 is shown in Fig. 4.78(b). Upon receipt of the Start pulse, successive bits are applied to the input of the internal 8-bit current D/A converter by the

TABLE 4.2 Successive-approximation process for $V_a = 10$ V

Step	V_b	B_3	B_2	B_1	B_0	Comparisons	Answer
1	8 V	1	0	0	0	Is $V_a \geq 8$ V?	Yes
2	12 V	1	1	0	0	Is $V_a \geq 12$ V?	No
3	10 V	1	0	1	0	Is $V_a \geq 10$ V?	Yes
4	11 V	1	0	1	1	Is $V_a \geq 11$ V?	No
	10 V	1	0	1	0	Read output	

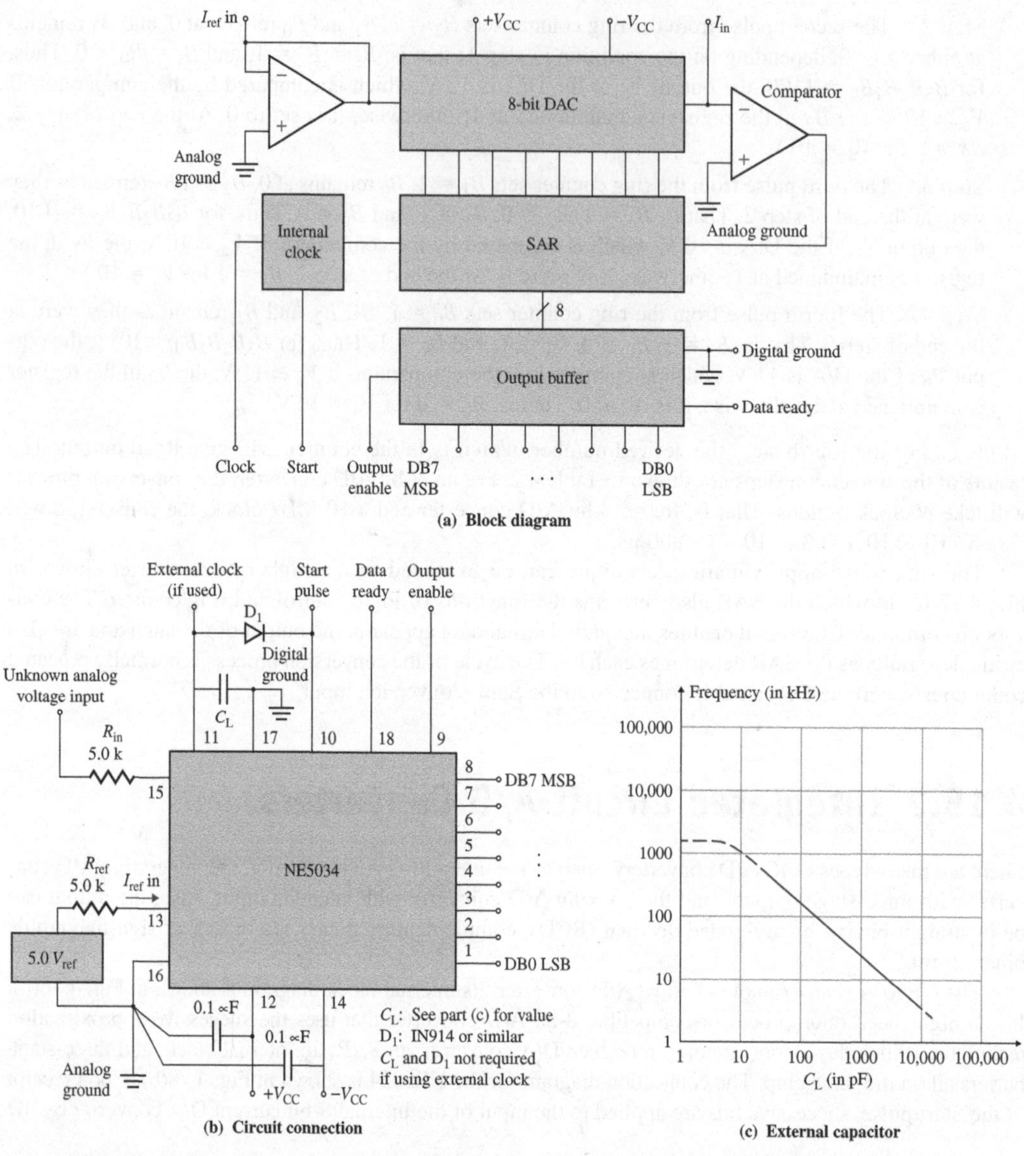

FIGURE 4.78 NE5034 8-bit A/D converter

I^2L SAR, beginning with the MSB (DB7). During the successive approximations, the sequence Data-Ready (DR) remains at 1. When the Output-Enable (OE) input is at logic 1, the data outputs assume a high-impedance status. When OE is at logic 0, the data are placed on the outputs. External capacitor C_L sets the internal clock frequency, as shown in Fig. 4.78(c). For C_L = 100 pF, for example, f_{clock} = 120 kHz.

The manufacturer's specifications for an A/D converter normally include the following parameters: *Input signal* is the maximum allowable analog input voltage range and may be unipolar or bipolar.

Conversion speed is the speed at which the A/D converter can make repetitive data conversions. The conversion time for successive-approximation converters ranges from 1 μs to 100 μs, whereas for an ultrafast parallel converter the time is in the range of 10 ns to 60 ns. *Quantizing error* is the error inherent in the conversion process because of the finite resolution of the discrete output. It is usually $\pm 1/2$ LSB. For a 10-bit converter with an analog input range of 0 to 10 V, the quantizing error will be $1/2^{10} \times 10 \text{ V} \approx 10 \text{ mV}$. *Accuracy* is the deviation of the actual bit transition value from the ideal transition value at any level over the range of the A/D converter. Accuracy includes errors from both the analog and the digital parts. With a digital error of 10 mV and a quantizing error of 10 mV, the overall error becomes 20 mV. With this amount of error, the converter will operate as a 9-bit A/D converter because a 9-bit converter has a quantizing error of $1/2^9 \times 10 \text{ V} \approx 20 \text{ mV}$.

KEY POINTS OF SECTION 4.15

- An A/D converter can convert an analog signal to a digital word that is proportional to the analog signal. Although there are many conversion techniques, the successive-approximation technique is the one most commonly used, mainly because of its excellent trade-offs in resolution, speed, accuracy, and cost.
- The specifications for an A/D converter include input signal range, conversion speed, quantizing error, and accuracy.

4.16 Voltage Regulators

A voltage regulator can provide a constant output voltage despite reasonable variations in the input supply voltage and the load. Most voltage regulators are designed to be relatively temperature insensitive and protect themselves and the load against overload. The output voltage of a power supply with a diode rectifier is shown in Fig. 4.79(a). The output voltage decreases as the load current increases as shown in Fig. 4.79(b) due to the internal resistance R_o of the power supply. The voltage regulation (VR), which is a measure of the output voltage degradation, is defined as

$$\text{VR} = \frac{V_{\text{no-load}} - V_{\text{Load}}}{V_{\text{no-load}}} \times 100\% \tag{4.107}$$

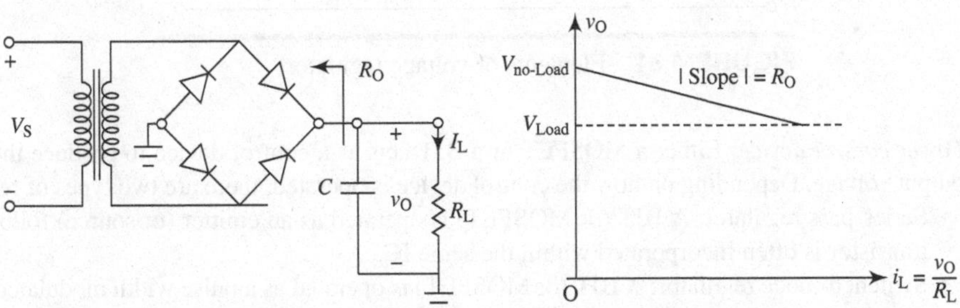

(a) Full-wave bridge **(b) Typical load characteristic**

FIGURE 4.79 Unregulated dc power supply

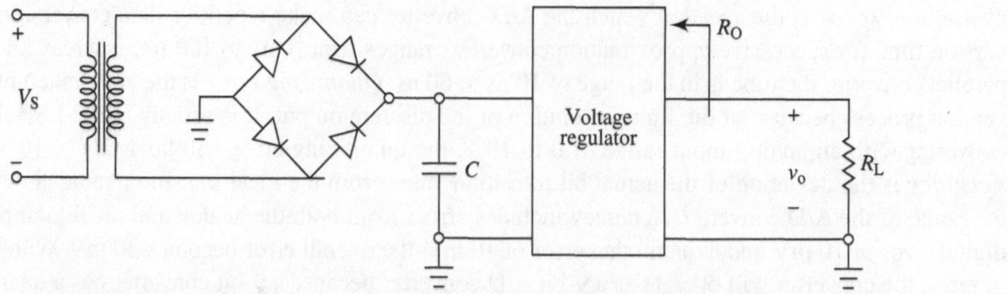

FIGURE 4.80 Full-wave bridge rectifier with a voltage regulator

The output resistance, which causes the voltage drop, should be as low as possible and there is a limit on the maximum load current $I_{L(max)}$. A voltage regulator is typically inserted between the output filter capacitor and the load as shown in Fig. 4.80. The regulator should maintain the output voltage V_O over a range of load current variations from 0 to $I_{L(max)}$.

Figure 4.81 shows the block diagram of a voltage regulator which consists of the following elements:

1. *Reference source*: The output voltage is compared with a reference signal. It is important that the reference source provide a voltage relatively independent of input source voltage and temperature changes over the operating range.
2. *Error detector*: The output voltage is compared to the reference signal and produces an error signal. This normally consists of a differential amplifier.
3. *Error amplifier*: The error signal is amplified to generate the control signal.
4. *Control signal generator*: The control circuit generates the gating signal for driving a power control device.

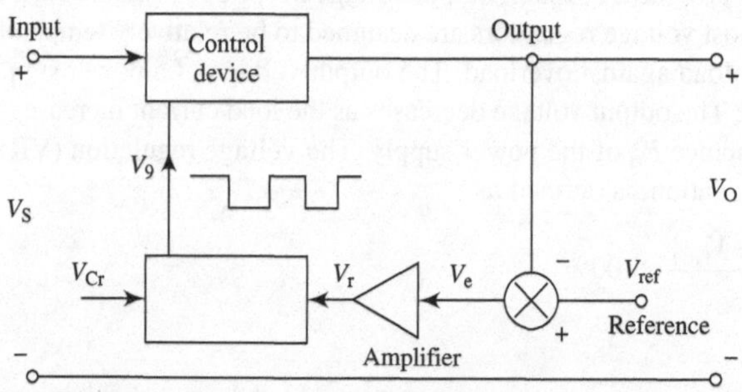

FIGURE 4.81 Elements of voltage regulators

5. *Power control device*: Either a MOSFET or a BJT acts as a control device to produce the desired output voltage. Depending on how the control device is operated, there are two types of regulators:
   - Series-pass regulator: A BJT (or MOSFET) is operated as an emitter (or source) follower. The transistor is often incorporated within the same IC.
   - Switched-mode regulator: A BJT (or MOSFET) is operated as a pulse width modulated (PWM) switch.
6. *Current limit*: This function is included for the output short-circuit protection within an internal operational amplifier.

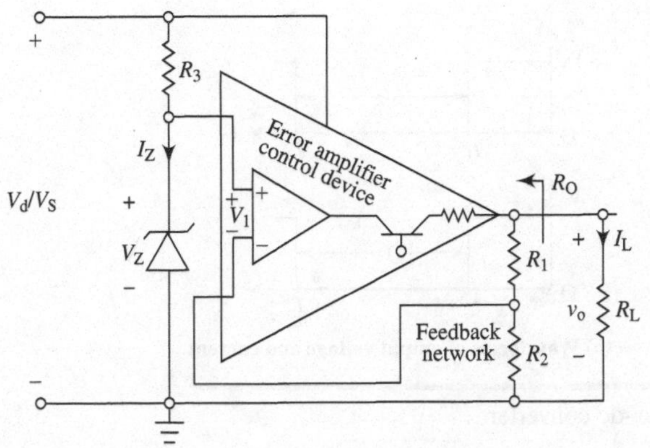

FIGURE 4.82 Schematic of series-pass voltage regulator

4.16.1 Series-Pass Voltage Regulator

The schematic of series-pass regulator is shown in Fig. 4.82. The Zener diode and resistance R_3 act as the reference source. Resistance R_{SC} ($<< R_L$) limits the load current. Resistances R_1 and R_2 provide the voltage-series feedback. For a high voltage gain (A) of the error amplifier, the closed-loop voltage gain is given by

$$A_f = \frac{V_O}{V_R} = \frac{A}{1+\beta A} \cong \frac{1}{\beta} = \frac{R_1 + R_2}{R_2} = 1 + \frac{R_1}{R_2} \tag{4.108}$$

Therefore, the output voltage depends on the Zener diode reference voltage and the ratio of two resistors. The principal disadvantage with series-pass regulators is relatively low efficiency as defined by

$$\eta = \frac{V_O I_O}{V_S I_S} = \frac{V_O}{V_S} \times 100\% \tag{4.109}$$

where $I_O = I_{IN}$ for a transistor current gain $\beta_F >> 1$. Since the power transistor is operating in the active region, the power dissipation in the output transistor is high in the range 30–50%. The typical efficiency is η = 50–70%. Heat sinking is required with an increased cost, weight, and size.

4.16.2 Fundamentals of Switching Mode Regulator

The switched-mode power supply (SMPS) or switching regulator controls the load power by varying the duty cycle of the series-pass transistor. The series-pass transistor is either turned ON or OFF .When switch SW, shown in Fig. 4.83(a), is closed for a time t_1, the input voltage V_S appears across the load. If the switch remains off for a time t_2, the voltage across the load is zero. The waveforms for the output voltage and load current are also shown in Fig. 4.83(b).

The switch can be implemented by using a BJT or a MOSFET. Neglecting the voltage drops of the power transistor, the average output voltage is given by

$$V_O = \frac{1}{T}\int_0^{t_1} v_o(t)dt = \frac{t_1}{T}V_S = t_1 f V_S = kV_S \tag{4.110}$$

where T is the switching period, $k = t_1/T$ is the duty cycle, and f is the switching frequency. The duty cycle k can be generated by comparing a dc reference signal v_r with a saw-tooth carrier signal v_{cr}. This is

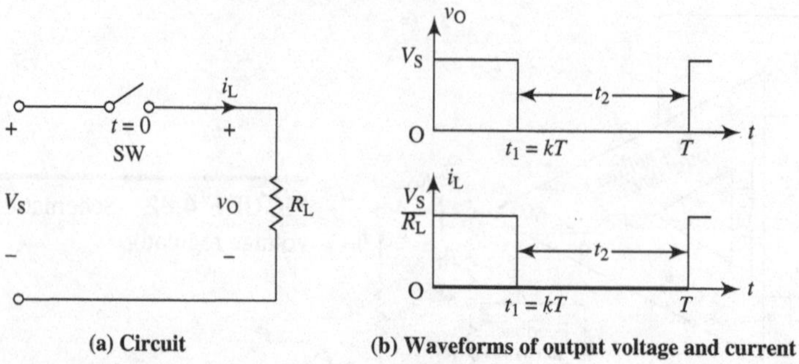

FIGURE 4.83 Step-down dc–dc converter

shown in Fig. 4.84, where V_r is the peak value of v_r and v_{cr} is the peak value of v_{cr}. The carrier signal is given by

$$v_{cr} = \frac{V_{cr}}{T}t \tag{4.111}$$

which must equal to the reference signal $v_r = v_{cr}$ (at t = kT) = V_r, i.e.,

$$V_r = \frac{V_{cr}}{T}kT \tag{4.112}$$

which gives the duty cycle k as

$$k = \frac{V_r}{V_{cr}} = M \tag{4.113}$$

where M is called the *modulation index*. By varying the reference signal from 0 to 1, the duty cycle k can be varied from 0 to 1.

The algorithm to generate the gating signal is as follows:

1. Generate a triangular waveform of period T as the carrier signal and a dc reference signal.
2. Compare these signals by a comparator to generate the difference and then a hard limiter to obtain a square-wave gate pulse of width kT, which must be applied to the switching device through an isolating circuit.
3. Any variation in the reference signal v_r varies linearly with the duty cycle k.

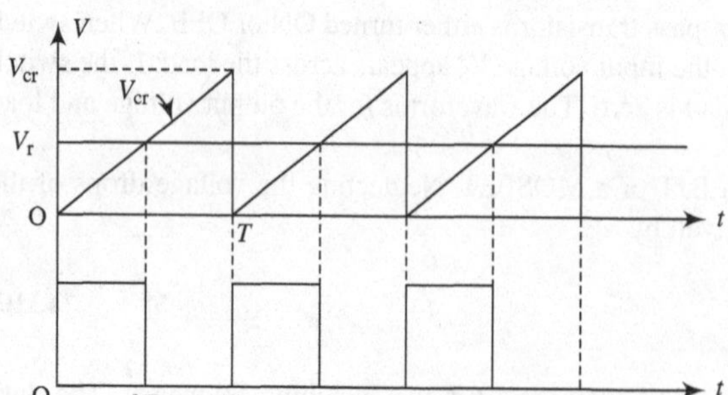

FIGURE 4.84 Generation of duty cycle

4.16.3 Functional Blocks of Switching Mode Regulator

We can notice from Fig. 4.83(b) that the output of dc–dc converters with a resistive load is discontinuous and this contains ripple harmonics. The ripple content is normally reduced by an LC filter. Switching regulators are commercially available as integrated circuits [9]. The designer can select the switching frequency by choosing the values of R and C of frequency oscillator. As a rule of thumb, to maximize efficiency, the minimum oscillator period should be about 100 times longer than the transistor switching time; for example, if a transistor has a switching time of 0.4 μs, the oscillator period would be 40 μs, which gives the maximum oscillator frequency of 25 kHz. This limitation is due to a switching loss in the transistor. The transistor switching loss increases with the switching frequency, and as a result the efficiency decreases. In addition, the core loss of inductors, which depends on the frequency, limits the high-frequency operation.

Figure 4.85 shows the functional bocks of a switch-mode power supply (SMPS). The dc output from the basic rectifier and capacitor filter powers the amplifiers and an oscillator operating at frequencies from several hundred hertz to 20 kHz. The comparator generates the PWM square-wave control signal. The PWM signal is used to switch an output transistor Q_1, between ON and OFF. This switched output is connected to the load R_L. To provide for a continuous power flow to the load, an LC filter acts as energy storage elements during the switch-off interval. A feedback circuit consisting of R_1 and R_2 is used to control the duty cycle. If the feedback voltage drops below the reference voltage V_{REF}, the error is amplified and used to change the transition point of the comparator and thus increases the duty cycle of Q_1. As the output increases to the desired level, the duty cycle stabilizes at some increased value. The frequency f_o of the rectangular wave is established by a fixed-frequency oscillator.

The output wave at the output of the dc–dc converter is a square wave of duration k_T as shown in Fig. 4.86(a). During the interval when the transistor is turned on, the inductor current rises almost linearly as shown in Fig. 4.86(b). During the interval when the transistor is turned off, the inductor current falls almost linearly. The capacitor voltage also varies with a 90° phase shift as shown in Fig. 4.86(c).

The peak-to-peak ripple current of the inductor can be found from [10]

$$\Delta I_L = \frac{V_S k(1-k)}{fL} \tag{4.114}$$

And the peak-to-peak ripple voltage of the capacitor can be found from [10].

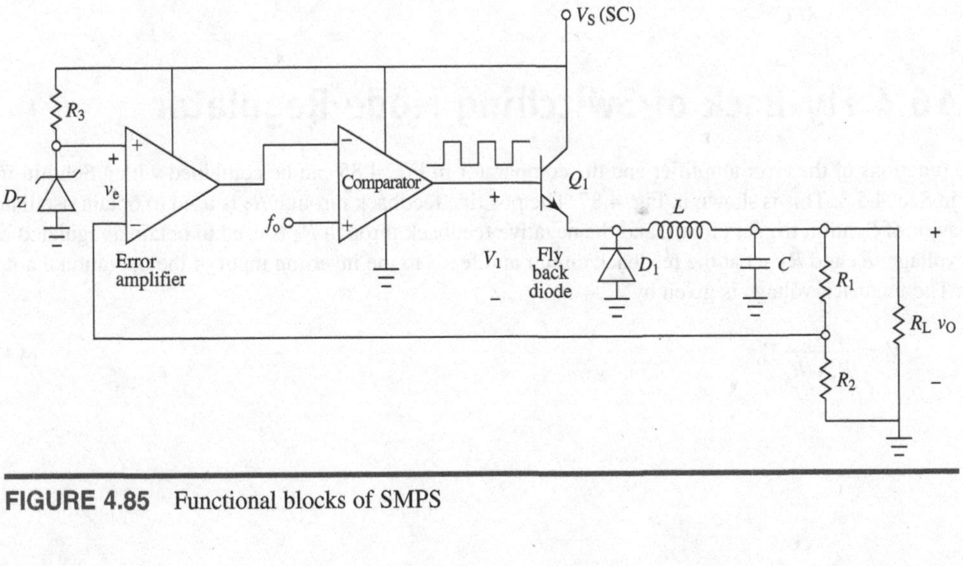

FIGURE 4.85 Functional blocks of SMPS

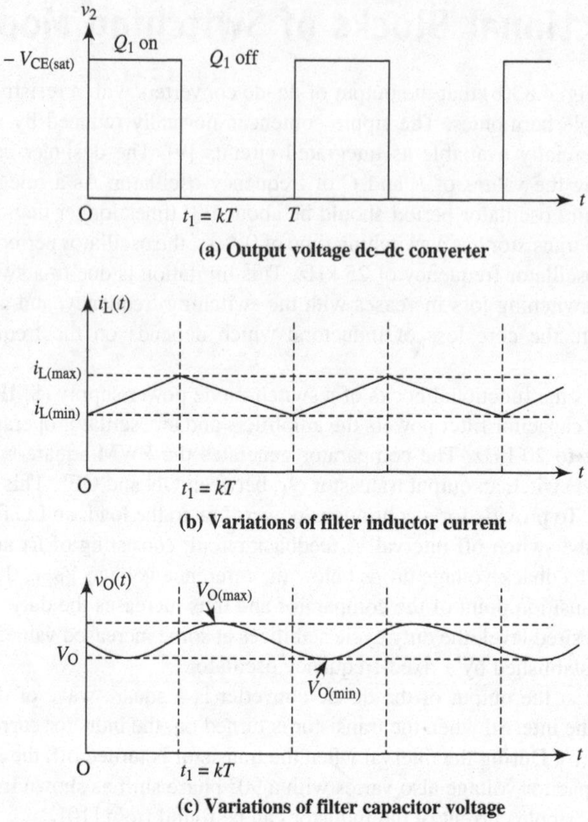

FIGURE 4.86 Waveforms of voltages and current

$$\Delta V_C = \frac{V_S k(1-k)}{8LCf^2} \tag{4.115}$$

4.16.4 Fly-Back of Switching Mode Regulator

The functions of the error amplifier and the comparator in Fig. 4.85 can be combined with a Schmitt trigger in Sec. 4.5.3. This is shown in Fig. 4.87. The positive feedback through R_2 is used to obtain oscillatory behavior of Schmitt trigger circuit, and the negative feedback through R_4 is used to obtain a regulated output voltage. R_3 and R_4 act at the feedback circuit and feeds to the inverting input of the operational amplifier. The feedback voltage is given by

$$V_- = \frac{R_4}{R_3 + R_4} V_O \tag{4.116}$$

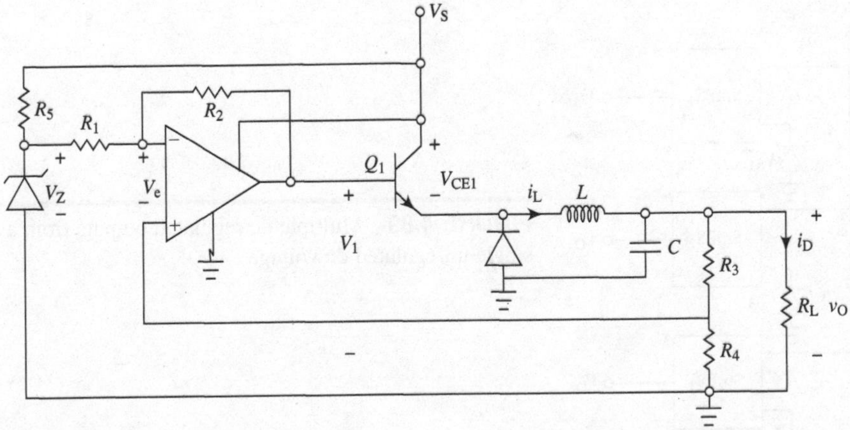

FIGURE 4.87 Fly-back dc-to-dc converter

The operational amplifier is configured as a comparator with hysteresis as shown in Fig. 4.88(a). The transfer function for this comparator is shown in Fig. 4.88(b). Applying Eqs. (4.21) and (4.22), we can get the voltage at which the comparator switches states as given by

$$V_{-} = V_R \pm \frac{R_1}{R_1 + R_2} V_1 \quad (4.117)$$

Therefore, if the hysteresis is kept at ±10 mV and $V_R = 5$ V, we can select the values of R_1 and R_2 so that the comparator will switch state at $V_{-} = 4.990$ V and $V_{-} = 5.010$ V. The oscillation frequency is given by [11]

$$f_o = \frac{1}{2\pi\sqrt{LC}} \sqrt{\frac{4}{\pi} \times \frac{V_R}{V_O} \times \frac{R_1 + R_2}{R_1}} \quad (4.118)$$

and the ripple factor RF can be approximated from [11].

$$\text{RF} = \frac{V_{O(max)} - V_{O(min)}}{V_O} = \frac{1}{15LCf_o^2} \times 100\% \quad (4.119)$$

Since the power dissipation of the switching transistor Q_1 is essentially zero, the overall regulator efficiency ranges between 70% and 90%. The SMPS can be configured to supply multiple dc voltages from a single dc source as shown in Fig. 4.89.

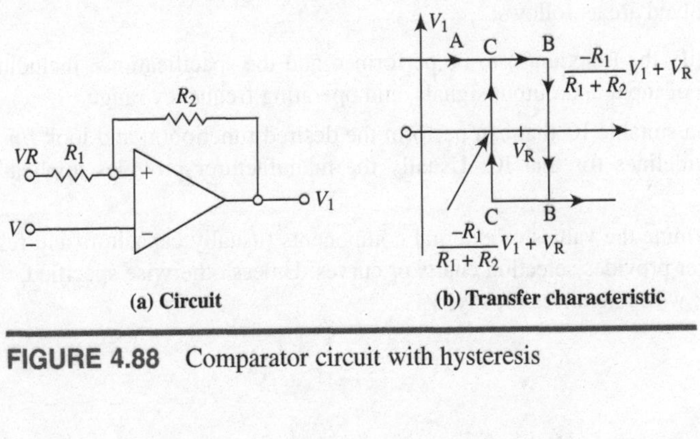

FIGURE 4.88 Comparator circuit with hysteresis

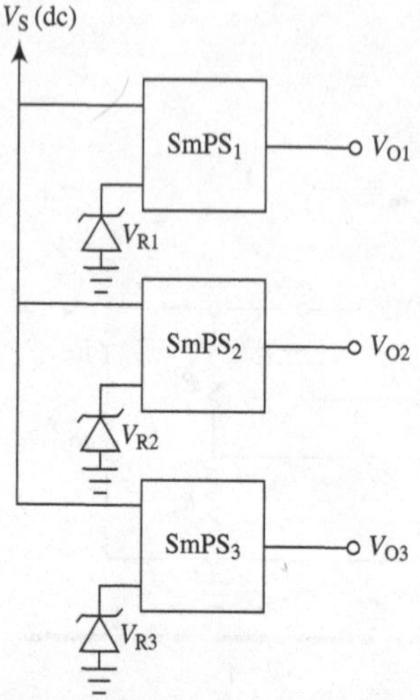

FIGURE 4.89 Multiple dc regulated outputs from a single unregulated dc voltage

KEY POINT OF SECTION 4.16

- A voltage regulator can provide a constant output voltage despite reasonable variations in input supply voltage and output load.
- There are two types of voltage regulators: series-pass regulator and switched-mode regulator. The switched-mode regulator has higher efficiency between 70% and 90%.
- An LC filter is generally connected to the output side of a voltage regulator for providing a continuous power flow to the load with low ripple content.

4.17 Circuit Design Using Analog Integrated Circuits

There are many analog ICs for general- and special-purpose applications. They include operational amplifiers, voltage comparators, instrument amplifiers, timers, buffers, interfacing circuits, voltage/frequency converters, data conversion circuits, power conversion and control circuits, and voltage regulators. The circuit design for an application using an IC is very simple and requires the selection of external components only. The steps involved are as follows:

Step 1. Identify the function(s) to be performed and the specifications, including available power supplies, range of input and output signals, and operating frequency range.

Step 2. Find a suitable IC that can perform the desired function(s) and look for application examples and/or guidelines for that IC. Usually the manufacturer provides application examples and guidelines.

Step 3. Determine the values of external components (usually capacitors and resistors). Generally, the manufacturer provides selection charts or curves. Unless otherwise specified, use standard values of components, with tolerances of, say, 5%.

Step 4. Simulate the circuit with a simulator such as PSpice/SPICE or Electronics Workbench, if the IC is supported by the simulator.

Step 5. Build and test the circuit, if possible.

Summary

Op-amp circuits with diodes can generate nonlinear functions for applications such as peak signal detectors, precision rectifiers, and comparator circuits. Op-amp circuits with diodes can reduce the effects of diode drop for low-voltage signals and are used for precision signal processing. The diode is placed in the feedback path of the op-amp and behaves as a superdiode with a negligible voltage drop on the order of microvolts.

Various waveforms are often required in electronic and control circuits. There are many integrated circuits that can be used to generate these waveforms. The design of wave generators is very simple and requires the selection of the external circuit components only. ICs such as op-amps, comparators, the NE/SE-566 VCO, the 555 timer, the NE/SE-565 PLL, and the 9400 series converters can be used to generate various waveforms.

A voltage regulator can provide a constant output voltage despite reasonable variations in input supply voltage and output load. The switched-mode regulators that have higher efficiency between 70% and 90% are commonly used.

References

1. J. R. Hufault, *Op-Amp Network Design*. New York: Wiley, 1986.
2. F. W. Hughes, *Op-Amp Handbook*. Englewood Cliffs, NJ: Prentice Hall, 1986.
3. C. F. Wojslow, *Operational Amplifiers*. New York: Wiley, 1986.
4. M. H. Rashid, *Introduction to PSpice Using OrCAD for Circuits and Electronics*. Englewood Cliffs, NJ: Prentice Hall, 2004, Chapter 10.
5. P. R. Gray, P. J. Hurst, S. H. Lewis, and R. G. Meyer, *Analysis and Design of Analog Integrated Circuits*. New York: Wiley, 2001.
6. J. M. Jacob, *Applications and Design with Analog Integrated Circuits*. Englewood Cliffs, NJ: Regents/Prentice Hall, 1993.
7. S. Soclof, *Design and Applications of Analog Integrated Circuits*. Englewood Cliffs, NJ: Prentice Hall, 1991, Chapters 15–17.
8. R. A. Gayakwad, *Op-Amps and Linear Integrated Circuits*. Englewood Cliffs, NJ: Prentice Hall, 1993.
9. Datasheet Catalog, http://www.DatasheetCatalog.com. A datasheet source for electronic components and semiconductors. Accessed November 2012.
10. M. H. Rashid, *Power Electronics—Circuits, Devices and Applications*. Upper Saddle River, NJ: Pearson Education, Inc, 3rd ed., 2004, Chapter 5.
11. S. G. Burns and P. R. Bond, *Principles of Electronic Circuits*. Boston, MA: International Thomson Publishing Company, 2nd ed., 1997.

Review Questions

1. What is a voltage limiter?
2. What is a soft limiter?
3. What is a clamper?
4. What is a comparator?
5. What is a limiting comparator?
6. What are the advantages of precision rectifiers?
7. What is a superdiode?
8. What are the differences between a comparator and an op-amp?
9. What is the principle of operation of a zero-crossing detector?
10. What is a Schmitt trigger?
11. What circuit parameters determine the upper and lower threshold voltages of a Schmitt trigger?
12. What are the effects of hysteresis on output voltage?
13. What is the principle of operation of a square-wave generator?
14. Why is a square-wave generator called an astable multivibrator?
15. What is saturation of an op-amp?
16. What is the purpose of series resistors in the input terminals of an op-amp in a square-wave generator?
17. What circuit parameters determine the frequency of a square wave?
18. What is the principle of operation of a triangular-wave generator?
19. What circuit parameters determine the frequency of a triangular wave?
20. What is the principle of operation of a sawtooth-wave generator?
21. What circuit parameters determine the frequency of a sawtooth wave?
22. What is the duty cycle of a sawtooth wave?
23. What is a voltage-controlled oscillator (VCO)?
24. What is the principle of operation of a VCO?
25. What circuit parameters determine the output frequency of a VCO?
26. What is the 555 timer?
27. What is a monostable multivibrator?
28. What advantages does the 555 timer connected as an astable multivibrator have over an op-amp astable multivibrator?
29. What circuit parameters determine the output frequency of the 555 timer connected as ramp generator?
30. What is a phase-lock loop (PLL)?
31. What are the main components of a PLL?
32. What is the principle of operation of a PLL?
33. What is the free-running mode of a PLL?
34. What is the capture mode of a PLL?
35. What is the phase-lock mode of a PLL?
36. What is the capture range of a PLL?
37. What is the lock range of a PLL?

38. What are the relationships among the free-running frequency, capture frequency, and lock frequency of a PLL?
39. What are some applications of the TelCom 9400 series converter?
40. What is a sample-and-hold circuit?
41. What are the main parts of a sample-and-hold circuit?
42. What is a digital-to-analog (D/A) converter?
43. What are the main parts of a D/A converter?
44. What is an analog-to-digital (A/D) converter?
45. What are the main parts of an A/D converter?
46. What is a voltage regulator?
47. What is voltage regulation?
48. What is a series-pass voltage regulator?
49. What is a switching-mode voltage regulator?
50. What is the function of a fly-wheel diode in a voltage regulator?
51. What is the function of a hysteresis loop in the comparator circuit of a switching voltage regulator?

Problems

The symbol D indicates that a problem is a design problem. The symbol P indicates that you can check the solution to a problem using PSpice/SPICE or Electronics Workbench.

4.2 *Circuits with Op-Amps and Diodes*

4.1 P **a.** The precision full-wave rectifier in Fig. 4.6(a) has $R_1 = R_2 = R_F = R = 10\ \text{k}\Omega$, $R_3 = 40\ \text{k}\Omega$, and $R_4 = 40\ \text{k}\Omega$. The input voltage is $v_S = 2 \sin(377\pi t)$. Plot the transfer characteristic and draw the waveform of the output voltage v_O.

b. Use PSpice/SPICE to plot the transfer characteristic and the output voltage.

4.2 D **a.** Design a precision full-wave rectifier as shown in Fig. 4.6(a) to provide a voltage gain of $A_f = v_O/v_S = 50$. The input voltage is $v_S = 0.01 \sin(200\pi t)$.

P **b.** Use PSpice/SPICE to check your design by plotting the output voltage.

4.3 D P **a.** Design a negative voltage limiter using the circuit in Fig. 4.10(a) by determining the values of R_1, R_2, R_3, and R_F. The supply DC voltage is $V_A = 15$ V. The circuit should limit the negative output voltage to $V_{O(min)} = -8$ V. The voltage gain without limiting is $A_f = -5$. The diode is fully turned on at a forward current of $i_D = 0.1$ mA, and its corresponding forward voltage drop is $V_D = 0.7$ V. The slope after the break point is to be limited to $S_1 = -1/30$.

b. Use PSpice/SPICE to plot the transfer characteristic for part (a).

4.4 The adjustable limiter in Fig. 4.11(a) has $R_1 = 12\ \text{k}\Omega$, $R_F = 70\ \text{k}\Omega$, $R_2 = 8\ \text{k}\Omega$, $R_3 = 1\ \text{k}\Omega$, $R_4 = 6\ \text{k}\Omega$, $R_5 = 1\ \text{k}\Omega$, $V_A = 15$ V, $-V_B = -15$ V, and $V_D = 0.7$ V. Determine **(a)** the positive clamping voltage $V_{O(max)}$ and the corresponding input voltage $V_{S(min)}$, **(b)** the negative clamping voltage $V_{O(min)}$ and the corresponding input voltage $V_{S(max)}$, and **(c)** the output voltage when the input voltage is $v_S = 5$ V.

4.5 D P **a.** Design an adjustable voltage limiter as shown in Fig. 4.11(a) to satisfy the following specifications: $V_{O(min)} = -5$ V, $V_{O(max)} = 5$ V, voltage gain $A_f = -10$. The slope is $S_1 = -1/20$ after break for $v_S > 0$, and the slope is $S_2 = -1/20$ after break for $v_S < 0$. The DC supply voltages are $V_A = 12$ V and

$-V_B = -12$ V. The on-state diode current is $i_D = 0.2$ mA, and the corresponding on-state diode voltage is $V_D = 0.7$ V.

b. Use PSpice/SPICE to plot the transfer characteristic. Assume $V_{CC} = 12$ V and $-V_{EE} = -12$ V. Use the PSpice/SPICE op-amp macromodel.

4.6 D P Design an output voltage–clamping circuit as shown in Fig. 4.14(a) so that the slope of the transfer characteristic is $S = v_O/v_S = 20$, $V_{O(max)} = 6.7$ V, and $V_{O(min)} = -8.7$ V. Determine the zener voltages V_{Z1} and V_{Z2}. Assume $V_D = 0.7$ V.

4.7 D P **a.** Design a hard limiter as shown in Fig. 4.15(a) by determining the values of R_1, R_2, R_3, R_4, and R_5. The circuit should limit the negative output voltage to $V_{O(min)} = -6$ V and the positive voltage to $V_{O(max)} =$ 6 V. The magnitude of the slopes after the break points should be less than or equal to 1/25. The diode drop is $V_D = 0.7$ V at $I_D = 0.1$ mA. The DC supplies are given by $V_A = -V_B = 15$ V.

b. Use PSpice/SPICE to plot the transfer characteristic. Assume $V_{CC} = 15$ V, $-V_{EE} = -15$ V, and $v_S = -5$ V to 5 V. Use the PSpice/SPICE op-amp macromodel.

4.8 D P **a.** Design a hard limiter as shown in Fig. 4.15(a) by determining the values of R_1, R_2, R_3, R_4, and R_5. The circuit should limit the negative output voltage to $V_{O(min)} = -5$ V and the positive voltage to $V_{O(max)} = 5$ V. The magnitude of the slopes after the break points should be less than or equal to 1/50. The diode drop is $v_D = 0.7$ V at $i_D = 0.1$ mA. The DC supplies are given by $V_A = -V_B = 15$ V.

b. Use PSpice/SPICE to plot the transfer characteristic. Assume $V_{CC} = 15$ V, $-V_{EE} = -15$ V, and $v_S = -5$ V to 5 V. Use the PSpice/SPICE op-amp macromodel.

4.9 D P **a.** Design a hard limiter as shown in Fig. 4.15(a) by determining the values of R_1, R_2, R_3, R_4, and R_5. The circuit should limit the negative output voltage to $V_{O(min)} = -7$ V and the positive voltage to $V_{O(max)} =$ 9 V. The magnitude of the slopes after the break points should be less than or equal to 1/50. The diode drop is $v_D = 0.7$ V at $i_D = 0.1$ mA. The DC supplies are given by $V_A = -V_B = 15$ V.

b. Use PSpice/SPICE to plot the transfer characteristic. Assume $V_{CC} = 15$ V, $-V_{EE} = -15$ V, and $v_S = -5$ V to 5 V. Use the PSpice/SPICE op-amp macromodel.

4.3 & 4.5 *Comparators and Schmitt Triggers*

For Probs. 4.10 through 4.14, use comparator LM111 and $v_S = 10 \sin(2000\pi t)$ to plot the hysteresis characteristic using PSpice/SPICE.

4.10 D P Design a Schmitt trigger as in Fig. 4.25(a) so that $V_{th} = |+V_{th}| = |-V_{th}| = 5$ V. Assume $V_{sat} = |-V_{sat}| = 12$ V.

4.11 P The parameters of the Schmitt trigger in Fig. 4.25(a) are $R_1 = 100\ \Omega$ and $R_F = 47$ kΩ. Calculate the threshold voltages $+V_{th}$ and $-V_{th}$. Assume $V_{sat} = |-V_{sat}| = 12$ V.

4.12 D P Design an inverting Schmitt trigger with the reference voltage of Fig. 4.25(a) so that $V_{Ht} = -8$ V and $V_{Lt} = -4$ V. Assume $V_{sat} = |-V_{sat}| = 12$ V.

4.13 D P Design a noninverting Schmitt trigger as in Fig. 4.28(a) so that $V_{th} = |+V_{th}| = |-V_{th}| = 5$ V. Assume $V_{sat} = |-V_{sat}| = 12$ V.

4.14 D P Design a noninverting Schmitt trigger with the reference voltage of Fig. 4.29(a) so that $V_{Ht} = 8$ V and $V_{Lt} = 4$ V. Assume $V_{sat} = |-V_{sat}| = 12$ V.

4.6–4.8 Square-, Triangular-, and Sawtooth-Wave Generators

For Probs. 4.15 through 4.19, use op-amp LF411 to plot the output using PSpice/SPICE.

4.15 Design the square-wave generator shown in Fig. 4.34(a) so that $f_o = 2$ kHz. Assume $V_{sat} = |-V_{sat}| = 10$ V.

D P

4.16 The parameters of the square-wave generator of Fig. 4.34(a) are $R_1 = 10$ kΩ, $R_F = 15$ kΩ, $R = 10$ kΩ, and $C = 0.047$ μF. Calculate the output frequency f_o.

P

4.17 Design the triangular-wave generator shown in Fig. 4.37(a) so that $f_o = 2$ kHz and $V_{th} = 5$ V. Assume $V_{sat} = |-V_{sat}| = 12$ V.

D P

4.18 The parameters of the triangular-wave generator of Fig. 4.37(a) are $R_1 = 10$ kΩ, $R_F = 40$ kΩ, $R = 10$ kΩ, and $C = 0.047$ μF. Calculate the output frequency f_o.

P

4.19 Design the sawtooth-wave generator shown in Fig. 4.40(a) so that $f_o = 5$ kHz, $V_{th} = 5$ V, and the circuit has a duty cycle of $k = t_1/T = 0.4$. Assume $V_{sat} = |-V_{sat}| = 12$ V.

D P

4.9 Voltage-Controlled Oscillators

4.20 a. Design a VCO as shown in Fig. 4.43(c) that has a nominal frequency of $f_o = 10$ kHz. Assume $V_{CC} = 15$ V.

D b. Calculate the modulation in the output frequencies if v_{CN} is varied by ±10%.

4.21 The parameters of the VCO in Fig. 4.43(c) are $R_A = 2.5$ kΩ, $R_1 = R_B = 10$ kΩ, and $C = 0.01$ μF.

a. Calculate the nominal frequency of the output waveform f_o.

b. Calculate the modulation in the output frequencies if v_{CN} is varied by ±10%. Assume $V_{CC} = 12$ V.

4.10 The 555 Timer

For Probs. 4.22 through 4.29, use the 555 timer to plot the output by PSpice/SPICE.

4.22 Design a monostable multivibrator as in Fig. 4.45(a) so that $t_p = 2$ ms. Assume $V_{CC} = 15$ V.

D P

4.23 Design an astable multivibrator as in Fig. 4.49(a) so that $k = 80\%$ and $f_o = 5$ kHz. Assume $V_{CC} = 15$ V.

D P

4.24 The parameters of the astable multivibrator in Fig. 4.49(a) are $R_A = 2.2$ kΩ, $R_B = 3.9$ kΩ, and $C = 0.1$ μF. Determine **(a)** the charging time t_c, **(b)** the discharging time t_d, and **(c)** the free-running frequency f_o.

P

4.25 Design a square-wave generator as in Fig. 4.52 so that $k = 50\%$ and $f_o = 5$ kHz. Assume $V_{CC} = 15$ V.

D P

4.26 The parameters of the square-wave generator in Fig. 4.52 are $R_A = 2.7$ kΩ, $R_B = 4.7$ kΩ, and $C = 1$ μF. Determine **(a)** the charging time t_c, **(b)** the discharging time t_d, and **(c)** the free-running frequency f_o.

P

4.27 Design a ramp generator as in Fig. 4.53(a) so that $k = 50\%$ and $f_o = 5$ kHz. Assume $V_{CC} = 15$ V, $V_{BE} = 0.7$ V, and a transistor of $\beta_F = 100$.

D P

4.28 The parameters of the ramp generator in Fig. 4.53(a) are $R = 10$ kΩ, $V_{CC} = 15$ V, $V_{BE} = 0.7$ V, and a transistor of $\beta_F = 100$. Determine the free-running frequency f_o.

P

4.29 D P Design the FSK modulator shown in Fig. 4.54(a) to produce frequencies of 1270 Hz and 1570 Hz corresponding to 1 (mark) and 0 (space), respectively.

4.11 *Phase-Lock Loops*

4.30 D Design a PLL as shown in Fig. 4.58(c) so that $f_o = 5$ kHz and $f_c = \pm 50$ Hz. Assume $V_{CC} = -V_{EE} = 15$ V.

4.31 The parameters of the PLL in Fig. 4.58(c) are $R_1 = 12\ \text{k}\Omega$, $C_1 = 0.01\ \mu\text{F}$, $C_2 = 10\ \mu\text{F}$, and $V_{CC} = -V_{EE} = 15$ V. Determine **(a)** the free-running frequency f_o, **(b)** the lock frequency f_L, and **(c)** the capture range f_c.

4.12 *Voltage-to-Frequency and Frequency-to-Voltage Converters*

4.32 D Design a V/F converter as shown in Fig. 4.64 so that $f_o = 2.5$ kHz at $v_I = 5$ V. The input voltage v_I can vary between 10 mV and 10 V. Assume $V_{DD} = -V_{SS} = 5$ V.

4.33 D Design an F/V converter as shown in Fig. 4.67 so that $V_O = 2.5$ V at $f_{in} = 10$ kHz. The input frequency f_{in} can vary between 0 and 20 kHz. Assume $V_{DD} = -V_{SS} = 5$ V.

4.13 *Sample-and-Hold Circuits*

4.34 D Design the SAH circuit shown in Fig. 4.71(a) so that the drop is within 0.5%. The leakage current in the hold mode is 1 nA, and the hold voltage is $V_h = 5$ V. The internal hold time is $t_h = 100\ \mu\text{s}$. Find the holding capacitance C_h.

4.35 D Design the SAH circuit shown in Fig. 4.71(a) so that the output tracks the input within 0.5%. The internal hold time is $t_h = 100$ ns. The input biasing current of the op-amp is $I_B = 10$ nA. Assume $R_F = R_1 = 20\ \text{k}\Omega$ and $v_I = 5$ V.

4.36 D Design the SAH circuit shown in Fig. 4.71(a) so that the drop is within 0.5%. The internal hold time is $t_h = 0.1$ ms, and the hold voltage is $V_h = 5$ V. The input biasing current of the op-amp is $I_B = 200$ nA. Assume $R_F = R_1 = 20\ \text{k}\Omega$.

4.14 *Digital-to-Analog Converters*

4.37 A 10-bit D/A converter of type 2*R* ladder, as shown in Fig. 4.73(a), has an input of 00 1001 1001, and the reference voltage is 5 V. Find the analog output voltage V_O.

4.38 A D/A converter is to have a full-scale output of 5 V and a resolution of less than 20 mV. What is the bit size?

4.39 The 8-bit D/A converter shown in Fig. 4.73(a) is to have a full-scale output of 10 V with a reference voltage of 5 V. Find the values of R_F and R.

4.15 *Analog-to-Digital Converters*

4.40 An 8-bit A/D converter has a reference voltage of 10 V. Find **(a)** the analog input corresponding to the binary outputs 1010 1010 and 0101 0101, **(b)** the binary output if $V_a = 3$ V, and **(c)** the resolution of the converter.

4.41 Construct a table similar to Table 4.2 for a 4-bit A/D converter if $V_a = 5$ V with a reference voltage of 16 V.

4.42 Construct a table similar to Table 4.2 for an 8-bit A/D converter if $V_a = 4$ V with a reference voltage of 10 V.

4.16 Voltage Regulators

4.43 The output voltage of a regulated power supply decreases from 9 V at no load to 8.95 V at a load current of 500 mA. Calculate the effective output resistance R_o and voltage regulation (VR).

4.44 A 12-V automobile battery decreases to 11.6 V under a starter current load of 200 A. Calculate the effective output resistance R_O and voltage regulation (VR).

4.45 The parameters of the circuit in Fig. 4.82 are $V_R = 4.7$ V, $8 \le V_S \le 12$, $I_{Z(min)} = 1$ mA, $P_{z(max)} = 250$ mW. Determine the minimum and maximum values for R_3 that will permit operation over all values of V_{IN}.

4.46 The parameters of the circuit in Fig. 4.82 are $V_R = 4.7$ V, $8 \le V_S \le 12$, $I_{Z(min)} = 1$ mA, $P_{z(max)} = 300$ mW. Determine the minimum and maximum values for R_3 that will permit operation over all values of V_{IN}.

4.47 The parameters of the circuit in Fig. 4.82 are $V_R = 1.63$ V and $V_O = 7.5$ V. Determine the values of R_1 and R_2 such that $R_1 \parallel R_2 = 2$ kΩ.

4.48 The switching regulator in Fig. 4.85 has an input voltage of $V_S = 12$ V. The required average output voltage $V_O = 5$ V at $R = 1$ kΩ and the peak-to-peak output ripple voltage $\Delta V_C = 20$ mV. The switching frequency $f_o = 26$ kHz. If the peak-to-peak ripple current of inductor is limited to $\Delta I_L = 04$ A, determine (a) the duty cycle k, (b) the filter inductance L, and (c) the filter capacitor C.

4.49 a. Design an SMPS in Fig. 4.87 by selecting the values of R_1, R_2, R_3, and R_4. The parameters are $V_S = 15$ V, comparator output $V_1 = \pm 13$ V, $V_O = 9$ V, $V_R = 4.7$ V, $L = 10$ mH, and $C = 10$ μF. Hysteresis ≤ 10 mV. (b) Calculate the resultant switching frequency f_o and the ripple factor (RF).

4.50 a. Design an SMPS in Fig. 4.87 by selecting the values of R_1, R_2, R_3, and R_4. The parameters are $V_S = 15$ V, comparator output $V_1 = \pm 13$ V, $V_O = 9$ V, $V_R = 4.7$ V, $L = 4.7$ mH, and $C = 10$ μF. Hysteresis ≤ 20 mV. (b) Calculate the resultant switching frequency f_o and the ripple factor (RF).

4.51 a. Design an SMPS in Fig. 4.87 by selecting the values of R_1, R_2, R_3, and R_4. The parameters are $V_S = 15$ V, comparator output $V_1 = \pm 13$ V, $V_O = 9$ V, $V_R = 4.7$ V, $L = 4.7$ mH, and $C = 22$ μF. Hysteresis ≤ 10 mV. (b) Calculate the resultant switching frequency f_o and the ripple factor (RF).

APPENDIX A

INTRODUCTION TO OrCAD

A.1 Introduction

This appendix provides an introduction to the electronic circuit simulation software OrCAD [1, 2]. The objectives of this appendix are the following:

- Becoming familiar with the electronic circuit simulation software OrCAD Capture
- Learning how to draw electrical circuits by OrCAD Capture

After installation instructions presented in Sec. A.2, Sec. A.3 provides an overview of the software package. Then Secs. A.4 to A.8 cover the basic steps in the circuit analysis process:

- Drawing the circuit
- Selecting the type of analysis
- Simulating the circuit
- Displaying the results of the simulation

The remaining sections, Secs. A.9 to A.13, deal with specific operations:

- Copying and capturing schematics
- Varying parameters
- Performing frequency response analysis
- Modeling devices and elements
- Creating netlists
- Adding library files

A.2 Installing the Software

The instructions for installing the OrCAD software are printed on the OrCAD installation disk or the CD-ROM. The software can be downloaded from the Cadence website at http://www.cadence.com/products/orcad/Pages/downloads.aspx

The following steps, which pertain to OrCAD Version 9.2, are applicable to other versions as well:

1. Place the schematics CD-ROM in the CD drive.
2. From Windows, enter the **File Manager** and click the left mouse button on the CD drive.
3. Click on **setup.exe**, **File**, **Run**, and **OK**.
4. Click on **OK** to select the products to install **Capture** and **PSpice**, as shown in Fig. A.1(a).
5. Click on **OK** to select the default **C:OrcadLite**.
6. Click on **Yes** to choose **Program Folder: Orcad Family Release 9.2 Lite Edition**, as shown in Fig. A.1(b).
7. Create the OrCAD Capture Lite and PSpice A/D icons from the Start menu; then go to **Programs**, **Orcad Family**, and **Capture Lite** menu. These icons are shown in Fig. A.2.
8. Click the left mouse button on the Capture Lite icon once, and the window of Capture Lite will open.

▶ **NOTE** Although there are later versions of OrCAD, the 9.2 version has proven to be bug free and works well.

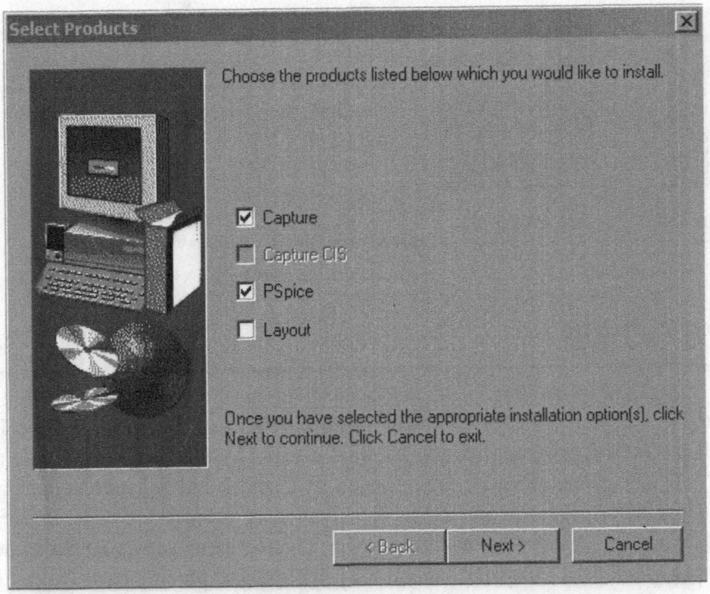

(a) Selecting products

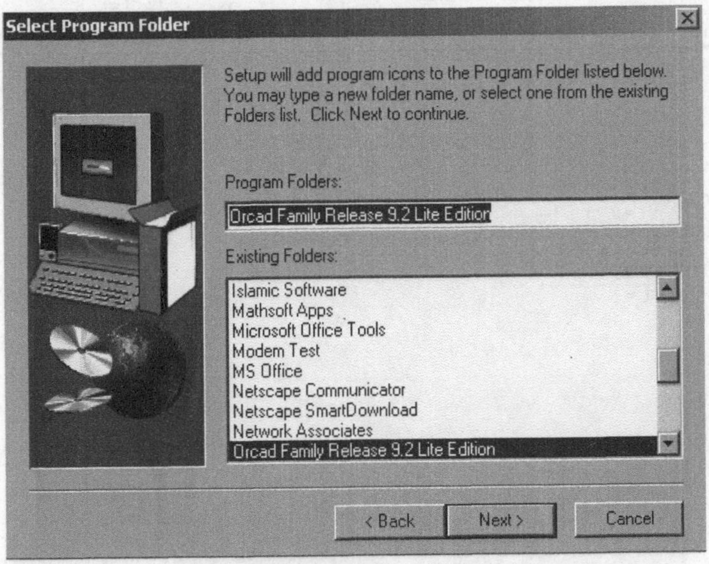

(b) Selecting the program folder

FIGURE A.1 Installation setup for OrCAD Capture

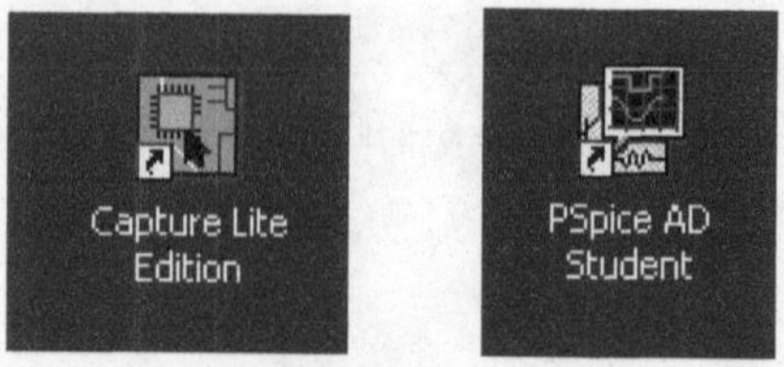

(a) **OrCAD Capture Lite** (b) **PSpice A/D**

FIGURE A.2 Icons for OrCAD Capture Lite and PSpice A/D

A.3 Overview

The OrCAD Capture software package has three major interactive programs: Capture, PSpice A/D, and Probe. *Capture* is a powerful program that lets us build circuits by drawing them within a window on the monitor. *PSpice A/D* lets us analyze the circuit created by Capture and generate voltage and current solutions. *Probe* is a graphic postprocessor and lets us display plots of voltages, currents, impedance, and power.

The general layout of Capture is shown in Fig. A.3. The top menu shows 10 main choices. The right-side menu shows the schematic "drawing" menu for selecting and placing parts. File, Edit, View, Place, and PSpice are most frequently used. For any help, click on the Help menu.

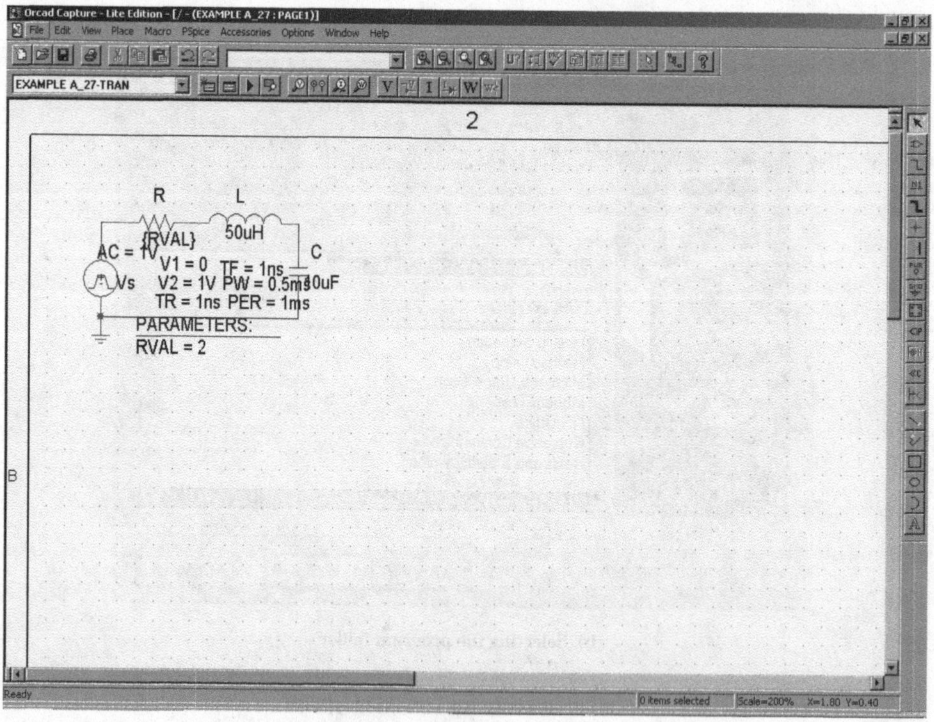

FIGURE A.3 General layout of OrCAD Capture

TABLE A.1 PSpice mouse operations

Button	Action	Function
Left	Single click	Select an item
	Double click	End a mode
	Double click on selected object	Edit a selection
	Single click on selected object and hold	Drag a selection
	Shift + single click	Extend a selection
Right	Single click	Abort the mode
	Double click	Repeat an action

The mouse follows an *object–action* sequence. First select an object and then perform an action, as in the following:

1. Click the mouse on a menu title so that it stays open. Then click on the command that you want.
2. A single click on the left mouse button selects an item.
3. A double click on the left mouse button performs an action such as to end a mode or edit a selection.
4. To drag a selected item, click the left button, hold it down, and move the mouse. Release the left button when placed.
5. To end a mode or edit a selection, click the right mouse button once.
6. To repeat an action, double click the right button.

These mouse operations are summarized in Table A.1.

A.4 The Circuit Analysis Process

As an example, let us draw and analyze the pulse response of an *RLC* circuit as shown in Fig. A.4. The steps to draw and analyze a circuit are as follows:

1. Draw the circuit under Capture.
2. Select the mode of analysis under PSpice.
3. Simulate the circuit under PSpice A/D.
4. Display the results under Probe.

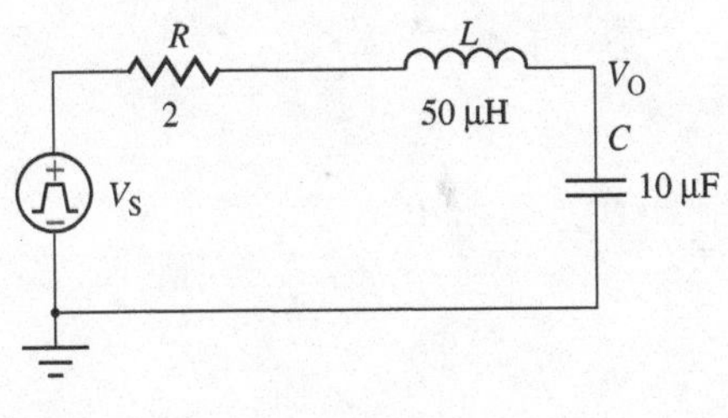

FIGURE A.4 An *RLC* circuit

A.5 Drawing the Circuit

The steps in drawing a circuit are as follows:

Step 1. Begin a new project.
Step 2. Get components from the Get New Part menu and place them on the drawing board.
Step 3. Rotate components as desired.
Step 4. Wire the components together.
Step 5. Label the components and add text as desired.
Step 6. Set attributes of the components.
Step 7. View the schematic.
Step 8. Save the circuit.

A.5.1 Beginning a New Project

Let us begin a new project to draw and analyze; call it Example B-3. Open the File menu, choose **New**, and then select **Project** as shown in Fig. A.5.

The New Project menu opens as shown in Fig. A.6. Give the file name of the new project (e.g., Example B-3), select **Analog or Mixed A/D**, and give the location of this new file (e.g., **C:\Rashid\PT3**). Next select **Create a blank project**, as shown in Fig. A.7.

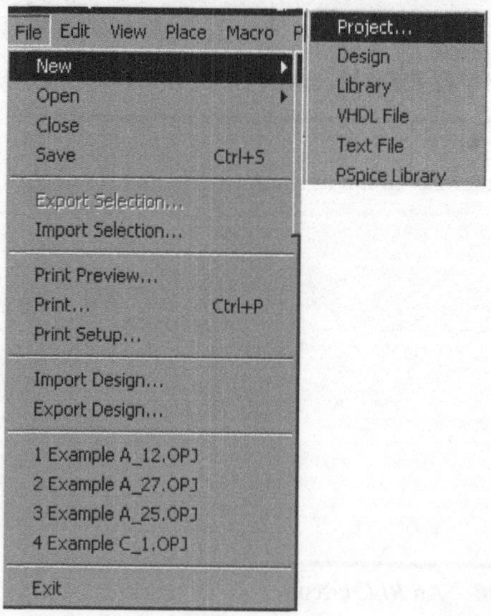

FIGURE A.5 File menu for a new project

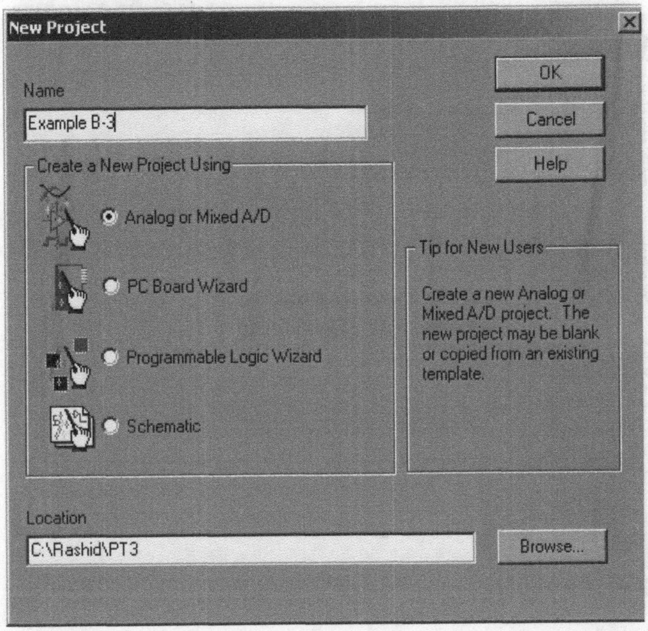

FIGURE A.6 Project menu

A.5.2 Getting and Placing Components

Let us start by placing a pulse source, a resistor, an inductor, a capacitor, and a ground on the drawing board. From the capture editor, you can place parts from the component libraries onto your schematic.

1. Use the Part command in the Place menu shown in Fig. A.8 or from the Place part menu on the right side. Alternately, choose the **Place part** command from the Draw menu on the right side. Choose **Browse** to browse the list of libraries or use the search command to enter the name of a known part, resistor R, as shown in Fig. A.9.
2. Choose **OK** or double click after selecting the part. The chosen part becomes the "current part" and is ready to be placed on your schematic.

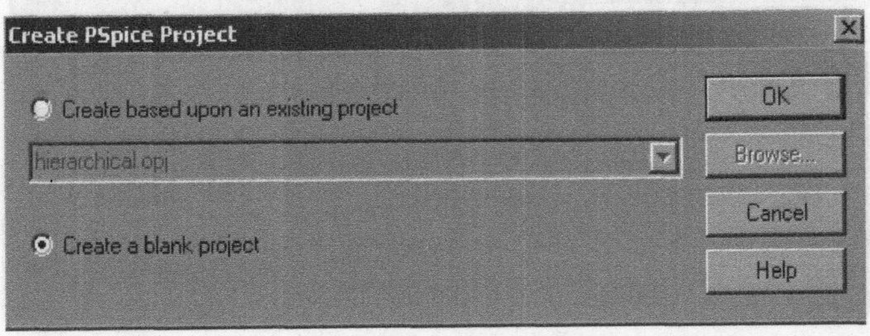

FIGURE A.7 Create a blank project

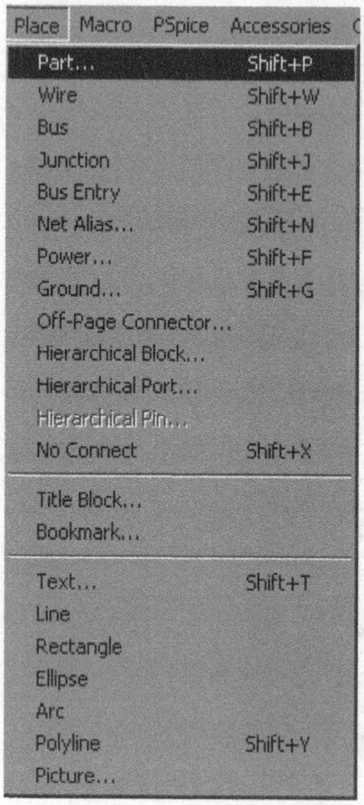

FIGURE A.8 Place menu

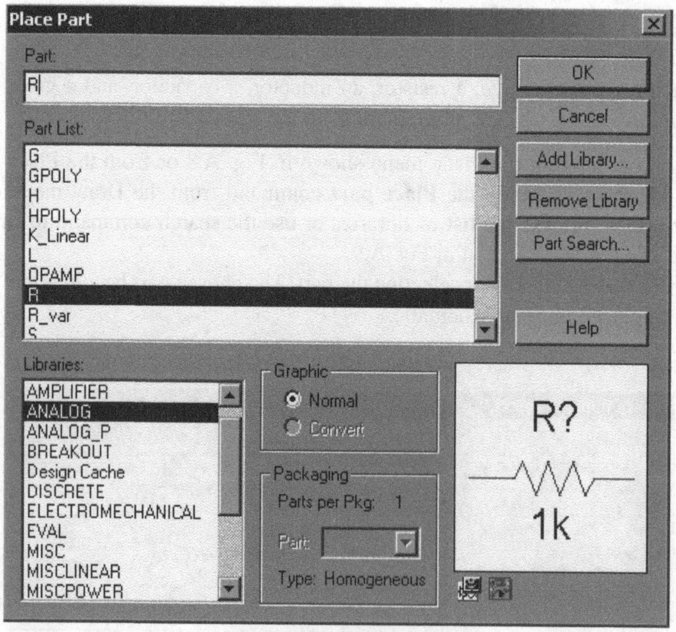

FIGURE A.9 Place part menu

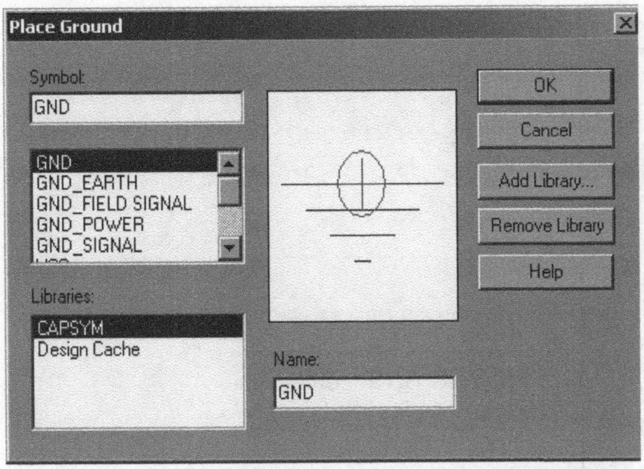

FIGURE A.10 Ground parts for an *RLC* circuit

3. The cursor is replaced by the shape of the part.
4. Click left to place an instance of the part; double click to place the part and end the mode or click right to end the mode without placing the part. To drag, point to an object, press and hold the left mouse button, and move the mouse. When the object is where you want it, release the mouse button.
5. To move a component, point to it, press and hold the left mouse button, and drag it to a new location.
6. To remove a component, select it, and choose **Delete** from the Edit menu.
7. Place a pulse source (VPULSE) from the **source.slb** library, a resistor (R), an inductor (L), and a capacitor (C) from the **analog.slb** library and a ground symbol **(GND)** from the Place menu or Place Ground (zero 0) menu on the right-side menu as shown in Fig. A.10. Arrange them as shown in Fig. A.11.

A.5.3 Rotating Components

Now rotate the capacitor so that it can be wired neatly into the circuit. Each time that you rotate a component, it turns clockwise 90°. The Flip command flips a selected object(s) to produce a mirror image of the object.

1. To rotate the capacitor (or other component), select it and choose **Rotate** from the Edit menu as shown in Fig. A.12. If you select an area of your schematic, the area is rotated around the center of the selection box. This is done as shown in Fig. A.13.
2. To flip a component, select it and choose **Flip** from the Edit menu.

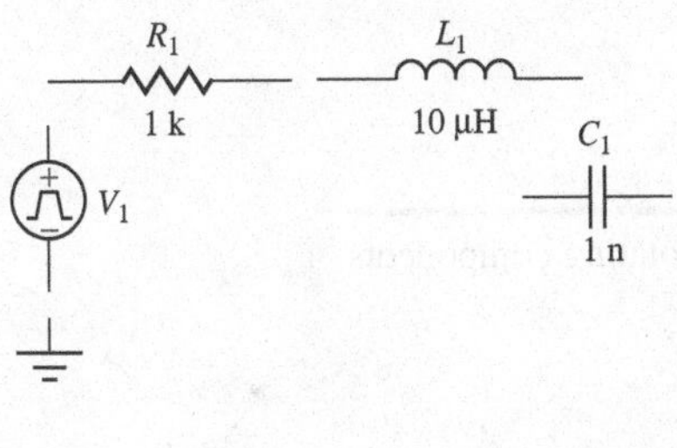

FIGURE A.11 Parts for an *RLC* circuit

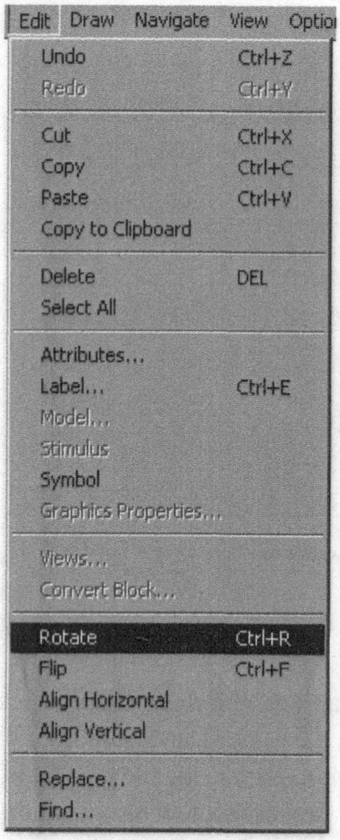

FIGURE A.12 Edit menu

▶ **HELP** If the Rotate command is dimmed, the capacitor is not selected. Try again by pointing to it so that the pointer becomes a hand and then click the left mouse button.

3. To deselect the selected capacitor (or other selected component), click it with the right mouse button or click an empty spot with the left mouse button.
4. To drag or rotate two or more components at once, first select them by drawing a rectangle around the components and then drag or rotate the rectangle.

▶ **HELP** To draw a rectangle around components, point above and beside one of the components that you want to select. Press and hold the left mouse button and drag diagonally until the rest of the components are in the rectangle that appears.

5. To deselect one of the selected components, click it with the right mouse button. To deselect everything on the marked rectangle, click an empty spot with the left mouse button.

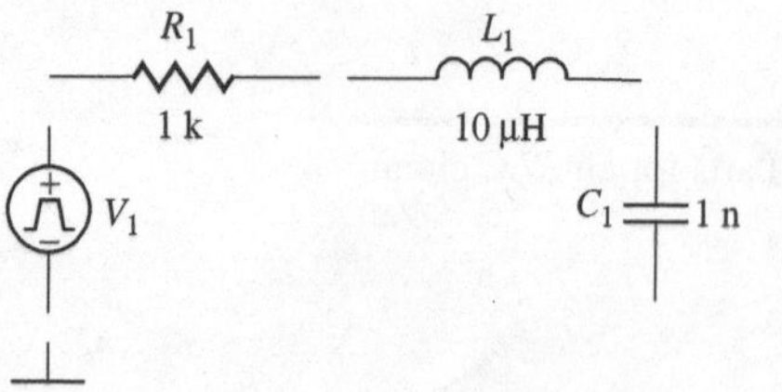

FIGURE A.13 Rotating components

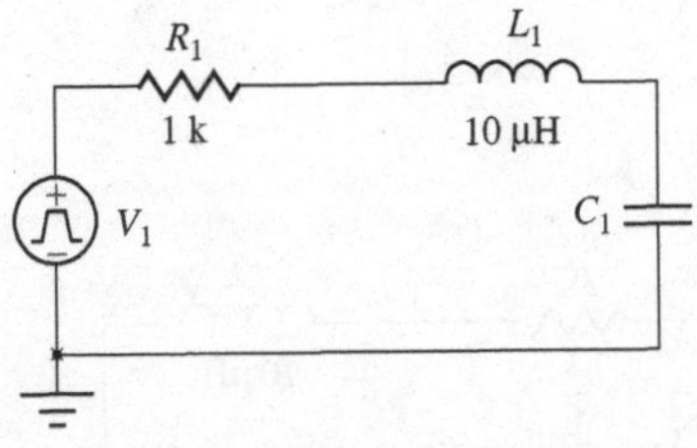

FIGURE A.14 Wiring circuit

A.5.4 Wiring Components

Once components are placed on the drawing board, you need to connect them. You can use the schematic editor to draw wires on your schematic and/or make vertices.

1. To draw a wire, choose **Wire** from the Place menu or Wire menu on the right side.
2. The cursor changes to a "wire" mode and is displayed as crosshairs. Click left to start drawing.
3. Move the mouse in any direction to extend the wire.
4. Click left again to end the wiring of a part. Click left once to start and click left again to stop. Continue until wiring of all parts is completed.
5. Press the **ESC** key or double click left or click right to end the wiring mode (the cursor will change from crosshairs to normal mode).

Now wire the circuit as shown in Fig. A.14.

▶ **TIP** To bring back the last command used (for wire), double click the right mouse button. In the lower right-hand corner, you will see the Wire command after the word *Cmd.*

A.5.5 Labeling Components and Adding Text

We can place a label on selected wires, bus segments, or ports. Wire and bus segments or ports may display multiple labels; however, all labels for a segment will contain the same text. Each component in a circuit can be labeled. We will assign levels R, L, C, V_S, and V_O.

1. To edit a label, select a wire, bus segment, or port for labeling.
2. Double click on the label to bring up the dialog box for **Display Properties**.
3. Enter the text for the label.
4. Click **OK**.
5. Change all levels R, L, C, V_S, and V_O as shown in Fig. A.15.

▶ **TIP** To move the level, left click on it and move to a desired location.

6. Next click on **Text** in the Place menu (see Fig. A.8).

You can place text anywhere on your schematic and size it to suit your needs.

1. To add text to your schematic, choose **Text** from the Place menu or **Place text** on the right-side menu.
2. In the dialog box, type in the desired text.
3. To change the font size, modify the **Font Size** shown in the dialog box to suit your needs.
4. Click **OK**.
5. Move the text to the desired location on the schematic and click left to place the text. Click right to end the mode.

Now place the output voltage, V_O as shown in Fig. A.16.

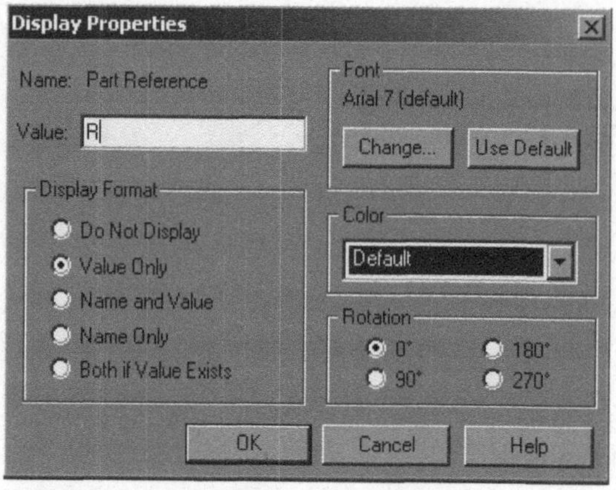

(a) **Display Properties menu**

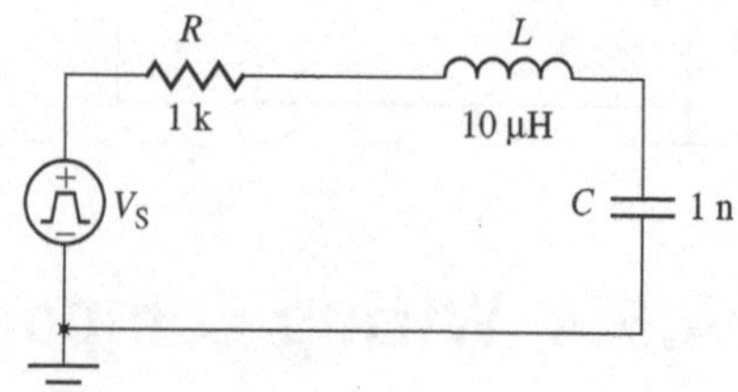

(b) ***RLC* circuit with desired labels**

FIGURE A.15 Labeling components

A.5.6 Setting Attributes

An attribute of a schematic item consists of a name and value pair.

1. To edit the properties of a selected object(s), select the object(s) to edit.
2. Choose **Properties** from the Edit menu or double click on the attribute text to bring up the Display Properties dialog box directly as shown in Fig. A.17.
3. Select an individual attribute: A dialog box appears in which you can enter a new value for the attribute. You can also select an entire part: A dialog box appears showing all attributes that can be edited for that part—for example, **Value Only**.

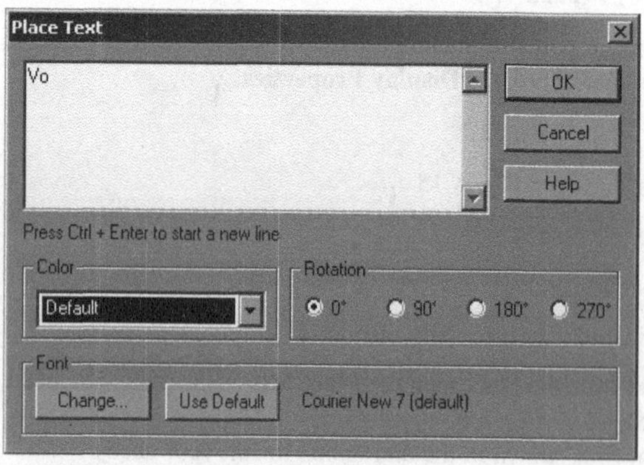

(a) **Place Text dialog box**

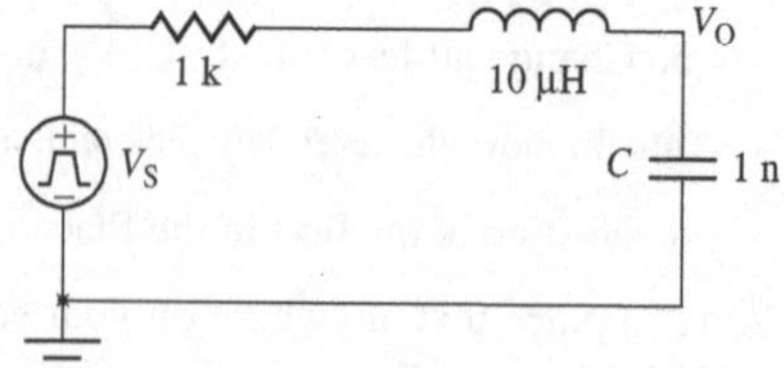

(b) ***RLC* circuit with desired text**

FIGURE A.16 Adding text

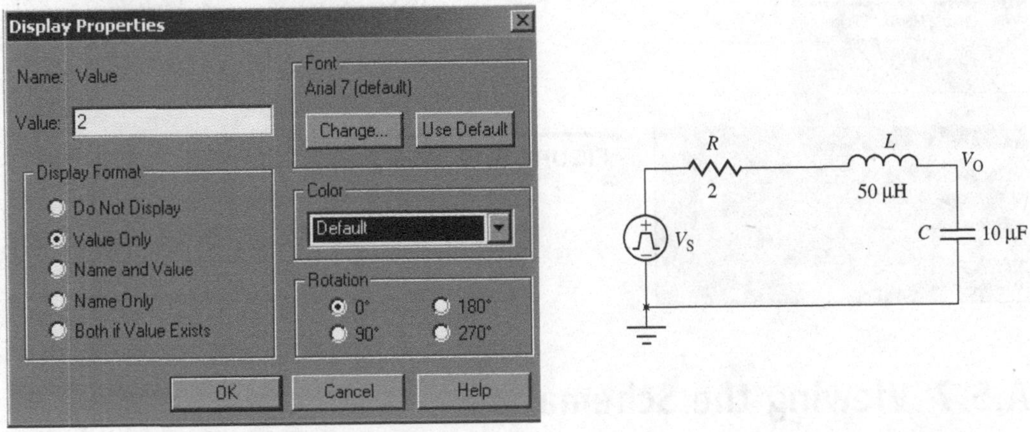

(a) **Display Properties dialog box** (b) ***RLC* circuit showing attributes**

FIGURE A.17 Setting attributes

4. Set $R = 2\ \Omega$, $L = 50\ \mu H$, and $C = 10\ \mu F$.

▶ **TIP** A quick way to set a component's value is to double click it.

▶ **HELP** Changes made to attributes in the schematic editor occur only on the particular part instance. Changes do not affect the underlying symbol in the library.

5. To change the value of an attribute, select it. The name and value should appear in the properties of the Edit menu.
   a. Change the value in the **Value** edit field and click **OK**.
   b. To delete an attribute, select it from the list and press **Delete**.
   c. To change whether the attribute name and/or value is shown on the schematic, select it and click **OK**.
   d. To add a new attribute, type the name and value in the **Edit** fields and press **OK**. Set the pulse source V_S as shown in Fig. A.18. Click on the attributes and then type V1 = 0, V2 = 1 V, TD (delay time) = 0, TR (rise time) = 1 ns, TF (fall time) = 1 ns, PW (pulse width) = 0.5 ms, PER (period) = 1 ms.
6. Click **OK** to incorporate the change.

▶ **TIP** Protected attributes, marked with an asterisk (*), cannot be changed using the Schematic Editor.

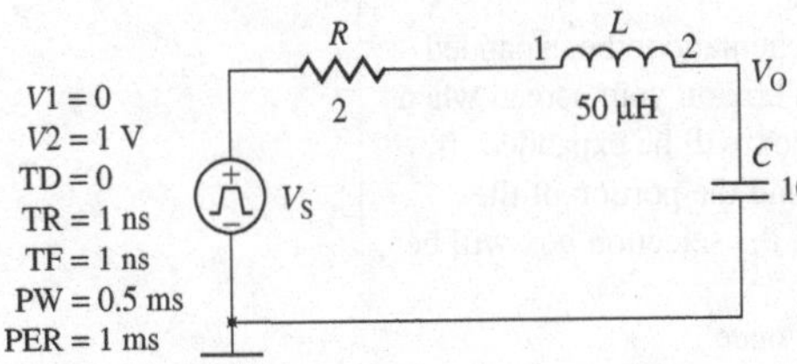

FIGURE A.18 Changing source attributes

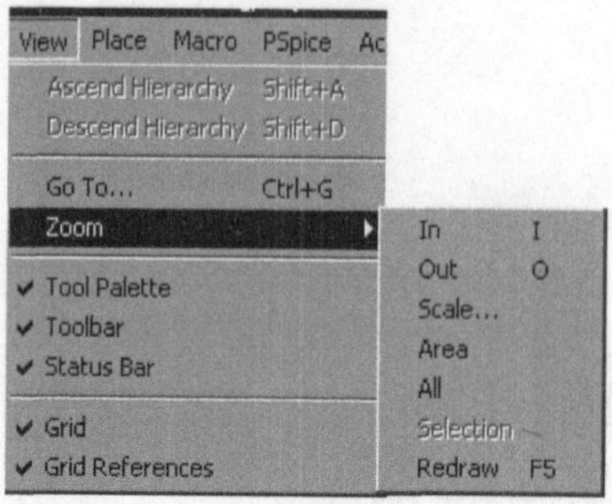

FIGURE A.19 View menu

A.5.7 Viewing the Schematic

You can change the viewing scale on your schematic via the View menu, shown in Fig. A.19. Table A.2 explains the functions of commands in the View menu.

A.5.8 Saving the Circuit File

Your schematic files are automatically saved into the current directory, unless otherwise specified.

- To save your schematic file, choose **Save** from the File menu, as shown in Fig. A.20, to save changes to the current file. You do not need to type a file name extension.

A circuit name must be a valid DOS file name. All schematic files are automatically given the file name extension OPJ. For example, the file name FIGB_3 will be saved as FIG B_3.OPJ.

TABLE A.2 View menu commands

Command	Function
Fit	Resets the viewing scale so that all parts, wires, and text can be seen on the screen.
In	Allows you to view an area on the schematic at closer range (i.e., magnify). After you select this command, a crosshair appears on the screen. Move the crosshair to the area you want to zoom in on.
Out	Changes the viewing scale so that you can view the schematic from a greater distance (i.e., view more of the schematic on the screen). After you select this command, a crosshair appears on the screen. Move the crosshair to define the center of the viewing area.
Area	Allows you to select a rectangular area on the schematic to be expanded to fill the screen. If you already have a selection box on your screen when you choose **Area**, the contents of the selection box will be expanded. If not, drag the mouse to form a selection box around the portion of the schematic you want to expand. The items within the selection box will be expanded to fill the screen.
Entire Page	Allows you to view the entire schematic page at once.

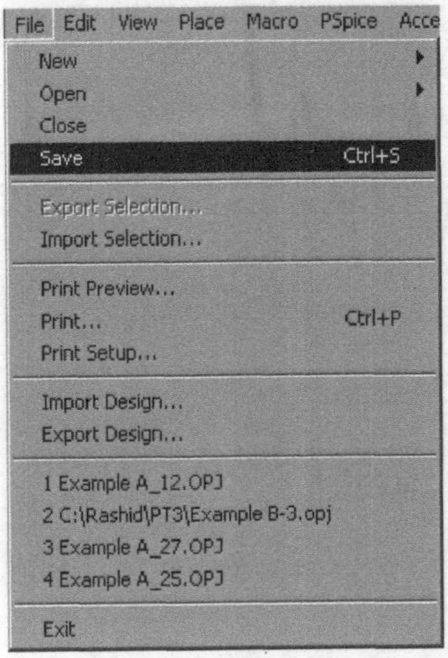

FIGURE A.20 File menu

A.6 Selecting the Type of Analysis

PSpice allows *DC sweep analysis, AC* (frequency response) *sweep analysis*, and *transient analysis*. The Setup command specifies which types of simulation analyses are enabled and allows the user to set up the parameters for selected analyses (described next).

When a signal is first applied to a circuit, there is a short-lived transient state before the circuit settles down to its usual responses. For the sample *RLC* circuit, we will conduct a transient analysis to examine the charging and discharging voltage of the capacitor.

1. Choose **New Simulation Profile** for a new simulation file or **Edit Simulation Profile** for an existing file from the PSpice menu, shown in Fig. A.21, and the Analysis Setup dialog box will open.
2. Enable transient analysis by clicking once in the **Enabled** space so that a check appears in the space, as shown in Fig. A.22.

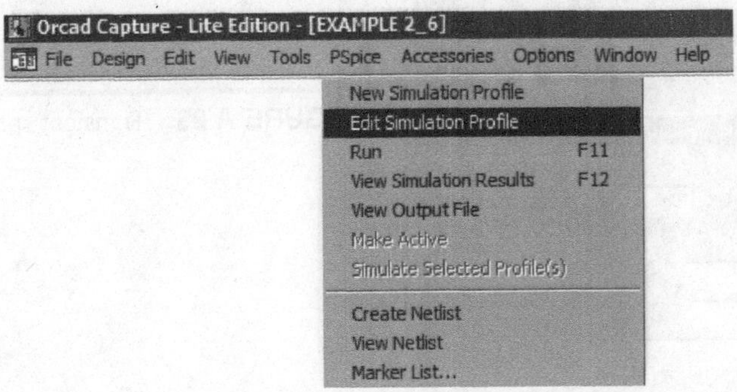

FIGURE A.21 PSpice setup for analysis

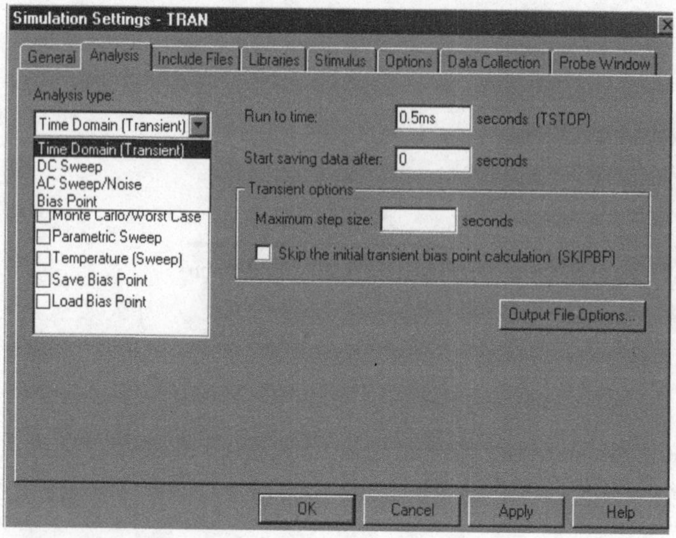

FIGURE A.22 Selecting Analysis type

3. Click on **Transient** to open the Transient specifications dialog box.
4. Type the print particulars, such as a **Print values** of 10 ns and a **Run to time** of 0.5 ms, as shown in Fig. A.23.
5. Click **OK**.

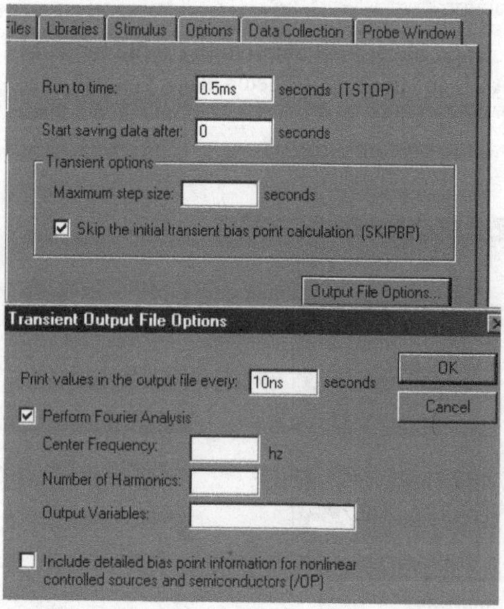

FIGURE A.23 Transient specifications

A.7 Simulation with PSpice

Now we are ready to simulate the circuit with PSpice. To begin the simulation, left click on **Simulate** in the Analysis menu.

During the circuit simulation process, PSpice creates and accesses a number of files. The first file is the *Schematics* file (*.SCH*), generated when a circuit drawn on the screen is saved. When the Schematics file is analyzed, three new files are generated: the *Circuit* file (*.CIR*), the *Netlist* file (*.NET*), and the *Alias* file (*.ALS*). The Circuit file (the master file) contains the *simulation directives* and references to the Netlist, Alias, and Model files. The Netlist file contains a Kirchhoff-like set of equations that lists parts and how they are connected; the equations relate the voltages and currents to the circuit elements through node numbers. The Alias file lists alternative names for circuit nodes. The Model file lists the characteristics and model statement of each component.

▶ **TIP** If there is any problem such as a missing attribute name or value, PSpice will indicate the error, and the simulation will be aborted. You can find the error message in the output file. For example, a message about an error in the file FIG1.1 will appear in FIG1_1.OUT.

When PSpice is run, each simulation directive in the Circuit (master) file specifies the information to be sent to the Output and Data files.

- The *Output* file (*.OUT*) is an ASCII file that holds the "audit trail" for the simulation. It contains a wide variety of information, including the original netlist, all output variables, and various tables.
- The *Data* file (*.DAT*) is sent to Probe, which uses the binary information to generate plots and graphs within the Probe window.

While a simulation is running, you can check the status of the simulation, as shown in Fig. A.24. When the simulation is completed, PSpice will display the message "Transient Analysis finished."

▶ **TIP** *Time step* is the internal simulation step for convergence to give a specified accuracy, whereas *Print step* (which was specified in the Transient specifications dialog box) is for printing or plotting the output variables.

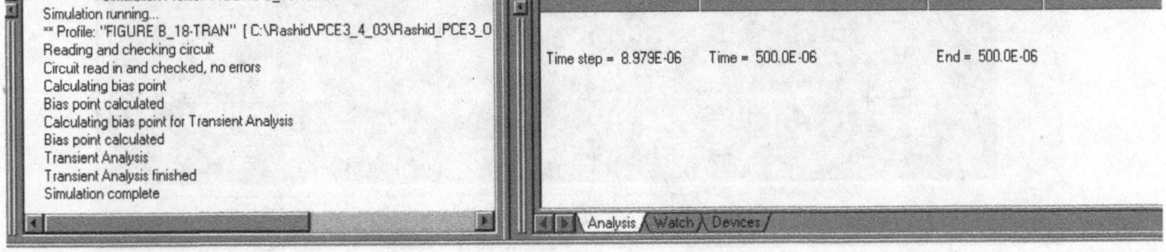

FIGURE A.24 PSpice analysis status box

A.8 Displaying the Results of a Simulation

Probe is a graphics postprocessor that allows the simulation results to be displayed in graphical form. After the calculations are completed, assuming no errors are found, PSpice sends the database it generated (FIG1_1.DAT) to Probe, which displays a graph.

1. To use Probe, choose **Run Probe** from the selecting Analysis type in Fig. A.22 menu (or use the F12 function key). As shown in Fig. A.25, Probe opens with an initial (default) graph in which the x-axis is automatically set to the transient variable, time.

▶ **TIP** To run Probe automatically after a simulation, choose **Automatically Run Probe After Simulation** from **Probe Setup** in the Analysis menu.

2. Choose **Trace** on the Probe menu. Within the large box at the top of the dialog box that appears, you will see the list of default *trace variables* that you can choose from, as shown in Fig. A.26. Before moving to the third step, specify the plot variables(s). For this example, we choose the following variables:
   a. V(R:2): V specifies the variable type (voltage, which is referenced to the ground), R specifies a component (resistor), and 2 specifies one end (node 2) of the component. A 1 indicates what was the left-hand side and a 2 what was the right-hand side when the component was initially placed horizontally. After one counterclockwise rotation, node 2 would be at the top.
   b. V(Vs: +): This is the voltage at the positive terminal of the voltage source V_S.
   c. I(R): This is the current flowing through resistance R from terminal 1 (left) to terminal 2 (right).

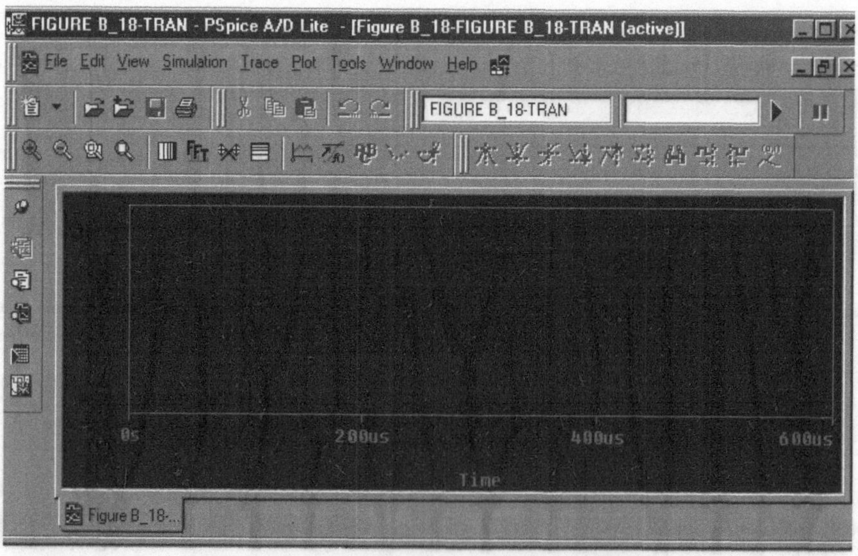

FIGURE A.25 Probe menu

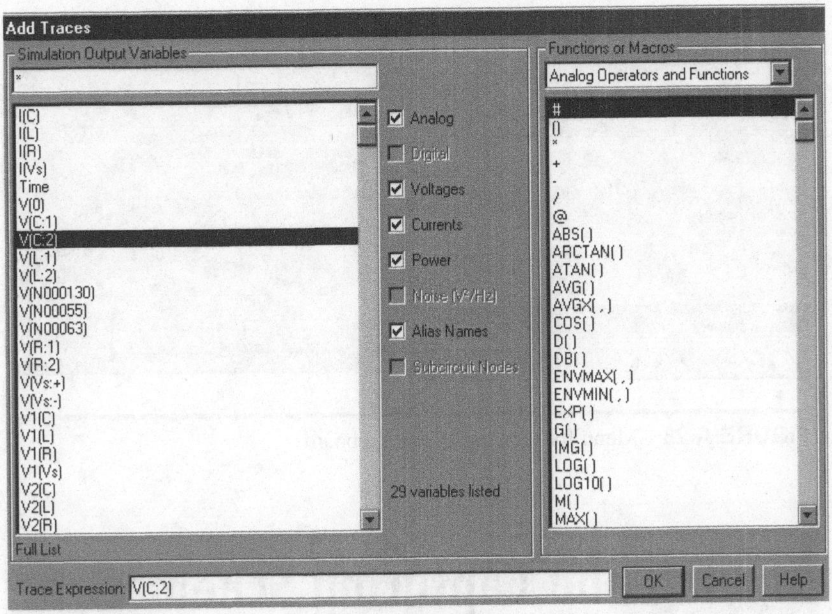

FIGURE A.26 Probe trace variables

3. After specifying the plot variables, type **V(C:2)** in the trace command box. This instructs Probe to display a graph of the voltage at terminal 2 of capacitor C, which is our desired output voltage. Click **OK** and you will see the plot of the output voltage, V(C:2), shown in Fig. A.27.
4. While copying to the clipboard and pasting to another document, the background and foreground colors can be changed as shown in the dialog box in Fig. A.28.

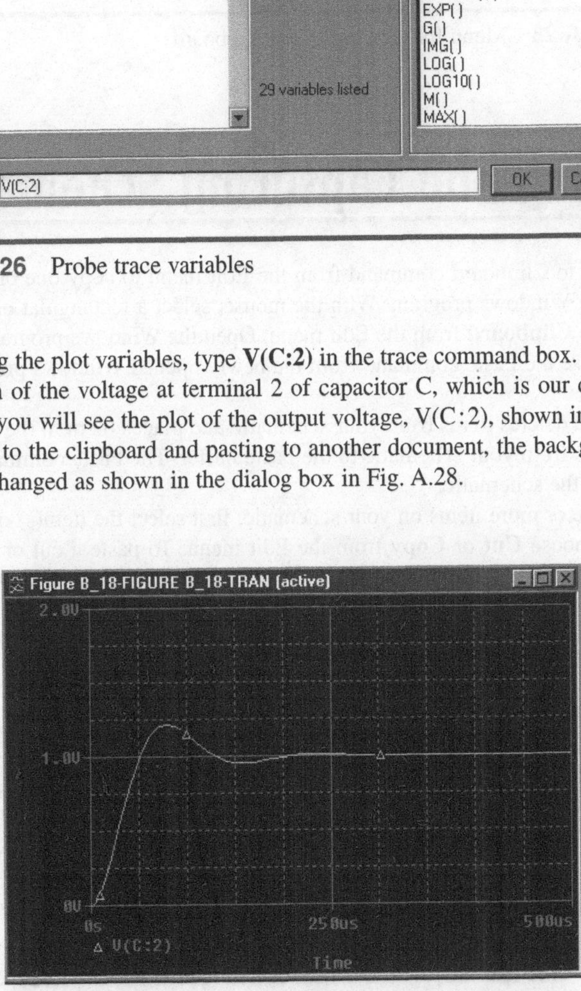

FIGURE A.27 Transient plot of the output voltage

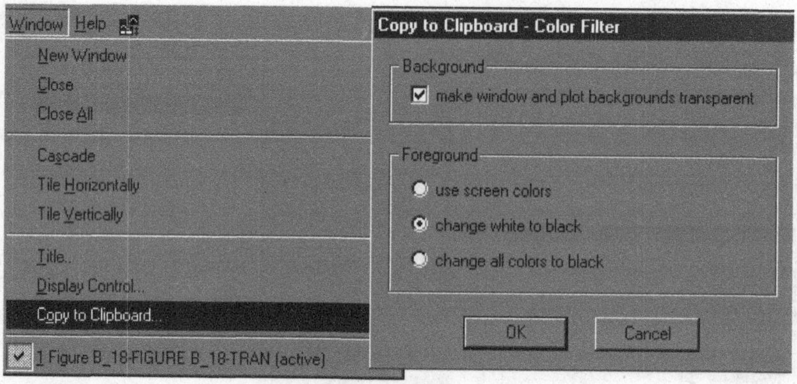

FIGURE A.28 Menu for copying to the clipboard

A.9 Copying and Capturing Schematics

You can use the Copy to Clipboard command from the Edit menu to copy one or more items from your schematic into another Windows program. With the mouse, select a rectangular area on the schematic to copy. Choose **Copy to Clipboard** from the Edit menu. Open the Windows program into which you want to paste the item(s). Use the Paste command from the newly opened Windows program to paste the item into the new file.

The Cut command deletes items from your schematic and places them in the Paste buffer. The Copy command copies items from your schematic to the Paste buffer. The Paste command places the contents of the Paste buffer on the schematic.

To cut or copy one or more items on your schematic, first select the item(s) on your schematic to be cut or copied. Then choose **Cut** or **Copy** from the Edit menu. To paste a cut or copied item onto your schematic, choose **Paste** from the Edit menu. Then place the cursor on the schematic where you'd like the cut or copied item to appear, and left click to place the item; right click to end the mode.

A.10 Varying Parameters

PSpice allows variation of component values or device parameters. We will plot the output voltage of our sample *RLC* circuit for three values: $R = 1\ \Omega$, $2\ \Omega$, and $10\ \Omega$.

1. To begin, get the part PARAM from the special.slb library file.
2. To change the value of *R* to a variable name such as RVAL, double click on the value of *R* and type **{RVAL}**.
3. To change the attribute of the part PARAM, double click on **PARAM** to open the Attributes dialog box, shown in Fig. A.29.
4. Choose **NAME1=RVAL** and **VALUE1=2** and click each time on **Save Attr**.

Parameters	
RVAL	10 k
CVAL	10 μF
LVAL	50 μH

(a) Symbol

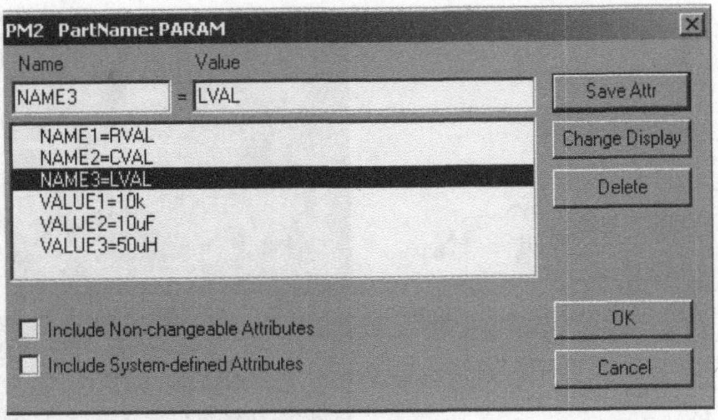

(b) Parameters

FIGURE A.29 Dialog box for PARAM

5. Choose **Setup** from the Analysis menu; then click once in the enabled space to enable parametric analysis.
6. Click on **Parametric** to open the box shown in Fig. A.30.
7. Enter or enable the Parametric specifications as follows: For Sweep variable, choose **Global parameter**; for Parameter name, type in **RVAL**; for Sweep Type, choose **Value list**; and for the values, type in 1 2 10.

▶ **TIP** You can define up to three variables in the same PARAM.

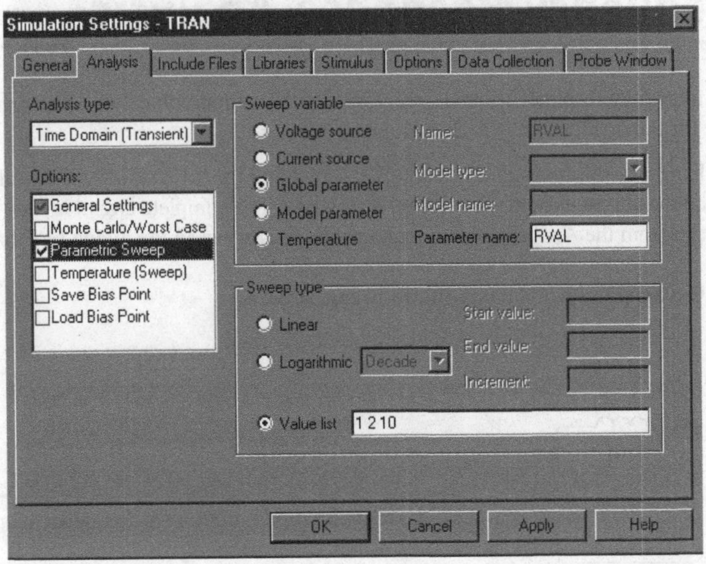

FIGURE A.30 Parametric specifications

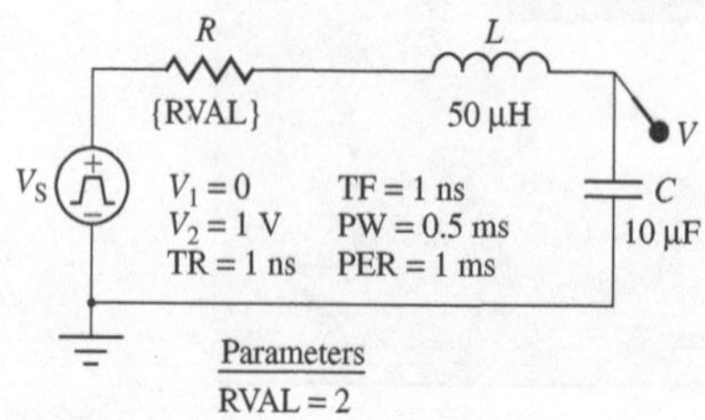

FIGURE A.31 *RLC* circuit with PARAM

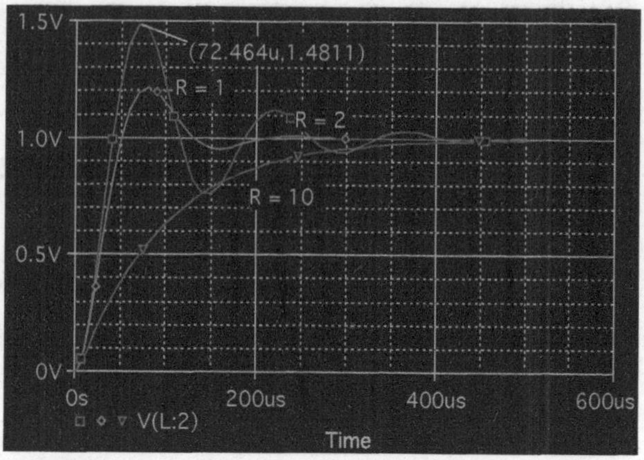

FIGURE A.32 Transient response for $R = 1\ \Omega$, $2\ \Omega$, $10\ \Omega$

8. Run the simulation for the complete circuit, shown in Fig. A.31. Use **Probe** to plot the output variable, V(C:2), as shown in Fig. A.32.

▶ **TIP** The labels R=1, R=2, and R=10 in Fig. A.32 were typed by choosing (in sequence) **Tools, Label**, and **Text** from the Probe menu. You can copy the Probe plot(s) to other Windows programs by choosing (in sequence) **Tools** and **Copy to Clipboard** from the Probe menu.

A.11 Frequency Response Analysis

As an example of frequency response analysis, we will plot the output voltage and phase angle of the example *RLC* circuit for three values: $R = 1\ \Omega$, $2\ \Omega$, and $10\ \Omega$.

1. To begin, change the attributes of the source V_S, shown in Fig. A.17, to 1 V AC by selecting the part and changing the attributes in the Part Name dialog box. The complete circuit is shown in Fig. A.33.
2. Choose **Setup** from the Analysis menu; then enable **AC Sweep** in the dialog box that appears (see Fig. A.22).
3. Click on **AC Sweep** to open the box shown in Fig. A.34.

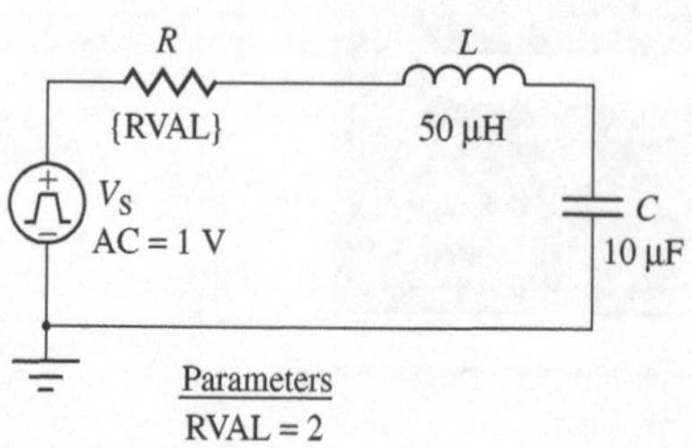

FIGURE A.33 *RLC* circuit for frequency response

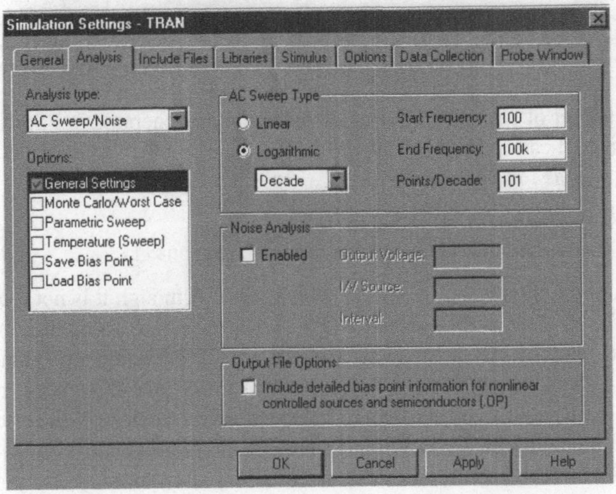

FIGURE A.34 Dialog box for frequency response analysis

4. Enter or enable the **AC Sweep** specifications as follows: For AC **Sweep Type**, choose **Decade**; for Start Frequency, type in 100; for End Frequency, type in 100k; for Points/Decade, type in 101.
5. Run the simulation for the complete circuit, shown in Fig. A.33. Use **Probe** to plot the output voltage, V(C:2), and the phase of the output voltage, VP(C:2), as shown in Fig. A.35.

▶ **TIP** If you add DB, as in VDB(C:2), you will get the output voltage in decibels.

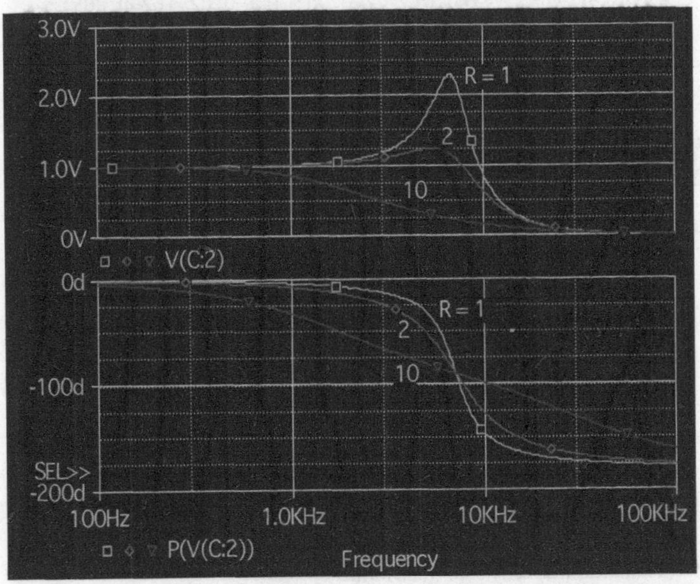

FIGURE A.35 Frequency response for $R = 1\ \Omega$, $2\ \Omega$, and $10\ \Omega$

A.12 Modeling Devices and Elements

A model that specifies a set of parameters for an element can be generated in PSpice using the .MODEL command. The same model can be used by one or more elements in the same circuit. The general form of the model statement [1] is as follows:

```
.MODEL  MNAME  TYPE (P1=A1 P2=A2
+       P3=A3 . . . PN=AN  [<tolerance specifications>])
```

MNAME is the name of the model and must start with a letter. Although it is not required, it is advisable to use the symbol of the element as the first letter (e.g., R for resistor, L for inductor). P1, P2, . . ., PN are the element parameters, and A1, A2, . . ., AN are their values. TYPE is the type name of the element and must be one of the types shown in Table A.3. An element must have the correct type name; that is, a resistor must have the type name RES, not IND or CAP. However, a circuit with several model names can include more than one model of the same type.

Tolerance specifications are used with .MC or .WORSE analysis only. They may be appended to each parameter using the following format:

```
[DEV/<distribution name> <value in % from 0 to 9>]
[LOT/<distribution name> <value in % from 0 to 9>]
```

where <distribution name> is either UNIFORM (in which case uniformly distributed deviations are generated over the range of $\pm$<value>) or GAUSS (in which case deviations with Gaussian distribution are generated over the range ± 4 and <value> specifies the ± 1 deviation).

Following are some sample model statements:

```
.MODEL RLOAD  RES (R=1 TC1=0.02 TC2=0.005)
.MODEL RLOAD  RES (R=1 DEV/GAUSS 0.5% LOT/UNIFORM 10%)
.MODEL CPASS  CAP (C=1 VC1=0.01 VC2=0.002 TC1=0.02 TC2=0.005)
```

TABLE A.3 Type names of elements

Type Name	Element
RES	Resistor
CAP	Capacitor
D	Diode
IND	Inductor
NPN	*npn* bipolar junction transistor
PNP	*pnp* bipolar junction transistor
NJF	*n*-channel junction FET
PJF	*p*-channel junction FET
NMOS	*n*-channel MOSFET
PMOS	*p*-channel MOSFET
GASFET	*n*-channel GaAs MESFET
VSWITCH	Voltage-controlled switch
ISWITCH	Current-controlled switch
CORE	Nonlinear magnetic core (transformer)

```
.MODEL    LFILTER    IND (L=1 IL1=0.1 IL2=0.002 TC1=0.02 TC2=0.005)
.MODEL    DNOM       D (IS=1E-9)
.MODEL    DLOAD      D (IS=1E-9 DEV 0.5% LOT 10%)
.MODEL    QMOD       NPN (BF=50 IS=1E-9)
```

A.12.1 Resistors

The symbol for a resistor is R. The name of a resistor must start with R, and the model statement takes the following general form:

```
R<name>  N+  N-  RNAME  RVALUE
```

A resistor does not have a polarity, and so the order of the nodes does not matter. However, the current is assumed to flow from the node designated by N+ as the positive node through the resistor to the node designated by N− as the negative node. RNAME is the model name that defines the parameters of the resistor. RVALUE is the nominal value of the resistance.

▶ **NOTE** Some versions of PSpice and SPICE do not make this assumption and thus do not allow you to refer to currents through a resistor. For example, such versions will not allow you to use a notation such as I(RL) to indicate the current through RL.

The model parameters for resistors are shown in Table A.4. If RNAME is omitted, RVALUE is the resistance in ohms; RVALUE can be positive or negative but must *not* be zero. If RNAME is included and TCE is *not* specified, the resistance as a function of temperature is calculated from

$$\text{RES} = \text{RVALUE} * \text{R} * [1 + \text{TC1} * (\text{T} - \text{T0}) + \text{TC2} * (\text{T} - \text{T0})^2]$$

If RNAME is included and TCE is specified, the resistance as a function of temperature is calculated from

$$\text{RES} = \text{RVALUE} * \text{R} * 1.01^{\text{TCE} * (\text{T}-\text{T0})}$$

where T and T0 are the operating temperature and the room temperature, respectively, in degrees Celsius.

Following are some sample resistor statements:

```
RL        5      6      5K
RLOAD     10     13     ARES 1MEG
.MODEL    RMOD          RES (R=1  TC1=0.02  TC2=0.005)
RINPUT    13     17     RRES  2K
.MODEL    ARES          RES (R=1  TCE=1.5)
```

TABLE A.4 Model parameters for resistors

Name	Meaning	Units	Default
R	Resistance multiplier		1
TC1	Linear temperature coefficient	$°C^{-1}$	0
TC2	Quadratic temperature coefficient	$°C^{-2}$	0
TCE	Exponential temperature coefficient	%/°C	0

TABLE A.5 Model parameters for capacitors

Name	Meaning	Units	Default
C	Capacitance multiplier		1
VC1	Linear voltage coefficient	V^{-1}	0
VC2	Quadratic voltage coefficient	V^{-2}	0
TC1	Linear temperature coefficient	$°C^{-1}$	0
TC2	Quadratic temperature coefficient	$°C^{-2}$	0

A.12.2 Capacitors

The symbol for a capacitor is C. The name of a capacitor must start with C, and the model statement takes the following general form:

```
C<name>   N+   N-   CNAME   CVALUE   IC=VO
```

N+ is the positive node, and N− is the negative node. The voltage of node N+ is assumed to be positive with respect to node N−, and the current flows from node N+ through the capacitor to node N−. CNAME is the model name, and CVALUE is the nominal value of the capacitor. IC defines the initial (time zero) voltage of the capacitor, VO.

The model parameters for capacitors are shown in Table A.5. If CNAME is omitted, CVALUE is the capacitance in farads; CVALUE can be positive or negative but must *not* be zero. If CNAME is included, the capacitance as a function of voltage and temperature is calculated from

$$CAP = CVALUE * C * (1 + VC1 * V + VC2 * V^2)[1 + TC1 * (T - T0) + TC2 * (T - T0)^2]$$

where T and T0 are the operating temperature and the room temperature, respectively, in degrees Celsius.

Following are some sample capacitor statements:

```
C1        2      6      0.01UF
CLOAD     10     13     10PF   IC=1.5V
CINPUT    14     16     DCAP   5PF
CX        10     25     DCAP   10NF   IC=3.5V
.MODEL    DCAP          CAP  (C=1  VC1=0.01  VC2=0.002  TC1=0.02  TC2=0.005)
```

▶ **TIP** The initial conditions (if any) apply only if you specify the UIC (use initial condition) option under the .TRAN command.

A.12.3 Inductors

The symbol for an inductor is L. The name of an inductor must start with L, and the model statement takes the following general form:

```
L<name>  N+   N-   LNAME   LVALUE   IC=IO
```

N+ is the positive node, and N− is the negative node. The voltage of N+ is assumed to be positive with respect to node N−, and the current flows from node N+ through the inductor to node N−. LNAME is the model name, and LVALUE is the nominal value of the inductor. IC defines the initial (time zero) current of the inductor IO.

TABLE A.6 Model parameters for inductors

Name	Meaning	Units	Default
L	Inductance multiplier		1
IL1	Linear current coefficient	A^{-1}	0
IL2	Quadratic current coefficient	A^{-2}	0
TC1	Linear temperature coefficient	$°C^{-1}$	0
TC2	Quadratic temperature coefficient	$°C^{-2}$	0

The model parameters for inductors are shown in Table A.6. If LNAME is omitted, LVALUE is the inductance in henries; LVALUE can be positive or negative but must *not* be zero. If LNAME is included, the inductance as a function of current and temperature is calculated from

$$\text{IND} = \text{LVALUE} * \text{L} * (1 + \text{IL1} * \text{I} + \text{IL2} * \text{I}^2)[1 + \text{TC1} * (\text{T} - \text{T0}) + \text{TC2} * (\text{T} - \text{T0})^2]$$

where T and T0 are the operating temperature and the room temperature, respectively, in degrees Celsius.

Following are some sample inductor statements:

```
LE        3      5     5MH
LLOAD     10     14    2UH   IC=0.1MA
LLINE     12     14    LMOD  2MH
LCHOKE    15     29    LMOD  5UH  IC=0.4A
.MODEL    LMOD         IND (L=1  IL1=0.1  IL2=0.002 TC1=0.02  TC2=0.005)
```

▶ **TIP** The initial conditions (if any) apply only if you specify the UIC (use initial condition) option using the .TRAN command.

A.12.4 Diodes

The model statement for a diode has the following general form:

```
.MODEL  DNAME  D (P1=A1 P2=A2 P3=A3 . . . PN=AN)
```

DNAME is the name of the model and can begin with any character, but its size is normally limited to eight characters. D is the type symbol for diodes. P1, P2, . . ., PN are the model parameters, and A1, A2, . . ., AN are their values. The model parameters are listed in Table A.7.

A.12.5 Bipolar Transistors

The model statement for *npn* transistors has the following general form:

```
.MODEL QNAME NPN (P1=A1 P2=A2 P3=A3 . . . PN=AN)
```

The general form of the model statement for *pnp* transistors is

```
.MODEL QNAME PNP (P1=A1 P2=A2 P3=A3 . . . PN=AN)
```

QNAME is the name of the BJT model, and NPN and PNP are the type symbols for *npn* and *pnp* transistors, respectively. QNAME can begin with any character, but its size is normally limited to eight characters.

TABLE A.7 Model parameters for diodes

Name	Model Parameter	Units	Default	Typical
IS	Saturation current (I_S)	A	1E-14	1E-14
RS	Parasitic resistance (R_S) (series lead and bulk resistance)	Ω	0	10
N	Emission coefficient (n)		1	1
TT	Transit time	s	0	0.1 ns
CJO	Zero-bias *pn* capacitance (C_{jo})	F	0	2 pF
VJ	Junction potential (V_j)	V	1	0.6
M	Junction grading coefficient		0.5	0.5
EG	Activation energy (0.67 for Shockley and 1.11 for silicon)	eV	1.11	1.11
XTI	IS temperature exponent		3	3
KF	Flicker noise coefficient		0	
AF	Flicker noise exponent		1	
FC	Forward-bias depletion capacitance coefficient		0.5	
BV	Reverse breakdown voltage	V	∞	50
IBV	Reverse breakdown current (reverse current at BV)	A	1E-10	

P1, P2, . . ., PN are the parameters, and A1, A2, . . ., AN are their values. Table A.8 shows the model parameters for BJTs.

A.12.6 JFETs

The model statement for an *n*-channel JFET has the following general form:

```
.MODEL JNAME NJF (P1=A1 P2=A2 P3=A3 . . . PN=AN)
```

The general form of the model statement for a *p*-channel JFET is

```
.MODEL JNAME PJF (P1=A1 P2=A2 P3=A3 . . . PN=AN)
```

JNAME is the name of the model, and NJF and PJF are the type symbols for *n*-channel and *p*-channel JFETs, respectively. P1, P2, . . ., PN are the parameters, and A1, A2, . . ., AN are their values. Table A.9 lists the model parameters for JFETs.

A.12.7 MOSFETs

The symbol for a metal oxide semiconductor field-effect transistor (MOSFET) is M. The name of a MOSFET must start with M, and the model statement takes the following general form:

```
M<name>  ND  NG  NS  NB  MNAME
+        [L=<value>] [W=<value>]
```

where ND, NG, NS, and NB are the drain, gate, source, and bulk (or substrate) nodes, respectively. MNAME is the model name. The positive current is the current flowing into a terminal; that is, the current flows from the drain node through the device to the source node in an *n*-channel MOSFET.

TABLE A.8 Model parameters for BJTs

Name	Model Parameter	Units	Default	Typical
IS	*pn* saturation current	A	1E-16	1E-16
BF	Ideal maximum forward beta		100	100
NF	Forward current emission coefficient		1	1
VAF(VA)	Forward Early voltage	V	∞	100
IKF(IK)	Corner for forward beta high-current roll-off	A	∞	10 MA
NE	Base–emitter leakage emission coefficient		1.5	2
BR	Ideal maximum reverse beta		1	0.1
NR	Reverse current emission coefficient		1	
IKR	Corner for reverse beta high-current roll-off	A	∞	100 MA
RB	Zero-bias (maximum) base resistance (base spreading resistance)	Ω	0	100
RE	Emitter ohmic resistance	Ω	0	1
RC	Collector ohmic resistance (collector lead and bulk resistance)	Ω	0	10
CJE	Base–emitter zero-bias *pn* capacitance (C_{jeo})	F	0	2 pF
VJE(PE)	Base–emitter built-in potential (V_{jbe})	V	0.75	0.7
MJE(ME)	Base–emitter *pn* grading factor		0.33	0.33
CJC	Base–collector zero-bias *pn* capacitance ($C_{\mu o}$)	F	0	1 pF
VJC(PC)	Base–collector built-in potential (V_{jc})	V	0.75	0.5
MJC(MC)	Base–collector *pn* grading factor		0.33	0.33
CJS(CCS)	Collector–substrate zero-bias *pn* capacitance	F	0	2 pF
MJS(MS)	Collector–substrate *pn* grading factor			0
FC	Forward-bias depletion capacitor coefficient		0.5	
TF	Ideal forward transit time	s	0	0.1 ns
TR	Ideal reverse transit time	s	0	10 ns
EG	Bandgap voltage (barrier height)	eV	1.11	1.11
XTI(PT)	IS temperature effect exponent (temperature coefficient for IS)		3	
KF	Flicker noise coefficient		0	6.6E-16
AF	Flicker noise exponent		1	1

TABLE A.9 Model parameters for JFETs

Name	Model Parameter	Units	Default	Typical
VTO	Threshold voltage	V	−2	−2
BETA	Transconductance coefficient	A/V^2	1E-4	1E-3
LAMBDA	Channel-length modulation	V^{-1}	0	1E-4
RD	Drain ohmic resistance	Ω	0	100
RS	Source ohmic resistance	Ω	0	100
IS	Gate *pn* saturation current	A	1E-14	1E-14
PB	Gate *pn* potential	V	1	0.6
CGD	Gate–drain zero-bias *pn* capacitance (C_{gdo})	F	0	5 pF
CGS	Gate–source zero-bias *pn* capacitance (C_{gso})	F	0	1 pF
FC	Forward-bias depletion capacitance coefficient		0.5	
KF	Flicker noise coefficient		0	
AF	Flicker noise exponent		1	

TABLE A.10 Model parameters for MOSFETs

Name	Model Parameter	Units	Default	Typical
LEVEL	Model type (Shichman-Hodges)		1	
L	Channel length	m	DEFL	
W	Channel width	m	DEFW	
VTO	Zero-bias threshold voltage	V	0	0.1
KP	Transconductance	A/V^2	2E-5	2.5E-5
GAMMA	Bulk threshold parameter	$V^{1/2}$	0	0.35
LAMBDA	Channel-length modulation (LEVEL=1)	V^{-1}	0	0.02
RD	Drain ohmic resistance	Ω	0	10
RS	Source ohmic resistance	Ω	0	10
IS	Bulk *pn* saturation current	A	1E-14	1E-15
JS	Bulk *pn* saturation current/area	A/m^2	0	1E-8
PB	Bulk *pn* potential	V	0.8	0.75
CBD	Bulk-drain zero-bias *pn* capacitance	F	0	5 pF
CBS	Bulk-source zero-bias *pn* capacitance	F	0	2 pF
CJ	Bulk *pn* zero-bias bottom capacitance/length	F/m^2	0	
MJ	Bulk *pn* bottom grading coefficient		0.5	
FC	Bulk *pn* forward-bias capacitance coefficient		0.5	
CGSO	Gate–source overlap capacitance/channel width	F/m	0	
CGDO	Gate–drain overlap capacitance/channel width	F/m	0	
CGBO	Gate-bulk overlap capacitance/channel length	F/m	0	
NSUB	Substrate doping density	$1/cm^3$	0	
TOX	Oxide thickness	m	∞	
UO	Surface mobility	$cm^2/V \cdot s$	600	
KF	Flicker noise coefficient		0	1E-26
AF	Flicker noise exponent		1	1.2

The model statement for an *n*-channel MOSFET has the following general form:

```
.MODEL  MNAME  NMOS (P1=A1 P2=A2 P3=A3 . . . PN=AN)
```

The general form of the model statement for a *p*-channel MOSFET is

```
.MODEL  MNAME  PMOS (P1=A1 P2=A2 P3=A3 . . . PN=AN)
```

NMOS and PMOS are the type symbols for *n*-channel and *p*-channel MOSFETs, respectively. MNAME can begin with any character, but its size is normally limited to eight characters. P1, P2, . . ., PN are the parameters, and A1, A2, . . ., AN are their values. Table A.10 lists the model parameters for MOSFETs.

A.13 Creating Netlists

Once you have drawn a schematic, you can create a netlist to use with other PSpice/SPICE software applications. After drawing your schematic, choose **Create Netlist** from the Analysis menu, shown in Fig. A.36. The netlist for Figs. A.31 and A.33 follows. It can be found in the file EXA-25.NET.

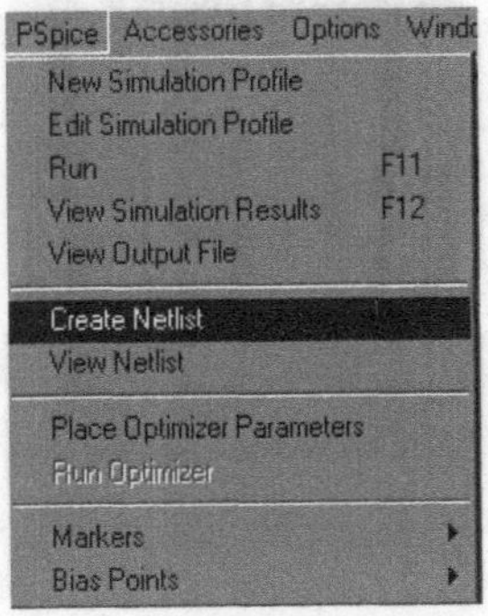

FIGURE A.36 Dialog box for creating netlist

```
* Schematics Netlist
C_C     0 $N_0001   10uF                  ; C is connected between nodes 0 and
                                          ; $N_0001
L_L     $N_0002 $N_0001   50uH            ; L is connected between nodes $N_0002
                                          ; and $N_0001
R_R     $N_0003 $N_0002 {RVAL}            ; R is connected between nodes $N_0003
                                          ; and $N_0002
V_Vs    $N_0003 0   AC   1V               ; Vs is connected between nodes $N_0003
                                          ; and 0
+PULSE 0 1V 0 1ns 1ns 0.5ms 1ms           ; Pulse voltage specifications
```

▶ **TIP** When you open a schematic file (e.g., EX5-1.SCH) for the first time, PSpice may indicate an error with the message "Not Finding Netlist." If this happens, first create the netlist by choosing **Create Netlist** from the Analysis menu, shown in Fig. A.36.

If the commands for transient analysis and AC analysis are included, the netlist for the PSpice circuit file becomes as follows:

```
Frequency Response ; The first line is the title line; PSpice always
                   ; ignores this statement
* Circuit Description
C_C     0 $N_0001   10uF                 ; C is connected between nodes 0 and
                                         ; $N_0001
L_L     $N_0002 $N_0001   50uH           ; L is connected between nodes $N_0002
                                         ; and $N_0001
R_R     $N_0003 $N_0002 {RVAL}           ; R is connected between nodes $N_0003
                                         ; and $N_0002
* Source Descriptions for both ac and pulse sources
V_Vs    $N_0003 0   AC   1V              ; Vs is connected between nodes $N_0003
                                         ; and 0
+ PULSE 0 1V 0 1ns 1ns 0.5ms 1ms         ; Pulse voltage specifications
* Analysis Descriptions for both ac and pulse sources
.AC   DEC 101 100HZ 100KHZ               ; ac analysis from 100 Hz to 100 kHz
                                         ; with decade increments of 101 points
                                         ; per decade
.TRAN   10ns 0.5ms                       ; Transient analysis from 0 to 0.5 ms
                                         ; with 10 ns printing/plotting interval
.PROBE                                   ; Graphics post-processor
.END                                     ; This is the last line and must always
                                         ; be included
```

TABLE A.11 Circuit elements and sources

Circuit Elements and Sources	First Letter	Model Type
Bipolar junction transistor	B	NPN/PNP
Capacitor	C	CAP
Current-controlled current source	F	
Current-controlled switch	W	VSWITCH
Current-controlled voltage source	H	
Diode	D	D
Exponential source	EXP	
GaAs MES field-effect transistor	B	GASFET
Ground	AGND	
Independent current source	I	
Independent DC voltage source	V	VDC
Inductor	L	IND/CORE
Junction field-effect transistor	J	NJF/PJF
MOS field-effect transistor	M	NMOS/PMOS
Mutual inductors (transformer)	K	
Piecewise linear source	PWL	
Polynomial source	POLY(n)	
Pulse voltage source	PULSE	VPULSE
Resistor	R	RES
Single-frequency frequency-modulation source	SFFM	
Sinusoidal voltage source	SIN	VSIN
Transmission line	T	
Voltage-controlled current source	G	
Voltage-controlled switch	S	VSWITCH
Voltage-controlled voltage source	E	EVALUE

The first letter and the model type for each element name are listed in Table A.11. The PSpice/SPICE commands are listed in Table A.12.

A.14 Adding Library Files

The directory of the PSpice circuit simulator (default name C:\MSIMEV61) has a subdirectory LIB, which contains library files of schematics and device models. This library has only a limited number of transistor schematics and models. If you want to use devices or elements with different model parameters, use elements or devices from the symbol library BREAKOUT.SLB, as shown in Fig. A.37. Use Notepad to modify the model statement in the BREAKOUT.LIB file—for example,

```
C:\OrcadLite\Capture\library\PSpice/Breakout.Lib
```

You can either use the library file RASHID.LIB that comes with the book or create your own library file containing the model statement of a device with the same model name as in the PSpice schematic—for example,

```
.MODEL      Q2N2222      Q (BF=100)                 ; BJT Q2N2222
.MODEL      D1N4148      D (IS=10E-15 BV=100)       ; Diode D1N4148
```

This way you keep the same model name and use the PSpice schematic symbols, but change the model parameters.

TABLE A.12 PSpice/SPICE commands

Analysis or Function	Command
Absolute value (operator)	ABS
AC/frequency analysis	.AC
DC operating point	.OP
DC sweep	.DC
Difference (operator)	DIF
End of subcircuit	.ENDS
Fourier analysis	.FOUR
Frequency response transfer function	FREQ
Function definition	.FUNC
Gain limit (operator)	GLIMIT
Global nodes	.GLOBAL
Graphics postprocessor	.PROBE
Include file	.INC
Initial conditions	.IC
Library file	.LIB
Model definition	.MODEL
Multiplier (operator)	MULTI
Node setting	.NODESET
Noise analysis	.NOISE
Options	.OPTIONS
Parameter definition	PARAM
Parameter variation	.PARAM
Parametric analysis	.STEP
Plot output	.PLOT
Print output	.PRINT
Probe	.PROBE
Sensitivity analysis	.SENS
Subcircuit call	X_Call
Subcircuit definition	.SUBCKT
Summation (operator)	SUM
Table (operator)	TABLE
Temperature	.TEMP
Transfer function	.TF
Transient analysis	.TRAN
Value	VALUE
Value of voltage-controlled voltage source	EVALUE
Width	.WIDTH

▶ **TIP** If you wish to change the model parameters of any device, use a word processing program in ASCII (DOS) text format (or DOS Editor or Notepad in PSpice) to edit the file EVAL.LIB and change the model parameters.

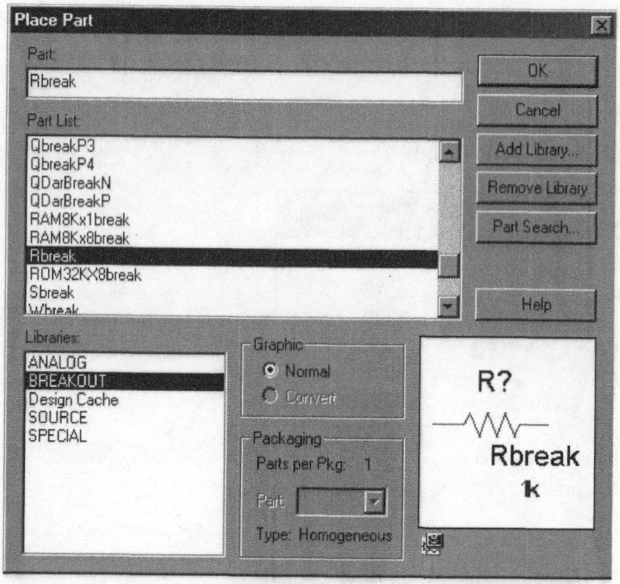

FIGURE A.37 Menu for adding parts

We can add the library file RASHID_MODEL.LIB while running the simulation. First, choose **Library and Include Files** from the Simulation Settings menu, shown in Fig. A.38. Then browse through the library files and select the file RASHID_MODEL.LIB. Click on **Add Library*** and then on **OK**. PSpice will look for device models in this library file while running the simulation. The Include Files dialog box is shown in Fig. A.39.

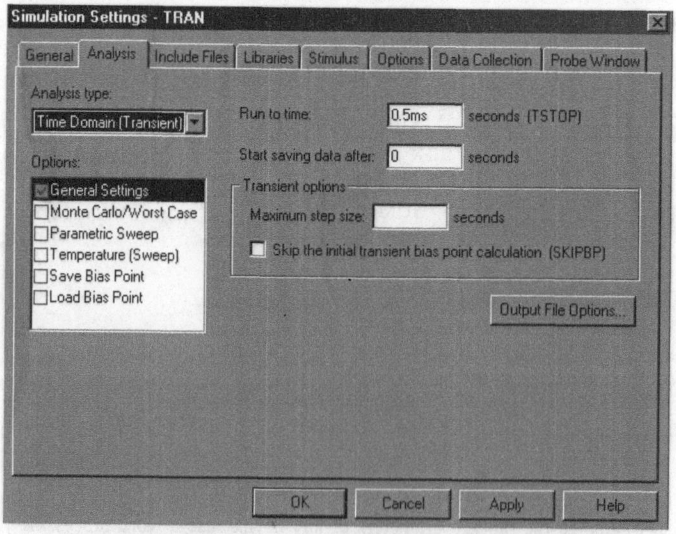

FIGURE A.38 Simulation Settings menu

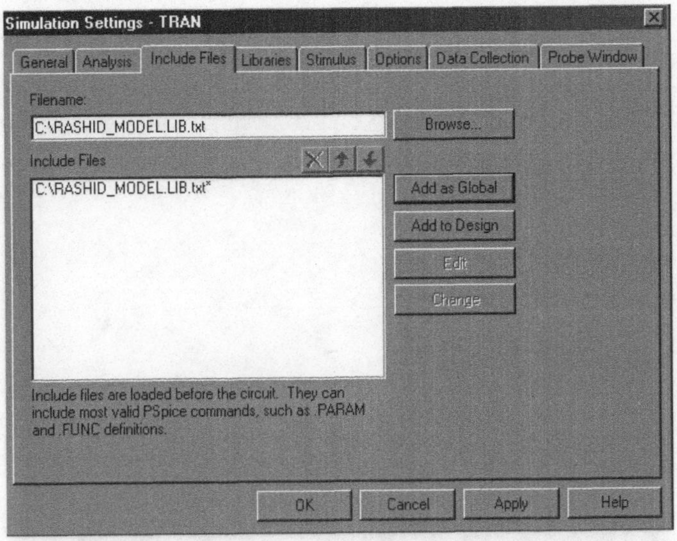

FIGURE A.39 Dialog box for adding library files

References

1. M. H. Rashid, *Introduction to PSpice Using OrCAD for Circuits and Electronics*. Upper Saddle River, NJ: Prentice Hall, 2003.
2. M. E. Herniter, *Schematic Capture with MicroSim PSpice*. Englewood Cliffs, NJ: Prentice Hall, 1996.

APPENDIX B

REVIEW OF BASIC CIRCUITS

B.1 Introduction

Electronic devices that are used as parts of electronic circuits are normally modeled by equivalent circuits. Doing a performance evaluation and design of any electronic circuit requires a knowledge of circuit analysis. This appendix reviews the basic circuit theorems and analysis techniques that are commonly used for electronic circuits.

B.2 Kirchhoff's Current Law

Kirchhoff's current law (KCL) states that the sum of all currents at a node must be zero; that is,

$$\sum I_n = 0$$

where I_n is the current flowing into nth node and $n = 1, 2, 3, \ldots, \infty$.

EXAMPLE B.1

Finding the currents in two parallel resistors For the circuit shown in Fig. B.1, find currents I_1 and I_2.

SOLUTION

Using KCL at node 1 gives

$$I_S - I_1 - I_2 = 0 \quad \text{or} \quad I_S = I_1 + I_2 \tag{B.1}$$

Since the voltage V_S across R_1 is the same as that across R_2, we can write $V_S = R_1 I_1 = R_2 I_2$, which gives $I_2 = R_1 I_1 / R_2$. Substituting for I_2, we get

$$I_S = I_1 + \frac{R_1}{R_2} I_1 = \frac{R_2 + R_1}{R_2} I_1$$

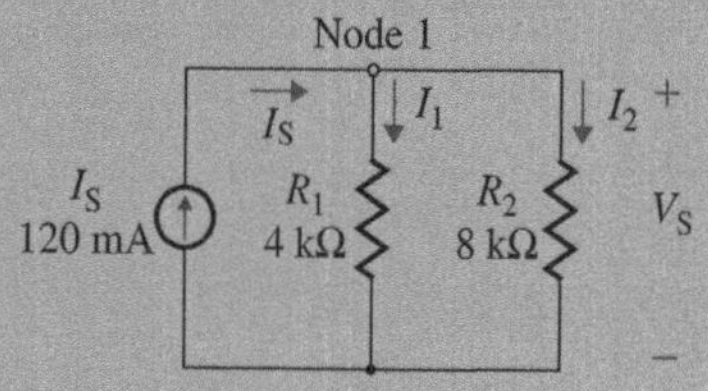

FIGURE B.1 Current distribution in two resistors

which gives the current I_1 through R_1 as

$$I_1 = \frac{R_2}{R_1 + R_2} I_S \qquad \text{(B.2)}$$

$$= \frac{8\ \text{k}\Omega}{4\ \text{k}\Omega + 8\ \text{k}\Omega} \times 120\ \text{mA} = 80\ \text{mA}$$

Similarly, substituting $I_1 = R_2 I_2 / R_1$ in Eq. (B.1) and simplifying, we get the current I_2 through R_2 as

$$I_2 = \frac{R_1}{R_1 + R_2} I_S \qquad \text{(B.3)}$$

$$= \frac{4\ \text{k}\Omega}{4\ \text{k}\Omega + 8\ \text{k}\Omega} \times 120\ \text{mA} = 40\ \text{mA}$$

The branch currents I_1 and I_2 can be found by applying Ohm's law as given by

$$V_S = I_S (R_1 \| R_2) = \frac{R_1 R_2}{R_1 + R_2} I_S$$

which can be used to find $I_1 = V_S/R_1$ as in Eq. (B.2) and $I_2 = V_S/R_2$ as in Eq. (B.3).

NOTE: Equations (B.2) and (B.3) give the current distribution in two resistors, and the two equations together are often known as the *current divider rule*.

EXAMPLE B.2

Finding the currents in three parallel resistors For the circuit shown in Fig. B.2, find the currents I_1, I_2, and I_3.

SOLUTION

The current source I_S is divided into I_1 through R_1, I_2 through R_2, and I_3 through R_3. Using KCL at node 1 gives

$$I_S - I_1 - I_2 - I_3 = 0 \quad \text{or} \quad I_S = I_1 + I_2 + I_3 \qquad \text{(B.4)}$$

Since the voltage V_S across R_1 is the same as that across R_2 and R_3, we can write $V_S = R_1 I_1 = R_2 I_2 = R_3 I_3$, which gives $I_2 = R_1 I_1 / R_2$ and $I_3 = R_1 I_1 / R_3$. Substituting for I_2 and I_3 in Eq. (B.4), we get

$$I_S = I_1 + \frac{R_1}{R_2} I_1 + \frac{R_1}{R_3} I_1 = \frac{R_1 R_2 + R_2 R_3 + R_3 R_1}{R_2 R_3} I_1$$

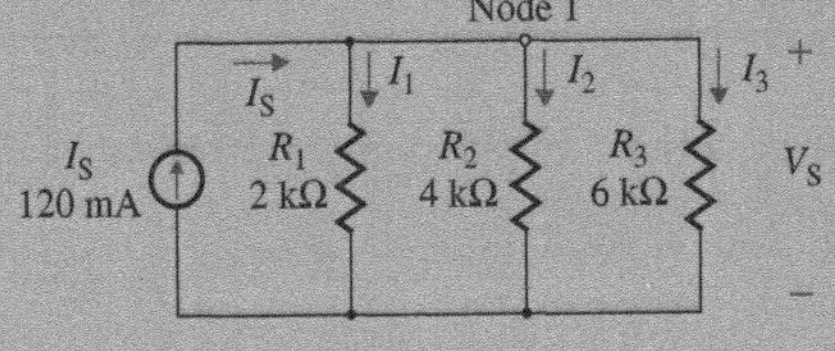

FIGURE B.2 Current distribution in three resistors

which gives the current I_1 through R_1 as

$$I_1 = \frac{1/R_1}{1/R_1 + 1/R_2 + 1/R_3} I_S = \frac{R_2R_3}{R_1R_2 + R_2R_3 + R_3R_1} I_S \tag{B.5}$$

$$= \frac{4\ \text{k}\Omega \times 6\ \text{k}\Omega}{2\ \text{k}\Omega \times 4\ \text{k}\Omega + 4\ \text{k}\Omega \times 6\ \text{k}\Omega + 6\ \text{k}\Omega \times 2\ \text{k}\Omega} \times 120\ \text{mA} = 65.45\ \text{mA}$$

Substituting $I_1 = R_2I_2/R_1$ and $I_3 = R_2I_2/R_3$ in Eq. (B.4) yields the current I_2 through R_2 as

$$I_2 = \frac{1/R_2}{1/R_1 + 1/R_2 + 1/R_3} I_S = \frac{R_1R_3}{R_1R_2 + R_2R_3 + R_3R_1} I_S \tag{B.6}$$

$$= \frac{2\ \text{k}\Omega \times 6\ \text{k}\Omega}{2\ \text{k}\Omega \times 4\ \text{k}\Omega + 4\ \text{k}\Omega \times 6\ \text{k}\Omega + 6\ \text{k}\Omega \times 2\ \text{k}\Omega} \times 120\ \text{mA} = 32.73\ \text{mA}$$

Similarly, substituting $I_1 = R_3I_3/R_1$ and $I_2 = R_3I_3/R_2$ in Eq. (B.4) yields the current I_3 through R_3 as

$$I_3 = \frac{1/R_3}{1/R_1 + 1/R_2 + 1/R_3} I_S = \frac{R_1R_2}{R_1R_2 + R_2R_3 + R_3R_1} I_S \tag{B.7}$$

$$= \frac{2\ \text{k}\Omega \times 4\ \text{k}\Omega}{2\ \text{k}\Omega \times 4\ \text{k}\Omega + 4\ \text{k}\Omega \times 6\ \text{k}\Omega + 6\ \text{k}\Omega \times 2\ \text{k}\Omega} \times 120\ \text{mA} = 21.82\ \text{mA}$$

The branch currents I_1, I_2, and I_3 can be found by applying Ohm's law as given by

$$V_S = I_S(R_1 \| R_2 \| R_3) = \frac{1}{(1/R_1) + (1/R_2) + (1/R_3)} I_S$$

which can be used to find $I_1 = V_S/R_1$ as in Eq. (B.5), $I_2 = V_S/R_2$ as in Eq. (B.6), and $I_3 = V_S/R_3$ as in Eq. (B.7).

NOTE: If there are n parallel resistances, the current I_n through the nth resistance R_n can be expressed in general form as

$$I_n = \frac{1/R_n}{(1/R_1) + (1/R_2) + \cdots + (1/R_m)} I_S$$

B.3 Kirchhoff's Voltage Law

Kirchhoff's voltage law (KVL) states that the sum of the voltages around any loop must be zero; that is,

$$\sum V_n = 0$$

where V_n is the voltage across the nth segment of a loop and $n = 1, 2, 3, \ldots, \infty$.

EXAMPLE B.3

Finding the voltage distribution in two resistors In Fig. B.3, the voltage V_S is divided into V_1 across R_1 and V_2 across R_2. Find the voltages V_1 and V_2.

SOLUTION

Using KVL around loop I gives

$$V_S - V_1 - V_2 = 0 \quad \text{or} \quad V_S = V_1 + V_2 \tag{B.8}$$

Since the current I_S through R_1 is the same as that through R_2, we can write $V_1 = R_1 I_S$ and $V_2 = R_2 I_S$. Substituting for V_1 and V_2 in Eq. (B.8), we get

$$V_S = V_1 + V_2 = R_1 I_S + R_2 I_S = (R_1 + R_2) I_S$$

which gives the current I_S as

$$I_S = \frac{V_S}{R_1 + R_2}$$

$$= \frac{24 \text{ V}}{4 \text{ k}\Omega + 8 \text{ k}\Omega} = 2 \text{ mA}$$

Therefore, the voltage V_1 across R_1 can be found from

$$V_1 = R_1 I_S = \frac{R_1}{R_1 + R_2} V_S \tag{B.9}$$

$$= \frac{4 \text{ k}\Omega}{4 \text{ k}\Omega + 8 \text{ k}\Omega} \times 24 \text{ V} = 8 \text{ V}$$

Similarly, the voltage V_2 across R_2 can be found from

$$V_2 = R_2 I_S = \frac{R_2}{R_1 + R_2} V_S \tag{B.10}$$

$$= \frac{8 \text{ k}\Omega}{4 \text{ k}\Omega + 8 \text{ k}\Omega} \times 24 \text{ V} = 16 \text{ V}$$

NOTE: The loop current I_S can be found directly by applying Ohm's law, $I_S = V_S/(R_1 + R_2)$. Equations (B.9) and (B.10) give the voltage distribution in two resistors only when the current through R_1 and R_2 is the same. This distribution of voltage is often known as the *voltage* (or *potential*) *divider rule*.

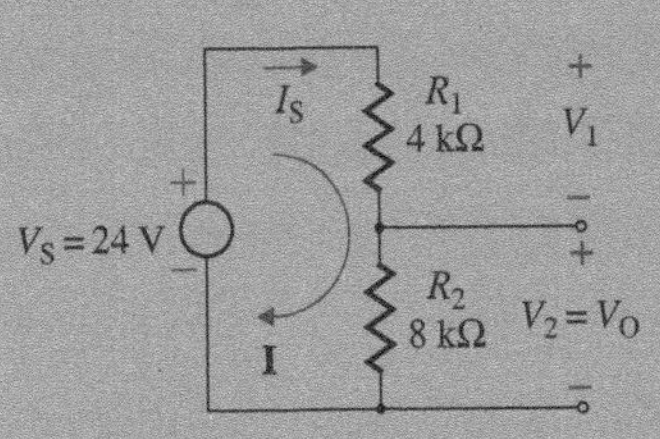

FIGURE B.3 Voltage distribution in two resistors

EXAMPLE B.4

Analyzing a circuit with a current-dependent current source For the circuit shown in Fig. B.4 with a current-dependent current source, find the currents I_B, I_C, and I_E and voltage V_C. Assume $R_{Th} = 15\ k\Omega$, $r_\pi = 1\ k\Omega$, $R_C = 2\ k\Omega$, $R_E = 500\ \Omega$, $\beta_F = 100$, $V_{CC} = 30$ V, and $V_{Th} = 5$ V.

SOLUTION

Using KCL at node 1, we get

$$I_E = I_B + I_C = I_B + \beta_F I_B = (1 + \beta_F)I_B \tag{B.11}$$

Using KVL around loop I, we get

$$V_{Th} = R_{Th}I_B + r_\pi I_B + R_E I_E = R_{Th}I_B + r_\pi I_B + R_E(1 + \beta_F)I_B$$

which gives I_B as

$$I_B = \frac{V_{Th}}{R_{Th} + r_\pi + R_E(1 + \beta_F)} \tag{B.12}$$

$$= \frac{5\ \text{V}}{15\ k\Omega + 1\ k\Omega + 500\ \Omega\ \times (1 + 100)} = 75.19\ \mu\text{A}$$

Current I_C, which is dependent only on I_B, can be found from

$$I_C = \beta_F I_B = \frac{\beta_F V_{Th}}{R_{Th} + R_\pi + R_E(1 + \beta_F)} \tag{B.13}$$

$$= \frac{100 \times\ 5\ \text{V}}{15\ k\Omega + 1\ k\Omega + 500\ \Omega\ \times (1 + 100)} = 7519\ \mu\text{A}$$

Then

$$I_E = I_B + I_C = 75.19\ \mu\text{A} + 7519\ \mu\text{A} = 7594\ \mu\text{A}$$

$$V_C = V_{CC} - I_C R_C = 30\ \text{V} - 7519\ \mu\text{A} \times 2\ k\Omega = 14.96\ \text{V}$$

Voltage V_E can be found from

$$V_E = R_E I_E = R_E(I_C + I_B) = R_E(\beta_F I_B + I_B) = R_E I_B(1 + \beta_F)$$

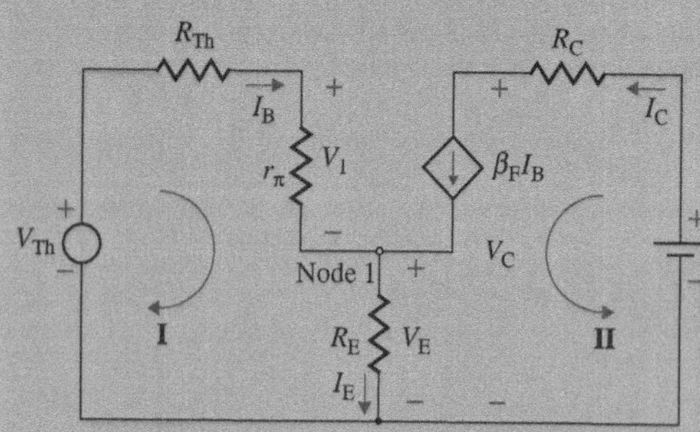

FIGURE B.4 Circuit with current-dependent current source

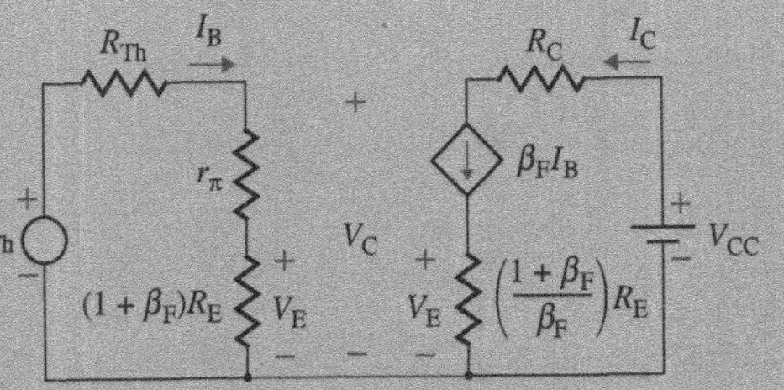

FIGURE B.5 Splitting of current R_E

Since $I_C = \beta_F I_B$, we get

$$V_E = R_E I_B(1 + \beta_F) = R_E I_C[(1 + \beta_F)/\beta_F)]$$

Thus, R_E offers a resistance of $R_E(1 + \beta_F)$ to current I_B in loop I and a resistance of $R_E(1 + \beta_F)/\beta_F$ to current I_C in loop II. Therefore, R_E can be split, or "reflected," into loop I and loop II by adjusting its value such that V_E is preserved on loop I and loop II. This arrangement is shown in Fig. B.5.

NOTE: The current-dependent current source $\beta_F I_B$ is often represented by a voltage-dependent voltage source $g_m V_1$ such that $\beta_F I_B = \beta_F V_1/r_\pi = (\beta_F/r_\pi)V_1$. g_m, which is known as the *transconductance*, is related to β_F by $g_m = \beta_F/r_\pi$. Therefore, by substituting $\beta_F = g_m r_\pi$, we can apply the above equations for a voltage-dependent voltage source that is often used as the small-signal model of bipolar and field-effect transistors.

B.4 Superposition Theorem

The superposition theorem states that the current through or voltage across any element in a linear network is equal to the algebraic sum of the currents or voltages produced independently by each source. To calculate the effect of one source, the other independent sources are removed by short-circuiting the voltage sources and open-circuiting the current sources. Any internal resistance associated with the removed voltage sources must be considered, however.

EXAMPLE B.5

Finding the output voltage using the superposition theorem The circuit in Fig. B.6 has a DC source $V_{DC} = 10$ V, an AC source $v_{ac} = 15 \sin(377\pi t)$, $R_1 = 2$ kΩ, and $R_2 = 3$ kΩ. Use the superposition theorem to determine the instantaneous output voltage v_O.

SOLUTION

$V_{DC} = 10$ V, $v_{ac} = 15 \sin(377\pi t)$, $R_1 = 2$ kΩ, and $R_2 = 3$ kΩ. The DC equivalent circuit with source V_{DC} only is shown in Fig. B.7(a); the output voltage due to V_{DC} is

$$V_{O1} = \frac{R_2}{R_1 + R_2} V_{DC} = \frac{3\text{ k}}{2\text{ k} + 3\text{ k}} \times 10 = 6\text{ V}$$

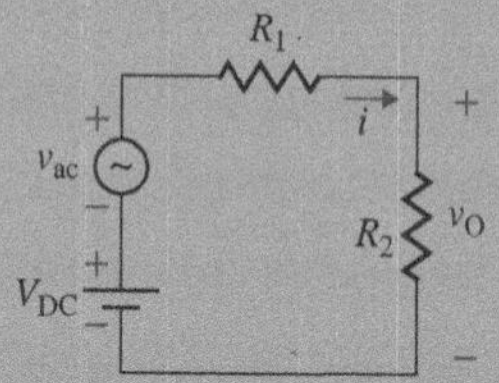

FIGURE B.6 Circuit for Example B.5

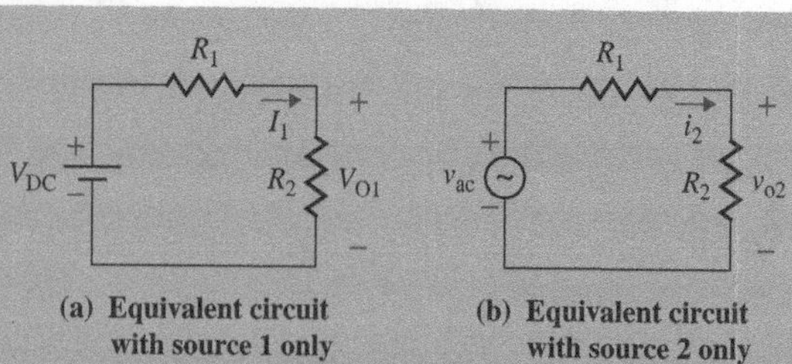

(a) Equivalent circuit with source 1 only

(b) Equivalent circuit with source 2 only

FIGURE B.7 Equivalent circuits for Example B.5

Figure B.7(b) shows the AC equivalent circuit with source v_{ac} only; the output voltage due to v_{ac} is

$$v_{o2} = \frac{R_2}{R_1 + R_2} v_{ac} = \frac{3\text{ k}}{2\text{ k} + 3\text{ k}} \times 15 \sin(377\pi t) = 9 \sin(377\pi t)$$

Therefore, the resultant output voltage v_O can be found by combining the output voltages due to individual sources; that is,

$$v_O = V_{O1} + v_{o2} = 6 + 9 \sin(377\pi t) = 3 \times [2 + 3 \sin(377\pi t)]$$

EXAMPLE B.6

Finding the output voltage using the superposition theorem A circuit with three input voltages V_{S1}, V_{S2}, and V_{S3} is shown in Fig. B.8. Use the superposition theorem to determine the output voltage V_O. Use $R_1 = 2\text{ k}\Omega$, $R_2 = 4\text{ k}\Omega$, $R_3 = 6\text{ k}\Omega$, $V_{S1} = 10\text{ V}$, $V_{S2} = 12\text{ V}$, and $V_{S3} = 15\text{ V}$.

SOLUTION

The equivalent circuit with source V_{S1} only is shown in Fig. B.9(a). Applying the voltage divider rule, we can find the output voltage due to V_{S1} as

$$V_{O1} = \frac{R_2 \| R_3}{R_1 + R_2 \| R_3} V_{S1} = \frac{4\text{ k} \| 6\text{ k}}{2\text{ k} + (4\text{ k} \| 6\text{ k})} \times 10 = 5.45\text{ V}$$

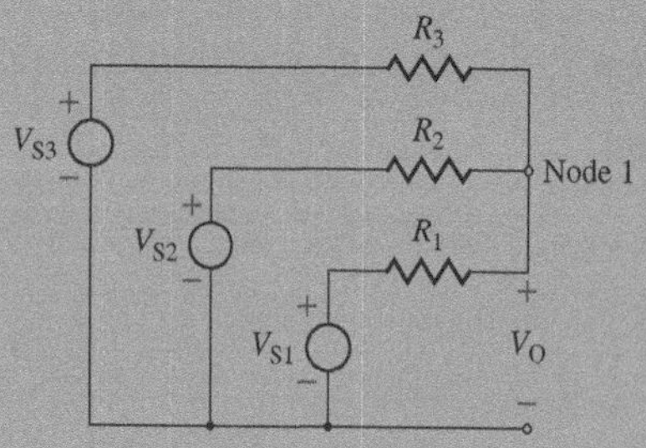

FIGURE B.8 Circuit for Example B.6

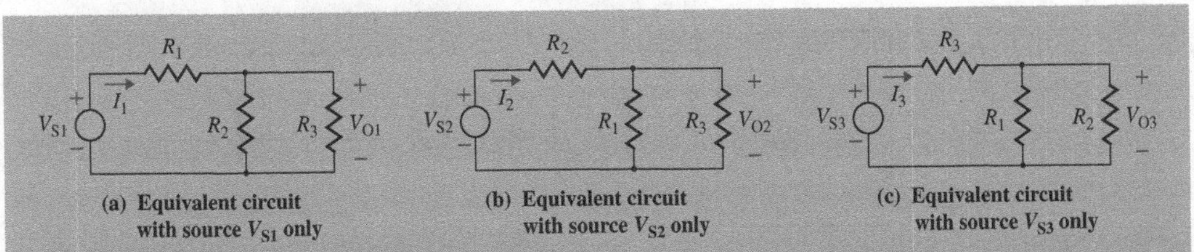

FIGURE B.9 Equivalent circuits for Example B.6

The equivalent circuit with source V_{S2} only is shown in Fig. B.9(b); the output voltage due to V_{S2} is

$$V_{O2} = \frac{R_1 \| R_3}{R_2 + R_1 \| R_3} V_{S2} = \frac{2\text{ k} \| 6\text{ k}}{4\text{ k} + (2\text{ k} \| 6\text{ k})} \times 12 = 3.27\text{ V}$$

The equivalent circuit with source V_{S3} only is shown in Fig. B.9(c); the output voltage due to V_{S3} is

$$V_{O3} = \frac{R_1 \| R_2}{R_3 + R_1 \| R_2} V_{S3} = \frac{2\text{ k} \| 4\text{ k}}{6\text{ k} + (2\text{ k} \| 4\text{ k})} \times 15 = 2.73\text{ V}$$

Therefore, the resultant output voltage V_O can be found by combining the output voltages due to individual sources; that is,

$$V_O = V_{O1} + V_{O2} + V_{O3} = 5.45 + 3.27 + 2.73 = 11.45\text{ V}$$

An alternative approach is to apply KVL at node 1 and look for V_O:

$$V_O = \frac{\text{Currents into node 1 if it were at ground potential}}{\text{Conductances radiating from node 1}}$$

$$= \frac{V_{S1}/R_1 + V_{S2}/R_2 + V_{S3}/R_3}{1/R_1 + 1/R_2 + 1/R_3} = \frac{10/2\text{ k} + 12/4\text{ k} + 15/6\text{ k}}{1/2\text{ k} + 1/4\text{ k} + 1/6\text{ k}} = 11.45\text{ V}$$

B.5 Thevenin's Theorem

Thevenin's theorem states that any two-terminal linear DC (or AC) network can be replaced by an equivalent circuit consisting of a voltage source and a series resistance (or impedance). This theorem is commonly used to find the voltage (or current) of a linear network with one or more sources. It allows us to concentrate on a specific portion of the network by replacing the remaining network by an equivalent circuit. In the case of sinusoidal AC circuits, the reactances are frequency dependent, and Thevenin's equivalent circuit is valid for only one frequency.

Figure B.10(a) shows a general DC network; Thevenin's equivalent circuit is shown in Fig. B.10(b). The steps in determining an equivalent voltage source V_{Th} and an equivalent resistance R_{Th} for Thevenin's equivalent circuit are as follows:

Step 1. Decide on the part of the network for which you desire a Thevenin's representation and mark the terminals, as shown in Fig. B.10(a).

Step 2. Remove that part of the network. In Fig. B.10(a), the load resistance R_L is to be removed.

Step 3. Mark the open-circuit terminals of the remaining network—namely, a and b.

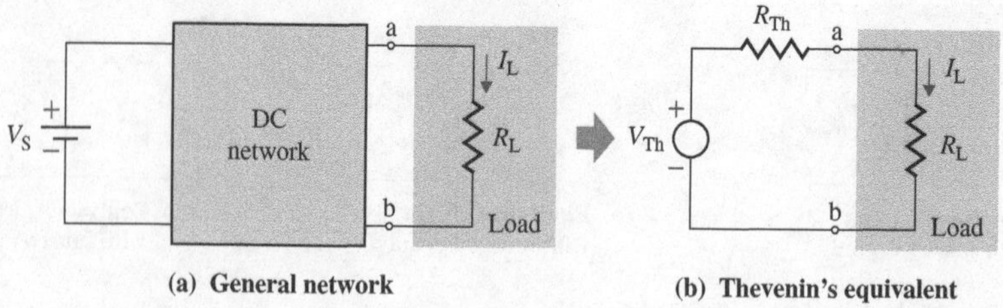

FIGURE B.10 Thevenin's equivalent circuit

Step 4. Determine the open-circuit voltage V_{Th} between terminals a and b.

Step 5. Set all independent sources to zero (voltage sources are replaced by short circuits and current sources by open circuits). Apply a test voltage V_X across terminals a and b. The ratio of V_X to its current I_X gives Thevenin's resistance R_{Th}.

EXAMPLE B.7

Finding Thevenin's equivalent circuit Represent the network shown in Fig. B.11(a) by Thevenin's equivalent, as shown in Fig. B.11(b). Assume $V_{CC} = 12$ V, $R_1 = 15$ kΩ, and $R_2 = 7.5$ kΩ.

SOLUTION

The open-circuit voltage, which is Thevenin's voltage between terminals a and b, can be found from the voltage divider rule in Eq. (B.10); that is,

$$V_{Th} = \frac{R_2}{R_1 + R_2} V_{CC} \tag{B.14}$$

$$= \frac{7.5\text{ k}}{15\text{ k} + 7.5\text{ k}} \times 12 = 4\text{ V}$$

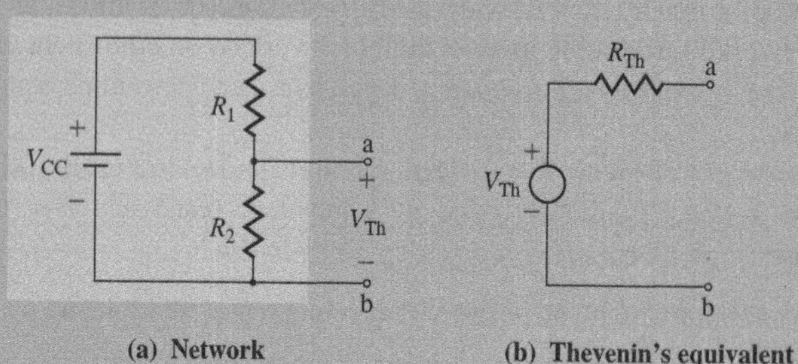

FIGURE B.11 Network for Example B.7

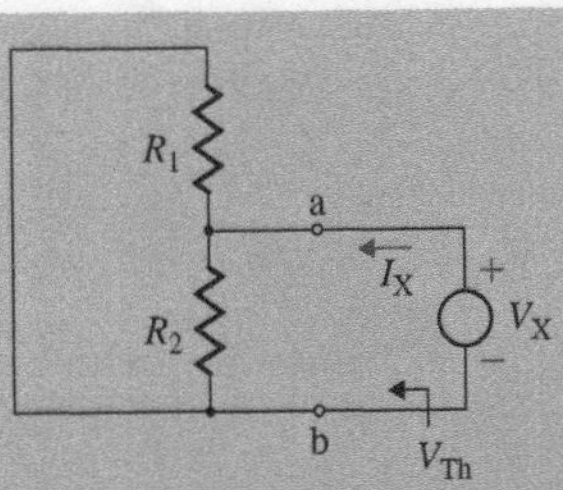

FIGURE B.12 Equivalent circuits

If source V_{CC} is set to zero and test voltage V_X is applied across terminals a and b, the circuit for determining R_{Th} is shown in Fig. B.12. R_{Th} becomes the parallel combination of R_1 and R_2. That is,

$$R_{Th} = \frac{V_X}{I_X} = R_1 \| R_2 \tag{B.15}$$

$$= 15\text{ k} \| 7.5\text{ k} = 5\text{ k}\Omega$$

EXAMPLE B.8

Finding Thevenin's equivalent circuit Represent the network shown in Fig. B.13(a) by Thevenin's equivalent, as shown in Fig. B.13(b). Assume $V_{CC} = 12$ V, $V_A = 9$ V, $R_1 = 15$ kΩ, and $R_2 = 7.5$ kΩ.

SOLUTION

Since there are two voltage sources V_{CC} and V_A, we will apply the superposition theorem to find V_{Th}. The equivalent circuit with source V_{CC} only is shown in Fig. B.14(a). Applying the voltage divider rule, we can find the output voltage due to V_{CC} only as

$$V_{O1} = \frac{R_2}{R_1 + R_2} V_{CC} = \frac{7.5\text{ k}}{15\text{ k} + 7.5\text{ k}} \times 12 = 4\text{ V}$$

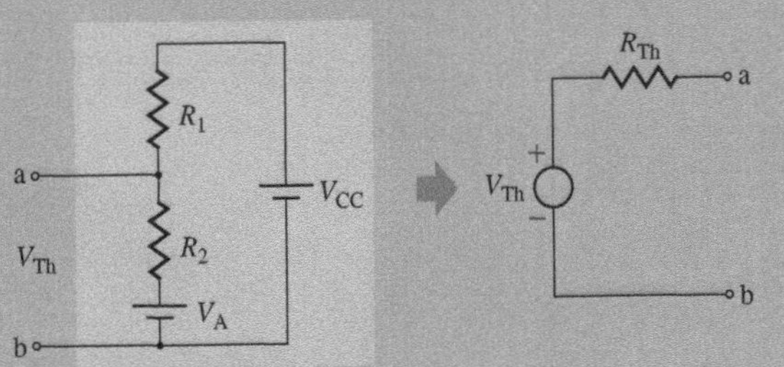

(a) Network (b) Thevenin's equivalent

FIGURE B.13 Network for Example B.8

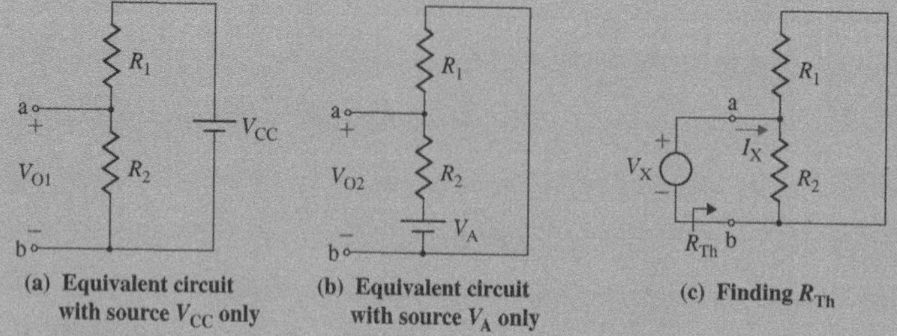

FIGURE B.14 Equivalent circuits

The circuit with source V_A only is shown in Fig. B.14(b); the output voltage due to V_A is

$$V_{O2} = \frac{R_1}{R_1 + R_2} V_A = \frac{15\text{ k}}{15\text{ k} + 7.5\text{ k}} \times 9 = 6\text{ V}$$

The resultant output voltage V_O, which equals V_{Th}, can be found by combining the output voltages due to individual sources; that is,

$$\begin{aligned} V_{Th} &= V_O = V_{O1} + V_{O2} \\ &= \frac{R_2}{R_1 + R_2} V_{CC} + \frac{R_1}{R_1 + R_2} V_A \qquad \text{(B.16)} \\ &= 4 + 6 = 10\text{ V} \end{aligned}$$

If sources V_A and V_{CC} are set to zero and test voltage V_X is applied across terminals a and b, the circuit for determining R_{Th} is shown in Fig. B.14(c). R_{Th} becomes the parallel combination of R_1 and R_2; that is,

$$\begin{aligned} R_{Th} &= \frac{V_X}{I_X} = R_1 \| R_2 \qquad \text{(B.17)} \\ &= 15\text{ k} \| 7.5\text{ k} = 5\text{ k}\Omega \end{aligned}$$

It is evident that Eqs. (B.15) and (B.17) are the same.

EXAMPLE B.9

Representing a network by Thevenin's equivalent circuit Represent the network shown in Fig. B.15 by Thevenin's equivalent circuit. The circuit values are $R_i = 1.5\text{ k}\Omega$, $R_C = 25\text{ k}\Omega$, $\beta_F = 50$, $h_r = 3 \times 10^{-4}$, and $V_s = 5\text{ mV}$.

(a) Calculate the parameters of Thevenin's equivalent circuit.

(b) Use PSpice/SPICE to check your results.

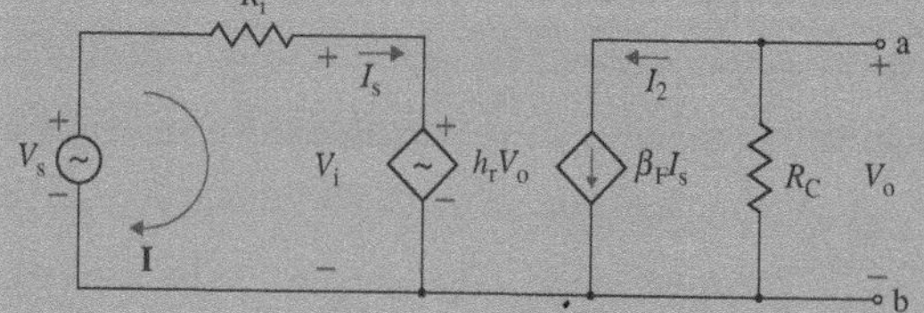

FIGURE B.15 Network for Example B.9

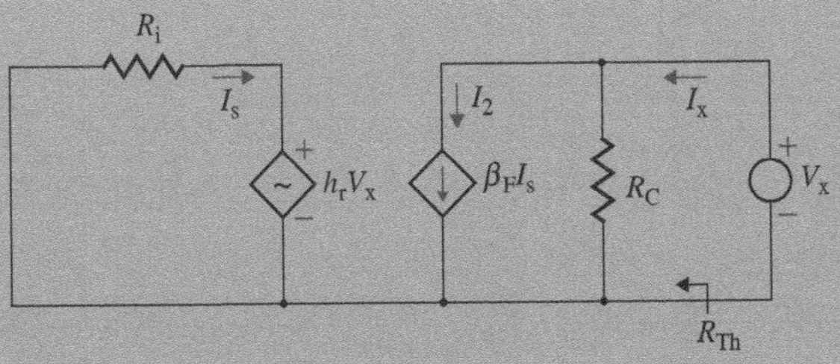

FIGURE B.16 Circuit for determining Thevenin's resistance

SOLUTION

(a) $R_i = 1.5\ \text{k}\Omega$, $R_C = 25\ \text{k}\Omega$, $\beta_F = 50$, $h_r = 3 \times 10^{-4}$, and $V_s = 5$ mV. The output voltage V_o between terminals a and b is

$$V_o = -I_2R_C = -\beta_F I_s R_C \tag{B.18}$$

The input current I_s can be found from loop I as

$$I_s = \frac{V_s - h_r V_o}{R_i} \tag{B.19}$$

Substituting I_s from Eq. (B.19) into Eq. (B.18) gives the output voltage V_o as

$$V_{Th} = V_o = \frac{-\beta_F R_C}{R_i - \beta_F h_r R_C} V_s \tag{B.20}$$

$$= \frac{-50 \times 25\ \text{k} \times 5\ \text{m}}{1.5\ \text{k} - 50 \times 3 \times 10^{-4} \times 25\ \text{k}} = -5.5556\ \text{V}$$

Thevenin's resistance R_{Th} can be determined from the circuit in Fig. B.16, which is obtained by short-circuiting the independent source V_s. Considering V_x and I_x as the test voltage and current, respectively, we get

$$I_s = -\frac{h_r V_x}{R_i} \tag{B.21}$$

$$I_x = \beta_F I_s + \frac{V_x}{R_C} \tag{B.22}$$

Substituting I_s from Eq. (B.21) into Eq. (B.22) gives

$$I_x = -\frac{\beta_F h_r V_x}{R_i} + \frac{V_x}{R_C} = \frac{R_i - \beta_F h_r R_C}{R_i R_C} V_x$$

which gives Thevenin's resistance R_{Th} as

$$R_{Th} = \frac{V_x}{I_x} = \frac{R_i R_C}{R_i - \beta_F h_r R_C} \tag{B.23}$$

$$= \frac{1.5\ \text{k} \times 25\ \text{k}}{1.5\ \text{k} - 50 \times 3 \times 10^{-4} \times 25\ \text{k}} = 33.33\ \text{k}\Omega$$

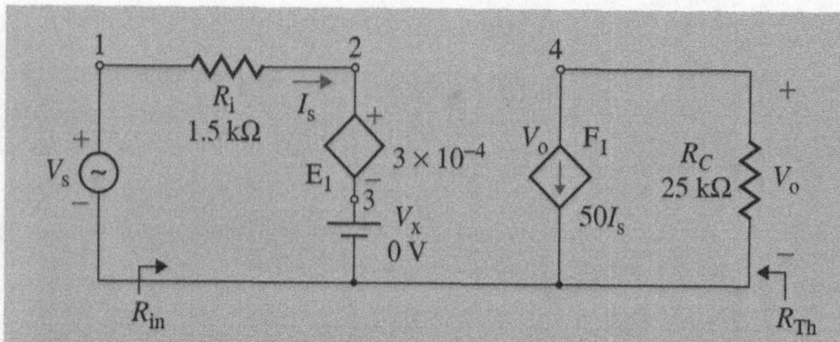

FIGURE B.17 Circuit for PSpice simulation

(b) The circuit for PSpice simulation is shown in Fig. B.17, which is run by OrCAD PSpice A/D. The results of the PSpice simulation are shown below:

```
NODE   VOLTAGE   NODE   VOLTAGE   NODE   VOLTAGE   NODE   VOLTAGE
( 1)    .0050    ( 2)   -.0017    ( 3)   0.0000    ( 4)   -5.5556
```

$$V_{Th} = V_o = V(4) = -5.5556 \text{ V}$$

```
****   SMALL-SIGNAL CHARACTERISTICS

V(4)/VS=-1.111E+03

INPUT RESISTANCE AT VS=1.125E+03

OUTPUT RESISTANCE AT V(4)=3.333E+04
```

Gain $A = V_o/V_s = -1111$

$R_{in} = V_s/I_s = 1.125 \text{ k}\Omega$

$R_{Th} = 33.33 \text{ k}\Omega$

B.6 Norton's Theorem

Norton's theorem states that any two-terminal linear DC (or AC) network can be replaced by an equivalent circuit consisting of a current source and a parallel resistance (or impedance). Norton's equivalent circuit can be determined from Thevenin's equivalent circuit; the relationship between them is shown in Fig. B.18. Norton's resistance R_N is identical to Thevenin's resistance R_{Th}, and Norton's current equals the short-circuit current at the terminals of interest.

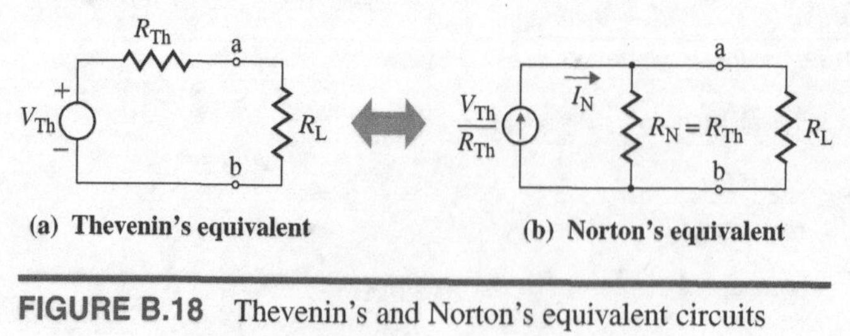

(a) Thevenin's equivalent **(b) Norton's equivalent**

FIGURE B.18 Thevenin's and Norton's equivalent circuits

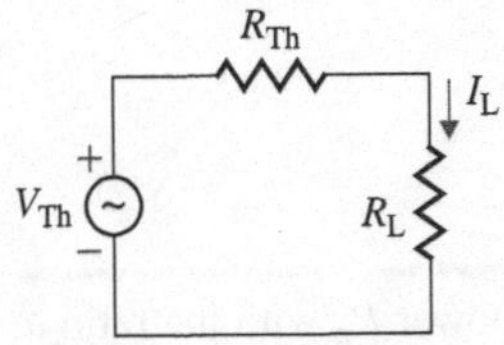

FIGURE B.19 Thevenin's equivalent circuit with a resistive load

B.7 Maximum Power Transfer Theorem

In electronic circuits, it is often necessary to transmit maximum power to the load. Let us consider the circuit in Fig. B.19, which could be Thevenin's equivalent circuit of a network. The power P_L delivered to load resistance R_L can be found from

$$P_L = I_L^2 R_L = \left[\frac{V_{Th}}{R_{Th} + R_L}\right]^2 R_L$$

$$= \frac{V_{Th}^2}{R_{Th}} \times \frac{1}{(1 + R_L/R_{Th})^2} \times \frac{R_L}{R_{Th}} \quad \text{(B.24)}$$

For a given circuit, V_{Th} and R_{Th} will be fixed. Therefore, the load power P_L depends on the load resistance R_L. If we set $R_L = uR_{Th}$, Eq. (B.24) becomes

$$P_L = \frac{V_{Th}^2}{R_{Th}} \frac{u}{(1 + u)^2}$$

$$= \frac{u}{(1 + u)^2} P$$

where $P = V_{Th}^2/R_{Th}$. Normalizing P_L with respect to P, we get the normalized power P_n as

$$P_n = \frac{P_L}{P} = \frac{u}{(1 + u)^2} \quad \text{(B.25)}$$

Figure B.20 shows the variation of normalized power P_n with u. The power P_n becomes maximum at $u = 1$; that is, $R_{Th} = uR_L = R_L$. The value of R_L for a maximum power transfer can also be determined from the condition $dP_L/dR_L = 0$. From Eq. (B.24),

$$\frac{dP_L}{dR_L} = V_{Th}^2\left[\frac{(R_{Th} + R_L)^2 - 2R_L(R_{Th} + R_L)}{(R_{Th} + R_L)^4}\right] = 0$$

or

$$(R_{Th} + R_L)^2 - 2R_L(R_{Th} + R_L) = 0$$

$$R_L = \pm R_{Th}$$

Since R_{Th} cannot be negative,

$$R_L = R_{Th} \quad \text{(B.26)}$$

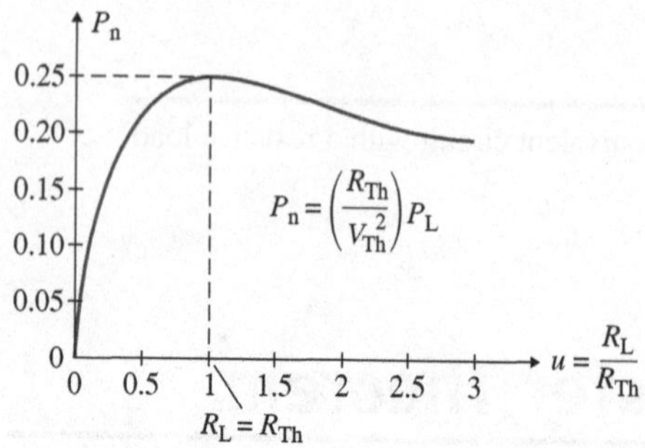

FIGURE B.20 Normalized power P_n with the ratio u

Thus, the maximum power transfer occurs when the load resistance R_L is equal to Thevenin's resistance R_{Th} of the network. For Norton's equivalent circuit of Fig. B.18(b), the maximum power will be delivered to the load when

$$R_N = R_L \tag{B.27}$$

Substituting R_L from Eq. (B.26) into Eq. (B.24) gives the maximum power P_{max} delivered to the load as

$$P_{max} = \frac{V_{Th}^2 R_L}{4R_L^2} = \frac{V_{Th}^2}{4R_L} \tag{B.28}$$

The input power P_{in} supplied by the source V_s is

$$P_{in} = \frac{V_{Th}^2}{R_{Th} + R_L} = \frac{V_{Th}^2}{2R_L} \tag{B.29}$$

Thus, the efficiency η at the maximum power transfer condition is

$$\eta = \frac{P_{max}}{P_{in}} \times 100\% = \frac{V_{Th}^2}{4R_L} \times \frac{2R_L}{V_{Th}^2} \times 100\% = 50\%$$

Therefore, the efficiency is always 50% at the maximum power transfer condition. In electronic circuits, the amount of power being transferred is usually small, and efficiency is often not of primary concern. However, efficiency is of major concern in circuits involving high power—for example, in power systems.

B.8 Transient Response of First-Order Circuits

The transient response gives the instantaneous value of an output voltage (or current) for a specified instantaneous input voltage (or current). The response due to a step-signal input is commonly used in evaluating electronic circuits because such a response allows us to predict the response due to other signals such as a pulse or square-wave input.

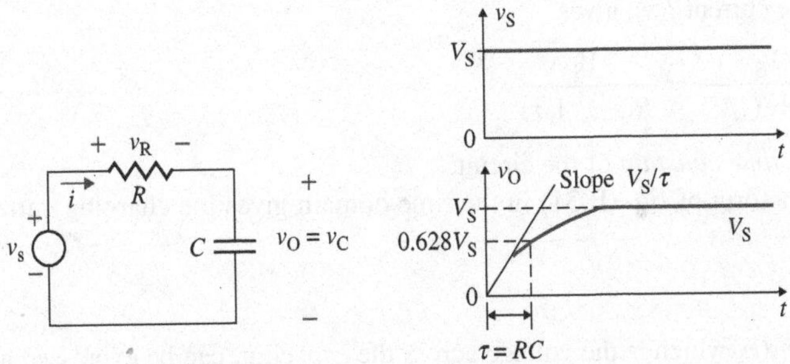

(a) **RC circuit** (b) **Step response**

FIGURE B.21 Series *RC* circuit

B.8.1 Step Response of Series *RC* Circuits

Consider the series *RC* circuit in Fig. B.21(a) with a step input voltage V_S. The output voltage v_O is taken across capacitor C. For $t \geq 0$, the charging current i of the capacitor can be found from

$$V_S = v_R + v_C = Ri + \frac{1}{C}\int i\,dt + v_C(t = 0) \tag{B.30}$$

with initial capacitor voltage $v_C(t = 0) = 0$.

Using the Laplace transformations in Table B.1, we can transform Eq. (B.30) into Laplace's domain of s as follows:

$$\frac{V_S}{s} = RI(s) + \frac{1}{Cs}I(s)$$

TABLE B.1 Some Laplace transformations

$f(t)$	$F(s)$
1	$\frac{1}{s}$
t	$\frac{1}{s^2}$
$e^{-\alpha t}$	$\frac{1}{s + \alpha}$
$\sin \alpha t$	$\frac{\alpha}{s^2 + \alpha^2}$
$\cos \alpha t$	$\frac{s}{s^2 + \alpha^2}$
$f'(t)$	$sF(s) - F(0)$
$f''(t)$	$s^2F(s) - sF(s) - F'(0)$

which, solved for the current $I(s)$, gives

$$I(s) = \frac{V_S}{sR + (1/C)} = \frac{V_S}{R(s + 1/\tau)} \tag{B.31}$$

where $\tau = RC$ is the *time constant* of the circuit.

The inverse transform of Eq. (B.31) in the time domain gives the charging current:

$$i(t) = \frac{V_S}{R} e^{-t/\tau} \tag{B.32}$$

The output voltage $v_O(t)$, which is the voltage across the capacitor, can be expressed as

$$v_O(t) = \frac{1}{C}\int_0^t i\,dt = \frac{1}{C}\int_0^t \frac{V_S}{R} e^{-t/\tau}\,dt = V_S(1 - e^{-t/\tau}) \tag{B.33}$$

In the steady state (at $t = \infty$), Eq. (B.32) gives

$$i(t = \infty) = 0$$

From Eq. (B.33),

$$v_O(t = \infty) = V_S \tag{B.34}$$

At $t = \tau$, Eq. (B.33) gives

$$v_O(t = \tau) = V_S(1 - e^{-1}) = 0.632V_S \tag{B.35}$$

The initial slope of the tangent to $v_O(t)$ can be found from Eq. (B.33):

$$\left.\frac{dv_O}{dt}\right|_{t=0} = \left.\frac{V_S}{\tau} e^{-t/\tau}\right|_{t=0} = \frac{V_S}{\tau} = \frac{V_S}{RC} \tag{B.36}$$

The transient response of $v_O(t)$, which rises expontentially due to a step input, is shown in Fig. B.21(b).

B.8.2 Step Response of Series *CR* Circuits

In a *CR* circuit, the output voltage is taken across resistance R instead of capacitance C, as shown in Fig. B.22(a). The output voltage v_O, which is the voltage across resistance R, can be found from Eq. (B.32); that is,

$$v_O(t) = Ri(t) = V_S e^{-t/\tau} \tag{B.37}$$

which, in the steady state (at $t = \infty$), gives

$$i(t = \infty) = 0$$

$$v_O(t = \infty) = 0$$

At $t = \tau$, Eq. (B.37) gives

$$v_O(t = \tau) = V_S e^{-1} = 0.368V_S \tag{B.38}$$

From Eq. (B.37), the initial slope of the tangent to $v_O(t)$ is

$$\left.\frac{dv_O}{dt}\right|_{t=0} = -\left.\frac{V_S}{\tau} e^{-t/\tau}\right|_{t=0} = -\frac{V_S}{\tau} = -\frac{V_S}{RC} \tag{B.39}$$

The response of $v_O(t)$, which falls exponentially due to a step voltage input, is shown in Fig. B.22(b).

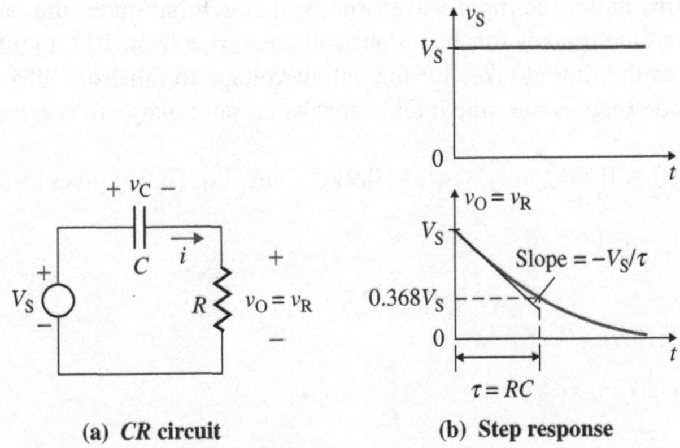

FIGURE B.22 Step response of a series *CR* circuit

B.8.3 Pulse Response of Series *RC* Circuits

An input pulse v_S of duration T, shown in Fig. B.23(a), is applied to the circuit of Fig. B.21(a). The response due to a pulse signal will depend on the ratio of time constant τ to duration T. We consider three cases: $\tau = T$, $\tau << T$, and $\tau >> T$.

In case 1, $\tau = T$, the output voltage $v_O(t)$ has just enough time to reach a near steady-state value of V_S. The capacitor C is charged exponentially to approximately voltage V_S. When the input voltage $v_S(t)$ falls to zero at $t = T$, the output (or capacitor) voltage $v_O(t)$ falls exponentially to zero, as shown

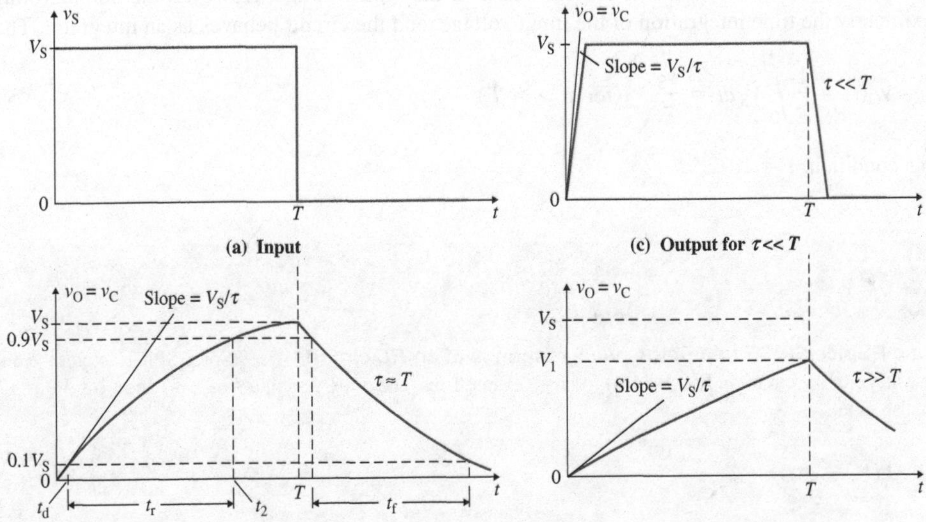

FIGURE B.23 Pulse response of an *RC* circuit

in Fig. B.23(b). The area under the input waveform must equal that under the output waveform. *Rise time* t_r is defined as the time it takes for the output voltage to rise from 10% to 90% of the final value. *Fall time* t_f is defined as the time it takes for the output voltage to fall from 90% to 10% of the initial value. *Delay time* t_d is defined as the time it takes for the output voltage to rise from 0% to 10% of the final value.

At $t = t_1 = t_d$, $v_O(t) = 0.1V_S$; at t_2, $v_O(t) = 0.9V_S$. Thus, Eq. (B.33) gives

$$0.1V_S = V_S(1 - e^{-t_1/\tau})$$
$$e^{-t_1/\tau} = 0.9$$
$$t_1 = -\tau \ln(0.9)$$

and

$$0.9V_S = V_S(1 - e^{-t_2/\tau})$$
$$e^{-t_2/\tau} = 0.1$$
$$t_2 = -\tau \ln(0.1)$$

The rise time t_r, which is equal to the fall time t_f, can be found as follows:

$$t_r = t_f = t_2 - t_1$$
$$= -\tau \ln(0.1) + \tau \ln(0.9) = \tau \ln(9) \approx 2.2\tau \quad \text{(B.40)}$$

In case 2, $\tau << T$, t_r and t_f are much smaller than T. The output voltage $v_O(t)$ represents the input signal more closely, as shown in Fig. B.23(c). This condition is generally satisfied by choosing circuit parameters such that $10\tau = T$.

In case 3, $\tau >> T$, there is not enough time for the output voltage $v_O(t)$ to reach the steady-state value of V_S. The output voltage at $t = T$ is V_1, which is much smaller than V_S, as shown in Fig. B.23(d). The output voltage starts to decay exponentially toward zero before reaching its maximum value. Thus, the output voltage will not be a true representative of the input voltage. However, the output voltage is approximately the time integration of the input voltage, and the circuit behaves as an integrator. That is,

$$v_O(t) = \frac{1}{\tau}\int_0^t V_S\, dt = \frac{V_S}{\tau} t \quad (\text{for } \tau >> T)$$

For this condition, $\tau = 10T$.

EXAMPLE B.10

Using PSpice/SPICE to plot the pulse response of an *RC* circuit Use PSpice/SPICE to plot the output voltage of the circuit in Fig. B.21(a) for $\tau = 0.1$ ms, 1 ms, and 5 ms. Assume $T = 2$ ms and $v_S = V_S = 1$ V of pulse input.

SOLUTION

For $\tau = 0.1$ ms, let $C = 0.1$ μF. Then

$$R = \frac{\tau}{C} = \frac{0.1 \text{ ms}}{0.1\ \mu\text{F}} = 1\ \text{k}\Omega$$

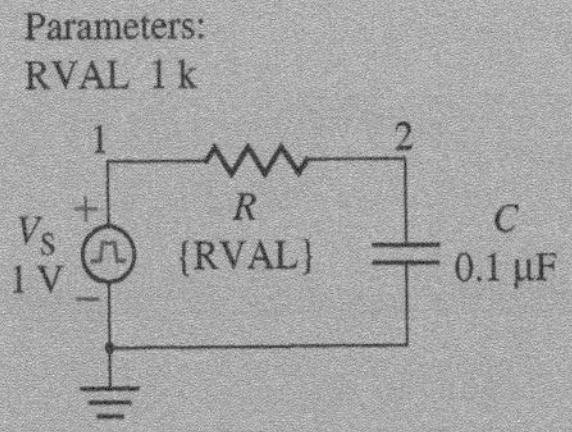

FIGURE B.24 *RC* circuit for PSpice simulation

For $\tau = 1$ ms, let $C = 0.1\ \mu$F. Then

$$R = \frac{\tau}{C} = \frac{1 \text{ ms}}{0.1\ \mu\text{F}} = 10 \text{ k}\Omega$$

For $\tau = 5$ ms, let $C = 0.1\ \mu$F. Then

$$R = \frac{\tau}{C} = \frac{5 \text{ ms}}{0.1\ \mu\text{F}} = 50 \text{ k}\Omega$$

The series *RC* circuit for PSpice simulation is shown in Fig. B.24 with a pulse input voltage. The list of the circuit file, which is run by OrCAD PSpice A/D, is as follows.

```
Example B.10  Pulse Response of Series RC Circuit
.PARAM RVAL=1K
.STEP PARAM RVAL LIST  1K  10K  50K
VS  1  0  PULSE (0V  1V  0  1NS  1NS  2MS  4MS)  ; Pulse input voltage
R  1  2  {RVAL}
C  2  0  0.1UF
.TRAN  0.1MS  4MS                                ; Transient analysis
.PROBE
.END
```

The PSpice plots of the output voltage $v_O(t)$ for three values of the time constant are shown in Fig. B.25. The lower the time constant τ, the faster the output voltage rises and falls.

NOTE: You can use the PSpice Parametric command for variable *R* to vary the time constant.

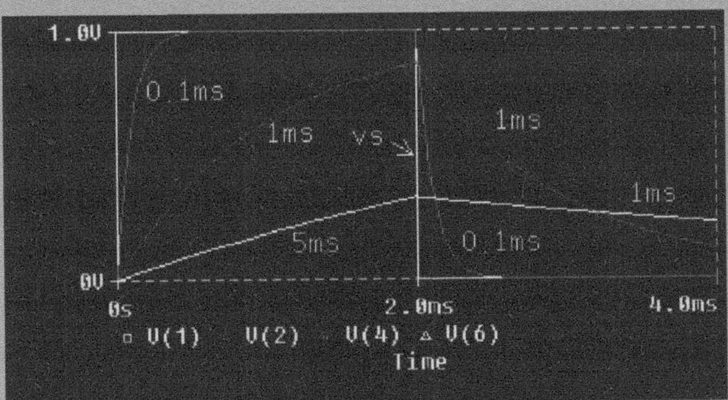

FIGURE B.25 Plots of $v_O(t)$ for Example B.10

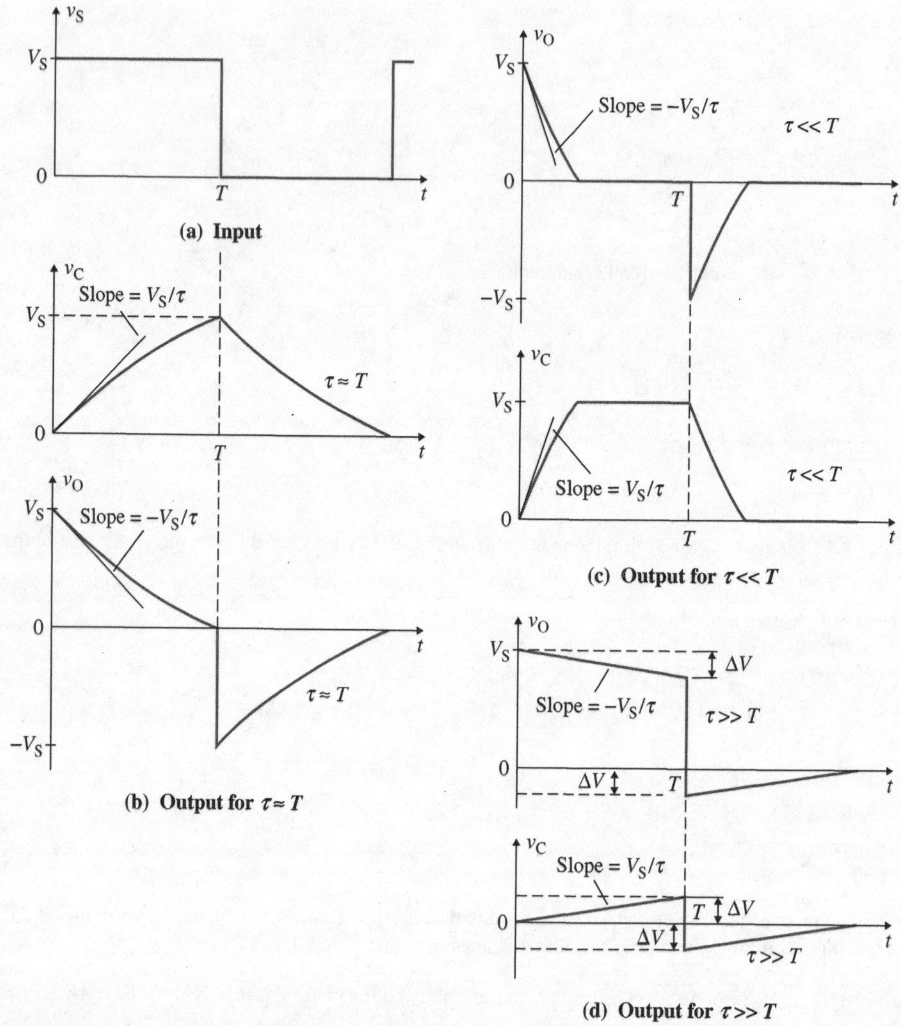

FIGURE B.26 Pulse response of a series *CR* circuit

B.8.4 Pulse Response of Series *CR* Circuits

An input pulse v_S of duration T, shown in Fig. B.26(a), is applied to the circuit of Fig. B.22(a). The response due to a pulse signal will depend on the ratio of time constant τ to duration T. Let us consider three cases: $\tau = T$, $\tau << T$, and $\tau >> T$.

In case 1, $\tau = T$, the capacitor voltage $v_C(t)$ starts to increase exponentially while the output voltage $v_O(t)$ starts to decay exponentially from V_S. This situation is shown in Fig. B.26(b). At $t = T$, the input signal v_S falls to zero and the capacitor discharges exponentially through the resistance R and the input source v_S. The output voltage v_O decays exponentially from a negative value toward zero.

In case 2, $\tau << T$, the output voltage $v_O(t)$ decays exponentially with a small time constant to zero. During the period $0 \le t \le T$, the capacitor charges exponentially to reach a steady-state value of V_S. For $t > T$, the capacitor discharges exponentially with a small time constant through the resistance R and the input source v_S. The output voltage $v_O(t)$ decays exponentially from a negative value toward zero. The waveforms for $v_O(t)$ and $v_C(t)$ are shown in Fig. B.26(c).

In case 3, $\tau >> T$, output voltage v_O falls only a small amount. The portion of the exponential curve v_O from $t = 0$ to $t = T$ will be almost linear, as shown in Fig. B.26(d). The fall in the output voltage v_O can be found approximately from Fig. B.26(d) as

$$\Delta V = \frac{V_S}{\tau} T \tag{B.41}$$

The *sag* S of the output voltage is defined as

$$S = \frac{\Delta V}{V_S} = \frac{V_S T/\tau}{V_S} = \frac{T}{\tau} = \frac{T}{RC} \tag{B.42}$$

EXAMPLE B.11

Using PSpice/SPICE to plot the pulse response of a *CR* circuit Use PSpice/SPICE to plot the output voltage $v_O(t)$ of the circuit in Fig. B.22(a) for $\tau = 0.1$ ms, 1 ms, and 5 ms. Assume $T = 2$ ms and $v_S = V_S = 1$ V of pulse input.

SOLUTION

For $\tau = 0.1$ ms, let $C = 0.1$ μF. Then

$$R = \frac{\tau}{C} = \frac{0.1\text{ ms}}{0.1\ \mu\text{F}} = 1\text{ k}\Omega$$

For $\tau = 1$ ms, let $C = 0.1$ μF. Then

$$R = \frac{\tau}{C} = \frac{1\text{ ms}}{0.1\ \mu\text{F}} = 10\text{ k}\Omega$$

For $\tau = 5$ ms, let $C = 0.1$ μF. Then

$$R = \frac{\tau}{C} = \frac{5\text{ ms}}{0.1\ \mu\text{F}} = 50\text{ k}\Omega$$

The series *CR* circuit for PSpice simulation is shown in Fig. B.27.

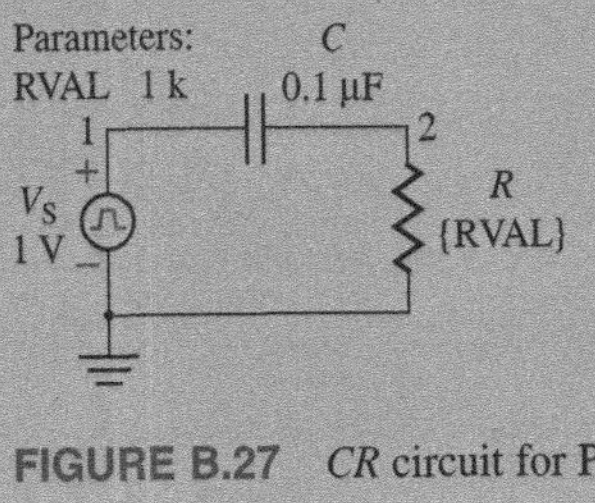

FIGURE B.27 *CR* circuit for PSpice simulation

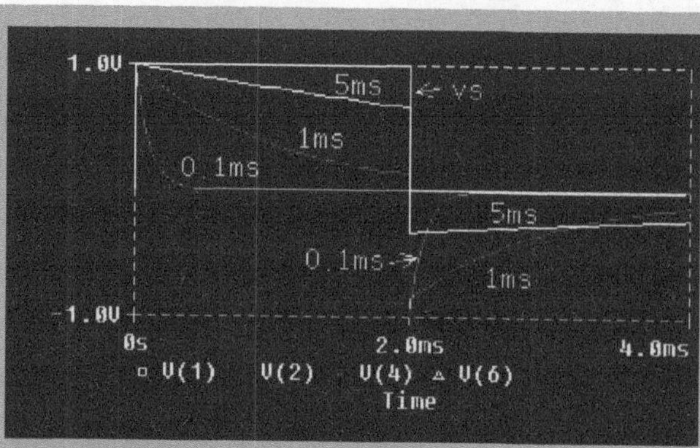

FIGURE B.28 Plots of $v_O(t)$ for Example B.11

The list of the circuit file is as follows.

```
Example B.11   Pulse Response of Series CR Circuit
.PARAM RVAL=1K
.STEP PARAM RVAL LIST  1K  10K  50K
VS  1  0  PULSE (0V  1V  0  1NS  1NS  2MS  4MS)     ; Pulse input voltage
R  2  0  {RVAL}
C  1  2  0.1UF
.TRAN  0.1MS  4MS                                    ; Transient analysis
.PROBE
.END
```

The plots of the output voltage $v_O(t)$ for three values of the time constant are shown in Fig. B.28.

EXAMPLE B.12

Pulse response of a parallel *RC* circuit A constant-current source $i_S = I_S$, shown in Fig. B.29(a), feeds a parallel *RC* circuit with $C = 0.1$ μF and $R = 100$ kΩ, as shown in Fig. B.29(b). The input is a pulse current of duration $T = 0.5$ ms. Determine **(a)** the instantaneous current $i_C(t)$ through capacitance *C*, **(b)** the instantaneous current $i_R(t)$ through resistance *R*, and **(c)** the sag *S* of the capacitor current.

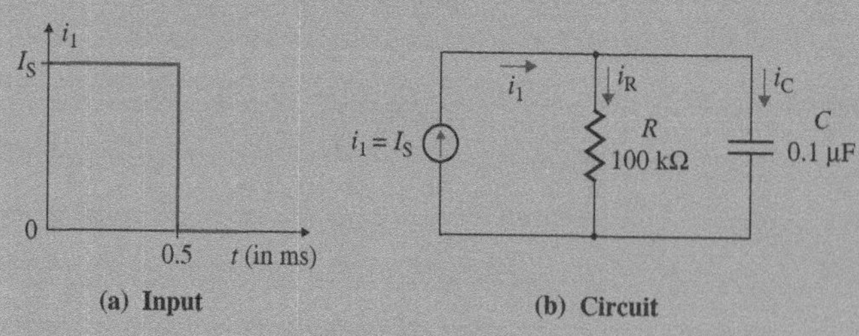

FIGURE B.29 Parallel *RC* circuit with constant current source I_S

SOLUTION

For a pulse signal, the input current in Laplace's domain is $I_S(s) = I_S/s$.

(a) Using the current divider rule, we can find the capacitor current I_C in Laplace's domain:

$$I_C(s) = \frac{R}{R + 1/Cs} I_S(s) = \frac{s}{s + 1/RC} I_S(s) = \frac{s}{s + 1/RC} \times \frac{I_S}{s} = \frac{I_S}{s + 1/RC}$$

$$= I_S \frac{I}{s + 1/RC} \tag{B.43}$$

The inverse transform of $I_C(s)$ in Eq. (B.43) gives

$$i_C(t) = I_S e^{-t/\tau} \tag{B.44}$$

where $\tau = RC$.

(b) The instantaneous current $i_R(t)$ through resistance R is

$$i_R(t) = I_S - i_C(t) = I_S(1 - e^{-t/\tau}) \tag{B.45}$$

(c) $\tau = RC = 100 \times 10^3 \times 0.1 \times 10^{-6} = 10$ ms, and $T = 0.5$ ms. Therefore, $\tau >> T$, and Eq. (B.42) gives

$$S = T/\tau = 0.5/10 = 5\%$$

B.8.5 Step Response of Series *RL* Circuits

A series RL circuit with a step-function input is shown in Fig. B.30(a). The output voltage v_O is taken across inductance L. The current i through the inductor can be deduced from

$$V_S = v_L + v_R = L\frac{di}{dt} + Ri \tag{B.46}$$

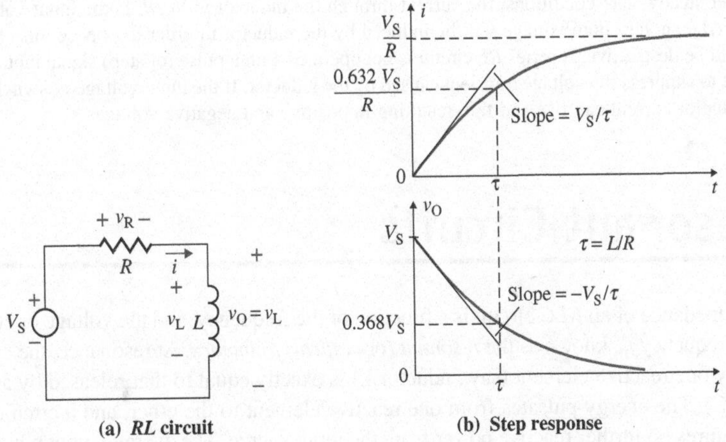

(a) *RL* circuit (b) Step response

FIGURE B.30 Step response of a series *RL* circuit

with an initial inductor current of zero, $i(t = 0) = 0$. In Laplace's domain of s, Eq. (B.46) becomes

$$\frac{V_S}{s} = LsI(s) + RI(s)$$

which, solved for $I(s)$, gives

$$I(s) = \frac{V_S}{s(sL + R)} = \frac{V_S}{Ls(s + 1/\tau)} = \frac{V_S}{R}\left[\frac{1}{s} - \frac{1}{(s + 1/\tau)}\right] \tag{B.47}$$

where $\tau = L/R$ is the *time constant* of an *RL* circuit. Taking the inverse transform of $I(s)$ in Eq. (B.47) gives the instantaneous current as

$$i(t) = \frac{V_S}{R}(1 - e^{-t/\tau}) \tag{B.48}$$

Using Eq. (B.48), we can find the voltage $v_O(t)$ across the inductance L:

$$v_O(t) = v_L(t) = L\frac{di}{dt} = V_S e^{-t/\tau} \tag{B.49}$$

In the steady state (at $t = \infty$),

$$v_O(t) = 0 \qquad \text{[from Eq. (B.49)]}$$

$$i(t) = \frac{V_S}{R} \qquad \text{[from Eq. (B.48)]}$$

If the output is taken across resistance R, the output voltage $v_O(t)$ becomes

$$v_O(t) = v_R(t) = Ri(t) = V_S(1 - e^{-t/\tau}) \tag{B.50}$$

In the steady state (at $t = \infty$), $v_R(t) = V_S$ and $i(t) = V_S/R$.

▶ **NOTE** Under steady-state conditions, the current through the inductor is V_S/R. If the input voltage v_S is turned off with a sharp edge, a very high voltage will be induced by the inductor in order to oppose this change of current. This voltage could be destructive. A series *RL* circuit is not operated with a pulse (or step) signal input unless there is a protection circuit to suppress the voltage transient caused by the inductor. If the input voltage v_S switches between $+V_S$ and $-V_S$, the inductor current can rise and fall, resulting in positive and negative voltages.

B.9 Resonant Circuits

The effective impedance of an *RLC* circuit is a function of the frequency, and the voltage or current becomes maximum at a frequency f_n, known as the *resonant* (or *natural*) *frequency*. At resonance, the energy absorbed at any instant by one reactive element (say, inductor L) is exactly equal to that released by another element (say, capacitor C). The energy pulsates from one reactive element to the other, and a circuit with no resistive element requires no further reactive power from the input source. The average input power, which is the power dissipated in the resistive element, becomes maximum at resonance. Resonant circuits are of two types: series resonant circuits and parallel resonant circuits.

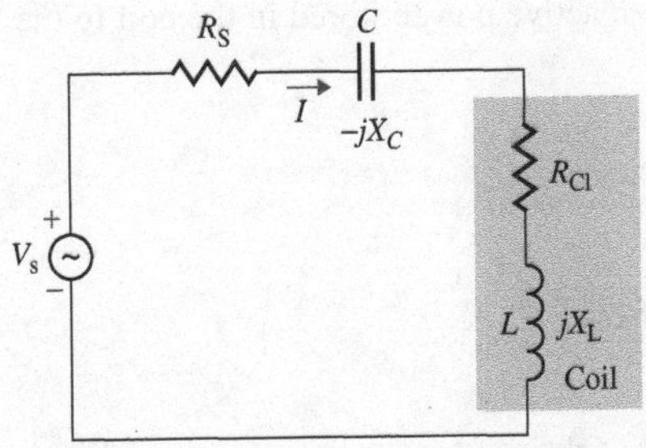

FIGURE B.31 Series resonant *RLC* circuit

B.9.1 Series Resonant Circuits

A series resonant *RLC* circuit is shown in Fig. B.31, where R_{Cl} is the internal resistance of the coil and R_S is the source resistance. If we define $R = R_{Cl} + R_S$, the total series impedance Z of the circuit is given by

$$Z = R + j(X_L - X_C) \tag{B.51}$$

The series resonance will occur at $f = f_n$ when

$$X_L = X_C \tag{B.52}$$

Equation (B.52) can also be stated as

$$\omega L = \frac{1}{\omega C} \quad \text{or} \quad 2\pi f_n L = \frac{1}{(2\pi f_n C)}$$

which gives the series resonant frequency f_n as

$$f_n = \frac{1}{2\pi\sqrt{LC}} \tag{B.53}$$

The impedance Z_n at resonance becomes

$$Z_n = Z = R \tag{B.54}$$

A series resonant circuit is normally defined by a *quality factor* Q_s, which is defined as the ratio of the reactive power stored in either the inductor or the capacitor to the average power dissipated in the resistor at resonance; that is,

$$Q_s = \frac{\text{Reactive power}}{\text{Average power}}$$

$$= \begin{cases} \dfrac{I^2 X_L}{I^2 R} = \dfrac{X_L}{R} = \dfrac{2\pi f_n L}{R} & \text{(for an inductive reactance)} \quad \text{(B.55)} \\[2ex] \dfrac{I^2 X_C}{I^2 R} = \dfrac{X_C}{R} = \dfrac{1}{2\pi f_n C R} & \text{(for a capacitive reactance)} \quad \text{(B.56)} \end{cases}$$

The quality factor Q_{Cl} of a coil is defined as the ratio of the reactive power stored in the coil to the power dissipated in the coil resistance R_{Cl}; that is,

$$Q_{Cl} = \frac{\text{Reactive power}}{\text{Power dissipated}} = \frac{X_L}{R_{Cl}}$$

The rms voltage V_L across inductor L at resonance can be found from

$$V_L = \frac{X_L V_s}{Z_n} = \frac{X_L V_s}{R} = Q_s V_s \tag{B.57}$$

The rms voltage V_C across capacitor C at resonance can be found from

$$V_C = \frac{X_C V_s}{Z_n} = \frac{X_L V_s}{R} = Q_s V_s \tag{B.58}$$

In many electronic circuits, the quality factor Q_s is high, in the range of 80 to 400. If $V_s = 30$ V and $Q_s = 80$, for example, then $V_C = V_L = 80 \times 30 = 2400$ V, and all electronic devices in the circuit will be subjected to this high voltage. Thus, a designer must be careful to protect the circuit from a high voltage across the inductor or the capacitor of a resonant circuit.

B.9.2 Parallel Resonant Circuits

A parallel resonant *RLC* circuit is shown in Fig. B.32(a). This circuit is also known as a *tank circuit*. The input signal to tank circuit is usually a current source. This type of circuit is frequently used with active devices such as transistors, which have the characteristic of a constant current source. By replacing the series *RL* combination of the inductor with a parallel combination, we can obtain the circuit in Fig. B.32(b), for which the admittance Y_{RL} is

$$Y_{RL} = \frac{1}{R_{Cl} + jX_L} = \frac{R_{Cl}}{R_{Cl}^2 + X_L^2} - j\frac{X_L}{R_{Cl}^2 + X_L^2} = \frac{1}{R_p} - j\frac{1}{X_p}$$

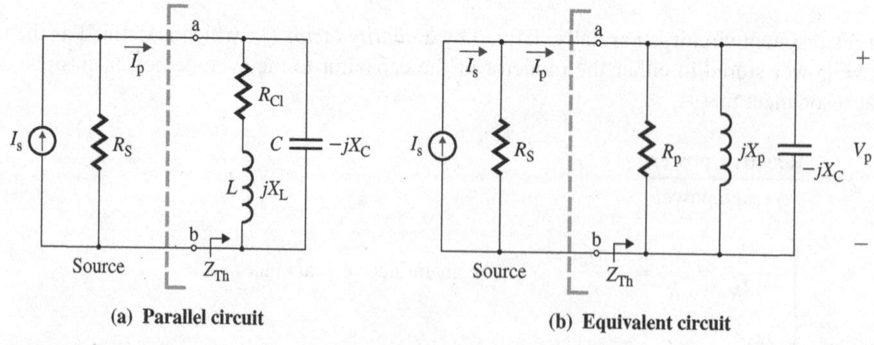

FIGURE B.32 Parallel resonant *RLC* circuit

$$\text{where} \quad R_p = \frac{R_{C1}^2 + X_L^2}{R_{C1}} \tag{B.59}$$

$$X_p = \frac{R_{C1}^2 + X_L^2}{X_L} \tag{B.60}$$

For the resonant condition,

$$X_p = X_C$$

Substituting X_p from Eq. (B.60) into the preceding equation, we get

$$\frac{R_{C1}^2 + X_L^2}{X_L} = X_C$$

$$R_{C1}^2 + X_L^2 = X_C X_L$$

$$X_L^2 = X_C X_L - R_{C1}^2$$

or

$$X_L^2 = \frac{L}{C} - R_{C1}^2$$

$$X_L = \left[\frac{L}{C} - R_{C1}^2\right]^{1/2} \tag{B.61}$$

which gives the parallel resonant frequency f_p as

$$f_p = \frac{1}{2\pi L}\left[\frac{L}{C} - R_{C1}^2\right]^{1/2} = \frac{1}{2\pi\sqrt{LC}}\left[1 - \frac{CR_{C1}^2}{L}\right]^{1/2} \tag{B.62}$$

$$= f_n\left[1 - \frac{CR_{C1}^2}{L}\right]^{1/2} \tag{B.63}$$

Thus, the parallel resonant frequency f_p, which is dependent on the coil resistance R_{Cl}, is less than the series resonant frequency f_n. For the conditions $(CR_{Cl}^2/L) << 1$ or $R_{Cl} << \sqrt{L/C}$, and for $R_{Cl} = 0$, Eq. (B.63) gives

$$f_p = f_n$$

The quality factor Q_p of the parallel resonant *RLC* circuit can be determined by the ratio of reactive power to the real power at resonance; that is,

$$Q_p = \frac{V_p^2/X_p}{V_p^2/(R_S \parallel R_p)} = \frac{R_S \parallel R_p}{X_p} \tag{B.64}$$

where V_p is the voltage across the parallel branches.

EXAMPLE B.13

Finding the parallel resonant frequency The parameters of the parallel resonant *RLC* circuit in Fig. B.32(a) are $R_{C1} = 47\ \Omega$, $L = 5$ mH, $C = 50$ pF, $R_S = 20$ kΩ, and the current source $I_s = 6$ mA. Calculate **(a)** the parallel resonant frequency f_p, **(b)** the voltage V_p across the resonant circuit at resonance, **(c)** the quality factor Q_{C1} of the coil, and **(d)** the quality factor Q_p of the resonant circuit.

SOLUTION

$R_{C1} = 47\ \Omega$, $L = 5$ mH, $C = 50$ pF, $R_S = 20$ kΩ, and $I_s = 6$ mA.

(a) From Eq. (B.53),

$$f_n = \frac{1}{2\pi \times \sqrt{5 \times 10^{-3} \times 50 \times 10^{-12}}} = 318.3 \text{ kHz}$$

From Eq. (B.63),

$$f_p = 318.3 \times 10^3 \times \left[1 - 50 \times 10^{-12} \times \frac{47^2}{5 \times 10^{-3}}\right]^{1/2} \approx 318.3 \text{ kHz}$$

(b) We know that

$$X_L = 2\pi f_p L = 2\pi \times 318.3 \times 10^3 \times 5 \times 10^{-3} = 9999.7\ \Omega$$

From Eq. (B.59), the effective resistance R_p of the parallel circuit is

$$R_p = \frac{47^2 + 9999.7^2}{47} = 2127.6 \text{ k}\Omega$$

Using the current divider rule, the rms current I_p through the parallel circuit is

$$I_p = \frac{R_S}{R_S + R_p} I_s = \frac{20 \text{ k}\Omega \times 6 \text{ mA}}{20 \text{ k}\Omega + 2127.6 \text{ k}\Omega} = 55.876\ \mu\text{A}$$

$$V_p = I_p R_p = 55.876\ \mu\text{A} \times 2127.6 \text{ k}\Omega = 118.88 \text{ V}$$

(c) We have

$$Q_{C1} = \frac{X_L}{R_{C1}} = \frac{9999.7}{47} = 212.8$$

(d) From Eq. (B.60), the effective inductive reactance X_p of the parallel circuit is

$$X_p = \frac{47^2 + 9999.7^2}{9999.7} = 9999.92\ \Omega$$

$$R_S \| R_p = \frac{20 \times 2127.6}{20 + 2127.6} = 19.81 \text{ k}\Omega$$

From Eq. (B.64), $Q_p = 19{,}810/9999.92 = 1.98$.

B.10 Frequency Response of First- and Second-Order Circuits

A sine wave (or sinusoid) is generally used to characterize electronic circuits, such as amplifiers and filters. The frequency response refers to the output characteristic for a sine-wave input. If a sine-wave input voltage

$$v_s(t) = V_m \sin \omega t \quad \text{(B.65)}$$

where V_m is the peak input voltage and ω is the frequency of the input voltage in radians per second, is applied to a circuit, the output voltage $v_o(t)$ can have a different amplitude and phase than the input voltage. The output voltage $v_o(t)$ will be of the form

$$v_o(t) = V_p \sin(\omega t - \phi) \quad \text{(B.66)}$$

where V_p is the peak output voltage. If f is the input frequency in hertz,

$$\omega = 2\pi f$$

A typical relationship between the input and output voltages of an amplifier is shown in Fig. B.33. An amplifier has in general a voltage gain of $A_v = V_p/V_m > 1$. Thus, the peak output V_p has a higher value than the input voltage V_m.

If $V_s(j\omega)$ and $V_o(j\omega)$ denote the rms values of input voltage and output voltage, respectively, as a function of frequency, the voltage gain $G(j\omega)$ is defined as

$$G(j\omega) = \frac{V_o(j\omega)}{V_s(j\omega)} \quad \text{(dimensionless)} \quad \text{(B.67)}$$

$G(j\omega)$ is a complex function with a magnitude and a phase. The magnitude $|G(j\omega)|$ gives the magnitude response, and the phase of $G(j\omega)$ gives the phase response. The magnitude and the phase are generally plotted against the frequency, with a logarithmic scale used for the frequency. The magnitude $|G(j\omega)|$ is normally expressed in decibels (dB):

$$\text{Magnitude in dB} = 20 \log_{10} |G(j\omega)|$$

We will review the frequency responses of the following circuits: first-order low-pass *RC* circuits, first-order high-pass *CR* circuits, second-order series *RLC* circuits, and second-order parallel *RLC* circuits.

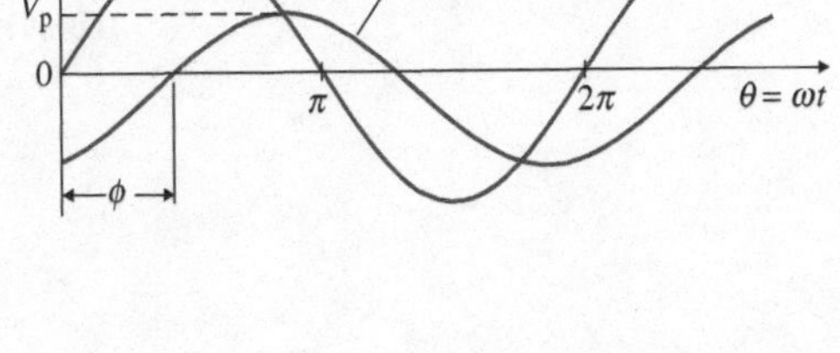

FIGURE B.33 Typical sinusoidal input and output voltages

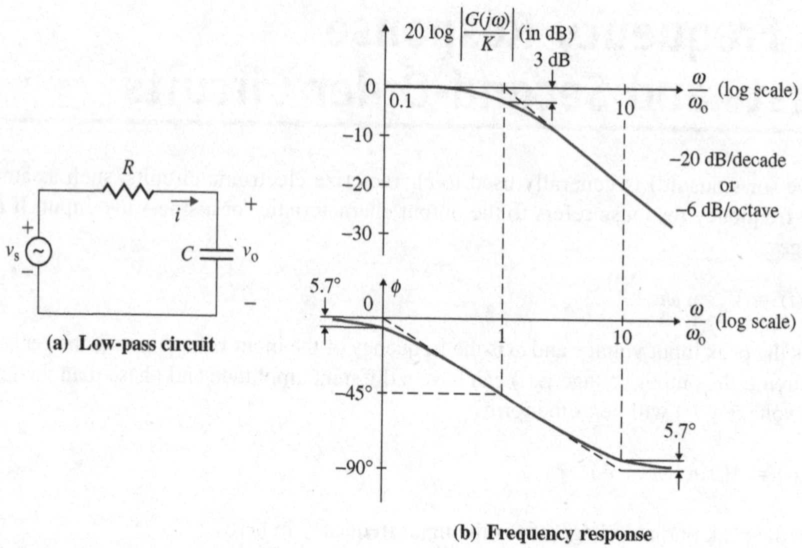

FIGURE B.34 First-order low-pass *RC* circuit

B.10.1 First-Order Low-Pass *RC* Circuits

A typical low-pass *RC* circuit is shown in Fig. B.34(a). The output voltage v_o is taken across capacitance *C*. The capacitor impedance in Laplace's domain is $1/Cs$. Using the voltage divider rule, we can find the voltage gain $G(s)$:

$$G(s) = \frac{V_o(s)}{V_S(s)} = \frac{1/Cs}{R + (1/Cs)} = \frac{1}{1 + sRC} \quad \text{(dimensionless)}$$

In the frequency domain, $s = j\omega$, and

$$G(j\omega) = \frac{1}{1 + j\omega RC} = \frac{1}{1 + j\omega\tau}$$

where $\tau = RC$. Thus, the magnitude $|G(j\omega)|$ of the voltage gain can be found from

$$|G(j\omega)| = \frac{1}{[1 + (\omega\tau)^2]^{1/2}} = \frac{1}{[1 + (\omega/\omega_o)^2]^{1/2}} \tag{B.68}$$

and the phase angle ϕ of $G(j\omega)$ is given by

$$\phi = -\tan^{-1}(\omega\tau) = -\tan^{-1}\left(\frac{\omega}{\omega_o}\right) \tag{B.69}$$

where $\omega_o = 1/RC = 1/\tau$.

For $\omega << \omega_o$,

$$|G(j\omega)| \approx 1$$

$$20 \log_{10}|G(j\omega)| = 0$$

$$\phi = 0$$

Therefore, at low frequency, the magnitude plot is a straight horizontal line at 0 dB. For $\omega >> \omega_o$,

$$|G(j\omega)| \approx \omega_o/\omega$$

$$20 \log_{10}|G(j\omega)| = 20 \log_{10}(\omega_o/\omega)$$

$$\phi = \pi/2$$

For $\omega = \omega_o$,

$$|G(j\omega)| = 1/\sqrt{2}$$

$$20 \log_{10}|G(j\omega)| = 20 \log_{10}(1/\sqrt{2}) = -3 \text{ dB}$$

$$\phi = \pi/4$$

Let us consider a high frequency such that $\omega_1 >> \omega_o$. At $\omega = \omega_1$, the magnitude is $20 \log_{10}(\omega_o/\omega_1)$. At $\omega = 10\omega_1$, the magnitude is $20 \log_{10}(\omega_o/10\omega_1)$. The change in magnitude between $\omega = \omega_1$ and $\omega = 10\omega_1$ becomes

$$20 \log_{10}(\omega_o/10\omega_1) - 20 \log_{10}(\omega_o/\omega_1) = 20 \log_{10}(1/10) = -20 \text{ dB}$$

If the frequency is doubled so that $\omega = 2\omega_1$, the change in magnitude becomes

$$20 \log_{10}(\omega_o/2\omega_1) - 20 \log_{10}(\omega_o/\omega_1) = 20 \log_{10}(1/2) = -6 \text{ dB}$$

The frequency response is shown in Fig. B.34(b). If the frequency doubled, the interval between the two frequencies is called an *octave* on the frequency axis. If the frequency is increased by a factor of 10, the interval between the two frequencies is called a *decade*. Thus, for a decade increase in frequency, the magnitude changes by −20 dB. The magnitude plot is a straight line with a slope of −20 dB/decade or −6 dB/octave. The magnitude curve is therefore defined by two straight-line asymptotes, which meet at the *corner frequency* (or *break frequency*) ω_o. The difference between the actual magnitude curve and the asymptotic curve is largest at the break frequency. The error can be found by finding the gain at $\omega = \omega_o$; that is, $|G(j\omega)| = 1/\sqrt{2}$, and $20 \log_{10}(1/\sqrt{2}) = -3$ dB. This error is symmetrical with respect to the break frequency. The break frequency is also known as the *3-dB frequency*.

The circuit in Fig. B.34(a) passes only the low-frequency signal, and the amplitude falls at higher frequencies. A circuit with this type of response is known as a *low-pass circuit*. The gain function (commonly known as the transfer function) of a low-pass circuit has the general form

$$G(s) = \frac{K}{1 + (s/\omega_o)} \tag{B.70}$$

where K is the magnitude of the gain function at $\omega = 0$ (or the *low-pass gain*). A low-pass circuit exhibits the characteristics of (a) finite output at a very low frequency, tending to zero, and (b) zero output at a very high frequency, tending to infinity.

EXAMPLE B.14

Using PSpice/SPICE to plot the frequency response of a low-pass *RC* circuit Use PSpice/SPICE to plot the frequency response of the low-pass *RC* circuit in Fig. B.34(a). Assume $V_m = 1$ V (peak AC), $R = 10$ kΩ, and $C = 0.1$ μF. The frequency f varies from 1 Hz to 100 kHz.

SOLUTION

The low-pass *RC* circuit for PSpice simulation is shown in Fig. B.35. The PSpice plot of the magnitude and phase angle are shown in Fig. B.36, which gives $f_o = 161$ Hz at -3 dB.

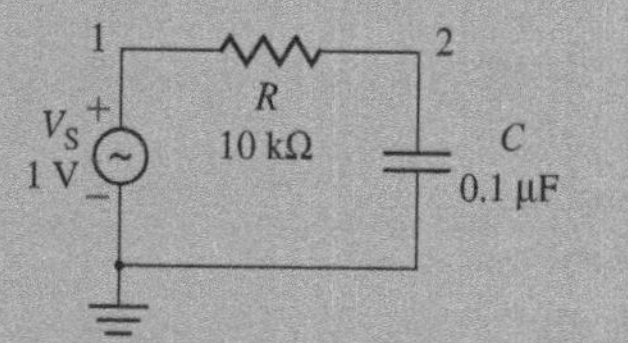

FIGURE B.35 Low-pass *RC* circuit for PSpice simulation

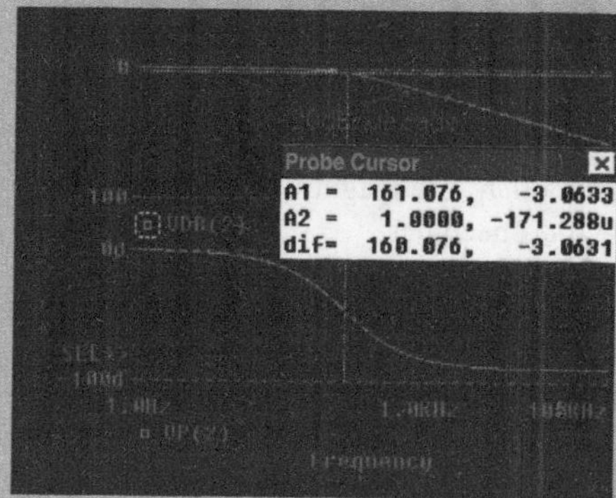

FIGURE B.36 Frequency response plots for Example B.14

B.10.2 First-Order High-Pass *CR* Circuits

A high-pass *CR* circuit is shown in Fig. B.37(a). The output voltage v_o is taken across resistance R. Using the voltage divider rule, we can find the voltage gain $G(s)$ in Laplace's domain:

$$G(s) = \frac{V_o(s)}{V_s(s)} = \frac{R}{R + (1/Cs)} = \frac{sRC}{1 + sRC} \quad \text{(dimensionless)}$$

In the frequency domain, $s = j\omega$, and

$$G(j\omega) = \frac{j\omega RC}{1 + j\omega RC} = \frac{j\omega\tau}{1 + j\omega\tau}$$

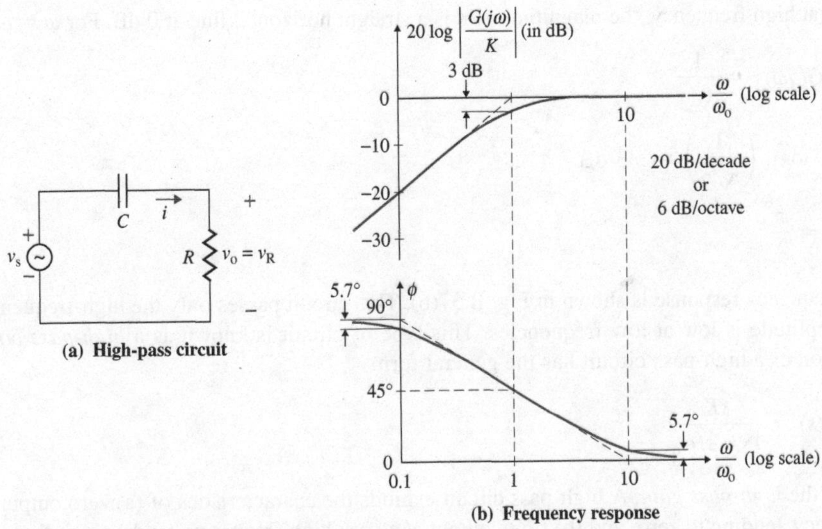

FIGURE B.37 First-order high-pass *CR* circuit

where $\tau = RC$. Thus, the magnitude $|G(j\omega)|$ of the voltage gain can be found from

$$|G(j\omega)| = \frac{\omega\tau}{[1 + (\omega\tau)^2]^{1/2}} = \frac{\omega/\omega_o}{[1 + (\omega/\omega_o)^2]^{1/2}} \tag{B.71}$$

and the phase angle ϕ of $G(j\omega)$ is given by

$$\phi = \frac{\pi}{2} - \tan^{-1}\left(\frac{\omega}{\omega_o}\right) \tag{B.72}$$

where $\omega_o = 1/RC = 1/\tau$.

For $\omega << \omega_o$,

$$|G(j\omega)| \approx \frac{\omega}{\omega_o}$$

$$20\log_{10}|G(j\omega)| = 20\log_{10}\left(\frac{\omega}{\omega_o}\right)$$

$$\phi = \frac{\pi}{2}$$

Therefore, for a decade increase in frequency, the magnitude changes by 20 dB. The magnitude plot is a straight line with a slope of +20 dB/decade or +6 dB/octave.

For $\omega >> \omega_o$,

$$|G(j\omega)| = 1$$

$$20\log_{10}|G(j\omega)| = 0$$

$$\phi \approx 0$$

Therefore, at high frequency, the magnitude plot is a straight horizontal line at 0 dB. For $\omega = \omega_o$,

$$|G(j\omega)| = \frac{1}{\sqrt{2}}$$

$$20\log_{10}\left(\frac{1}{\sqrt{2}}\right) = -3 \text{ dB}$$

$$\phi = \frac{\pi}{4}$$

The frequency response is shown in Fig. B.37(b). This circuit passes only the high-frequency signal, and the amplitude is low at low frequencies. This type of circuit is known as a *high-pass circuit*. The gain function of a high-pass circuit has the general form

$$G(s) = \frac{sK}{1 + s/\omega_o} \tag{B.73}$$

where K is the *high-pass gain*. A high-pass circuit exhibits the characteristics of (a) zero output at a very low frequency, tending to zero, and (b) finite output at a very high frequency, tending to infinity.

EXAMPLE B.15

Using PSpice/SPICE to plot the frequency response of a high-pass *CR* circuit Use PSpice/SPICE to plot the frequency response of the high-pass *CR* circuit in Fig. B.37(a). Assume $V_m = 1$ V (peak AC), $R = 10$ kΩ, and $C = 0.1$ µF. The frequency f varies from 1 Hz to 100 kHz.

SOLUTION

The high-pass *CR* circuit for PSpice simulation is shown in Fig. B.38. The PSpice plots of the magnitude and phase angle are shown in Fig. B.39, which gives $f_o = 157$ Hz at -3 dB.

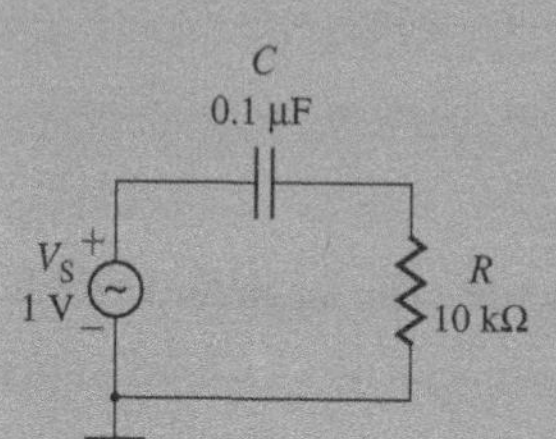

FIGURE B.38 High-pass *CR* circuit for PSpice simulation

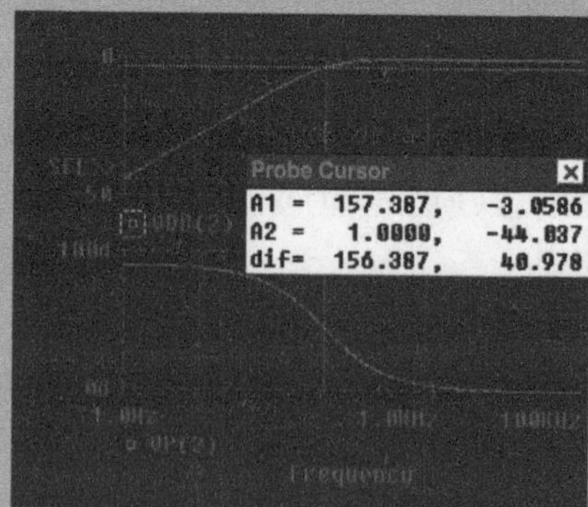

FIGURE B.39 Frequency response plots for Example B.15

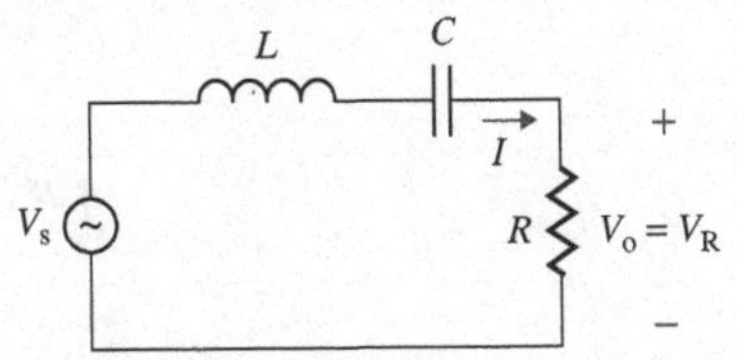

FIGURE B.40 Series *RLC* circuit

B.10.3 Second-Order Series *RLC* Circuits

A series *RLC* circuit is shown in Fig. B.40. The output voltage v_o is taken across resistance R. Using the voltage divider rule, we can find the voltage gain (or the transfer function) in Laplace's domain of s:

$$G(s) = \frac{V_o(s)}{V_s(s)} = \frac{R}{R + sL + (1/Cs)} = \frac{sR/L}{s^2 + (sR/L) + (1/LC)} \quad \text{(dimensionless)} \tag{B.74}$$

Defining $\omega_n = 1/\sqrt{LC}$ as the *natural frequency* in rad/s and $\alpha = R/(2L)$ as the *damping factor*, we can write Eq. (B.74) as

$$G(s) = \frac{2\alpha s}{s^2 + 2\alpha s + \omega_n^2} \tag{B.75}$$

Let us define

$$\delta = \frac{\alpha}{\omega_n} = \frac{R}{2L}\sqrt{LC} = \frac{R}{2}\sqrt{\frac{C}{L}}$$

as the *damping ratio*. Then Eq. (B.75) becomes

$$G(s) = \frac{2\delta\omega_n s}{s^2 + 2\delta\omega_n s + \omega_n^2} \quad \text{(dimensionless)} \tag{B.76}$$

where $\delta < 1$. (Note that δ is not necessarily less than 1, but has been assumed at less than 1 for this discussion.) In the frequency domain, $s = j\omega$. Thus,

$$G(j\omega) = \frac{2\delta\omega_n j\omega}{(j\omega)^2 + 2\delta\omega_n(j\omega) + \omega_n^2} = \frac{j2\delta\omega/\omega_n}{-(\omega/\omega_n)^2 + (j2\delta\omega/\omega_n) + 1}$$

$$= \frac{j2\delta\omega/\omega_n}{1 + (j2\delta\omega/\omega_n) - (\omega/\omega_n)^2} \tag{B.77}$$

Let us define $u = \omega/\omega_n$ as the *frequency ratio* or the *normalized frequency*. Then Eq. (B.77) can be simplified to

$$G(j\omega) = \frac{j2\delta u}{1 + j2\delta u - u^2}$$

The magnitude $|G(j\omega)|$ can be found from

$$|G(j\omega)| = \frac{2\delta u}{[(1 - u^2)^2 + (2\delta u)^2]^{1/2}} \tag{B.78}$$

The phase angle ϕ of $G(j\omega)$ can be found from

$$\phi = \frac{\pi}{2} - \tan^{-1}\left(\frac{2\delta u}{1 - u^2}\right) \tag{B.79}$$

For low frequencies $u << 1$,

$$|G(j\omega)| \approx 2\delta u$$

$$20 \log_{10} |G(j\omega)| \approx 20 \log_{10} (2\delta u)$$

$$\phi \approx \frac{\pi}{2}$$

Therefore, at low frequencies, the magnitude plot is a straight line with a slope of +20 dB/decade or +6 dB/octave. For $u = 1$, $|G(j\omega)| \approx 1$ only if

$$\delta \approx 1$$

$$20 \log_{10} |G(j\omega)| = 0 \text{ dB}$$

$$\phi = 0$$

For $u >> 1$,

$$|G(j\omega)| \approx \frac{2\delta u}{u^2} = \frac{2\delta}{u}$$

$$20 \log_{10} |G(j\omega)| \approx 20 \log_{10} (2\delta) - 20 \log_{10} (u) \approx -20 \log_{10} (u)$$

$$\phi \approx -\frac{\pi}{2}$$

Therefore, at high frequencies, the magnitude plot is a straight line with a slope of −20 dB/decade or −6 dB/octave. The actual characteristic will differ considerably from the asymptotic lines, and the error will depend on the damping ratio δ. The magnitude and frequency plots of the series *RLC* circuit are shown in Fig. B.41. The magnitude peaks at $u = 1$.

If the output voltage of the series *RLC* circuit falls below 70% of its maximum value, the output is not considered the significant value. The *cutoff frequency* is defined as that value of frequency for which the magnitude of the gain drops to 70.7% of its maximum value $|G(j\omega)|_{max} = 1$. Thus, at cutoff frequencies, Eq. (B.78) gives

$$|G(j\omega)| = \frac{2\delta u}{[(1 - u^2)^2 + (2\delta u)^2]^{1/2}} = 0.707 = \frac{1}{\sqrt{2}}$$

or $$\sqrt{2}(2\delta u) = [(1 - u^2)^2 + (2\delta u)^2]^{1/2}$$

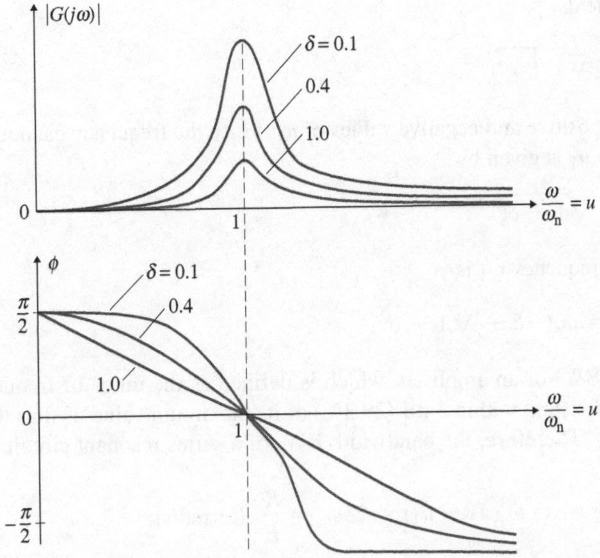

FIGURE B.41 Frequency response of a series *RLC* circuit

Squaring both sides yields

$$2(2\delta u)^2 = (1 - u^2)^2 + (2\delta u)^2$$

or

$$(2\delta u)^2 = (1 - u^2)^2 \tag{B.80}$$

The possible solutions of Eq. (B.80) are

$$2\delta u_1 = 1 - u_1^2$$

$$u_1^2 + 2\delta u_1 - 1 = 0 \tag{B.81}$$

$$2\delta u_2 = -(1 - u_2^2) = u_2^2 - 1$$

$$u_2^2 - 2\delta u_2 - 1 = 0 \tag{B.82}$$

Solving Eq. (B.82) yields

$$u_2 = \delta \pm \sqrt{1 + \delta^2}$$

Since the frequency cannot be negative, the upper cutoff frequency ratio u_2 is given by

$$u_2 = \delta + \sqrt{1 + \delta^2} \tag{B.83}$$

and the upper cutoff frequency ω_2 is

$$\omega_2 = u_2\omega_n = \omega_n(\delta + \sqrt{1 + \delta^2}) \tag{B.84}$$

Solving Eq. (B.81) yields

$$u_1 = -\delta \pm \sqrt{1 + \delta^2}$$

which will give both positive and negative values of u_1. Since the frequency cannot be negative, the lower cutoff frequency ratio u_1 is given by

$$u_1 = -\delta + \sqrt{1 + \delta^2} \tag{B.85}$$

and the lower cutoff frequency ω_1 is

$$\omega_1 = u_1\omega_n = \omega_n(-\delta + \sqrt{1 + \delta^2}) \tag{B.86}$$

The *bandwidth* (BW) of an amplifier, which is defined as the range of frequencies over which the gain remains almost constant within 3 dB (29.3%) of its maximum value, is thus the difference between the cutoff frequencies. Therefore, the bandwidth $\mathrm{BW_s}$ of a series resonant circuit can be found from

$$\mathrm{BW_s} = \omega_2 - \omega_1 = \omega_\mathrm{n}(u_2 - u_1) = 2\delta\omega_\mathrm{n} = \frac{R}{L} \quad \text{(in rad/s)} \tag{B.87}$$

$$\mathrm{BW_s} = f_2 - f_1 = \frac{1}{2\pi}\frac{R}{L} \quad \text{(in Hz)} \tag{B.88}$$

From Eq. (B.55), $R/L = 2\pi f_\mathrm{n}/Q_\mathrm{s}$. Thus, Eq. (B.88) can be rewritten as

$$\mathrm{BW_s} = \frac{1}{2\pi}\frac{R}{L} = \frac{1}{2\pi}\frac{2\pi f_\mathrm{n}}{Q_\mathrm{s}} = \frac{f_\mathrm{n}}{Q_\mathrm{s}} \tag{B.89}$$

which shows that the larger the value of Q_s, the smaller the value of bandwidth $\mathrm{BW_s}$, and vice versa. It can be shown that Eq. (B.89) can be also applied to calculate the bandwidth $\mathrm{BW_p}$ of a parallel resonant circuit; that is,

$$\mathrm{BW_p} = \frac{f_\mathrm{p}}{Q_\mathrm{p}} \tag{B.90}$$

where f_p is the parallel resonant frequency in Eq. (B.63) and Q_p is the quality factor of a parallel resonant circuit in Eq. (B.64).

EXAMPLE B.16

Finding the frequency response of a series *RLC* circuit The series *RLC* circuit in Fig. B.40 has $R = 50\ \Omega$, $L = 4$ mH, and $C = 0.15\ \mu$F.

(a) Determine the series resonant frequency f_n, the damping ratio δ, the quality factor Q_s, the cutoff frequencies, and the bandwidth $\mathrm{BW_s}$.

(b) Use PSpice/SPICE to plot the magnitude and phase angle of the output voltage for $R = 50\ \Omega$, $100\ \Omega$, and $200\ \Omega$. The frequency f varies from 100 Hz to 1 MHz. Assume $V_\mathrm{m} = 1$ V peak AC.

SOLUTION

(a) $R = 50\ \Omega$, $L = 4$ mH, and $C = 0.15\ \mu$F, so

$$\omega_n = \frac{1}{\sqrt{LC}} = \frac{10^5}{\sqrt{4 \times 1.5}} = 40{,}825 \text{ rad/s}$$

The series resonant frequency is

$$f_n = \frac{\omega_n}{2\pi} = \frac{40{,}825}{2\pi} = 6497.5 \text{ Hz}$$

Since $\alpha = R/(2L) = 50/(2 \times 4 \times 10^{-3}) = 6250$, the damping ratio is

$$\delta = \frac{\alpha}{\omega_n} = \frac{6250}{40{,}825} = 0.1531$$

From Eq. (B.55),

$$Q_s = \omega_n \frac{L}{R} = 40{,}825 \times 4 \times \frac{10^{-3}}{50} = 3.266$$

For the lower cutoff frequency, Eqs. (B.85) and (B.86) give

$$u_1 = -\delta + \sqrt{1 + \delta^2} = -0.1531 + \sqrt{1 + 0.1531^2} = 0.85855$$

$$\omega_1 = u_1\omega_n = 0.85855 \times 40{,}825 = 35{,}050.4 \text{ rad/s}$$

Thus, $f_1 = 35{,}050.4/2\pi = 5578$ Hz. For the upper cutoff frequency, Eqs. (B.83) and (B.84) give

$$u_2 = \delta + \sqrt{1 + \delta^2} = 0.1531 + \sqrt{1 + 0.1531^2} = 1.16475$$

$$\omega_2 = u_2\omega_n = 1.16475 \times 40{,}825 = 47{,}551 \text{ rad/s}$$

Thus, $f_2 = 47{,}551/2\pi = 7568$ Hz. From Eq. (B.89), the bandwidth is

$$BW_s = f_2 - f_1 = \frac{f_n}{Q_s} = \frac{6497.5}{3.266} = 1989.4 \text{ Hz}$$

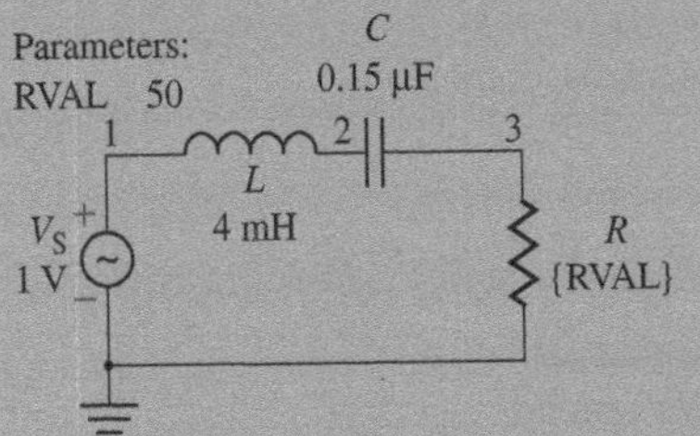

FIGURE B.42 Series *RLC* circuit for PSpice simulation

(b) The series *RLC* circuit for PSpice simulation is shown in Fig. B.42. The list of the circuit file is as follows.

```
Example B.16  Frequency Response of Series RLC Circuit
.PARAM RVAL = 50
.STEP PARAM RVAL LIST  50  100  200
Vm  1  0  AC  1V                    ; AC input of 1 V peak
L  1  2  4MH
C  2  3  0.15UF
R  3  0  {RVAL}
.AC DEC 100 100HZ 1MEGHz            ; AC analysis from f = 100 Hz to 1 MHz
                                    ; with a decade change and 100 points per decade
.PROBE
.END
```

The PSpice plots of the magnitude and phase angle (using EXB-16.SCH) are shown in Fig. B.43. The plot for $R = 50\ \Omega$ gives $f_1 = 5578$ Hz, $f_2 = 7568$ Hz, $f_n = 6457$ Hz, and $BW_s = f_2 - f_1 = 1990$ Hz.

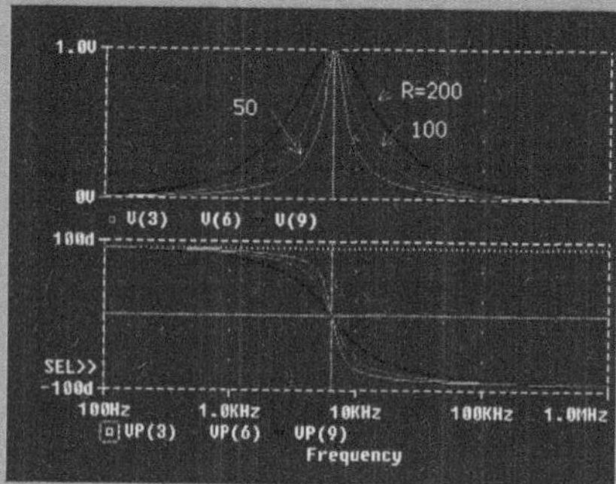

FIGURE B.43 Frequency response plots for Example B.16

B.10.4 Second-Order Parallel *RLC* Circuits

A parallel *RLC* circuit is shown in Fig. B.44. The output voltage v_o is taken across the parallel combination of *R*, *L*, and *C*. The transfer function $G(s) = V_o(s)/I_s(s)$ in Laplace's domain of *s* is the equivalent impedance *Z*(*s*).

The function

$$\frac{1}{Z(s)} = \frac{1}{R} + \frac{1}{sL} + sC = \frac{sL + R + s^2LCR}{sRL}$$

$$= \frac{s^2 + (s/RC) + (1/LC)}{s/C} \quad \text{(in siemens, or mhos)}$$

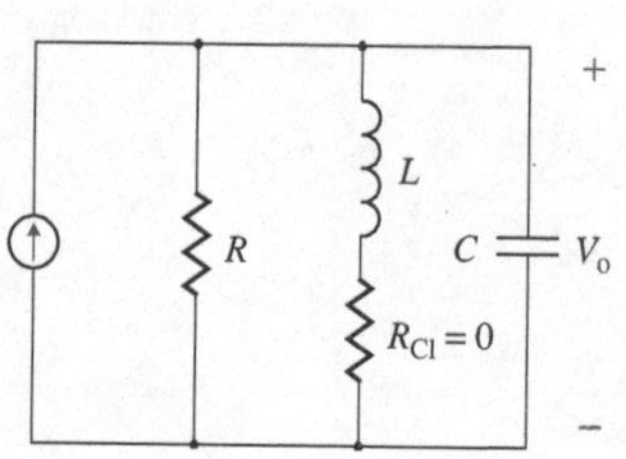

FIGURE B.44 Parallel *RLC* circuit

gives the transfer function $G(s)$ as

$$G(s) = \frac{V_o(s)}{I_s(s)} = Z(s) = \frac{s/C}{s^2 + (s/RC) + (1/LC)}$$

$$= R\frac{s/RC}{s^2 + (s/RC) + (1/LC)} \quad \text{(in ohms)} \tag{B.91}$$

Defining $\omega_n = 1/\sqrt{LC}$ as the *resonant frequency* in rad/s and $\alpha = 1/(2RC)$ as the *damping factor*, we can write Eq. (B.91) as

$$G(s) = R\frac{2\alpha s}{s^2 + 2\alpha s + \omega_n^2} \quad \text{(in ohms)} \tag{B.92}$$

Let us define

$$\delta = \frac{\alpha}{\omega_n} = \frac{1}{2RC} \times \sqrt{LC} = \frac{1}{2R}\sqrt{\frac{L}{C}} \tag{B.93}$$

as the *damping ratio*. Then Eq. (B.92) becomes

$$G(s) = R\frac{\delta\omega_n s}{s^2 + 2\delta\omega_n s + \omega_n^2} \quad \text{(in ohms)} \tag{B.94}$$

where $\delta < 1$. (Note that δ is not necessarily less than 1, but has been assumed less than 1 for this discussion.) The right-hand side of Eq. (B.94) is $R/2$ multiplied by Eq. (B.76). Following the development of Eqs. (B.78) and (B.79), we find the magnitude $|G(j\omega)|$ as

$$|G(j\omega)| = \frac{2\delta uR}{[(1-u^2)^2 + (2\delta u)^2]^{1/2}} \quad \text{(in ohms)} \tag{B.95}$$

and the phase angle ϕ of $G(j\omega)$ as

$$\phi = \frac{\pi}{2} - \tan^{-1}\left(\frac{2\delta u}{1-u^2}\right) \tag{B.96}$$

The magnitude and frequency plots of a parallel *RLC* circuit are shown in Fig. B.45. The maximum value $|G(j\omega)|_{max} = |Z(j\omega)|_{max} = 1$. At the cutoff frequencies, the magnitude of the gain drops to 70.7% of its maximum value R. Thus, Eq. (B.95) gives

$$|G(j\omega)| = \frac{2\delta uR}{[(1-u^2)^2 + (2\delta u)^2]^{1/2}} = 0.707R = \frac{R}{\sqrt{2}}$$

or $$\sqrt{2}(2\delta u) = [(1-u^2)^2 + (2\delta u)^2]^{1/2}$$

Squaring both sides yields

$$2(2\delta u)^2 = (1-u^2)^2 + (2\delta u)^2$$

or $$(2\delta u)^2 = (1-u^2)^2 \tag{B.97}$$

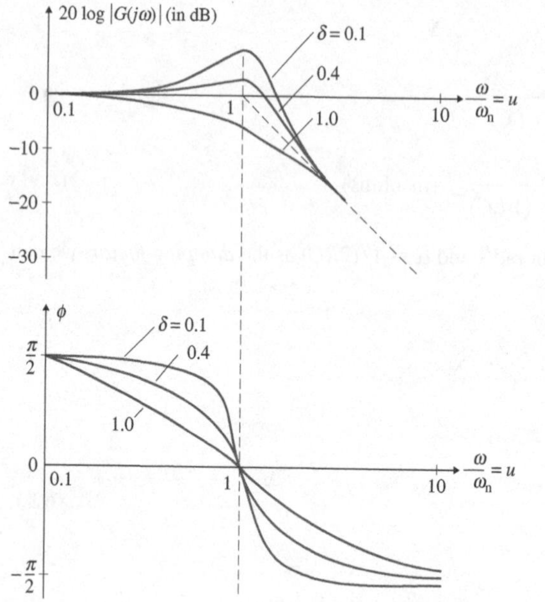

FIGURE B.45 Frequency response of parallel *RLC* circuit

which is the same as Eq. (B.80). Equations (B.81) to (B.86) can be applied to find ω_1 and ω_2. Then the bandwidth BW_p of the parallel resonant circuit can be found from

$$\begin{aligned} BW_p &= \omega_2 - \omega_1 = \omega_n(u_2 - u_1) = 2\delta\omega_n \\ &= 2\frac{1}{2R}\sqrt{\frac{L}{C}} \times \frac{1}{\sqrt{LC}} = \frac{1}{RC} \quad \text{(in rad/s)} \end{aligned} \tag{B.98}$$

▶ **NOTE** For a parallel circuit, $BW_p = 1/RC$ only; for a series circuit, $BW_s = R/L$.

EXAMPLE B.17

Finding the frequency response of a parallel *RLC* circuit The parallel *RLC* circuit in Fig. B.44 has $R = 50\ \Omega$, $L = 4$ mH, and $C = 0.15\ \mu$F.

(a) Determine the parallel resonant frequency f_p, the damping ratio δ, the cutoff frequencies, the bandwidth BW_p, and the quality factor Q_p of the circuit.

(b) Use PSpice/SPICE to plot the magnitude and phase angle of the output voltage for $R = 50\ \Omega$, $100\ \Omega$, and $200\ \Omega$. The frequency f varies from 100 Hz to 100 kHz. Assume $I_m = 1$ A peak AC.

SOLUTION

(a) $R = 50\ \Omega$, $L = 4$ mH, $C = 0.15\ \mu$F, and $I_m = 1$ A peak AC, so

$$\omega_n = \frac{1}{\sqrt{LC}} = \frac{10^5}{\sqrt{4 \times 1.5}} = 40{,}825 \text{ rad/s}$$

The parallel resonant frequency is

$$f_p = \frac{\omega_n}{2\pi} = \frac{40{,}825}{2\pi} = 6497.5 \text{ Hz}$$

Since $\alpha = 1/(2RC) = 1/(2 \times 50 \times 0.15 \times 10^{-6}) = 66.667 \times 10^3$, the damping ratio is

$$\delta = \frac{\alpha}{\omega_n} = 66.667 \times \frac{10^3}{40{,}825} = 1.633$$

For the lower cutoff frequency, Eqs. (B.85) and (B.86) give

$$u_1 = -\delta + \sqrt{1 + \delta^2} = -1.633 + \sqrt{1 + 1.633^2} = 0.28186$$

$$\omega_1 = u_1\omega_n = 0.28186 \times 40{,}825 = 11{,}507 \text{ rad/s}$$

Thus, $f_1 = 11{,}507/2\pi = 1831$ Hz. For the upper cutoff frequency, Eqs. (B.83) and (B.84) give

$$u_2 = \delta + \sqrt{1 + \delta^2} = 1.633 + \sqrt{1 + 1.633^2} = 3.54786$$

$$\omega_2 = u_2\omega_n = 3.54786 \times 40{,}825 = 144{,}841 \text{ rad/s}$$

Thus, $f_2 = 144{,}841/2\pi = 23{,}052$ Hz. From Eq. (B.98), the bandwidth is

$$\text{BW}_p = f_2 - f_1 = \frac{1}{RC} = \frac{1}{50 \times 0.15 \times 10^{-6}} = 133{,}333.3 \text{ rad/s, or } 21{,}220 \text{ Hz}$$

Using Eq. (B.90), we can find the quality factor:

$$Q_p = \frac{f_p}{BW_p} = \frac{6497.5}{21{,}220} = 0.3062$$

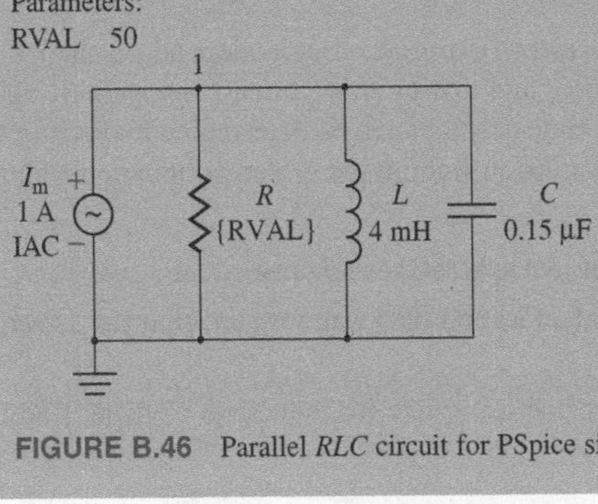

FIGURE B.46 Parallel *RLC* circuit for PSpice simulation

(b) The parallel *RLC* circuit for PSpice simulation is shown in Fig. B.46. The list of the circuit file is as follows.

```
Example B.17 Frequency Response of a Parallel RLC Circuit
.PARAM RVAL = 50
.STEP PARAM RVAL LIST  50  100  200
IM  0  1  AC  1A        ; ac input of 1 V peak
L   1  0  4MH
C   1  0  0.15UF
R   1  0  {RVAL}
.AC  DEC  100  100HZ  1MEGHZ
.PROBE
.END
```

The PSpice plots of the magnitude and phase angle (using EXB-17.SCH) are shown in Fig. B.47. The plot for $R = 50\ \Omega$ gives $f_1 = 1834$ Hz, $f_2 = 22.56$ kHz, $f_p = 6457$ Hz, and $BW_p = f_2 - f_1 = 20{,}726$ Hz.

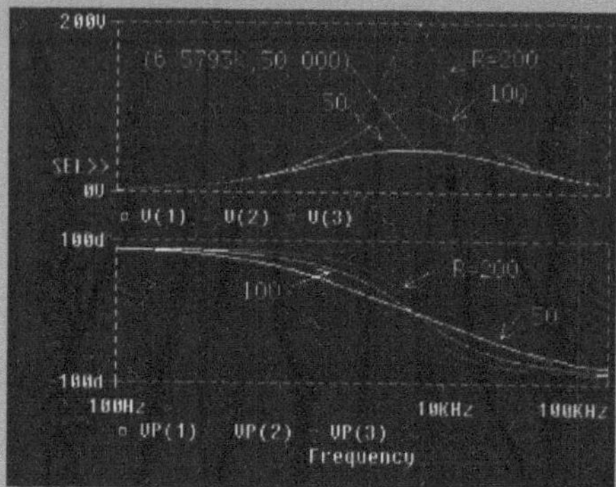

FIGURE B.47 Frequency response plots for Example B.17

B.11 Time Constants of First-Order Circuits

We have seen that the transient and frequency responses of first-order circuits depend on their time constants. The time constant of an *RC* network is $\tau = RC$, and that of an *RL* circuit is $\tau = L/R$. Many circuits have more than two components. The effective time constant can be determined by finding the effective resistance and capacitance of the circuit. The steps in finding the effective time constant are as follows:

Step 1. Set the voltage source(s) to zero and the current source(s) as an open circuit.

Step 2. If there is more than one capacitor (or inductor) and only one resistor, find the effective capacitance or inductance seen by the resistor.

Step 3. If there is more than one resistor and only one capacitive (or inductive) element, find the effective resistance seen by the capacitor (or inductor).

EXAMPLE B.18

Finding the effective time constant The circuit in Fig. B.48 has $R_1 = R_2 = R_3 = 6\ \text{k}\Omega$ and $C = 0.1\ \mu\text{F}$. Determine **(a)** the effective time constant τ, **(b)** the cutoff frequency ω_o, and **(c)** the bandwidth BW.

SOLUTION

If the source is shorted, the effective resistance seen by capacitor C is the parallel combination of R_1, R_2, and R_3. The effective resistance R is given by

$$\frac{1}{R} = \frac{1}{R_1} + \frac{1}{R_2} + \frac{1}{R_3} \quad \text{or} \quad R = \frac{R_1}{3} = \frac{6\ \text{k}}{3} = 2\ \text{k}\Omega$$

(a) The effective time constant is

$$\tau = CR = 2\ \text{k}\Omega \times 0.1\ \mu\text{F} = 0.2\ \text{ms}$$

(b) The cutoff frequency is

$$\omega_o = \frac{1}{\tau} = \frac{1}{0.2\ \text{ms}} = 5000\ \text{rad/s, or } 795.8\ \text{Hz}$$

(c) At $\omega = 0$, capacitor C is open-circuited, and the output voltage has a finite value. At a high frequency, tending to infinity ($\omega = \infty$), capacitor C is short-circuited, and the output voltage becomes zero. This is a low-pass circuit with $f_1 = 0$ and $f_2 = f_o = 795.8$ Hz. Thus, the bandwidth is

$$\text{BW} = f_2 - f_1 = 795.8\ \text{Hz}$$

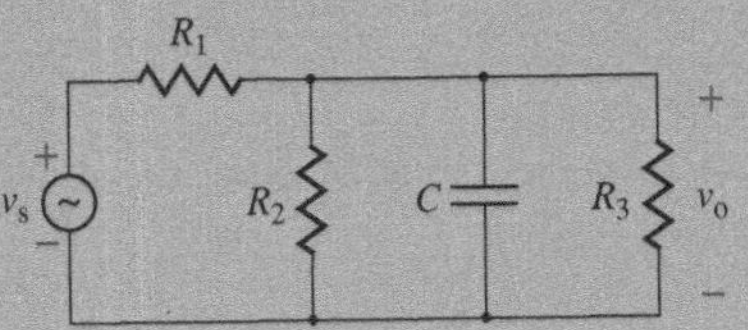

FIGURE B.48 Circuit for Example B.18

EXAMPLE B.19

Finding the effective time constant The circuit in Fig. B.49 has $R_1 = R_2 = R_3 = 10\ \text{k}\Omega$ and $C_1 = 0.1\ \mu\text{F}$. Determine **(a)** the effective time constant τ and **(b)** the cutoff frequency ω_o.

SOLUTION

If the source is shorted, the effective resistance is the sum of R_1 and $(R_2 \| R_3)$; that is,

$$R = R_1 + (R_2 \| R_3) = 10\ \text{k} + 10\ \text{k} \| 10\ \text{k} = 15\ \text{k}\Omega$$

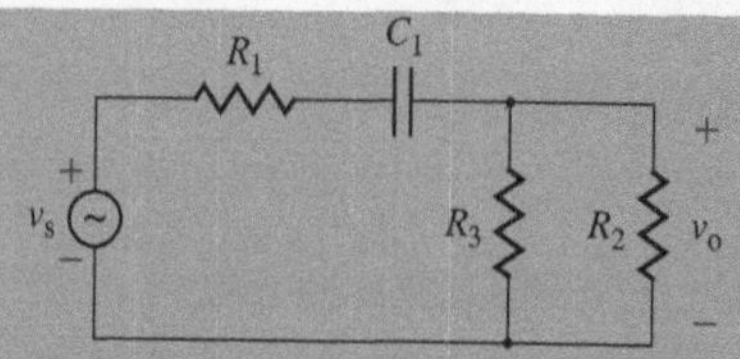

FIGURE B.49 Circuit for Example B.19

(a) The effective time constant is

$$\tau = CR = 15 \text{ k}\Omega \times 0.1 \ \mu\text{F} = 1.5 \text{ ms}$$

(b) The cutoff frequency is

$$\omega_o = \frac{1}{\tau} = \frac{1}{1.5 \text{ ms}} = 667 \text{ rad/s, or } 106 \text{ Hz}$$

APPENDIX C

LOW-FREQUENCY HYBRID BJT MODEL

A bipolar transistor can be represented by hybrid (h) parameters. If i_b, v_{be}, i_c, and v_{ce} are the small-signal variables of a transistor, as shown in Fig. C.1(a), they are related to the hybrid parameters as follows:

$$v_{bc} = h_{ie} i_b + h_{re} v_{ce} \tag{C.1}$$

$$i_c = h_{fe} i_b + h_{oe} v_{ce} \tag{C.2}$$

where h_{ie} is the *short-circuit input resistance* (or simply the *input resistance*), defined by

$$h_{ie} = \left.\frac{v_{be}}{i_b}\right|_{v_{ce}=0} \quad \text{(in ohms)} \tag{C.3}$$

h_{re} is the *open-circuit reverse voltage ratio* (or the *voltage-feedback ratio*), defined by

$$h_{re} = \left.\frac{v_{be}}{v_{ce}}\right|_{i_b=0} \quad \text{(dimensionless)} \tag{C.4}$$

h_{fe} is the *short-circuit forward-transfer current ratio* (or the *small-signal current gain*), defined by

$$h_{fe} = \left.\frac{i_c}{i_b}\right|_{v_{ce}=0} \quad \text{(dimensionless)} \tag{C.5}$$

and h_{oe} is the *open-circuit output admittance* (or simply the *output admittance*), defined by

$$h_{oe} = \left.\frac{i_c}{v_{ce}}\right|_{i_b=0} \quad \text{(in siemens)} \tag{C.6}$$

The low-frequency hybrid model is shown in Fig. C.1(b). The input has a voltage-controlled voltage source in which the controlling voltage is the output voltage. The output circuit has a current-controlled current source in which the controlling current is the input current. The subscript e on the h parameters indicates that these hybrid parameters are derived for a common-emitter configuration. It is also possible to have common-base and common-collector configurations.

BJT manufacturers specify the common-emitter hybrid parameters. The parameter h_{re}, which takes into account the effect of v_{CE} on i_B, is very small, with a typical value of 0.5×10^{-4}. The parameter h_{oe}, which represents the admittance of the CE junction, is also very small; its value is typically 10^{-6} S. The parameters h_{re} and h_{oe} can often be omitted from the circuit model without significant loss of accuracy, especially for hand calculations.

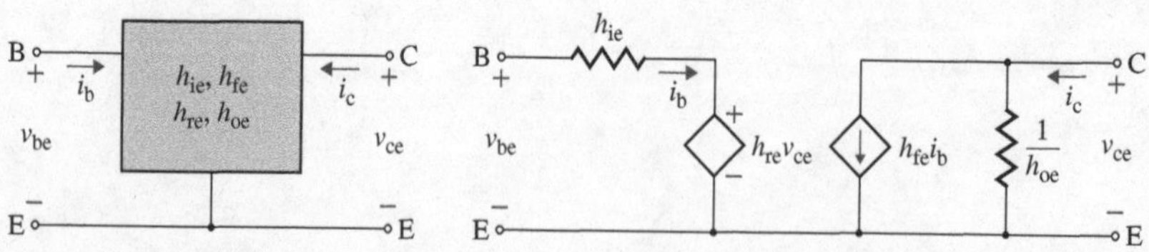

(a) Transistor variables (b) Hybrid model

FIGURE C.1 Low-frequency hybrid model

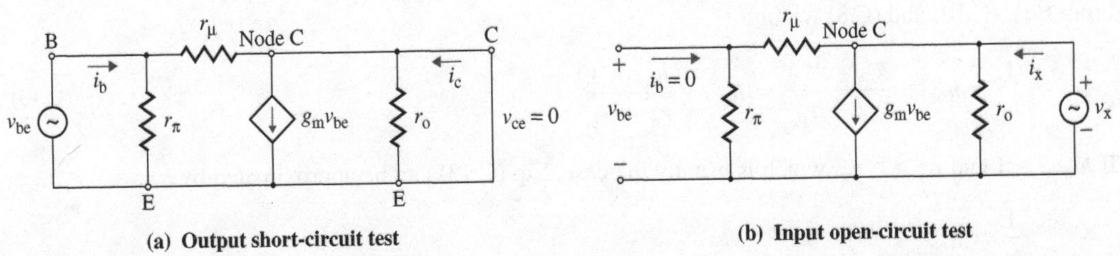

FIGURE C.2 Equivalent circuits for deriving h parameters

The π-model parameters can be related to the h parameters by applying short-circuit and open-circuit tests. The parameters h_{ie} and h_{fe}, which are short-circuit parameters, can be derived in terms of π-model parameters from Fig. C.2(a) by shorting the collector-to-emitter terminals.

$$h_{ie} = \left.\frac{v_{be}}{i_b}\right|_{v_{ce}=0} = r_\pi \| r_\mu = \frac{r_\pi r_\mu}{r_\pi + r_\mu} \tag{C.7}$$

$$h_{fe} = \left.\frac{i_c}{i_b}\right|_{v_{ce}=0} = \frac{g_m v_{be}}{v_{be}/r_\pi} = g_m r_\pi \tag{C.8}$$

As shown in Fig. C.2(b), h_{oe} and h_{re} can be determined by open-circuiting the base-to-emitter terminals and then applying the voltage divider rule.

$$h_{re} = \left.\frac{v_{be}}{v_x}\right|_{i_b=0} = \frac{r_\pi}{r_\pi + r_\mu} \tag{C.9}$$

Summing the currents at the collector node in Fig. C.2(b), we get

$$i_x = \frac{v_x}{r_o} + g_m \frac{r_\pi v_x}{r_\pi + r_\mu} + \frac{v_x}{r_\pi + r_\mu} = v_x\left[\frac{1}{r_o} + \frac{g_m r_\pi + 1}{r_\pi + r_\mu}\right]$$

which gives

$$h_{oe} = \frac{i_x}{v_x} = \frac{1}{r_o} + \frac{g_m r_\pi + 1}{r_\pi + r_\mu} \tag{C.10}$$

Using Eq. (C.8), we get

$$r_\pi = \frac{h_{fe}}{g_m} \tag{C.11}$$

Using Eq. (C.7) and $h_{fe} = g_m r_\pi$ from Eq. (C.8), we get

$$r_\mu = \frac{r_\pi + r_\mu}{r_\pi} h_{ie} = \frac{h_{ie}}{h_{re}} \tag{C.12}$$

From Eqs. (C.10) and (C.8) we find

$$\frac{1}{r_o} = h_{oe} - \frac{g_m r_\pi + 1}{r_\pi + r_\mu} = h_{oe} - \frac{h_{fe} + 1}{r_\pi + r_\mu} \tag{C.13}$$

If $h_{fe} >> 1$ and $r_\mu >> r_\pi$, which is usually the case, Eq. (C.13) can be approximated by

$$\frac{1}{r_o} \approx h_{oe} - \frac{h_{fe}}{r_\mu}$$

which gives

$$r_o = \left[h_{oe} - \frac{h_{fe}}{r_\mu}\right]^{-1} \tag{C.14}$$

$$= \frac{V_A}{I_C}$$

r_μ is very large, and h_{re} is negligibly small. The π model becomes similar to the h model in Fig. C.2(b) such that $r_\pi \equiv h_{ie}$, $\beta_f \equiv h_{fe}$, and $r_o \equiv 1/h_{oe}$. Being open-circuited, r_μ has a high value and can often be neglected.

EXAMPLE C.1

Converting hybrid parameters to π-model parameters The h parameters of a transistor are $h_{ie} = 1\ \text{k}\Omega$, $h_{fe} = 100$, $h_{re} = 0.5 \times 10^{-4}$, and $h_{oe} = 10 \times 10^{-6}$ S. The collector biasing current is $I_C = 11.62$ mA. Calculate the small-signal π-model parameters of the transistor. Assume $V_T = 25.8$ mV at 25°C.

SOLUTION

$$g_m = \frac{I_C}{V_T} = \frac{11.62\ \text{mA}}{25.8\ \text{mV}} = 0.4504\ \text{A/V}$$

From Eq. (C.11),

$$r_\pi = \frac{h_{fe}}{g_m} = \frac{100}{0.4504} = 222\ \Omega$$

From Eq. (C.12),

$$r_\mu = \frac{h_{ie}}{h_{re}} = \frac{1\ \text{k}\Omega}{0.5 \times 10^{-4}} = 20\ \text{M}\Omega$$

From Eq. (C.14),

$$r_o = \left[h_{oe} - \frac{h_{fe}}{r_\mu}\right]^{-1} = \left[10 \times 10^{-6} - \frac{100}{20 \times 10^{6}}\right]^{-1} = 200\ \text{k}\Omega$$

EXAMPLE C.2

Finding the hybrid-model parameters of the circuit shown in Fig. C.3 The circuit values are $R_1 = 4\ \text{k}\Omega$, $R_2 = 10\ \text{k}\Omega$, and $R_f = 16\ \text{k}\Omega$. Determine the hybrid parameters h_{ie}, h_{re}, h_{fe}, and h_{oe}.

SOLUTION

$R_1 = 4\ \text{k}\Omega$, $R_2 = 6\ \text{k}\Omega$, and $R_f = 20\ \text{k}\Omega$. Using Eq. (C.3) and the equivalent circuit as shown in Fig. C.4(a), h_{ie} can be determined from

$$h_{ie} = \left.\frac{v_S}{i_S}\right|_{v_O=0} = R_1 \| R_f = 4\ \text{k} \| 16\ \text{k} = 3.2\ \text{k}\Omega$$

Using Eq. (C.4) and the equivalent circuit shown in Fig. C.4(b), h_{re} can be determined from

$$h_{re} = \left.\frac{v_S}{v_O}\right|_{i_S=0} = \frac{R_1}{R_1 + R_f} = \frac{4\ \text{k}}{4\ \text{k} + 16\ \text{k}} = 0.2\ \text{V/V}$$

Using Eq. (C.5) and the equivalent circuit shown in Fig. C.4(c), h_{fe} can be determined from

$$h_{re} = \left.\frac{i_O}{i_S}\right|_{v_O=0} = \frac{-R_1}{R_1 + R_f} = \frac{-4\ \text{k}}{4\ \text{k} + 16\ \text{k}} = -0.2\ \text{A/A}$$

Using Eq. (C.6) and the equivalent circuit shown in Fig. C.4(d), h_{oe} can be determined from

$$h_{oe} = \left.\frac{i_O}{v_O}\right|_{i_S=0} = \frac{1}{R_1 \| (R_1 + R_f)} = \frac{1}{4\ \text{k} \| (4\ \text{k} + 16\ \text{k})} = 1.5 \times 10^{-4}\ \mho$$

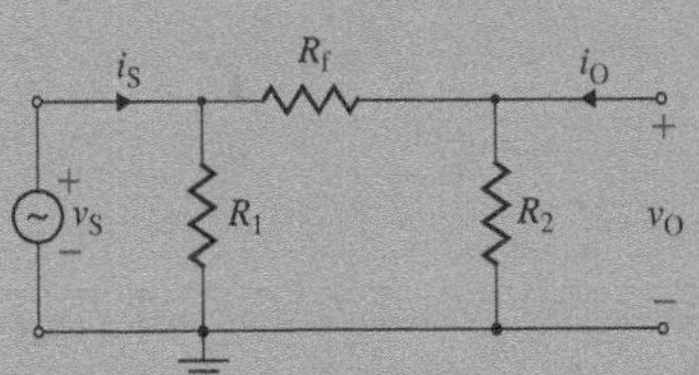

FIGURE C.3 Circuit for Example C.2

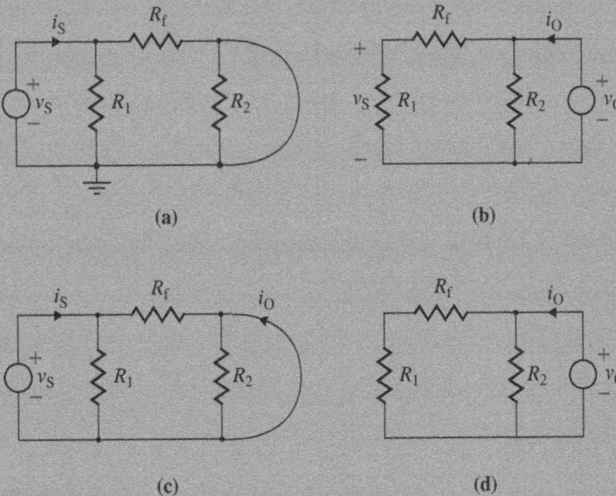

FIGURE C.4 Equivalent circuits for determining hybrid parameters for Example C.2

APPENDIX D

EBERS–MOLL MODEL OF BIPOLAR JUNCTION TRANSISTORS

A transistor is a nonlinear device that can be modeled using the nonlinear characteristics of diodes. The Ebers–Moll model is a large-signal model commonly used for modeling BJTs. One version of the model is based on the assumption of one forward-biased diode and one reverse-biased diode. This arrangement is shown in Fig. D.1 for an *npn* transistor. This model, referred to as the *injection version* of the Ebers–Moll model, is valid for active, saturation, and cutoff regions. Under normal operation in the active region, one junction of the BJT is forward biased and the other is reverse biased.

The emitter–base and collector–base diodes can be described using the Shockley diode characteristic:

$$I_F = I_{ES}\left(\exp\frac{V_{BE}}{V_T} - 1\right) \tag{D.1}$$

$$I_R = I_{CS}\left(\exp\frac{V_{BC}}{V_T} - 1\right) \tag{D.2}$$

where $V_T = kT/q = 25.8$ mV at 25°C
I_{ES} = reverse saturation current of base–emitter diode
I_{CS} = reverse saturation current of base–collector diode

Both I_{ES} and I_{CS} are temperature dependent. If $V_{BE} > 0$, diode D_F is forward biased and its current I_F causes a corresponding current $\alpha_F I_F$. If $V_{BC} > 0$, diode D_R is reverse biased. The subscripts F and R are used to designate forward and reverse conditions, respectively. Using Kirchhoff's current law (KCL) at the emitter and collector terminals, we can write the emitter current I_E as

$$\begin{aligned} I_E &= -I_F + \alpha_R I_R \\ &= -I_{ES}\left(\exp\frac{V_{BE}}{V_T} - 1\right) + \alpha_R I_{CS}\left(\exp\frac{V_{BC}}{V_T} - 1\right) \end{aligned} \tag{D.3}$$

and the collector current I_C as

$$\begin{aligned} I_C &= \alpha_F I_F - I_R \\ &= \alpha_F I_{ES}\left(\exp\frac{V_{BE}}{V_T} - 1\right) - I_{CS}\left(\exp\frac{V_{BC}}{V_T} - 1\right) \end{aligned} \tag{D.4}$$

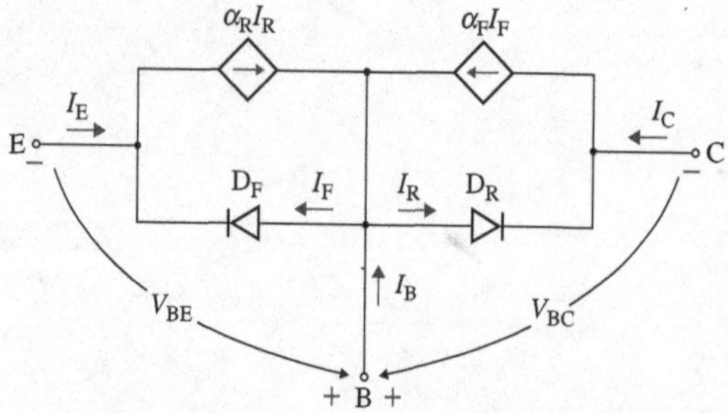

(a) Injection version model

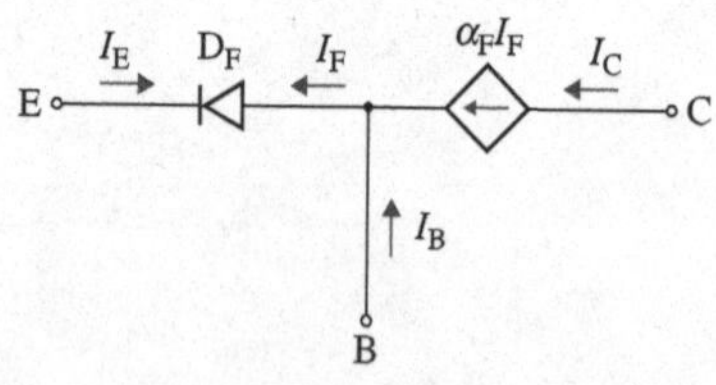
(b) Approximate injection version model

FIGURE D.1 Ebers–Moll injection version model for *npn* transistor

If $V_{BE} = 0$, $\alpha_R I_{CS} = I_S$ represents the reverse saturation leakage current of diode D_R. Similarly, if $V_{BE} = 0$, $\alpha_F I_{ES} = I_S$ represents the reverse saturation leakage current of diode D_F. If we assume ideal diodes, the forward and reverse saturation leakage currents are related by

$$\alpha_R I_{CS} = \alpha_F I_{ES} = I_S \tag{D.5}$$

where I_S is known as the *transistor saturation current.*

The current from the collector to the base with the emitter open-circuited can be found by letting $I_C = I_{CBO}$ and $I_E = 0$. Since the collector–base junction is normally reverse biased, $V_{BC} < 0$ and $|V_{BC}| >> V_T$, $\exp(V_{BC}/V_T) << 1$. With these conditions in Eqs. (D.3) and (D.4), we get

$$0 = -I_{ES}\left(\exp\frac{V_{BE}}{V_T} - 1\right) - \alpha_R I_{CS}$$

$$I_{CBO} = \alpha_F I_{ES}\left(\exp\frac{V_{BE}}{V_T} - 1\right) + I_{CS}$$

Solving these two equations for I_{CBO} yields

$$I_{CBO} = -\alpha_F\alpha_R I_{CS} + I_{CS} = (1 - \alpha_R\alpha_F)I_{CS} = I_{CS} - \alpha_F I_S \tag{D.6}$$

The current from the emitter to the base with the collector open-circuited can be found by letting $I_E = I_{EBO}$ and $I_C = 0$. Since the emitter junction is reverse biased, $V_{BE} < 0$ (i.e., $V_{EB} > 0$) and $|V_{BE}| >> V_T$, $\exp(V_{BE}/V_T) << 1$. Equations (D.3) and (D.4) give

$$I_{EBO} = I_{ES} + \alpha_R I_{CS}\left(\exp\frac{V_{BC}}{V_T} - 1\right)$$

$$0 = -\alpha_F I_{ES} - I_{CS}\left(\exp\frac{V_{BC}}{V_T} - 1\right)$$

Solving these two equations for I_{EBO} gives

$$I_{EBO} = I_{ES} - \alpha_R\alpha_F I_{ES} = (1 - \alpha_R\alpha_F)I_{ES} \tag{D.7}$$

From Eqs. (D.5), (D.6), and (D.7), we get

$$\alpha_F I_{EBO} = I_{ES}(1 - \alpha_R\alpha_F)\alpha_F = \alpha_R I_{CS}(1 - \alpha_R\alpha_F) = \alpha_R I_{CBO} \tag{D.8}$$

Since diode D_F is forward biased and diode D_R is reverse biased, $V_{EB} < V_{BC}$. Thus, I_{EBO} is less than I_{CBO}, and α_F is greater than α_R. In the active region, diode D_R is reverse biased, and $I_R \approx 0$; that is, $I_E = -I_F$, and $I_C = \alpha_F I_F = -\alpha_F I_E$. Thus, Fig. D.1(a) can be approximated by Fig. D.1(b).

The circuit model of Fig. D.1(a) relates the dependent sources to the diode currents. In circuit analysis, it is convenient to express the current source in a form that is controlled by the terminal currents. Eliminating $[\exp(V_{BE}/V_T) - 1]$ from Eqs. (D.3) and (D.4) and then using Eq. (D.6), we get

$$I_C = -\alpha_F I_E - (1 - \alpha_F\alpha_R)I_{CS}\left(\exp\frac{V_{BC}}{V_T} - 1\right)$$

$$= -\alpha_F I_E - I_{CBO}\left(\exp\frac{V_{BC}}{V_T} - 1\right) \tag{D.9}$$

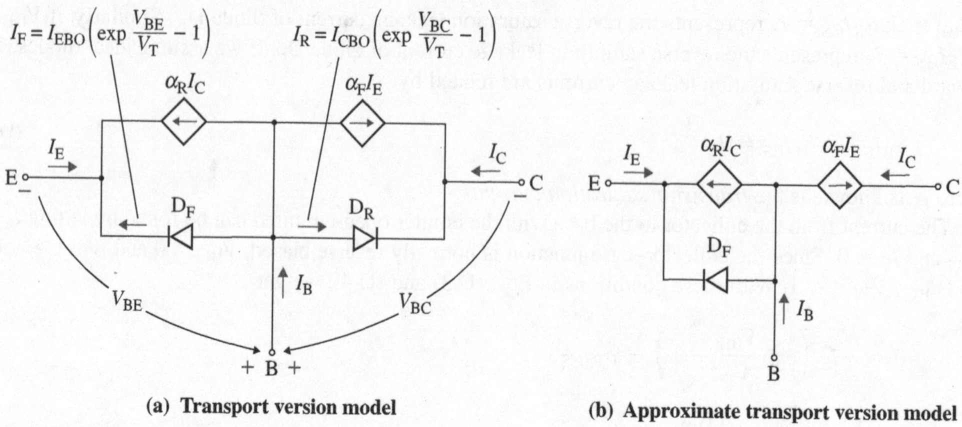

FIGURE D.2 Ebers–Moll transport version model

Similarly, eliminating [exp (V_{BC}/V_T) − 1] from Eqs. (D.3) and (D.4) and then using Eq. (D.7) gives

$$I_E = -\alpha_R I_C - (1 - \alpha_R\alpha_F)I_{ES}\left(\exp\frac{V_{BE}}{V_T} - 1\right)$$

$$= -\alpha_R I_C - I_{EBO}\left(\exp\frac{V_{BE}}{V_T} - 1\right) \quad \text{(D.10)}$$

The circuit model corresponding to Eqs. (D.9) and (D.10) is shown in Fig. D.2(a). The current sources are controlled by collector current I_C and emitter current I_E. This model, referred to as the *transport version* of the Ebers–Moll model, is normally used in computer simulations with PSpice/SPICE. In fact, the linear models are the approximate versions of the Ebers–Moll model in Fig. D.2(a).

Assuming that $V_{BC} > 0$ and $I_R \approx 0$, $I_C = -\alpha_F I_E$ and Fig. D.2(a) can be approximated by Fig. D.2(b). If we substitute $I_C = -\alpha_F I_E$, Eq. (D.10) becomes

$$I_E = \alpha_R\alpha_F I_E - I_{EBO}\left(\exp\frac{V_{BE}}{V_T} - 1\right)$$

which relates I_E to α_R, α_F, and V_{BE} by

$$I_E = \frac{I_{EBO}\exp\left(V_{BE}/V_T - 1\right)}{1 - \alpha_R\alpha_F} \quad \text{(D.11)}$$

I_{ES} and I_{CS} are also known as the *short-circuit saturation currents*, whereas I_{CBO} and I_{EBO} are known as the *open-circuit saturation currents*. Note that the injection and transport versions of the Ebers–Moll model are interchangeable. Once the parameters of one version are known, the parameters of the other version can be found.

EXAMPLE D.1

Finding the currents for the Ebers–Moll model of an *npn* transistor An *npn* transistor is biased so that $V_{BE} = 0.3$ V and $V_{CE} = 6$ V. If $\alpha_F = 0.99$, $\alpha_R = 0.90$, and $I_{CBO} = 5$ nA, find all the currents for the injection version of the Ebers–Moll model of Fig. D.1(a). Assume thermal voltage $V_T = 25.8$ mV.

SOLUTION

$V_T = 25.8 \text{ mV} = 0.0258$ V. Since V_{BE} is positive, the base–emitter junction is forward biased. The voltage at the collector–base junction is

$$V_{CB} = V_{CE} + V_{EB} = V_{CE} - V_{BE} = 6 - 0.3 = 5.7 \text{ V}$$

or $$V_{BC} = -5.7 \text{ V}$$

Since V_{CB} is positive, the collector–base junction is reverse biased, and the transistor operates in the active region. From Eq. (D.8),

$$I_{EBO} = \frac{\alpha_R I_{CBO}}{\alpha_F} = 0.9 \times \frac{5 \text{ nA}}{0.99} = 4.545 \text{ nA}$$

From Eq. (D.6),

$$I_{CS} = \frac{I_{CBO}}{1 - \alpha_R \alpha_F} = \frac{5 \text{ nA}}{1 - 0.9 \times 0.99} = 45.87 \text{ nA}$$

From Eq. (D.7),

$$I_{ES} = \frac{I_{EBO}}{1 - \alpha_R \alpha_F} = \frac{4.545 \text{ nA}}{1 - 0.9 \times 0.99} = 41.697 \text{ nA}$$

From Eq. (D.1), the forward diode current is

$$I_F = 41.697 \times 10^{-9} \times \left(\exp\frac{0.3}{0.0258} - 1\right) \approx 4.678 \text{ mA}$$

From Eq. (D.2), the reverse diode current is

$$I_R = 45.87 \times 10^{-9} \times \left(\exp\frac{-5.7}{0.0258} - 1\right) \approx -45.87 \text{ nA}$$

and $$\alpha_R I_R \approx -0.9 \times 45.87 \text{ nA} = -41.28 \text{ nA}$$

$$\alpha_F I_F \approx 0.99 \times 4.678 \text{ mA} = 4.63 \text{ mA}$$

$$I_E = -I_F + \alpha_R I_R = -4.678 \times 10^{-3} - 41.28 \times 10^{-6} \approx -4.719 \text{ mA}$$

$$I_C = \alpha_F I_F - I_R = 4.63 \times 10^{-3} + 45.87 \times 10^{-9} \approx 4.63 \text{ mA}$$

$$I_B = -(I_E + I_C) = -(-4.719 \times 10^{-3} + 4.63 \times 10^{-3}) = 89 \text{ } \mu\text{A}$$

EXAMPLE D.2

Finding the currents for the Ebers–Moll model of a *pnp* transistor The model parameters of the *pnp* transistor are $\alpha_R = 0.9$, $\alpha_F = 0.99$, $I_{ES} \approx I_{CS} = 45$ nA, $V_{EB} = 0.4$ V, and $V_{CB} = 0.3$ V. Determine **(a)** the currents for the transport version of the Ebers–Moll model in Fig. D.2(a) and **(b)** $\beta_F = \beta_{forced}$. Assume thermal voltage $V_T = 25.8$ mV.

SOLUTION

For $V_{EB} = 0.4$ V and $V_{CB} = 0.3$ V, the collector–base and emitter–base junctions are forward biased. Therefore, the transistor is operated in the saturation region.

(a) For a *pnp* transistor, all the polarities of voltages and currents are reversed; thus, Eqs. (D.3) and (D.4) become

$$I_E = I_{ES}\left(\exp\frac{V_{EB}}{V_T} - 1\right) - \alpha_R I_{CS}\left(\exp\frac{V_{CB}}{V_T} - 1\right) \tag{D.12}$$

$$= 45 \times 10^{-9}\left(\exp\frac{0.4}{25.8\text{ m}} - 1\right) - 0.9 \times 45 \times 10^{-9}\left(\exp\frac{0.3}{25.8\text{ m}} - 1\right)$$

$$= 243.48\text{ mA} - 4.54\text{ mA} = 238.94\text{ mA}$$

and
$$I_C = -\alpha_F I_{ES}\left(\exp\frac{V_{EB}}{V_T} - 1\right) + I_{CS}\left(\exp\frac{V_{CB}}{V_T} - 1\right) \tag{D.13}$$

$$= -0.99 \times 45 \times 10^{-9}\left(\exp\frac{0.4}{25.8\text{ m}} - 1\right) + 45 \times 10^{-9}\left(\exp\frac{0.3}{25.8\text{ m}} - 1\right)$$

$$= -241.04\text{ mA} + 5.05\text{ mA} = -235.99\text{ mA}$$

Thus, $I_B = -(I_E + I_C) = -(238.94 - 235.99) = -2.95$ mA

(b) Since the transistor is operated in the saturation region, the value of the forward current gain becomes less than that of the active region. The forward current gain is known as the *forced current gain* β_{forced}:

$$\beta_{forced} = \frac{I_C}{I_B} = \frac{235.99\text{ mA}}{2.95\text{ mA}} = 80$$

EXAMPLE D.3

Finding the collector–emitter saturation voltage of a BJT An *npn* transistor is operated in the saturation region, and its parameters are $\alpha_R = 0.9$, $\alpha_F = 0.989$, $\beta_F = 89.91$, and $V_T = 25.8$ mV. Calculate the collector–emitter saturation voltage $V_{CE(sat)}$.

SOLUTION

$$\beta_F = \frac{\alpha_F}{1 - \alpha_F} = \frac{0.989}{1 - 0.989} = 89.91$$

We know that

$$V_{BC} = V_{BE} + V_{EC} = V_{BE} - V_{CE}$$

In the saturation region, $V_{BE} > 0$ and $V_{BC} > 0$; that is,

$$\exp\left(\frac{V_{BE}}{V_T}\right) >> 1 \quad \text{and} \quad \exp\left(\frac{V_{BC}}{V_T}\right) >> 1$$

Using Eqs. (D.3) and (D.5), we get

$$I_E = -I_{ES}e^{V_{BE}/V_T} + a_F I_{ES}e^{V_{BC}/V_T}$$

$$\approx -I_{ES}e^{V_{BE}/V_T} + \alpha_F I_{ES}e^{(V_{BE}-V_{CE})/V_T}$$

$$\approx -I_{ES}e^{V_{BE}/V_T}[1 - \alpha_F e^{-V_{CE}/V_T}]$$

But $I_E + I_C + I_B = 0$, so $-I_E = (I_C + I_B)$; that is,

$$I_C + I_B = I_{ES}e^{V_{BE}/V_T}[1 - \alpha_F e^{-V_{CE}/V_T}] \tag{D.14}$$

Similarly, using Eqs. (D.4) and (D.5), we get

$$I_C = \alpha_F I_{ES}\left(\exp\frac{V_{BE}}{V_T} - 1\right) - I_{CS}\left(\exp\frac{V_{BC}}{V_T} - 1\right)$$

$$= \alpha_F I_{ES}e^{V_{BE}/V_T}\left[1 - \frac{1}{\alpha_R}e^{-V_{CE}/V_T}\right] \tag{D.15}$$

Dividing Eq. (D.14) by Eq. (D.15) yields

$$\frac{I_C + I_B}{I_C} = 1 + \frac{I_B}{I_C} = \frac{1 - \alpha_F e^{-V_{CE}/V_T}}{\alpha_F[1 - (e^{-V_{CE}/V_T})\alpha_R]}$$

which, after simplification, becomes

$$V_{CE} = V_T \ln\left\{\frac{\alpha_F[1 + I_C/I_B\,(1 - \alpha_R)]}{\alpha_R[\alpha_F + I_C/I_B\,(\alpha_F - 1)]}\right\} \tag{D.16}$$

In the active region, the base current I_B is related to the collector current I_C by $I_B = I_C/\beta_F$. The saturation region is considered to begin at the point where the forward current gain β_F is 90% of the value in the active region. That is, $\beta_{sat} = \beta_{forced} = 0.9\beta_F$ and $I_C = 0.9I_B\beta_F$. Equation (D.16) gives the collector–emitter saturation voltage as

$$V_{CE(sat)} = V_T \ln\left\{\frac{\alpha_F[1 + 0.9\beta_F(1 - \alpha_R)]}{\alpha_R[\alpha_F + 0.9\beta_F(\alpha_F - 1)]}\right\}$$

Substituting $\beta_F = \alpha_F/(1 - \alpha_F)$ in the denominator gives

$$V_{CE(sat)} = V_T \ln\left[\frac{1 + 0.9\beta_F(1 - \alpha_R)}{\alpha_R(1 - 0.9)}\right] \tag{D.17}$$

For $V_T = 0.0258$ V, $\alpha_R = 0.9$, and $\beta_F = 89.91$, we get

$$V_{CE(sat)} = 0.0258 \times \ln\left[\frac{1 + 0.9 \times 89.91 \times (1 - 0.9)}{0.9 \times (1 - 0.9)}\right] = 0.119 \text{ V}$$

APPENDIX E

PASSIVE COMPONENTS

E.1 Resistors

TABLE E.1 Standard values (in ohms) for metal film resistors (tolerance ±1%) and carbon resistors (tolerance ±5% and ±10%)

1%	5%	10%	1%	5%	10%	1%	5%	10%	1%	5%	10%
10.0	10	10	20.0	20		40.2			61.9	62	
10.2			20.5			41.2			63.4		
10.5			21.0			42.2			64.9		
10.7			21.5			43.2	43		66.5		
11.0	11		22.1	22	22	44.2			68.1	68	68
11.3			22.6			45.3			69.8		
11.5			23.2			46.4			71.5		
11.8	12	12	23.7			47.5	47	47	73.2		
12.1			24.3	24		48.7			75.0	75	
12.4			24.9			49.9			76.8		
12.7			25.5			51.1	51		78.7		
13.0	13		26.1			52.3			80.6		
13.3			26.7	27	27	53.6			82.5	82	82
13.7			27.4			54.9			84.5		
14.0			28.0			56.2	56	56	86.6		
14.3			28.7			57.6			88.7		
14.7			29.4			59.0			90.9	91	
15.0	15	15	30.1	30		60.4			93.1		
15.4			30.9						95.3		
15.8			31.6						97.6		
16.2	16		32.4	33	33						
16.5			33.2								
16.9			34.0								
17.4			34.8								
17.8			35.7	36							
18.2	18	18	36.5								
18.7			37.4								
19.1			38.3								
19.6			39.2	39	39						

▶ **NOTE** The available values for carbon are $1\ \Omega \leq R \leq 100\ \text{M}\Omega$ and for metal film are $10\ \Omega \leq R \leq 10\ \text{M}\Omega$. The available values are obtained by multiplying the sequence number by a power of 10 (i.e., 10^{-1}, 10^{0}, 10^{1}, 10^{2}, 10^{3}, and so on).

TABLE E.2 Standard values (in ohms) for wire-wound resistors (tolerance ±5%)

0.008	0.75	7.5	27	62	180	450	1 k	4 k
0.01	1.0	8	30	70	200	470	1.2 k	5 k
0.02	1.5	10	33	75	220	500	1.3 k	10 k
0.03	2.0	12	35	80	250	560	1.5 k	15 k
0.05	2.5	15	40	82	270	600	1.8 k	20 k
0.1	3.0	16	45	100	300	680	2 k	25 k
0.15	3.3	20	47	110	330	700	2.2 k	40 k
0.2	4.0	22	50	120	390	750	2.5 k	50 k
0.26	5.0	22.5	56	150	400	910	3 k	100 k
0.3	6	25	60	160	430		3.5 k	150 k
0.5	7							

TABLE E.3 Power ratings for resistors

Type	Tolerance	Power Rating
Carbon resistors	5% and 10%	$\frac{1}{8}$ W
		$\frac{1}{4}$ W
		$\frac{1}{2}$ W
		1 W
		2 W
Metal film resistors	1%	$\frac{1}{8}$ W
		$\frac{1}{4}$ W
		$\frac{1}{2}$ W
Wire-wound resistors	5%	5 W
		12 W
		25 W
		50 W
		100 W
		225 W

TABLE E.4 Color code for resistors

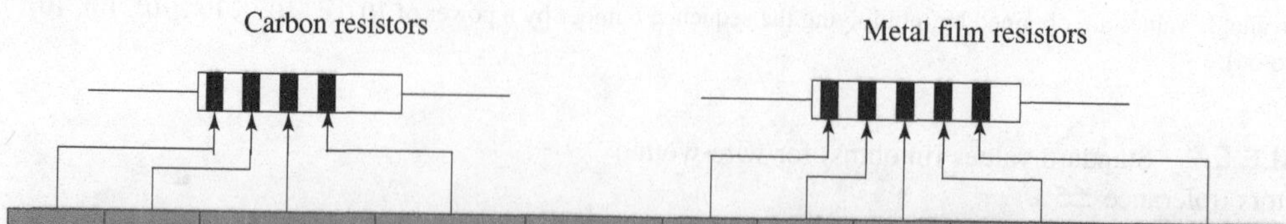

1st Digit	2nd Digit	Multiplier	Tolerance	Color	1st Digit	2nd Digit	3rd Digit	Multiplier	Tolerance
0	0	$10^0 = 1$	—	Black	0	0	0	$10^0 = 1$	—
1	1	$10^1 = 10$	—	Brown	1	1	1	$10^1 = 10$	±1%
2	2	$10^2 = 100$	—	Red	2	2	2	$10^2 = 100$	—
3	3	$10^3 = 1$ k	—	Orange	3	3	3	$10^3 = 1$ k	—
4	4	$10^4 = 10$ k	—	Yellow	4	4	4	$10^4 = 10$ k	—
5	5	$10^5 = 100$ k	—	Green	5	5	5	$10^5 = 100$ k	—
6	6	$10^6 = 1$ M	—	Blue	6	6	6	$10^6 = 1$ M	—
7	7	$10^7 = 10$ M	—	Violet	7	7	7	$10^7 = 10$ M	—
8	8		—	Gray	8	8	8		—
9	9		—	White	9	9	9		—
—	—		±5%	Gold	—	—	—		—
—	—		±10%	Silver	—	—	—		—
—	—		±20%	No band	—	—	—		—

E.2 Potentiometers

TABLE E.5 Standard values (in ohms) for carbon composition, linear taper potentiometers (tolerance ±10%; power rating 2.25 W)

50	1 k	10 k	100 k	1 M
150	1.5 k	15 k	150 k	1.5 M
200	2 k	20 k	200 k	2 M
250	2.5 k	25 k	250 k	2.5 M
350	3.5 k	35 k	350 k	3.5 M
500	5 k	50 k	500 k	5 M
750	7.5 k	75 k	750 k	

TABLE E.6 Standard values (in ohms) for conductive plastic potentiometers (tolerance ±10%; power rating $\frac{1}{2}$ W)

250			
1 k	10 k	100 k	1 M
2.5 k	25 k	250 k	2.5 M
5 k	50 k	500 k	5 M

TABLE E.7 Standard values (in ohms) for CERMET potentiometers (tolerance ±10%; power rating $\frac{1}{2}$ W)

50				
100	1 k	10 k	100 k	1 M
200	2 k	20 k	200 k	2 M
500	5 k	50 k	500 k	

E.3 Capacitors

TABLE E.8 Standard values for polarized aluminum electrolytic capacitors (tolerance −10% to +50%)

Voltage (V)	Capacitance (μF)	Voltage (V)	Capacitance (μF)	Voltage (V)	Capacitance (μF)
10	22	25	10	50	0.1
	33		22		0.22
	47		33		0.33
	100		47		0.47
	220		100		1.0
	330		220		2.2
	470		330		3.3
	1000		470		4.7
	2200		1000		10
	3300		2200		22
	4700		3300		33
	6800		4700		47
	10,000				100
					220
					330
					470
					1000
					2200

TABLE E.9 Standard values for ceramic disc capacitors (tolerance ±10%)

Voltage (V)	Capacitance (pF)
200	10
	15
	22
	33
	47
	68
	100
	150
	220
	330
	470
	680
	1000
	1500
	2200
	3300
	4700
	6800
	10,000
	15,000

TABLE E.10 Standard values for mylar polyester capacitors (tolerance ±10%)

Voltage (V)	Capacitance (μF)
100	0.001
	0.0015
	0.0022
	0.0033
	0.0047
	0.0068
	0.0082
	0.01
	0.015
	0.022
	0.027
	0.033
	0.039
	0.047
	0.056
	0.068
	0.082
	0.1
	0.12
	0.15
	0.18
	0.22
	0.27
	0.33
	0.39
	0.47
	0.56
	0.68
	0.82
	1

TABLE E.11 Standard values for ceramic variable capacitors (tolerance ±10%)

	Capacitance (pF)	
Voltage (V)	Min	Max
250	1	4.5
	2.5	10
	4	18
	6	35
	7	40
	8	50

APPENDIX F

DESIGN PROBLEMS

Mini Design Projects

F.1 Design a circuit that will sum the sensing signals generated by the probes of an electrocardiogram. The output can be expressed in terms of the input signals (V_A, V_B, V_C, and V_D) as follows:

$$V_O = 1.4V_A + 5V_B + 3V_C + 0.6V_D$$

The accuracy should be better than 2%. The DC power supplies are ± 15 V.

F.2 Design a BJT buffer amplifier to give a midfrequency voltage gain of $|A_v| = v_L/v_s \approx 1$ and an input resistance of $R_{in} = v_s/i_s \geq 50$ kΩ. The load resistance is $R_L = 10$ kΩ. Assume a source resistance of $R_S = 500$ Ω and $V_{CC} = 15$ V.

F.3 Design an NMOS buffer amplifier to give a midfrequency voltage gain of $|A_v| = v_L/v_s \approx 1$ and an input resistance of $R_{in} = v_s/i_s \geq 500$ kΩ. The load resistance is $R_L = 10$ kΩ. Assume a source resistance of $R_S = 500$ Ω and $V_{DD} = 15$ V.

F.4 Design a BJT buffer amplifier with an active load to give a midfrequency voltage gain of $|A_v| = v_L/v_s \approx 1$ and an output resistance of $R_{in} = v_s/i_s \geq 50$ kΩ. The load resistance is $R_L = 10$ kΩ. Assume a source resistance of $R_S = 500$ Ω and $V_{CC} = V_{EE} = 15$ V.

Medium Design Projects

F.5 Design a DC power supply to electronic equipment from an AC supply of 120 V (rms) ± 10%, 60 Hz. The load requires ± 12 V ± 5% at 0.5 A. Use discrete devices only and design for minimum expense.

F.6 a. Design a BJT amplifier to give a midfrequency voltage gain of $|A_v| = v_L/v_s = 20 \pm 5\%$ at a load resistance of $R_L = 10$ kΩ. Assume a source resistance of $R_S = 500$ Ω and $V_{DD} = 15$ V.

b. Modify the design so that the amplifier operates in the frequency range from 10 Hz to 100 kHz.

F.7 a. Design a depletion NMOS amplifier to give a midfrequency voltage gain of $|A_v| = v_L/v_s = 20 \pm 5\%$ at a load resistance of $R_L = 10$ kΩ. Assume a source resistance of $R_S = 1.5$ kΩ and $V_{DD} = 15$ V.

b. Modify the design so that the amplifier operates in the frequency range from 5 Hz to 50 kHz.

F.8 a. Design a MOSFET amplifier to give a midfrequency voltage gain of $|A_v| = v_L/v_s = 10 \pm 5\%$ at a load resistance of $R_L = 10$ kΩ. Assume a source resistance of $R_S = 1.5$ kΩ and $V_{DD} = 12$ V.

b. Modify the design so that the amplifier operates in the frequency range from 10 Hz to 50 kHz.

F.9 a. Design an NMOS amplifier with an active load to give a midfrequency voltage gain of $|A_v| = v_L/v_s \geq 250 \pm 5\%$ at a load resistance of $R_L = 20$ kΩ. Assume a source resistance of $R_S = 1.5$ kΩ and $V_{DD} = 15$ V.

b. Modify the design so that the amplifier operates in the frequency range from 10 Hz to 50 kHz.

Large Design Projects

F.10 The input signal to an amplifier is $v_s = 2$ mV, and the amplifier has a source resistance of $R_S = 1$ kΩ.

a. Design a BJT amplifier to give a midfrequency voltage gain A_v ($=v_L/v_s$ with a load resistance $R_L = 10$ kΩ) of greater than 650. The input resistance R_{in} of the amplifier should be greater than 70 kΩ, and the output resistance R_{out} should be less than 250 Ω. The DC supply voltage is $V_{CC} = 12$ V.

b. Modify the design so that the amplifier operates in the frequency range from 10 kHz to 80 kHz.

c. Apply feedback and modify the design. The input resistance must be increased by 20 (i.e., $R_{if} \geq 20\ R_{in}$), and the output resistance must be decreased by 20 (i.e., $R_{of} = R_o/20$).

d. Apply feedback and modify the design so that the amplifier oscillates at a frequency of $f_o = 20$ kHz.

F.11 The input signal to an amplifier is $v_s = 2$ mV, and the amplifier has a source resistance of $R_S = 1$ kΩ.

a. Design an NMOS amplifier to give a midfrequency voltage gain A_v ($= v_L/v_s$ with a load resistance $R_L = 10$ kΩ) of greater than 450. The input resistance R_{in} of the amplifier should be greater than 500 kΩ, and the output resistance R_{out} should be less than 250 Ω. The DC supply voltage is $V_{DD} = 12$ V.

b. Modify the design so that the amplifier operates in the frequency range from 20 kHz to 60 kHz.

c. Apply feedback and modify the design. The input resistance must be increased by 10 (i.e., $R_{if} \geq 10\ R_{in}$), and the output resistance must be decreased by 10 (i.e., $R_{of} \leq R_o/10$).

d. Apply feedback and modify the design so that the amplifier oscillates at a frequency of $f_o = 20$ kHz.

F.12 Design a BJT operational amplifier to give a large-signal differential gain of $10^3 \pm 10\%$. The input resistance should be higher than 100 kΩ ± 5%, and the output resistance should be less than 210 Ω ± 5%. The op-amp should have a common-mode rejection ratio of CMRR = $10^4 \pm 10\%$. The unity-gain bandwidth must be better than $10^4 \pm 10\%$.

F.13 Design a multistage BJT amplifier to meet the following specifications: voltage gain, $|A_v| = v_L/v_s = 600$ ± 5% (with load); input resistance, $R_i = v_s/i_s \geq 25$ kΩ; output resistance, $R_o \leq 300$ Ω; load resistance, $R_L = 25$ kΩ; source resistance, $R_s = 1$ kΩ; DC supply, $V_{CC} = 15$ V; input signal, $v_s = 1$ mV to 2 mV (peak sinusoidal), 1 kHz. The amplifier should operate in the frequency range from 10 Hz to 30 kHz.

F.14 Design a multistage MOS amplifier to meet the following specifications: voltage gain, $|A_v| = v_L/v_s = 400$ ± 5% (with load); input resistance, $R_i = v_s/i_s \geq 25$ kΩ; output resistance, $R_o \leq 300$ Ω; load resistance, $R_L = 25$ kΩ; source resistance, $R_s = 1$ kΩ; DC supply, $V_{DD} = 15$ V; input signal, $v_s = 1$ mV to 2 mV (peak sinusoidal), 1 kHz. The amplifier should operate in the frequency range from 100 Hz to 25 kHz.

F.15 Design a differential amplifier to satisfy the following conditions: a differential voltage gain of $A_o \geq 2500$; common-mode rejection ratio of CMRR ≥ 4500; load of $R_L = 50$ kΩ; coupling capacitor of $C_L = \infty$. Use all BJTs and only one resistance. Use PSpice to verify your design with $v_d = 5$ μV and $v_c = 10$ mV. Supply voltages are $V_{CC} = V_{EE} = 15$ V. The amplifier should operate in the frequency range up to 25 kHz.

F.16 Design a differential amplifier to satisfy the following conditions: a differential voltage gain of $A_o \geq 2500$; common-mode rejection ratio of CMRR ≥ 4500; load of $R_L = 50$ kΩ; coupling capacitor of $C_L = \infty$. Use all MOSFETs. Use PSpice to verify your design with $v_d = 5$ μV and $v_c = 10$ mV. Supply voltages are $V_{DD} = V_{SS} = 15$ V. The amplifier should operate in the frequency range up to 25 kHz.

APPENDIX G

MILLER'S CAPACITOR METHOD

Let us consider the small-signal equivalent of a transistor amplifier as shown in Fig. G.1. C and C_i are the internal capacitances of the transistor. Assuming that the current through capacitor C is very small compared to the voltage dependent current source $g_m V_i$, the output voltage in frequency domain is

$$V_o = -g_m V_i (R_o \| R_L) \tag{G.1}$$

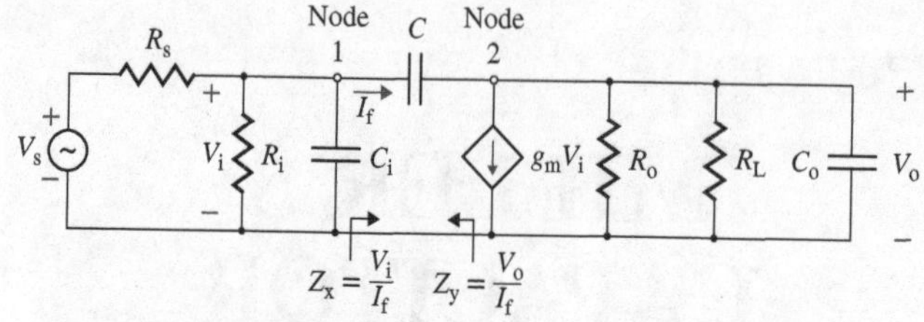

FIGURE G.1 Small signal equivalent of a transistor amplifier

Using Eq. (G.1), the current I_f flowing through C (from the left side to the right side) in frequency domain is given by

$$\begin{aligned} I_f &= j\omega C(V_i - V_o) = j\omega C[V_i + g_m V_i (R_o \| R_L)] \\ &= j\omega C[1 + g_m (R_o \| R_L)]V_i = j\omega C_m V_i \end{aligned} \tag{G.2}$$

where $C_m = C[1 + g_m (R_o \| R_L)]$

Using Eq. (G.1), the current $-I_f$ flowing through C (from the right side to the left side) is given by

$$\begin{aligned} -I_f &= j\omega C(V_o - V_i) = j\omega C\left[V_o + \frac{V_o}{g_m (R_o \| R_L)}\right] \\ &= j\omega C\left[1 + \frac{1}{g_m (R_o \| R_L)}\right]V_o = j\omega C_n V_o \end{aligned} \tag{G.4}$$

$$\text{where } C_n = C\left[1 + \frac{1}{g_m (R_o \| R_L)}\right] \tag{G.5}$$

Thus, capacitor C, which is connected between the input and output terminals of a high-gain amplifier with 180° phase reversal, can be replaced by a shunt capacitor C_m on the input side and another capacitor C_n on the output side. This arrangement is shown in Fig. G.2. The value of C is seen on the input side as a multiplying factor almost equal to the voltage gain $g_m(R_o \| R_L)$. This effect, known as *Miller's effect*, is

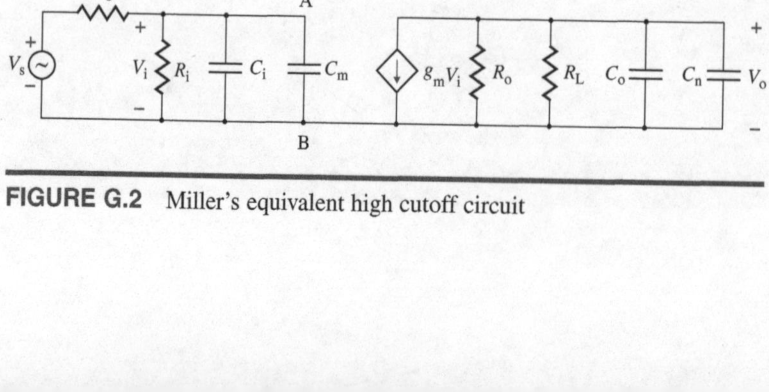

FIGURE G.2 Miller's equivalent high cutoff circuit

dominant in amplifiers with a high voltage gain and a phase reversal. Therefore, the high-frequency poles can be found from

$$f_{H1} = \frac{1}{2\pi(C_i + C_m)(R_s \| R_i)} \tag{G.6}$$

$$f_{H2} = \frac{1}{2\pi(C_o + C_n)(R_o \| R_L)} \tag{G.7}$$

Since $A_v = -g_m(R_o \| R_L)$ is the voltage gain, we can rewrite Eqs. (G.3) and (G.5) as

$$C_m = C(1 - A_v) \tag{G.8}$$

$$C_n = C\left(1 - \frac{1}{A_v}\right) \tag{G.9}$$

Therefore, if the voltage A_v is negative, then $C_m > C$ and there is a capacitance multiplication. Thus, the effective capacitance C_m between nodes 1 and 2 (in Fig. G.1) can be increased by a voltage amplifier. That is, a small capacitance connected between the input and output terminals of a voltage amplifier will have a much larger effective capacitance between the input terminals A and B (in Fig. G.2). C_m is called the *Miller capacitance*. The Miller capacitance plays an important role in designing the high-frequency response of amplifiers and in the design of active filters. For example, f_{H1} in Eq. (G.6), which is contributed mainly by C_m, will be much lower than f_{H2}, and thus f_{H1} will set the dominant high frequency.

G.1 MOSFET Drain Current

The conductance of the channel between the drain and the source can be found from

$$g_{DS} = \frac{1}{r_{DS}} = \frac{W}{L}\mu_n Q_n \tag{G.10}$$

where μ_n is the mobility of the electrons in the reverse-biased (also called the inversion) layer under the oxide layer, Q_n is the magnitude of the reverse-biased layer charge per unit area, W is the channel width, and L is the channel length

Let V_t be the MOSFET threshold voltage. If we consider a small incremental drain-to-source voltage v along the channel, we can find Q_n from the gate oxide capacitance C_{on} and the voltage difference $(v_{GS} - V_t - v)$ as given by

$$Q_n(v) = C_{on}(v_{GS} - V_t - v) \tag{G.11}$$

which can be applied to obtain the drain current as given by

$$i_D = \frac{W}{L}\mu_n C_{on}\int_0^{v_{DS}} (v_{GS} - V_t - v)dv = \frac{W}{L}\mu_n C_{on}\left[\left(v_{GS} - V_t\right)v_{DS} - v_{DS}^2\right] \tag{G.12}$$

which can be expressed in a more general form in terms of external voltages v_{GS} and v_{DS}

$$i_D = K_n[(v_{GS} - V_t)v_{DS} - v_{DS}^2] \tag{G.13}$$

Substituting $v_{DS} = V_{GS} - V_t$ at which the saturation occurs gives the drain current in the saturation region:

$$i_D = K_n(v_{GS} - V_t)^2 \tag{G.14}$$

For a depletion type MOSFET, the threshold voltage V_t is replaced by the pinch-off voltage V_p. Equation (G.14) becomes

$$i_D = K_n(v_{GS} - V_p)^2 \tag{G.15}$$

where $K_n = I_{DSS}/V_p^2$ and I_{DSS} is the maximum steady-state drain current at $v_{GS} = 0$. Equation (G.11) can also be written as

$$i_D = I_{DSS}\left(1 - \frac{v_{GS}}{V_p}\right)^2 \tag{G.16}$$

G.2 Early Effects

The collector current of a BJT can be expressed in general form as

$$i_C = I_{SC}e^{v_{BE}/V_T} \tag{G.17}$$

The slope introduced by the Early effect is almost linear with i_C and v_{CE} characteristics. If the collector current characteristics are extrapolated to zero collector current, the curves intercept the voltage axis at a point known as the *Early voltage*. The Early voltage is positive for an npn transistor and negative for a pnp transistor. From Fig. G.3, we can write the output conductance as

$$g_o = \frac{1}{r_o} = \frac{di_C}{dv_{CE}} = \frac{I_C}{v_{CE} + V_A} \tag{G.18}$$

where r_o is the output resistance of the transistor. If we include the finite slope of i_C versus v_{CE} characteristics due to the Early effect, the collector current in Eq. (G.17) can be modified to

$$i_C = I_{SC}e^{v_{BE}/V_T}\left(\frac{I_C}{v_{CE} + V_A}\right) \tag{G.19}$$

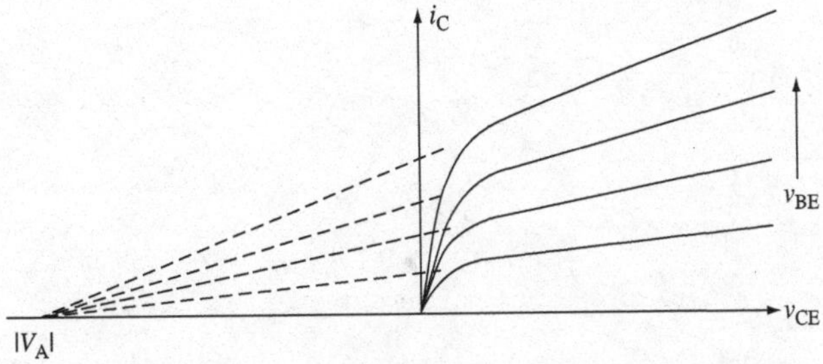

FIGURE G.3 Early voltage

APPENDIX H

HIGH-FREQUENCY TRANSFER FUNCTION METHOD

The high-frequency behavior is determined solely by the internal capacitors of the amplifier. The equivalent circuit for determining the high break frequencies is shown in Fig. H.1. Capacitance C, between the input and output terminals of the amplifier, can be replaced by Miller's equivalent capacitances. Therefore, we can find the frequency response by s-domain analysis or by Miller's capacitor method.

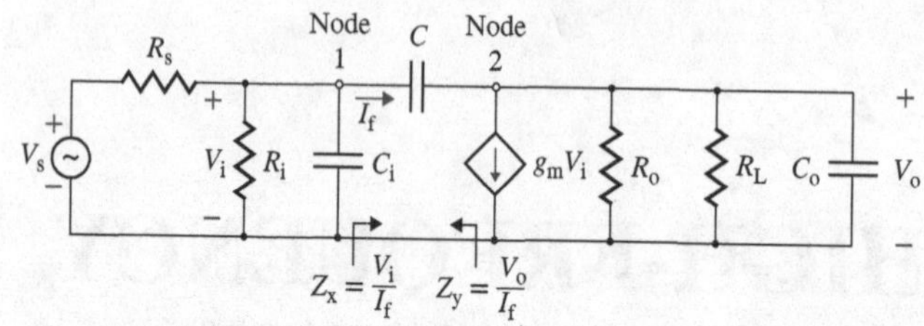

FIGURE H.1 Small signal equivalent high cutoff circuit

As we did for the low break frequencies, we will derive the transfer function for high break frequencies. Applying Kirchhoff's current law at nodes 1 and 2, we get the following equations in Laplace's domain of s:

$$\frac{V_s - V_i}{R_s} = \frac{V_i}{R_i} + V_i C_i s + (V_i - V_o)Cs \tag{H.1}$$

$$g_m V_i + \frac{V_o}{R_o} + \frac{V_o}{R_L} + V_o C_o s + (V_o - V_i)Cs = 0 \tag{H.2}$$

Equations (H.1) and (H.2) can be solved to give the voltage transfer function:

$$\frac{V_o(s)}{V_s(s)} = \frac{-(g_m - Cs)R_1R_2/R_s}{1 + s[R_1(C_i + C) + R_2(C_o + C) + g_m CR_1R_2] + s^2R_1R_2(C_iC_o + C_iC + C_oC)} \tag{H.3}$$

where $R_1 = (R_s \parallel R_i)$ and $R_2 = (R_o \parallel R_L)$. The denominator of Eq. (H.3) has two poles. If p_1 and p_2 are the two poles, the denominator can be written as

$$D(s) = \left(1 + \frac{s}{p_1}\right)\left(1 + \frac{s}{p_2}\right) = 1 + s\left(\frac{1}{p_1} + \frac{1}{p_2}\right) + \frac{s^2}{p_1 p_2} \tag{H.4}$$

If the poles are widely separated, which is generally the case, and p_1 is assumed to be the dominant pole, while p_2 is assumed large, then Eq. (H.4) can be approximated by

$$D(s) = 1 + \frac{s}{p_1} + \frac{s^2}{p_1 p_2} \tag{H.5}$$

Equating the coefficients of s in Eq. (H.3) to those in Eq. (H.5) yields

$$p_1 = \frac{1}{R_1(C_i + C) + R_2(C_o + C) + g_m CR_1R_2} \tag{H.6}$$

Equating the coefficients of s^2 in Eq. (H.3) to those in Eq. (H.5) yields

$$p_2 = \frac{R_1(C_i + C) + R_2(C_o + C) + g_m C R_1 R_2}{R_1 R_2(C_i C_o + C_i C + C_o C)} \tag{H.7}$$

In practice, the value of C is higher than that of C_i and C_o, and Eqs. (H.6) and (H.7) can be simplified further as follows:

$$p_1 \approx \frac{1}{g_m C R_1 R_2} \tag{H.8}$$

$$p_2 = \frac{g_m C}{C_i C_o + C_i C + C_o C} \tag{H.9}$$

Here p_1 and p_2 correspond to the break frequencies ω_1 and ω_2 in frequency domain.

▶ NOTES

1. The dominant pole p_1 decreases as C increases, whereas p_2 increases as C increases. Therefore, increasing C causes the poles to split apart, possibly making p_1 the dominant pole.
2. If $C >> C_i$ and $C >> C_o$, Eq. (H.9) can be approximated as

$$p_2 \approx \frac{g_m C}{C(C_i + C_o)} = \frac{g_m}{C_i + C_o} \tag{H.10}$$

3. If there is no feedback capacitance ($C = 0$), Eq. (H.3) gives the following poles:

$$p_1 = \frac{1}{C_i R_1} \tag{H.11}$$

$$p_2 = \frac{1}{C_o R_2} \tag{H.12}$$

ANSWERS TO SELECTED PROBLEMS

Chapter 1

1.13 $W/L = 39$, $V_t = 5.06$ V

1.19 **(a)** $I_{D1,2} = 0.168$ mA
(b) $A_d = -1.832$, $A_c = 9.945 \times 10^{-3}$, CMRR = 184.2, $v_O = -54.96$ mV

1.21 **(a)** 96.46 μA, 103.54 μA

1.22 **(a)** 0.715 mA **(b)** 139.8 kΩ
(c) 100 V **(d)** 1.2909

1.24 **(a)** 1.36 mA **(b)** 73.53 kΩ
(c) 100 V **(d)** 1.282

1.26 **(a)** 4.152 mA **(b)** 24 kΩ
(c) 100 V **(d)** 1.142

1.28 $R_2 = 11.88$ kΩ, $R_o = 55.69$ MΩ

1.30 **(a)** $R_1 = 2.86$ MΩ **(b)** $R_o = 760$ MΩ

1.31 1.231 mA, 9.26 MΩ

1.35 For $V_A = 100$ and $\beta_F = 100$, $r_o = 100$ kΩ; for $V_{CC} = 15$ V, $R_1 = 13.6$ kΩ

1.37 **(a)** 2.955 mA, 2.011 mA **(b)** −961.5, −0.25, 38.44×10^6, 20.8 Ω, 6.04 MΩ

1.41 769.2, 2.08 MΩ, 8 MΩ, 587.9

1.45 −28.57, 0, 0

1.47 −60.14, 0, 0

1.49 $A_{do} = -961$, $\omega_H = 51.1$ krad/s

Chapter 2

2.3 26 mV

2.5 1 pV

2.9 **(a)** ±2.7 mV **(b)** −0.5 V ± 2.7 mV

2.12 **(a)** 0.6 mV **(b)** 2.6 mV

2.17 **(a)** 835.96 kΩ **(b)** 7.75 mA/V
(c) 6480 **(d)** 125 kΩ

2.20 **(a)** 0.2 μA **(b)** 258 kΩ
(c) 6.16 MHz

2.21 **(a)** 0.1 μA **(b)** 516 kΩ

2.23 521,963

2.24 347,104

2.28 **(a)** ±66.66 V/μs **(b)** 205.6 MHz

Chapter 3

3.1 **(a)** 2 kΩ, 3.28 kΩ **(b)** 94

3.2 **(a)** 89.25 ps **(b)** 4.635 mW

3.4 27.38 mA/V^2

3.7 **(a)** 3 V **(b)** 0.4 V, 3.04 mA

3.9 **(a)** 3.5 V, 0.39 V **(b)** 1.5 V, 3.37 mA, 8.438 mW, 0 W

3.11 4.2 V

3.16 **(a)** 2.5 V, 4 V, 1 V **(b)** 0

3.17 **(a)** 4.914 V, 1.914 V

3.18 **(a)** 2.5 V, 3.5 V, 1.5 V

3.19 0.464

3.22 **(a)** 2.5 V, 3.5 V, 1.5 V **(b)** 4.95 V, 0.25 mA, 0.05 V, 0.25 mA

3.23 **(a)** 1/1.5 **(b)** 2.1 V, 2.276 V
(c) 1.02 ns **(d)** 0.51 pJ

3.28 **(a)** 17.55 kΩ **(b)** 2
(c) 0.049 mW

3.30 **(a)** 1175 Ω, 3384 Ω **(b)** 1.006, 1.63 V
(c) 42

3.32 **(a)** 4.8 kΩ, 14.4 kΩ **(b)** $N = 1$

3.36 0.2 V, $I_B = 2.102$ mA, $I_3 = 1.92$ mA, $N = 58$

3.37 $I_1 = 0.58$ mA, $I_2 = 1.64$ mA, $I_{B3} = 2.102$ mA, $N = 64$

3.38 $R_1 = 4$ kΩ, $I_{B1} = 725$ μA, $I_{LO} = 410$ μA

3.41 21.15 mA, 1.15 mA, 0.846 mA

3.43 **(a)** $v_{O1} = -0.8$ V, $R_{C1} = 400$ Ω, $R_4 = 4.5$ kΩ

3.44 **(a)** $v_A = 2.1$ V, $R_E = 2.3$ kΩ, $v_{O1} = 2.2$ V, $R_4 = 3.7$ kΩ, $V_{B6} = 0.2$ V

3.45 $R_3 = 1.75$ kΩ, $R_E = 1.75$ kΩ, $R_4 = 1250$ Ω

3.47 **(a)** 0.106 ns

Chapter 4

4.4 (a) 3.32 V, 0.57 V (b) −2.66 V, 0.456 V
(c) −3.1 V

4.10 For $R_1 = 10\ \text{k}\Omega$, $R_F = 14\ \text{k}\Omega$,
$R_x = 5833\ \Omega$

4.14 For $R_1 = 10\ \text{k}\Omega$, $R_F = 50\ \text{k}\Omega$,
$V_{ref} = 4.8$ V

4.18 2127.6 Hz

4.21 (a) 16 kHz (b) 12.8 kHz

4.26 (c) 196 Hz

4.31 (a) 2.5 kHz (b) ±200/3 kHz
(c) 543 Hz

4.37 0.747 V

4.38 8

INDEX